INTERNATIONAL SEMINAR ON
NUCLEAR WAR AND PLANETARY EMERGENCIES
43rd Session

The Role of Science in the Third Millennium.
Energy: Pollution: Oil Spills – Air, Water and Waste Criteria / Climate: Nuclear Power Plants—Sustainable Systems / Paleoclimate Natural Variations and Implications – Sensitivity, Climategate, USA Legislation / Cosmic Objects: Countermeasures / Mitigation of Terrorist Acts: Scientific Contributions to Biosecurity / Lectio Magistralis: Science in Everyday Life / Information Security: Cyber Conflicts and Stability / The Erice Science for Peace Prize Award

THE SCIENCE AND CULTURE SERIES
Nuclear Strategy and Peace Technology*

Series Editor: Antonino Zichichi

1981 — International Seminar on Nuclear War — 1st Session: The World-wide Implications of Nuclear War

1982 — International Seminar on Nuclear War — 2nd Session: How to Avoid a Nuclear War

1983 — International Seminar on Nuclear War — 3rd Session: The Technical Basis for Peace

1984 — International Seminar on Nuclear War — 4th Session: The Nuclear Winter and the New Defence Systems: Problems and Perspectives

1985 — International Seminar on Nuclear War — 5th Session: SDI, Computer Simulation, New Proposals to Stop the Arms Race

1986 — International Seminar on Nuclear War — 6th Session: International Cooperation: The Alternatives

1987 — International Seminar on Nuclear War — 7th Session: The Great Projects for Scientific Collaboration East-West-North-South

1988 — International Seminar on Nuclear War — 8th Session: The New Threats: Space and Chemical Weapons — What Can be Done with the Retired I.N.F. Missiles-Laser Technology

1989 — International Seminar on Nuclear War — 9th Session: The New Emergencies

1990 — International Seminar on Nuclear War — 10th Session: The New Role of Science

1991 — International Seminar on Nuclear War — 11th Session: Planetary Emergencies

1991 — International Seminar on Nuclear War — 12th Session: Science Confronted with War (unpublished)

1991 — International Seminar on Nuclear War and Planetary Emergencies — 13th Session: Satellite Monitoring of the Global Environment (unpublished)

1992 — International Seminar on Nuclear War and Planetary Emergencies — 14th Session: Innovative Technologies for Cleaning the Environment

1992 — International Seminar on Nuclear War and Planetary Emergencies — 15th Session (1st Seminar after Rio): Science and Technology to Save the Earth (unpublished)

1992 — International Seminar on Nuclear War and Planetary Emergencies — 16th Session (2nd Seminar after Rio): Proliferation of Weapons for Mass Destruction and Cooperation on Defence Systems

1993 — International Seminar on Planetary Emergencies — 17th Workshop: The Collision of an Asteroid or Comet with the Earth (unpublished)

1993 — International Seminar on Nuclear War and Planetary Emergencies — 18th Session (4th Seminar after Rio): Global Stability Through Disarmament

1994 — International Seminar on Nuclear War and Planetary Emergencies — 19th Session (5th Seminar after Rio): Science after the Cold War

1995 — International Seminar on Nuclear War and Planetary Emergencies — 20th Session (6th Seminar after Rio): The Role of Science in the Third Millennium

1996 — International Seminar on Nuclear War and Planetary Emergencies — 21st Session (7th Seminar after Rio): New Epidemics, Second Cold War, Decommissioning, Terrorism and Proliferation

1997 — International Seminar on Nuclear War and Planetary Emergencies — 22nd Session (8th Seminar after Rio): Nuclear Submarine Decontamination, Chemical Stockpiled Weapons, New Epidemics, Cloning of Genes, New Military Threats, Global Planetary Changes, Cosmic Objects & Energy

1998 — International Seminar on Nuclear War and Planetary Emergencies — 23rd Session (9th Seminar after Rio): Medicine & Biotechnologies, Proliferation & Weapons of Mass Destruction, Climatology & El Nino, Desertification, Defence Against Cosmic Objects, Water & Pollution, Food, Energy, Limits of Development, The Role of Permanent Monitoring Panels

1999 — International Seminar on Nuclear War and Planetary Emergencies — 24th Session: HIV/AIDS Vaccine Needs, Biotechnology, Neuropathologies, Development Sustainability — Focus Africa, Climate and Weather Predictions, Energy, Water, Weapons of Mass Destruction, The Role of Permanent Monitoring Panels, HIV Think Tank Workshop, Fertility Problems Workshop

2000 — International Seminar on Nuclear War and Planetary Emergencies — 25th Session: Water — Pollution, Biotechnology — Transgenic Plant Vaccine, Energy, Black Sea Pollution, Aids — Mother–Infant HIV Transmission, Transmissible Spongiform Encephalopathy, Limits of Development — Megacities, Missile Proliferation and Defense, Information Security, Cosmic Objects, Desertification, Carbon Sequestration and Sustainability, Climatic Changes, Global Monitoring of Planet, Mathematics and Democracy, Science and Journalism, Permanent Monitoring Panel Reports, Water for Megacities Workshop, Black Sea Workshop, Transgenic Plants Workshop, Research Resources Workshop, Mother–Infant HIV Transmission Workshop, Sequestration and Desertification Workshop, Focus Africa Workshop

2001 — International Seminar on Nuclear War and Planetary Emergencies — 26th Session: AIDS and Infectious Diseases — Medication or Vaccination for Developing Countries; Missile Proliferation and Defense; Tchernobyl — Mathematics and Democracy; Transmissible Spongiform Encephalopathy; Floods and Extreme Weather Events — Coastal Zone Problems; Science and Technology for Developing Countries; Water — Transboundary Water Conflicts; Climatic Changes — Global Monitoring of the Planet; Information Security; Pollution in the Caspian Sea; Permanent Monitoring Panels Reports; Transmissible Spongiform Encephalopathy Workshop; AIDS and Infectious Diseases Workshop; Pollution Workshop

2002 — International Seminar on Nuclear War and Planetary Emergencies — 27th Session: Society and Structures: Historical Perspectives — Culture and Ideology; National and Regional Geopolitical Issues; Globalization — Economy and Culture; Human Rights — Freedom and Democracy Debate; Confrontations and Countermeasures: Present and Future Confrontations; Psychology of Terrorism; Defensive Countermeasures; Preventive Countermeasures; General Debate; Science and Technology: Emergencies; Pollution, Climate — Greenhouse Effect; Desertification, Water Pollution, Algal Bloom; Brain and Behaviour Diseases; The Cultural Emergency: General Debate and Conclusions; Permanent Monitoring Panel Reports; Information Security Workshop; Kangaroo Mother's Care Workshop; Brain and Behaviour Diseases Workshop

2003 — International Seminar on Nuclear War and Planetary Emergencies — 29th Session: Society and Structures: Culture and Ideology — Equity — Territorial and Economics — Psychology — Tools and Countermeasures — Worldwide Stability — Risk Analysis for Terrorism — The Asymmetric Threat — America's New "Exceptionalism" — Militant Islamist Groups Motives and Mindsets — Analysing the New Approach — The Psychology of Crowds — Cultural Relativism — Economic and Socio-economic Causes and Consequences — The Problems of American Foreign Policy — Understanding Biological Risk Chemical Threats and Responses — Bioterrorism — Nuclear Survivial Criticalities — Responding to the Threats — National Security and Scientific Openness — Working Groups Reports and Recommendations

2003 — International Seminar on Nuclear War and Planetary Emergencies — 30th Session: Anniversary Celebrations: The Pontifical Academy of Sciences 400th — The 'Ettore Majorana' Foundation and Centre for Scientific Culture 40th — H.H. John Paul II Apostolate 25th — Climate/Global Warming: The Cosmic Ray Effect; Effects on Species and Biodiversity; Human Effects; Paleoclimate Implications; Evidence for Global Warming — Pollution: Endocrine Disrupting Chemicals; Hazardous Material; Legacy Wastes and Radioactive Waste Management in USA, Europe; Southeast Asia and Japan — The Cultural Planetary Emergency: Role of the Media; Intolerance; Terrorism; Iraqi Perspective; Open Forum Debate — AIDS and Infectious Diseases: Ethics in Medicine; AIDS Vaccine Strategies — Water: Water Conflicts in the Middle East — Energy: Developing Countries; Mitigation of Greenhouse Warming — Permanent Monitoring Panels Reports — Workshops: Long-Term Stewardship of Hazardous Material; AIDS Vaccine Strategies and Ethics

2004 — International Seminar on Nuclear War and Planetary Emergencies — 31st Session: Multidisciplinary Global Approach of Governments and International Structures: Societal Response — Scientific Contributions to Policy — Economics — Human Rights — Communication — Conflict Resolution — Cross-Disciplinary Responses to CBRN Threats: Chemical and Biological Terrorism — Co-Operation Between Russia and the West — Asymmetrical Conflicts — CBW Impact — Cross-Disciplinary Challenges to Emergnecy Management, Media Information and Communication: Role of Media in Global Emergencies — Emergency Responders — Working Groups' Reports and Recommendations

2004 — International Seminar on Nuclear War and Planetary Emergencies — 32nd Session: Limits of Development: Migration and Cyberspace; in Europe; Synoptic European Overview; From and Within Asia; Globalization — Climate: Global Warming; a Chronology; Simple Climate Models; Energy and Electricity Considerations — T. S. E.: CJD and Blood Transfusion; BSE in North America; Gerstmann-Straussler-Scheinker Disease — The Cultural Emergency: Innovations in Communications and IT — Cosmic Objects: Impact Hazard; Close Approaches; Asteroid Deflection; Risk Assessment and Hazard Reduction; Hayabusa and Follow Up — Aids and Infectious Diseases: Ethics in Medicine; International Co-operation; Laboratory Biosecurity Guidelines; Georgian Legislation; Biosecurity Norms and International Organizations, Legal Measures Against Biocrimes — Water and Pollution: Cycle Overview; Beyond Cost and Price; Requirements in Rural Iran; Isotope Techniques; Clean and Reliable Water for the 21st Century — Permanent Monitoring Panels Reports — Workshops: Global Biosecurity; Cosmic Objects

2005 — International Seminar on Nuclear War and Planetary Emergencies — 34th Session: Energy: Nuclear and Renewable Energy; Energy Technologies for the 21st Century; Repositories Development; Nuclear Power in Europe and in Asia; The Future of Nuclear Fusion — Climate: Global Warming; Celestial Climate Driver; Natural and Anthropogenic Contributions; Climate Data and Comparison with Models; Understanding Common Climate Claims — AIDS and Infectious Diseases: New Threats from Infectious Agents — SARS Epidemic; Vaccines Development; Transmissible Spongiform Encephalopathies Update — Limits of Development: International Points of View on Migration — Pollution: Science and Technology; Subsurface Laser Drilling — Desertification: A Global Perspective; Integrated Approaches — Disarmament and Cultural Emergencies: A WFS Achievement in China; Non-Proliferation — Permanent Monitoring Panel Reports — Workshops: Energy; Information Security; Building Resilence Associated with the Third Meeting on Terrorism

2006 — International Seminar on Nuclear War and Planetary Emergencies — 36th Session: Energy: Global Nuclear Power Future; Global Monitoring of the Planet Proliferation: Nuclear Weapons; AIDS and Infectious Diseases: Avian Flu — Global Health; Climatology: Global Warming/Aerosols and Satellites; Pollution: Plastic Contaminants in Water; Information Security: Relevance of Cyber Security; Limits of Development: Development of Sustainability; Defence Against Cosmic Objects; WFS General Meeting: Cultural Energy-Focus: Terrorism; Permanent Monitoring Panel Reports; Limits of Development Permanent Monitoring Panel Meeting; World Energy Monitoring Workshop.

2007 — International Seminar on Nuclear War and Planetary Emergencies — 38th Session: World Energy Crisis; Managing Climate Change; Mitigation of Greenhouse Gases; Geoengineering & Adaptation; Theoretical Alternatives to Climate Modelling; US Missile Defence Shield; Global Monitoring of the Planet; Life Cycle Nuclear Energy Environmental Issues; The Epidemic of Alzheimer; Infectious Agents and Cancer

2008 — International Seminar on Nuclear War and Planetary Emergencies — 40th Session: Energy: Nuclear Power Present and Future; Sustainability of Biofuels; Resolving the Nuclear Waste – Climatology Model and Statistics; Ozone and Climate Change Interaction; Spatio-Temporal Field of Atmospheric CO_2; Forest Policies – Medicine: Vector-Borne Diseases; Screening Technology – Pollution: Air-Borne Particulates – Global Monitoring of the Planet: Disarmament and Non-Proliferation Regime; The Crisis in Internet Security; The Northern Sea Route – The Erice Science for Peace Award Scientific Session

2009 — International Seminar on Nuclear War and Planetary Emergencies — 42nd Session: World Energy Crisis—Energy & Pollution: Essential Technologies for Managing the Coupled Challenges of Climate Change and Energy Security, Energy, Water, Climate, Pollution & Limits of Development in Asian Countries; Global Monitoring of the Planet—Sensitivity of Climate to Additional CO_2 as Indicated by Water Cycle Feedback Issues, Climate Uncertainties Addressed by Satellites, The Basic Mathematics Needed for all Models; Pollution and Medicine—The Revolution in the Environmental Health Sciences and the Emergence of Green Chemistry; Information Security—Cyber Conflict and Cyber Stability: Finding a Path to Cyber Peace; Cultural Pollution—The Erice Science for Peace Award Scientific Session

2010 — International Seminar on Nuclear War and Planetary Emergencies — 43rd Session: The Role of Science in the Third Millennium Energy: Nuclear Power Plants—Sustainable Systems / Pollution: Oil Spills – Air, Water and Waste Criteria / Climate: Paleoclimate Natural Variations and Implications – Sensitivity, Climategate, USA Legislation / Cosmic Objects: Countermeasures / Mitigation of Terrorist Acts: Scientific Contributions to Biosecurity / Lectio Magistralis: Science in Everyday Life / Information Security: Cyber Conflicts and Stability / The Erice Science for Peace Prize Award

*For the complete list of titles in this series, please go to:
http://worldscibooks.com/series/scsnsp_series.shtml

THE SCIENCE AND CULTURE SERIES

Nuclear Strategy and Peace Technology

INTERNATIONAL SEMINAR ON NUCLEAR WAR AND PLANETARY EMERGENCIES

43rd Session

The Role of Science in the Third Millennium.
Energy: Pollution: Oil Spills – Air, Water and Waste Criteria / Climate: Nuclear Power Plants—Sustainable Systems / Paleoclimate Natural Variations and Implications – Sensitivity, Climategate, USA Legislation / Cosmic Objects: Countermeasures / Mitigation of Terrorist Acts: Scientific Contributions to Biosecurity / Lectio Magistralis: Science in Everyday Life / Information Security: Cyber Conflicts and Stability / The Erice Science for Peace Prize Award

"E. Majorana" Centre for Scientific Culture
Erice, Italy, 19 – 24 August 2010

Series Editor and Chairman: A. Zichichi

Edited by R. Ragaini

World Scientific

NEW JERSEY • LONDON • SINGAPORE • BEIJING • SHANGHAI • HONG KONG • TAIPEI • CHENNAI

Published by

World Scientific Publishing Co. Pte. Ltd.
5 Toh Tuck Link, Singapore 596224
USA office: 27 Warren Street, Suite 401-402, Hackensack, NJ 07601
UK office: 57 Shelton Street, Covent Garden, London WC2H 9HE

British Library Cataloguing-in-Publication Data
A catalogue record for this book is available from the British Library.

INTERNATIONAL SEMINAR ON NUCLEAR WAR AND PLANETARY EMERGENCIES — 43RD SESSION: THE ROLE OF SCIENCE IN THE THIRD MILLENNIUM ENERGY: NUCLEAR POWER PLANTS—SUSTAINABLE SYSTEMS / POLLUTION: OIL SPILLS—AIR, WATER AND WASTE CRITERIA / CLIMATE: PALEOCLIMATE NATURAL VARIATIONS AND IMPLICATIONS—SENSITIVITY, CLIMATEGATE, USA LEGISLATION / COSMIC OBJECTS: COUNTERMEASURES / MITIGATION OF TERRORIST ACTS: SCIENTIFIC CONTRIBUTIONS TO BIOSECURITY / LECTIO MAGISTRALIS: SCIENCE IN EVERYDAY LIFE / INFORMATION SECURITY: CYBER CONFLICTS AND STABILITY / THE ERICE SCIENCE FOR PEACE PRIZE AWARD

ISBN-13 978-981-4365-92-5
ISBN-10 981-4365-92-0

Printed in Singapore by B & Jo Enterprise Pte Ltd

CONTENTS

SECTION 1

SESSION 1

SCIENCE IN THE THIRD MILLENNIUM

WHY SCIENCE IS NEEDED FOR THE CULTURE OF THE THIRD MILLENNIUM: THE MOTOR FOR PROGRESS

PROFESSOR ANTONINO ZICHICHI
CERN, Geneva, Switzerland and University of Bologna, Italy and Centro Enrico Fermi, Italy

Dear Colleagues, Ladies and Gentlemen,

I welcome you all to this 43rd Session of the International Seminars on Nuclear War and Planetary Emergencies and declare the Session to be open.

Once again, as every year since 1981, a distinguished group of Interdisciplinary Scientists is gathered, here at the Ettore Majorana Foundation and Centre for Scientific Culture in Erice, to participate in the fight against the consequences of the 63 Planetary Emergencies identified, 30 years ago, by the World Federation of Scientists.

During these 30 years, we were successful in avoiding a planetary nuclear holocaust and conducted 100 pilot projects in 50 developing countries on a fully voluntary basis, promoting and implementing a close co-operation between scientists of the East and West, North and South, thus contributing to the fall of the Berlin Wall. We are now still striving to avoid an environmental and cultural planetary holocaust by addressing issues such as;

- World Energy Crisis
- Global Monitoring of the Planet
- Pollution
- Information Security
- Defence against Cosmic Objects
- Mitigation of Terrorists Acts

One of our biggest challenges nowadays is to convince the great public that the best way to study these problems, with clarity and rigour, is through Science. We have been teaching the public to identify the great achievements of our intellect: Language, Rigorous Logic and Science and to distinguish them from their applications.

In my talk, last year, I gave many details and examples to enlarge our understanding of the problem. In view of its importance, I will repeat some of my arguments for those of you who were not present last year but they will not appear in this year's proceedings (readers are invited to refer to the proceedings of the 42nd Session of the International Seminars on Nuclear War and Planetary Emergencies).

As mentioned before, the topics of this 43rd Session of our Seminars deal with:

- **World Energy Crisis**
- **Global Monitoring of the Planet**
- **Pollution**
- **Information Security**
- **Defence Against Cosmic Objects**
- **Mitigation of Terrorists Acts**

Let me show you the synthesis of all the Planetary Emergencies grouped in 15 classes. The details of these Emergencies are in the Tables I, II, III, IV, V, VI, VII, VIII, IX, X, XI, XII, XIII, XIV, and XV.

	SUMMARY	number of P.E.
I	*WATER*	= 4
II	*SOIL*	= 3
III	*FOOD*	= 5
IV	*ENERGY*	= 5
V	*POLLUTION*	= 6
VI	*LIMITS OF DEVELOPMENT*	= 3
VII	*CLIMATIC CHANGES*	= 1
VIII	*GLOBAL MONITORING OF THE PLANET*	= 6
IX	*NEW MILITARY THREATS IN THE MULTIPOLAR WORLD*	= 3
X	*SCIENCE AND TECHNOLOGY FOR DEVELOPING COUNTRIES TO AVOID A NORTH-SOUTH ENVIRONMENTAL HOLOCAUST*	= 1
XI	*THE PROBLEM OF ORGAN SUBSTITUTION*	= 5
XII	*INFECTIOUS DISEASES*	= 7
XIII	*CULTURAL POLLUTION*	= 7
XIV	*COMMON DEFENCE AGAINST COSMIC OBJECTS*	= 2
XV	*THE HUGE MILITARY INVESTMENTS*	= 5
	Total	**= 63**

Synthesis of Planetary Emergencies.

I *Total: 4*

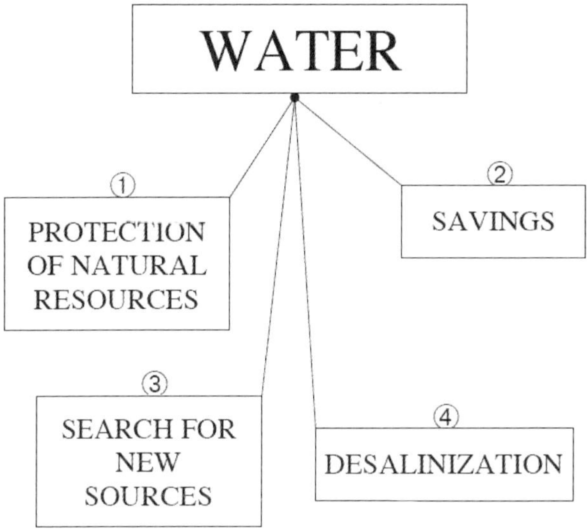

II *Total: 3*

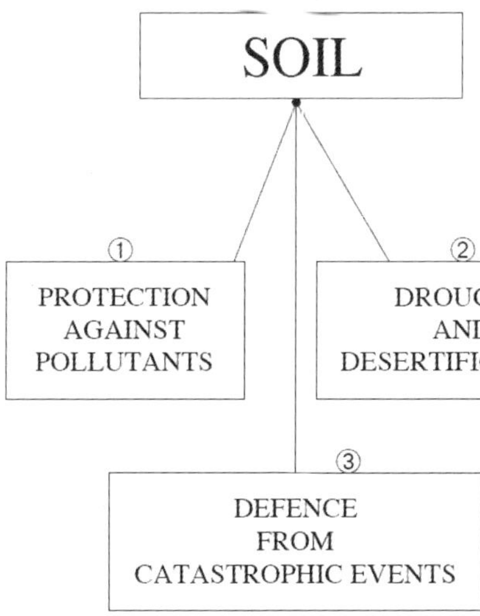

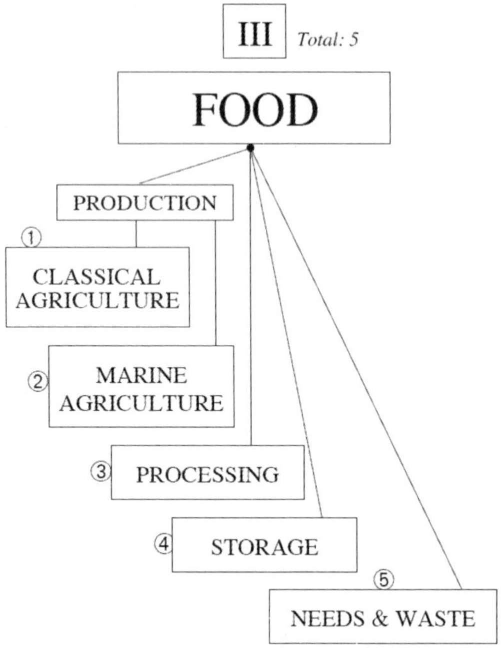
III
Total: 5
FOOD
PRODUCTION
1 CLASSICAL AGRICULTURE
2 MARINE AGRICULTURE
3 PROCESSING
4 STORAGE
5 NEEDS & WASTE

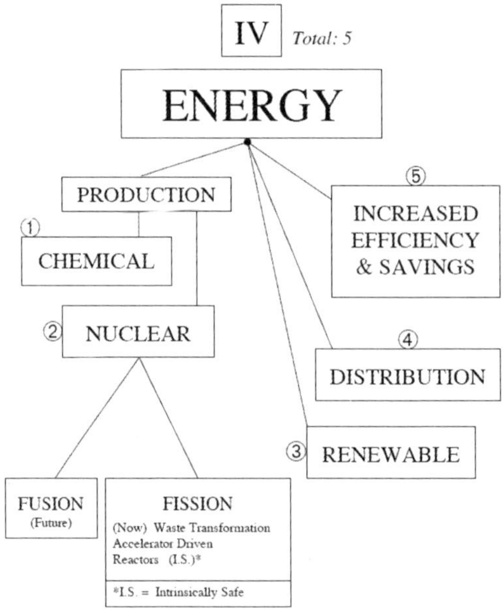
IV
Total: 5
ENERGY
PRODUCTION
1 CHEMICAL
2 NUCLEAR
FUSION
(Future)
FISSION
(Now) Waste Transformation
Accelerator Driven
Reactors (I.S.)*
*I.S. = Intrinsically Safe
3 RENEWABLE
4 DISTRIBUTION
5 INCREASED EFFICIENCY & SAVINGS

V *Total: 6*

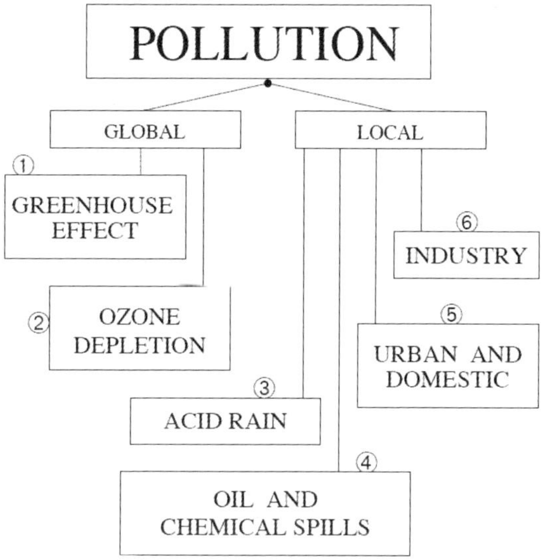

VI *Total: 3*

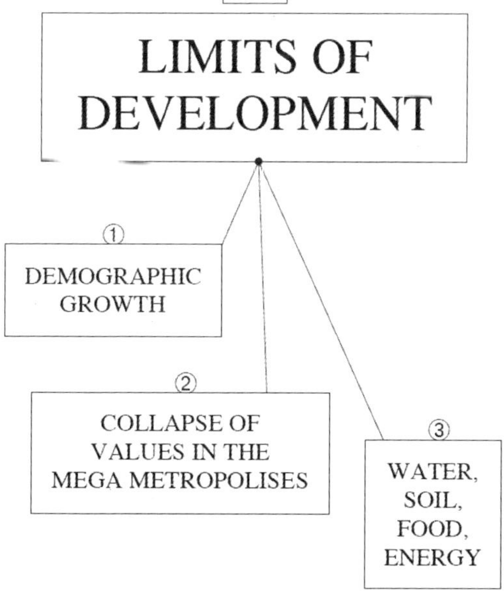

VII *Total: 1*

CLIMATIC CHANGES

①

AND THE SCIENTIFIC PROBLEMS ASSOCIATED WITH PREDICTION

VIII *Total: 6*

GLOBAL MONITORING OF THE PLANET

① POLLUTION MONITORING

② ATMOSPHERIC MONITORING

③ DEFENCE AGAINST FLOODS AND OTHER UNEXPECTED EXTREME METEOROLOGICAL EVENTS

④ EARTHQUAKES

⑤ VOLCANOES & CALDERA

⑥ DEFENCE AGAINST COSMIC OBJECTS

IX *Total: 3*

NEW MILITARY THREATS IN THE MULTIPOLAR WORLD

① GLOBAL DEFENCE SYSTEM

② DANGER OF PROLIFERATION OF WMD*

③ TERRORISM

*WMD = Weapons of Mass Destruction

X *Total: 1*

SCIENCE AND TECHNOLOGY FOR DEVELOPING COUNTRIES TO AVOID A NORTH-SOUTH ENVIRONMENTAL HOLOCAUST

①

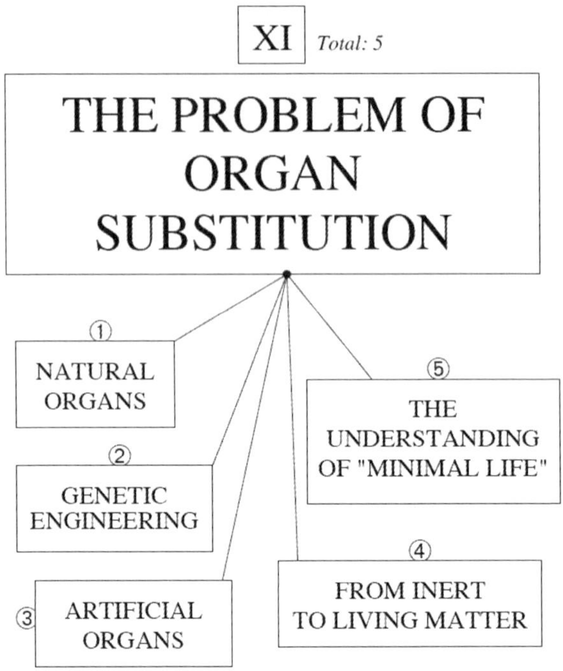
XI
Total: 5
THE PROBLEM OF ORGAN SUBSTITUTION
① NATURAL ORGANS
② GENETIC ENGINEERING
③ ARTIFICIAL ORGANS
④ FROM INERT TO LIVING MATTER
⑤ THE UNDERSTANDING OF "MINIMAL LIFE"

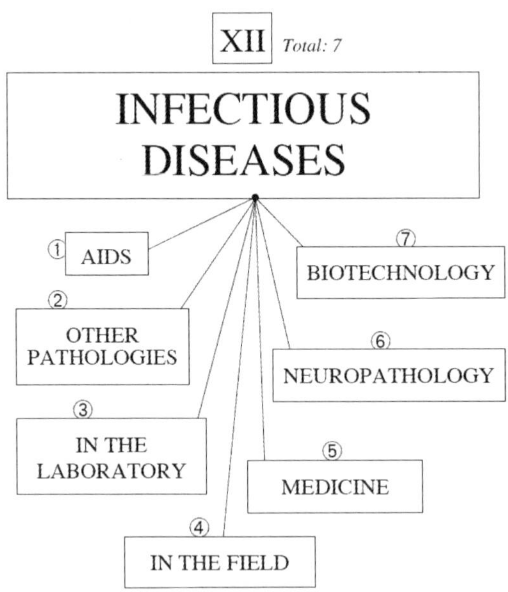
XII
Total: 7
INFECTIOUS DISEASES
① AIDS
② OTHER PATHOLOGIES
③ IN THE LABORATORY
④ IN THE FIELD
⑤ MEDICINE
⑥ NEUROPATHOLOGY
⑦ BIOTECHNOLOGY

XIII *Total: 7*

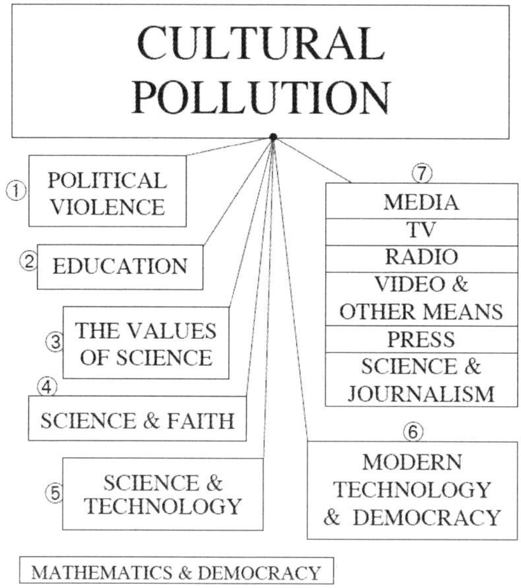

XIV *Total: 2*

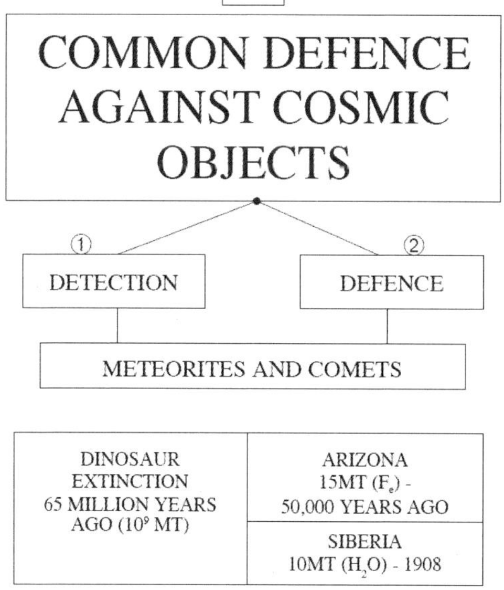

SCHOLARSHIPS PROGRAMME

To illustrate what WFS has been accomplishing concretely over the years, a few words about our WFS National Scholarship Programmes. As you all know, the WFS launched the National Scholarship Programmes in Newly Emerging Countries, East European Countries and Developing Countries.

Many of their young scientists require, upon graduation from their Universities, a year or more of training, under the supervision of an experienced scientist, before they are able to start a successful career in science.

Opportunities for obtaining post-graduate fellowships, to study in foreign institutes and laboratories, are scarce. In addition, their economic situation is such as to forbid them to continue their academic careers in their own country and forces them to choose jobs which are not related to their higher studies.

The World Federation of Scientists and the World Laboratory therefore instituted a National Scholarship Programme, which allows these young graduates to conduct scientific research activities in their own country, under the supervision of the best and most experienced national scientists. The duration of the scholarship depends on the field of Science chosen and otherwise is typically for one year - A few results:

National Scholarship Programmes were established in 33 countries, which are: Albania, Azerbaijan, Belarus, Bulgaria, Burundi, China, Ecuador, Estonia, Georgia, Ghana, India, Indonesia, Jordan, Kazakhstan, Korea, Kyrgyzstan, Lithuania, Macedonia, Madagascar, Malaysia, Mexico, Moldova, Pakistan, Palestine, the Philippines, Romania, Senegal, Slovenia, Tajikistan, the Ukraine, Vietnam, Uruguay and Uzbekistan.

THE ERICE PRIZE

In order to promote the values of scientific culture worldwide and following a proposal by the WFS, on the occasion of the 25th anniversary of the foundation of the EMFCSC, a special law was voted unanimously by the Sicilian Parliament to establish the 'Ettore Majorana Prize – Erice – Science for Peace'.

Every year, the Prize is awarded to distinguished scientists and world leaders who have contributed to the promotion of the values of scientific culture.

In the past decades, 58 prizes were awarded to:

P.A.M. Dirac, P.L. Kapitza, A.D. Sakharov, E. Teller, V.F. Weisskopf, J.B.G. Dausset, S.D. Drell, M. Gell-Mann, H.W. Kendall, L.C. Pauling, A. Salam, C. Villi, R. Doll, J.C. Eccles, T.D. Lee, L. Montagnier, Qian Jiadong, J.S. Schwinger, U. Veronesi, G.M.C. Duby, R.L. Garwin, S.L. Glashow, D.C. Hodgkin, R.Z. Sagdeev, K.M.B. Siegbahn, Y.P. Velikhov, J. Karle, J.-M.P. Lehn, A. Magnéli, N.F. Ramsey, H. Rieben, J.J. van Rood, C.S. Wu, R.L. Mössbauer, A. Müller, H. Kohl, M.S. Gorbachev, H.H. John Paul II, R. Clark, M. Cosandey, A. Peterman, R. Wilson, Lord J. Alderdice, J.I. Friedman, M. Koshiba, S. Coleman.

The 2007 prize was awarded to A.N. Chilingarov, P.C.W. Chu, L. Esaki, W.N. Lipscomb Jr., J. Szyszko and M-K. Wu.

The 2008 prize was awarded to Herbert Aaron Hauptman, David Hunter Hubel, Robert Huber, Bengt Samuelsson, Honglie Sun and Ada Yonath.

Let me close with a testimony of how much we believe in the Fermi Statement:

NEITHER SCIENCE NOR CIVILIZATION COULD EXIST WITHOUT MEMORY

Giancarlo Wick was a great physicist and strong supporter of the Erice Centre. This is the reason why the **Giancarlo Wick Gold Medal Award** has been established by the WFS.

It is a great pleasure for me to let you know that the Scientific Committee for the Gian Carlo Wick Gold Medal Award, chaired by T.D. Lee, has unanimously decided that the recipient for 2010 should be Professor Toichiro Kinoshita of Cornell University, with the following citation:

For his fundamental and very accurate calculations in Quantum Electrodynamics with deep consequences for the Physics of Fundamental Particles

Tonight we will have the celebration ceremony for the Gian Carlo Wick Gold Medal.

GIAN CARLO WICK
GOLD MEDAL AWARD CEREMONY
Eugene P. Wigner Institute – Sandro Pertini Cloister
20.00 – 20.30

And now, I invite Professor Toichiro Kinoshita to join me in order to give us a lecture on Quantum Electrodynamics, Electron Anomalous Magnetic Moment and Fine Structure Constant.

QUANTUM ELECTRODYNAMICS, ELECTRON ANOMALOUS MAGNETIC MOMENT AND FINE STRUCTURE CONSTANT

TOICHIRO KINOSHITA
Cornell University
Ithaca, USA

INTRODUCTION

Quantum mechanics[1,2] is the basis of our understanding of the material world and the foundation of many modern technologies.

However, it was known from the beginning that it is applicable only to a slowly moving object whose velocity v is small compared with the velocity c of light in vacuum.

In 1928 Dirac discovered an equation, called Dirac equation, which is consistent with special relativity and works at all velocities.[3] One of many successes of Dirac equation is its prediction that the gyromagnetic ratio of the electron g is equal to 2, in good agreement with the measurement then available.

However, the Dirac equation had a puzzling feature that it appears to predict presence of negative energy states. This was resolved by the reinterpretation that these states actually represent positive energy states of a positron (which has the same mass but electric charge opposite to that of the electron). Theoretically this is equivalent to treating the electron and the positron as operators of quantized Dirac field. The validity of this interpretation was given a firm experimental support by the discovery of the positron in cosmic ray experiment.[4]

By 1929 **quantum electrodynamics (QED)**, which describes the interaction of the electron (positron) field and the electromagnetic field in a way consistent with quantum mechanics and relativity, was formulated by Heisenberg and Pauli[5] and Dirac.[6]

Predictions of this theory in the lowest order of perturbation theory agreed with observations within the precision of few percents. However, if one tries to improve the prediction of theory further by including **higher order effects**, one runs into a peculiar situation that the result of calculation becomes **infinitely large**. This remained a big puzzle of QED for 15 years.

The key to solve this puzzle was found in 1947 when two experiments[7,8] showed that the prediction of Dirac equation requires a tiny but **non-vanishing** correction.

One is **the Lamb shift of the hydrogen atom**: $2S1/2$ state is about 1050 MHz higher than $2P1/2$ state, in disagreement with the Dirac equation which predicts that they are at the same level (namely degenerate).[7]

The other is **the anomalous magnetic moment of the electron** which was found in the measurement of the Zeeman splitting of Ga atom. Namely, the g-factor of the electron is about 0.1 % larger than the prediction of Dirac equation[8]:

$$a_e \equiv (g - 2)/2 = 0.001\ 15(4).$$

These experiments played crucial roles in solving the puzzle by helping establish the concept of **renormalization**, which is an assertion that, if the parameters of QED describing the rest mass and electric charge of an electron are identified with the observed mass and charge, these divergences disappear from calculated physical processes.

Bethe applied this concept within the **nonrelativistic framework** and obtained the Lamb shift in rough agreement with the measurement.[9]

For an unambiguous treatment of renormalization, however, it is necessary to have a relativistic formulation. Such a formulation was being developed by Tomonaga and Schwinger, unknown to each other.[10–12] Schwinger applied his theory of renormalized QED to calculate a_e and obtained[13]

$$a_e = \frac{\alpha}{2\pi} = 0.001161....$$

in excellent agreement with the measurement.

Together with Bethe's work on the Lamb shift, this provided firm experimental foundation to the **renormalized QED**.

Progress in g–2 measurement over 60 years since 1947 (spin precession[14], Penning trap[15]) has improved precision of measurement by **eight orders of magnitude**. Latest innovations and improvements of the Penning trap method have led to the precision of 0.24 ppb[16]:

$$a_e\,(\text{exp}) = 1\ 159\ 652\ 180.73\ (0.28) \times 10^{-12} \quad [0.24\ \text{ppb}].$$

In view of the simple theoretical structure and the availability of precise measurements the electron g–2 is particularly suitable for high precision test of the validity of QED. However, the (old) computational method employed by Tomonaga and Schwinger would have been too cumbersome to handle the theory to comparable precision.

It is the **diagrammatic method and use of Feynman propagators** invented by Feynman[17], developed by Dyson[18], and supplemented by Wick's theorem[19], that simplifies the calculation enormously and enables us to evaluate g–2 to any order in a systematic manner.

THEORY OF a_e UP TO ORDER α^4

Following Dyson we can write a_e as a_e = a_e (QED) + a_e (hadron) + a_e (electroweak), where

$$a_e(QED) = A_1 + A_2(m_e/m_\mu) + A_2(m_e/m_\tau) + A_3(m_e/m_\mu, m_e/m_\tau),$$

$$A_i = A_i^{(2)}\left(\frac{\alpha}{\pi}\right) + A_i^{(4)}\left(\frac{\alpha}{\pi}\right)^2 + A_i^{(6)}\left(\frac{\alpha}{\pi}\right)^3 + \dots, \qquad i = 1,2,3.$$

The first four A_1 terms are:

$A_1^{(2)} = 0.5$ 1 Feynman diagram (analytic)[13]

$A_1^{(4)} = -0.328\,478\,965\ldots$ 7 Feynman diagrams (analytic)[20,21]

$A_1^{(6)} = 1.181\,241\,456\ldots$ 72 Feynman diagrams (numerical, analytic)[22,23]

$A_1^{(8)} = -1.914\,4\,(35)$ 891 Feynman diagrams (numerical)[24,25]

$A2$ term is small but not negligible: $\sim 2.72 \times 10^{-12}$.

$A3$ term is completely negligible: $\sim 2.4 \times 10^{-21}$.

Non-QED contribution (Standard Model) are:

$$a_e\,(\text{hadron}) = 1.689\,(20) \times 10^{-12}\ {}^{(26,\,27)}$$

$$\text{and, } a_e\,(\text{electroweak}) = 0.030 \times 10^{-12}\ {}^{(28)}$$

From these results, and assuming $|A_1^{(10)}| < 4.6$ for the unknown 10th-order term[29], one finds:

$$a_e\,(\text{Rb}) = 1\,159\,652\,182.79\,(0.11)(0.37)(7.72) \times 10^{-12},$$

$$a_e\,(\text{exp}) - a_e\,(\text{Rb}) = -2.06\,(7.72) \times 10^{-12}.$$

where

$$\alpha{-1}(\text{Rb}) = 137.035\,998\,84\,(91) \qquad [6.7\ \text{ppb}],$$

is the value obtained by an optical lattice method.[30]

Note that uncertainty 0.11 of $A_1^{(8)}$ and *guesstimated error* 0.37 of $A_1^{(10)}$ are much smaller than 7.72 of the best measured α available[30]. In other words non-QED α, even the best one, is too crude to test QED to the precision achieved by theory and measurement of a_e.

Thus it makes more sense to test QED by an alternative approach:

Get α from theory and measurement of a_e
and compare it with other α's.

This yields

$$\alpha{-1}(a_e) = 137.035\,999\,085\,(12)(37)(33) \qquad [0.37\ \text{ppb}],$$

where 12, 37, 33 are the uncertainties of 8th-order, 10th-order, and a_e (exp), respectively.

High precision α's obtained by various means are compared in Figure 1. Note that $\alpha(a_e) \equiv \alpha(a_e\ [\mathrm{HV}\ 08])$ is 18 times more accurate than α(Rb).

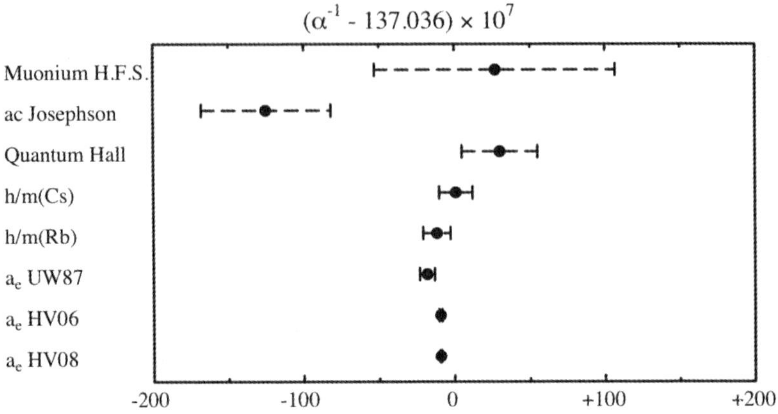

Fig. 1: Comparison of various α^{-1} of high precision.

THE TENTH-ORDER TERM

Is it possible to make $\alpha(a_e)$ even more precise? On the experimental side the Harvard group is working to reduce the measurement error further. On the theoretical side further progress is impossible unless $A^{(10)}$ is evaluated explicitly.

Thus we began working on 10th-order terms more than 7 years ago anticipating future improvement in the measurement of a_e.

12672 Feynman diagrams contribute to $A_1^{(10)}$.
9080 Feynman diagrams contributing to $A_2^{(10)}$.

Obviously this is a gigantic project, requiring systematic and highly automated approach.

First step: Classify them into gauge-invariant sets. There are 32 gauge-invariant sets within 6 supersets.

Actually some of these sets are not too difficult. By 2006 we have evaluated 17 of 32 sets.[33] Since then we have also evaluated and published three more sets.[34,35,36]

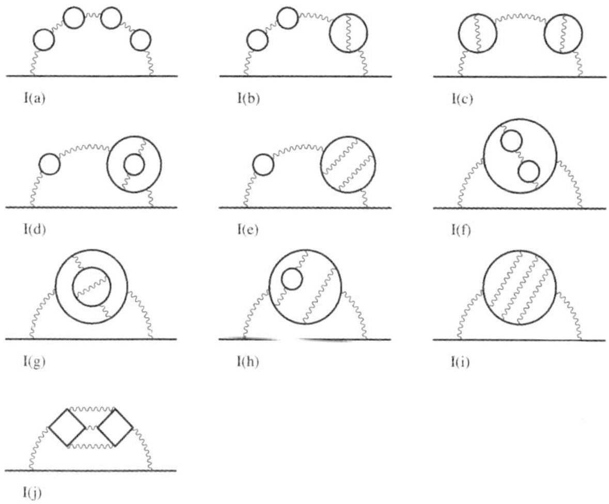

Fig. 2: Diagrams of Superset I. This set consists of 10 subsets, all built from a second-order vertex. Solid lines represent electron propagating in the magnetic field. Wavy lines represent photons. 208 diagrams contribute to $A_1^{(10)}$. *498 contribute to* $A_2^{(10)}(m_e/m_\mu)$.

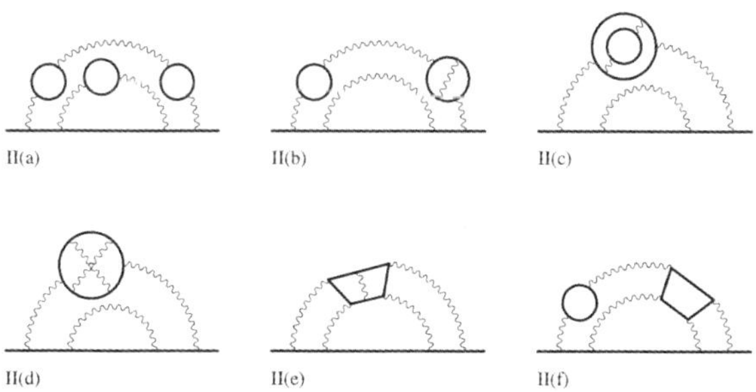

Fig. 3: Diagrams of Superset II. This set is built from fourth-order proper vertices. 600 diagrams contribute to $A_1^{(10)}$. *1176 contribute to* $A_2^{(10)}(m_e/m_\mu)$.

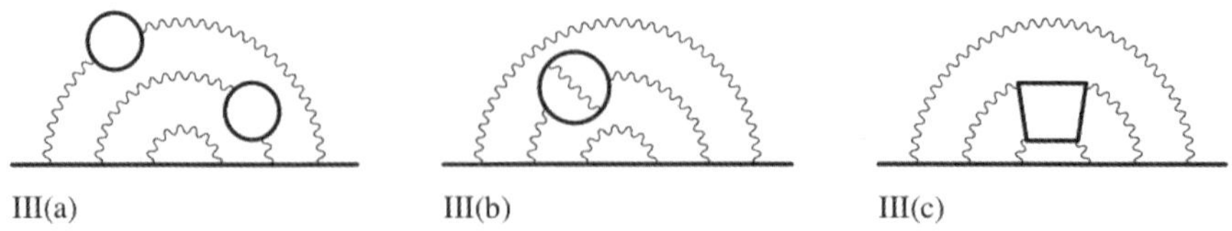

Fig. 4: Diagrams of Superset III. This set is built from sixth-order proper vertices. 1140 diagrams contribute to $A_1^{(10)}$. 1740 contribute to $A_2^{(10)}(m_e/m_\mu)$.

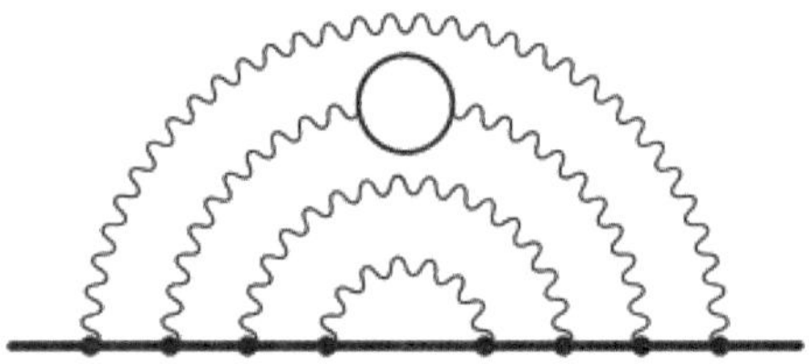

Fig. 5: Diagrams of Superset IV. This set is built from eighth-order proper vertices. 2072 diagrams contribute to $A_1^{(10)}$ and $A_2^{(10)}(m_e/m_\mu)$.

The remaining sets are more difficult to handle. To deal with these sets we developed an **auomatic code generator.**[31,32] Integrals thus obtained are evaluated by **VEGAS, adaptive-iterative Monte-Carlo routine.**[37] By now we have FORTRAN codes of all the remaining sets, which have been evaluated numerically at least once. Particularly difficult is thc Set V whose codes are huge and is the source of largest numerical uncertainty. After many thousand hours of computation, the uncertainty in $A_1^{(10)}$ has been reduced to less than 0.8, which is still preliminary and crude but is already about 6 times more precise than the previous guesstimate $|A_1^{(10)}| < 4.6$.[29] The uncertainty in $A_1^{(10)}$ is being reduced further.

We are working to reduce the uncertainty of $A_1^{(8)}$, too. At present the uncertainty has been reduced to less than 80% of the previous result.[25]

In the near future we will obtain $\alpha(a_e)$ which will be about 30 times more precise than $\alpha(Rb)$. What this means is that we cannot test the validity of QED in terms of a_e unless some way is found to measure an independent α to a precision comparable to that of $\alpha(a_e)$. Until then, $\alpha(a_e)$ serves as a standard by which the validity of other measurements and their theories are examined.

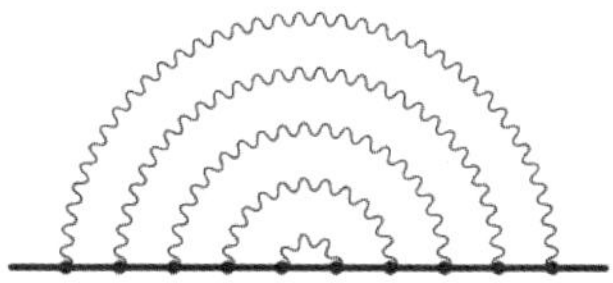

Fig. 6: Diagrams of Superset V. This set consists of 10th-order proper vertices with no closed lepton loop. 6354 diagrams contribute to $A_1^{(10)}$. *No contribution to* $A_2^{(10)}$ (m_e/m_μ).

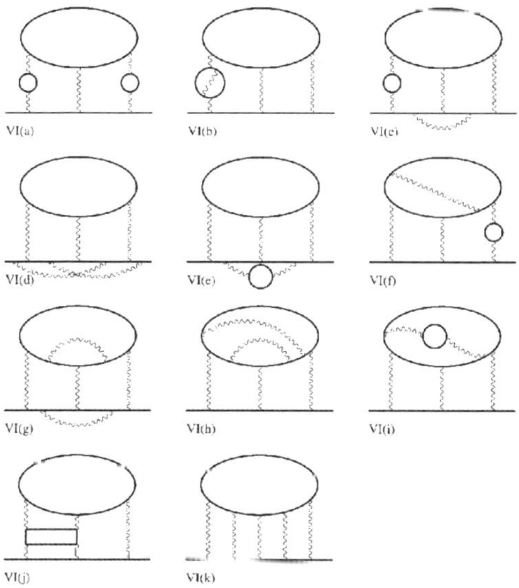

Fig. 7: Diagrams of Superset VI. This set consists of 11 subsets, which contain light-by-light- scattering *subdiagrams. 2298 diagrams contribute to* $A_1^{(10)}$. *3594 contribute to* $A_2^{(10)}$ (m_e/m_μ).

DISCUSSION

The discoverers of QED regarded the renormalization procedure as a *jelly-built* temporary fix to be replaced by something better.[38,39] In fact it was soon found that, as the theory of fundamental physics, QED must be enlarged to include hadronic and weak interactions, which led to the Standard Model (SM). But the jelly-built structure itself has remained as the basic framework of SM.

SM itself is generally regarded as temporary measure which requires further modification to accommodate new physics. Such modification is most likely to come from experiments at high energy accelerators such as LHC.

However, the impact of new physics on a_e may not necessarily be straightforward. As a matter of fact, a possibility exists that it has no detectable effect on $g-2$.

Reason: Independent α's measured by non-QED means must also include the effect of new physics.

Recall that masses and charges involved in ordinary QM cannot be correctly identified as physical mass and charge to the precision that requires radiative corrections.

For proper interpretation ordinary QM, by which these measurements of α are interpreted, must be the nonrelativistic limit of explicitly renormalized QED (or SM)[40], unfortunately misnamed NRQED.

It is amazing that the jelly-built structure still looks good at the precision exceeding 1 part in 109. If disagreement is detected at the next level of precision it might indicate that the jelly-built structure is not as sturdy as has been thought thus far. This might imply that breakdown of SM comes, not necessarily from high energy region, but from an entirely unexpected cause.

ACKNOWLEDGMENTS

This work is supported in part by the U.S. National Science Foundation Grant NSF-PHY-0757868. The author thanks M. Nio for assistance in preparation of this article.

REFERENCES

1. W. Heisenberg, *Z. Physik* 33:879 (1925).
2. E. Schro¨dinger, *Ann. der Phys.* 81:109 (1926).
3. P.A.M. Dirac, *Proc. Roy. Soc.* (London), A117, 610 (1928).
4. C.D. Anderson, *Phys. Rev.* 41:405 (1932).
5. W. Heisenberg and W. Pauli, *Z. Physik* 56:1 (1925).
6. P.A.M. Dirac, *Proc. Roy. Soc.* (London), A136, 453 (1932).
7. W.E. Lamb and R.C. Retherford, *Phys. Rev.* 72:241 (1947).
8. P. Kusch and H.M. Foley, *Phys. Rev.* 72:1256 (1947).
9. H.A. Bethe, *Phys. Rev.* 72:339 (1947).
10. S. Tomonaga, RIKEN-IHO 22, 545 (1943) in Japanese; *Prog. Theor. Phys.* 1:27 (1946).
11. Z. Koba and S. Tomonaga, *Prog. Theor. Phys.* 2:218 (1947).
12. J. Schwinger, *Phys. Rev.* 73:416L (1948); 74:1439 (1948); 75:651 (1949); 76:790 (1949).
13. J. Schwinger, *Phys. Rev.* 73:416L (1948); 75:898 (1949).
14. J.C. Wesley and A. Rich, *Phys. Rev.* A 4:1341 (1971).
15. R.S. Van Dyck, Jr., P.B. Schwinberg, and H.G. Dehmelt, *Phys. Rev. Lett.* 59:26 (1987).
16. D. Hanneke, S. Fogwell, and G. Gabrielse, *Phys. Rev. Lett.* 100:120801 (2008).
17. R.P. Feynman, *Phys. Rev.* 74:1430 (1948); 76:769 (1949).
18. F.J. Dyson, *Phys. Rev.* 75:486 (1949); 75:1736 (1949).
19. G.C. Wick, *Phys. Rev.* 80:268 (1950).

20. A. Petermann, *Helv. Phys. Acta* 30:407 (1957).
21. C.M. Sommerfield, *Phys. Rev.* 107:328 (1957).
22. S. Laporta and E. Remiddi, *Phys. Lett. B*, 379:283 (1996).
23. T. Kinoshita, *Phys. Rev. Lett.* 75:4728 (1995).
24. T. Kinoshita and M. Nio, *Phys. Rev.* D 73:013003 (2006).
25. T. Aoyama, M. Hayakawa, T. Kinoshita, and M. Nio, *Phys. Rev.* D 77:053012 (2008).
26. B. Krause, *Phys. Lett.* B, 390:392 (1997).
27. A. Nyfeller, arXiv:0901.1172 [hep-ph].
28. A. Czarnecki, B. Kraus, and W.J. Marciano, *Phys. Rev. Lett.* 76:3267 (1996).
29. P.J. Mohr, B.N. Taylor, D.B. Newell, *Rev. Mod. Phys.* 80:633 (2008).
30. P. Clad´e, et al., *Phys. Rev.* A 74:052109 (2006).
31. T. Aoyama, M. Hayakawa, T. Kinoshita, and M. Nio, *Nucl. Phys.* B740:138 (2006).
32. T. Aoyama, M. Hayakawa, T. Kinoshita, and M. Nio, *Nucl. Phys.* B796,:184 (2008).
33. T. Kinoshita and M. Nio, *Phys. Rev.* D 73:053007 (2006).
34. T. Aoyama, M. Hayakawa, T. Kinoshita, M. Nio, and N. Watanabe, *Phys. Rev.* D 78:053005 (2008).
35. T. Aoyama, M. Hayakawa, T. Kinoshita, and M. Nio, *Phys. Rev.* D 78:113006 (2008).
36. T. Aoyama, K. Asano, M. Hayakawa, T. Kinoshita, M. Nio, and N. Watanabe, *Phys. Rev.* D 78:113006 (2008).
37. G.P. Lepage, *J. Comput. Phys.* 27:192 (1978).
38. S. Tomonaga, private communication.
39. F.J. Dyson, in *Physics Today*, 15, August 2006.
40 W.E. Caswell and G.P. Lepage, *Phys. Lett.* 167B:437 (1986).

THE POTENTIAL IMPACT OF THE IRRATIONAL ON THE SUCCESS OF SCIENCE

THE LORD ALDERDICE FRCPsych
Chairman, PMP on Motivations for Terrorism
House of Lords, London UK

According to a Pew Research survey of 3,003 Americans, conducted in recent days (but just before 13 August 2010 when President Obama defended Muslims' right to build an Islamic centre near Ground Zero in New York) a growing number of Americans incorrectly believe that President Barack Obama is a Muslim. Some 18% said the President was a Muslim (up from 11% in March 2009) and among Republican supporters, that number was 34%. Just less than that (a third of those quizzed) correctly identified Mr. Obama as a Christian, despite his regular public observations of his faith. Forty-three percent of those questioned said they did not know what Mr. Obama's religion was.

This finding is even more startling than the regular observation of the high percentages of people in the Muslim world who believe that the 9/11 attacks were conducted either by the CIA or by Israel in order to implicate Muslims and justify the subsequent attacks on Afghanistan and Iraq and, possibly in the future, on Iran.

It is clear that whole populations, including in some of the most highly educated parts of the world, can be held in the grip of wholly false beliefs that have potentially enormous and damaging implications for international relations and world peace. It is also clear that such false beliefs have a number of specific characteristics. They usually have some piece of material that can be misinterpreted—the President is called Barack Hussein Obama. His name is unusual and his middle name is a common Muslim name. The misinterpretation has negative connotations—in the aftermath of 9/11, the implication that he is a Muslim is not just randomly negative, it implies that he is an enemy or in league with the enemy of the country. There is a negative and clearly paranoid colouring to the delusion, and it is not reducing in strength through exposure to reality—on the contrary as his overall popularity rating slides, so the idea that he is a Muslim gains strength. The false belief fulfils a function. It expresses feelings of fear and anger in an American community which sees the eclipse of its pre-eminent place in the world, the demise of its remarkable prosperity, the failure in a number of military adventures and the sense that it is losing the so-called 'wars' on terrorism, drugs, poverty and so on. The delusionary belief that their President is actually a closet enemy of the state, out to bring it to its knees expresses profound fears and anger about these threats and losses to an America which has positively defined its very identity in the 'American Dream' of success and a good life based on its political and economic freedoms, religious faith and educated population. One should not ignore the fact that many people in the country do not share the enthusiasm of the rest of the world that the United States has elected a black President, but to express criticism of the President on account of his skin colour or race is not acceptable in a country which has had such profound struggles, and indeed a civil war, on this very question of the place of Americans of African origin in U.S. society.

It is crucial to observe that this irrational belief is emerging in a country with a free press and an educated population which has access to more information than at any time in history. One is reminded that the Germany in which National Socialism emerged was also one of the most highly educated and advanced countries in the world at the time and had a profoundly democratic Weimar constitution, but its population felt deeply the sense of loss and humiliation represented by the defeat in the First World War, the reparations of the Treaty of Versailles and the economic crisis of the 1930's and the last Great Depression. In such circumstances education and democratic freedoms are not a sufficient protection against the rise of powerful irrational and dangerous paranoid beliefs, indeed the educated population in Germany turned their abilities to the use and abuse of science itself with the catastrophic consequences which still leave a profound legacy on our world two generations on.

Professor Zichichi, in his important presentation today, *Why Science is needed for the Culture of the Third Millennium*, has emphasized that without the development of language and subsequently of logic it would not have been possible for human beings to move on to the science on which so much of the progress of the modern world is based. Language enabled mankind to step away from the rest of the animal kingdom by being able to exert some more control, inhibition, delay and transformation of his physical emotions of hunger, aggression, anxiety, terror, envy and the sexual drives. The development of language also facilitated communication between individuals and within the self by enabling the capacities of reflection and memory. In addition, language made communication and learning possible between groups of people and succeeding generations, especially when it became possible to record thoughts and ideas in written form. Language made possible the growth of logic and the rationality through which it was possible to distinguish between what one wished or feared on the one hand, and external reality on the other. Science took this further by testing thinking against reality in a very powerful way.

It is important to appreciate however that the evolution of language and the development of logic did not remove emotion, nor did it remove the possibility of a falling back or regression into emotionally driven mental functioning and behaviour. In the individual we see dissolution of the more advanced thinking capacities in mental illness and the re-appearance of what we call 'primary process thinking'. Individuals can also lose the capacity to think rationally and logically as a result of organic damage or emotional overload. All of us as individuals can find our capacity to think logically being transiently affected by anger, through a reversal in fortunes or a humiliation, or more positively by falling in love. However, when someone falls mentally ill for organic or psychological reasons such a disturbance of their thinking may not be transient, rarely has positive consequences, and is not usually resolved merely by improving the person's external circumstances.

One of the disturbances of individual thinking which can appear in psychotic mental illness and in organic dementias is the *'delusion'*. This is a fixed false belief, out of line with the person's cultural context, and from which the person cannot be persuaded by rational argument or reasonable evidence. Their appreciation of reality is distorted to fit their delusionary belief, rather than their delusion being susceptible to disproof by confronting reality. Their behaviour is then disturbed as they act on the basis of their delusionary ideas. As I noted earlier, however, fixed false beliefs can also arise in groups where they are widely shared by many people with a common cultural context. This is not a sign of widespread individual psychotic illness or organic disturbance. It is a group phenomenon; a disturbance of the functioning of the 'mind' of the group. It does indeed appear that while the group does not have a brain in the way we understand individuals to have it can, nevertheless, have a mind that operates through the relationships between human beings. What is it that makes whole groups or communities hold to fixed false beliefs contrary to reasonable evidence? Why do communities hold to what we would regard as non-scientific thinking even though individuals within the group may be functioning within normal limits?

From the time of the Enlightenment it was generally assumed that such irrational beliefs were simply a matter of ignorance and that a few generations of education, especially scientifically informed education, would relieve communities of irrationality. However, while some progress was made in this regard it was not anything like as widespread as scientific thinkers had hoped, nor was rational scientific thinking a permanent achievement. It is clear that it is not only in the mental life of individuals that irrationality can take over, but also in the thinking of a group or community. Our research in the World Federation of Scientists has shown us that such a descent into irrational or non-scientific thinking is often associated with deep group anxiety. Examples of this arise when a community feels it is in danger of economic chaos, physical external attack, or threats to religious faith, culture or other aspects of the group's identity. Science and scientific thinking, which one can see as the characteristic of healthy, rational functioning in a group or community, can be damaged when communities descend into fundamentalism which is anti-scientific and produces a primitive form of religion or of group thinking in general. It is important to note that all religious thinking does not fall into this category. I am referring specifically to that form of thinking which we describe as 'fundamentalist'. Any religious family of belief may be held either in a fundamentalist form, or in a more advanced form. Indeed a descent into the irrational is not a specifically religious issue at all and is not only seen where there is a direct attack on scientific thinking from anti-science religious fundamentalists. Even some scientists can become scientific fundamentalists. The scientific approach can be corrupted from 'inside the camp' as it were. A properly scientific approach demands that in the face of uncertainty we take a rational approach which accepts that there is uncertainty, and that we may have to live with 'not knowing' for a long time until the truth becomes clear. However when there is a serious threat or uncertainty for a group it may seek to remove the uncomfortable anxiety provoked by this uncertainty by jumping to conclusions which are given a cloak of scientific respectability by the fact that it is held by people who are called scientists, rather than by the scientific rigor of the approach. For example, the uncertainty and fear over climate change and environmental degradation can lead communities to jump to illusory conclusions about how far humanity can control the forces of nature rather than living with the uncertainty which a rigorous and long-term scientific investigation requires.

Unfortunately the levers of rational enquiry are not strong in the face of passionate anxiety and the resultant regression into a form of thinking dominated by emotion rather than logic. There are a number of problematic implications. Firstly, such thinking is by its very nature not very susceptible to rational argument and debate, not least because part of its function is not to facilitate understanding but to express anxiety.

Secondly, it tends to be held aggressively (though not usually violently) resulting in attacks on those who insist on holding on to scientific uncertainty. The problem of predicting outcomes in certain circumstances, such as where there are a very large number of variables, or where for other reasons prediction is simply not possible, is very frightening to many people who would rather have false certitude than honest uncertainty. The increasing adoption of irrational beliefs in the United States, for example, is not restricted to the religion of the President. School curricula are being changed to set creationism alongside evolution in science classes. In some cases creationism is actually replacing teaching about evolution. Not alone can we not be sure that education can entirely protect us from unscientific thinking, but even worse education can be made to promote irrational anti-scientific thinking.

How can we ensure that our societies are not overwhelmed by fear-driven irrationality? Some of the answer is directly political with a large 'P' and as such may seem outside of the remit of the WFS. What is however within our purview is to apply our scientific approach to understanding the nature of irrationality itself, not only in the individual (as we do in psychiatry), but also in the group. Only if we focus upon and increase our understanding of group irrationality and how and why it occurs can we hope to prevent our global community from falling victim to ignorance and misinformation which pretends to be knowledge, and an appearance or 'gloss' of science contributing to the worsening cycle of panic rather than building a truly scientific approach that may be able to help us out of our problems. This is what we have been trying to do through the Permanent Monitoring Panel on Motivations for Terrorism. We believe we have made significant progress in our understanding of irrational group thinking and behaviour, but as Galileo found more than 400 years ago, new scientific understandings which threaten current views and structures of power are usually not well-received and adopted especially when they are presented to a population whose form of thinking is not dominated by rationality. In the same way when we find ways of understanding how irrational beliefs come about we are often accused by the fundamentalists on one side of sympathizing with the fundamentalists on the other side.

An example, from another field, of how apparent progress can actually lead to a worsening reality is provided by the growth of some of the structures of democracy in Africa. In the past, sub-Saharan African society was largely tribal and its technology was modest. The judgement of the tribal chief was often reached after some consultation with the elders but was final, however in the larger scheme of things it was also of limited effect because the capacity to wage war and cause destruction was modest. Law and the institutions of western style democracy accompanied by the technology of security, defence and the weapons of war were rapidly imported into Africa during the twentieth century without a concomitant fundamental change and development of the indigenous culture and in particular without the time for such processes to properly take place. This meant that old-style chiefs could now use electoral processes and law, not to create liberal democracies but to maintain old tribal attitudes, and in addition to enforce them not with the limited capacities of spears and foot-soldiers, but with a fearsome battery of modern weapons of war, which they could not themselves produce, but which are now available for them to use. It had been believed that if the institutional and physical tools of modernity were provided, a tolerant liberal culture would inevitably develop and in short order. It is by no means clear that this is true. The technology made possible by science may well be used by a society whose culture is enriched by a highly developed and positive view of humanity. In this case we can have some chance (though no certainty) of relative peace, stability and mutual benefit and understanding. On the other hand, without the culture of science, the technology developed out of the science of recent centuries can contribute instead to fear, economic chaos, war, and social destruction. Terrorism is a symptom of such cultural regression of the group through fundamentalist ways of thinking, and in addition political radicalization.

The recent economic chaos is another case where the gloss of economic 'science' was used to bolster the wish for cost-free prosperity, but actually led to catastrophe. The illusion of the 'science' of economics lies in its failure to recognize that with so many variables, and a very high level of complexity of function, accurate prediction in economics is simply not possible. In this case the irrational way of thinking was not a paranoid one developed in the service of fear, but an illusion in the service of a different emotion—the wish for easy wealth. Economic markets are largely a function of group psychology rather than mathematical calculus, and so psychological contrarians are as likely to be able predict outcomes as the mathematical economic analysts. Markets are not random, but nor (except in certain limited contexts) are they predictable, and to believe otherwise is not rational belief, it is an expression of a wish.

In each of these three areas—the paranoid functioning of a group in a climate of fear and uncertainty; the challenge to a group to adapt quickly to new and unfamiliar technological developments; and the seductiveness of the wish to believe in the rapid, easy creation of wealth—we need the culture of science to protect us from disaster. However since we cannot avoid the development of irrational group thinking we must work on a more scientific understanding of man's irrationality, especially irrationality in group functioning, to help us protect science and the scientific method from being dismissed by religious fundamentalists who have always been suspicious of it, or abused by some who claim to be adherents of science but are actually themselves fundamentalists who mislead many people and potentially bring on cultural catastrophe.

Let me finish by noting that I have concentrated entirely on the potential negative impact of the irrational on the success of science, however this focus is a function of the shortage of time and the immediate dangers of the negative consequences. Irrationality is also crucial for creativity—the capacity to imagine, to dream, to fantasize and to go beyond and sometimes against the current perceived rational view of reality is as essential for scientific as it is for artistic creativity. But my time is gone and the consideration of these positive aspects of the irrational must wait for another day.

SCIENCE AND THE FUTURE OF CYBERSPACE: AN ITU PERSPECTIVE

DR. HAMADOUN I. TOURÉ
Secretary-General, International Telecommunication Union
Geneva, Switzerland

Professor Zichichi, Distinguished Guests, Ladies and Gentlemen,

It is a great pleasure to be here with you in Erice today to discuss ITU's perspective on *Science and the Future of Cyberspace*. Given that they are now so completely interdependent, however, I could just as easily say that I am here to discuss ITU's perspective on *Cyberspace and the Future of Science*.

The sum of scientific knowledge and understanding has increased exponentially over the past two decades, and I believe that this is in very great part due to the massive uptake and proliferation of information and communication technologies (ICTs).

We have seen the number of mobile cellular subscriptions grow almost five-hundred-fold, from 11 million worldwide at the end of 1990 to over five billion by the end of this year.

Twenty years ago, virtually nobody was online. Today, almost two billion people have access to the Internet, and by the end of this year there will be close to 900 million mobile broadband subscriptions worldwide.

At ITU, we believe that this will have a profound affect on the future of each and every person on the planet, and that this will have huge implications for the future of science.

We are already seeing many examples of this, across multiple sectors, and I firmly believe that over the coming years we will see many more.

Take the field of medicine, where massively powerful computers and the power of online collaboration have enabled huge advances in our understanding of human biology and chemistry, helping us unlock the secrets of life itself.

ICTs are also driving a revolution in the provision and delivery of healthcare—through applications as simple as using text messages to remind patients to take their medicine or as complex as surgery performed remotely using telemedicine.

Looking at the field of education—which of course is the key to all scientific progress—we are now seeing a dramatic shift away from classrooms and into cyberspace.

Distance learning is not a new phenomenon, but its uptake—as more and more people get online—has recently mushroomed, and the world's biggest universities are now online establishments. The two largest are the Indira Gandhi National Open University in India, which has an astonishing three million students enrolled, and the Allama Iqbal Open University in Pakistan, which has 1.8 million students enrolled.

This will change everything. For the first time in human history, the potential of individuals to realise their dreams and ambitions will not be limited by the place they were born, by their status in the family, or by their wealth. As long as they are able to get online, and into cyberspace, anybody in the world will be able to get a decent education.

Distinguished guests,

ICTs will also help us address the most pressing issues of our time—including, most notably, climate change.

Take the example of satellites, which will be increasingly important in the future in helping to boost food output and reduce emissions—with satellite monitoring producing 98% fewer emissions than ordinary ground monitoring. Satellite-based intelligence services for farmers, for their part, cost less than U.S. $15 per hectare annually and can increase yields by as much as 10%.

Satellites also play a vital role when natural disasters and catastrophes occur, often providing the only reliable communications links when other networks have been destroyed. ITU is proud to play a very active role wherever we can: in Haiti and Chile, earlier this year, for example, where we were quick to provide emergency communications services.

ICTs will also play an increasingly important role in helping to reduce energy consumption. While technology itself contributes around 2.5% of global greenhouse gas emissions, smart use of modern technologies could cut global power consumption by 15%.

ICTs help to reduce waste, cut business travel and make industry more efficient. And new technologies being developed within ITU—such as Next-Generation Networks—can reduce network and data centre power consumption by up to 40%.

In the future I am confident that the smart application of science will see us achieve much more in cyberspace than we have ever been able to achieve in the real world alone.

I expect transport, for example, to become much safer than it is today, with cars—equipped with smart ICTs and communicating with one another in real-time—becoming able to avoid having accidents altogether.

And I expect networked ICTs to dramatically improve the lives of the disabled, the elderly and the infirm.

Underpinning all of this progress will be broadband networks—and equally, without ubiquitous broadband networks, we will be able to make none of this progress.

It is therefore absolutely essential that we create the right conditions to allow the rapid rollout and proliferation of broadband networks, and that access to broadband becomes much more affordable around the world. This point is especially important, as today there are quite extraordinary disparities in the affordability of broadband access.

Ladies and gentlemen,

Let me give you some dramatic figures from ITU's 'Measuring the Information Society 2010' report.

In the top 21 countries included in the fixed broadband Internet sub-basket, published in the report, broadband subscriptions cost less than 1% of average monthly income—and under 3% of average monthly income in a further 22 countries. Here in Italy, for example, the figure is just 0.98%.

At the other end of the scale, in the most expensive 28 countries in ITU's list—most of which are UN-designated Least Developed Countries, LDCs—a monthly broadband subscription costs over 100% of average monthly income.

What a terrible irony that is!

The people who can least afford access to broadband are being asked to pay the most, relative to their income.

This is scandalous. But there are grounds for optimism, and broadband access is in fact getting more affordable every year. Worldwide, broadband prices dropped by 42% between 2008 and 2009, and broadband became more affordable in almost every market across the globe last year.

There are many reasons for this, and I am confident that broadband will continue to get more affordable as time goes on.

One of the most important drivers in terms of bringing down costs in both the developed and the developing world has been an ongoing increase in capacity, itself driven by a newly competitive environment.

In Africa, for example, we are seeing new broadband capacity coming on-stream fast, and I was delighted to be personally present in Kenya when a new submarine cable was brought onshore in March, and then to be in French Guyana earlier this month to see new satellites for Africa being launched.

By increasing the size and scale of cyberspace, these are truly inspirational events, and signs of the very positive times in which we live.

It is vital that we keep up the momentum, and that we do everything in our power to ensure that the digital divide does not become a broadband divide.

This is why ITU launched our 'Build on Broadband' initiative earlier in the year, and why we are so proud to be playing a key leadership role in the 'Broadband Commission for Digital Development'.

The Broadband Commission is chaired by President Kagame of Rwanda and Carlos Slim Helú, Honorary Lifetime Chairman of Grupo Carso. The Vice-Chairs are myself, and Irina Bokova, Director General of UNESCO, and we have a total of over 50 very high-profile Commissioners from the public and private sectors.

Our Commissioners come from many different walks of life and many different backgrounds, but they all share a common vision: the desire to see the vast benefits of broadband brought to all the world's people.

Together, we are determined to demonstrate broadband's truly transformational nature. And together, we believe we have a unique, once-in-a-generation opportunity to drive social and economic progress.

The Commission will be delivering its report to the UN Secretary-General, Ban Ki-moon, on 19 September, just ahead of the 2010 MDG Summit in New York.

Distinguished guests,

The future of science and the future of cyberspace are clearly inter-linked and inter-dependent, and will rely on ubiquitous broadband networks.

But while broadband can bring the world's riches within reach of all the world's people, it also opens up new opportunities for wrong-doers, and in particular cybercriminals.

It is estimated that each year cybercriminals steal up to a trillion dollars worth of intellectual property from businesses worldwide. In addition, many millions of individuals have their privacy violated; suffer identity theft or have their hard-earned savings stolen from them.

Governments constantly face cyberattacks—and terrorists increasingly rely not just on their weapons, but on the power of cyberspace technologies like GPS and VoIP to sow destruction.

As a result, the next major war will probably begin in cyberspace rather than on the ground.

Given the scale of the threat—and the phenomenal harm that can be caused by even a single cyber attack—we cannot rely on ad hoc solutions or hope to survive by strengthening our defences only after attacks have occurred.

No, we must work together, to ensure a coordinated response.

This is why ITU is playing a lead role in coordinating global efforts in this area, and why we launched the Global Cybersecurity Agenda (GCA) in 2007.

The GCA is now in its operational phase, with a physical home in Malaysia at the headquarters of IMPACT—the International Multilateral Partnership Against Cyber-Threats.

If users lack faith in the security of the Internet, it will not continue to flourish as a facilitator of learning, as a platform for telemedicine, or as a tool for more efficient and accountable government. Or indeed as a key driver of trade and commerce, as a global communications channel, or as a vital research tool.

So we need to be sure we have an environment where criminals cannot hide behind legal loopholes and regulatory inconsistencies. Nations with less well-developed ICT legislation should no longer find themselves host to villainous online activities. And all states—from the most prosperous to the most disadvantaged—need to have an effective shield with which to safeguard themselves.

We must therefore take action to enhance security in cyberspace, and increase cooperation and coordination at the global level.

Existing frameworks, while good, are not in fact global. The best example is the Budapest Convention on Cybercrime, which has been ratified by only 26 countries. The nature of the convention makes it hard for other countries to join.

ITU therefore offers to host international discussion: identifying aspects we can agree on (there are many); recognizing that there are many different viewpoints; and working with all stakeholders, from government, industry and civil society.

Ladies and gentlemen,

We cannot predict exactly how the development of the information society, built on broadband, will enrich the lives of everyone on the planet. But I believe it will be as transformational—and as surprising—as the widespread use of electricity.

Originally aimed at providing better lighting, electricity totally transformed our world, powering large-scale industrialization, stimulating national and international commerce, and allowing skyscrapers and mass urban transport systems to be built.

I am absolutely confident that we will be astonished by what broadband networks help us achieve in the next twenty years—which will see scientific advances beyond our wildest dreams, and quite extraordinary social and economic development.

I believe it is perhaps the greatest opportunity we have ever known for human progress.

So in closing let me encourage each and every one of you to act as an 'ICT advocate', and in particular as a broadband advocate. For the future of science and the future of mankind.

Thank you.

THE EVOLVING NUCLEAR WEAPON THREAT TO SOCIETY

RICHARD L. GARWIN
IBM Fellow Emeritus, Thomas J. Watson Research Center,
IBM Research Division, Yorktown Heights, NY, USA

ABSTRACT

Vast reductions in deployed nuclear weapons have taken place, and the international consensus against nuclear weapon proliferation is strong. However, world security is threatened by emergent nuclear powers and especially by the growing threat of non-state deliberate unauthorized use of state-owned nuclear weapons and by the threat of improvised nuclear explosives fashioned from stocks of weapon-usable fissionable materials—highly-enriched uranium or plutonium

THE NATURE AND NUMBER OF NUCLEAR WEAPONS

I begin by reminding you of the destructiveness of nuclear weapons, and their current numbers around the world. Then I discuss a bit what it takes to build and detonate a nuclear weapon and the change in perception of the nuclear weapon threat over the decades since the two used in war at Hiroshima and Nagasaki August, 1945.

Fig. 1: Hiroshima, August 1945.

Since then, almost 1000 nuclear weapons have been exploded in tests in the atmosphere, mostly by the United States and the former Soviet Union. U.S., Soviet, and U.K. atmospheric testing was brought to an end by the Limited Test Ban Treaty of 1963. Since 1992, the United States has not tested a nuclear weapon underground, and similarly with Russia.

The bombs used in 1945 are shown in Figure 2.

Fig. 2: Little Boy and Fat Man—Hiroshima and Nagasaki bombs. ~13 and 20 kilotons explosive yield, ~ 4 tons weight.

They weighed more than 8000 lbs and had yields of 13 and 20 kt respectively. In contrast, the 3 W-78/Mk-12A warheads on the U.S. Minuteman III missile (Figure 3), each weighs about 700-800 lbs and has a yield of 335 kt. The warhead is 71 inches long and 21 inches in base diameter. It was placed into service in late 1979.

Fig. 3: Three multiple independently targeted reentry vehicles for the Minuteman III.

The Comprehensive Test Ban Treaty was opened for signature in 1996, and since then there have been nuclear tests underground only by India and Pakistan in 1998 and by North Korea in 2006 and 2009.

All nuclear weapons begin with the assembly of a supercritical mass of fissile material that will support an exponentially growing neutron chain reaction with a generation time of 10-20 nanoseconds. A single fission is nothing; it would kill a few cells in the body. But a microsecond's worth of chain reaction in the case of the Nagasaki bomb fissions a kilogram of plutonium and liberates energy within the bomb equivalent to 17,000 tons of high explosive. This produces an enormous fireball, and the heated air rises to create the typical "mushroom cloud."

Fig. 4: 15-kiloton surface burst. Nevada Test Site, 25 May 1953.

Nuclear weapons are again in the news. Their control and even potential elimination is a priority with the Obama Administration and with President Obama himself. Following President Obama's speech in Prague April 5, 2009, we have had the signature in Prague April 8, 2010, by Presidents Obama and Medvedev of the "New START" Treaty to replace the 1991 Strategic Arms Reduction Treaty—START—that expired December 5, 2009. Furthermore, on April 6, 2010 the U.S. government issued the Nuclear Posture Review report—NPR[1], a totally unclassified document with the purpose to specify what the U.S. intends to do about its nuclear weapons over the next five to ten years, and how it views nuclear weapons in the world. On April 12-13, President Obama

[1] http://www.defense.gov/npr/docs/2010%20nuclear%20posture%20review%20report.pdf.

presided in Washington over the Nuclear Security Summit, concentrating on rendering more secure the material that might be used to make nuclear weapons, transforming the administration goal of securing all weapon-usable material within four years to a global goal to do the same thing.

By "weapon usable material" is meant highly enriched uranium (by IAEA definition >20% U-235 in U-238, but in reality probably >80% U-235). The other major material for building nuclear weapons is plutonium, of which the Pu-239 isotope is most useful. However, even the plutonium separated from power reactor fuel in some countries for recycle into power reactors as Mixed OXide fuel (MOX) can be used by knowledgeable weapon builders to make a nuclear weapon without yield penalty or, by relative novices, with no less than a 1-2 kiloton yield (i.e., the content of 1000 trucks loaded with 2 tons of HE). Even a 1 kiloton nuclear explosion could kill many tens of thousands of people in a city within days by exposure to radiation fallout.

For an excellent account, see *The Making of the Atomic Bomb* by Richard Rhodes. The wartime Manhattan Project to develop and produce the nuclear weapons cost the United States some $2 billion, and for many years that cost of "a billion dollars per bomb" and a feeling of American superiority persuaded many that only America could possess the superweapon. By 1949, though, the Soviet Union detonated its first nuclear weapon—a copy of the plutonium design tested July 16, 1945 near Alamogordo, New Mexico. The plutonium was created in production reactors fueled by natural uranium; the necessary 6 kg of Pu required 6000 megawatt-days of reactor power—in the case of the 200 MW (thermal) Hanford reactor, this was 30 full-power days. For a modern power reactor of one-million kWe (3000 MWt) 2 days of operation would suffice.

The Hiroshima gun-type weapon used about 60 kg of HEU, produced by a particularly inefficient "gaseous diffusion" isotopic separation process from the 0.71% U-235 content in natural uranium. Now, most of the world uses the gas centrifuge in an extremely clever design that consumes only about 2% of the energy needed by gaseous diffusion for the same product. The common "light water" power reactor, of which the United States operates 104 at present, each produces about a million kWe of electrical power and consumes annually about 25 tons of low-enriched uranium fuel, fissioning a ton per year of U-235. The 25 tons of spent fuel downloaded annually contains about 150 kg of Pu-239 and almost 100 kg of Pu-240 and heavier Pu isotopes.

Most of the "separative work" required to produce a ton of HEU is already invested in the ton of LEU (4% U-235) fed annually to each power reactor. Each "unit of separative work" (SWU) costs on the world market about $100; 151 SWU are invested in each kg of U-235 in 4.4% fuel, so the SWU content of 1 kg of U-235 content costs about $15,100 and the SWUs in the ton of U-235 annual feed represents about $15.1 million in enrichment cost.

At 60 kg of HEU per gun-type weapon and a SWU content of 232 SWU/kg, the SWU cost per kg is $23,200 and the enrichment cost of a weapon is $1.39 million. Naturally, a centrifuge plant that is built to make a few weapons will have higher cost than one that is amortized over 10 years of commercial fuel production, but the cost is negligible for a state of any size.

And unfortunately, with the passage of time and the availability of the internet, the original "secrets" of the atomic bomb have become relatively common knowledge.

Among useful books to help understand both nuclear weapons and nuclear power is "Megawatts and Megatons," by Richard L. Garwin and Georges Charpak (2001).

On January 4, 2007 four U.S. leaders in the security field—George P. Shultz, Sam Nunn, Henry A. Kissinger, and William J. Perry, following a well-prepared meeting at Stanford University, published an Op-Ed in the Wall Street Journal to proclaim their goal of the elimination of nuclear weapons and, in the nearer term, massive reductions in nuclear weaponry. President Obama has adopted this goal, and the April, 2010, NPR details some of the early conditions that must be met in order that a world without nuclear weapons might come about.

Although the Department of Defense was responsible for the NPR, the Department of Energy, the State Department and the National Security Council (NSC) all contributed heavily, and it the NPR is a policy document that guides the entire government. Fortunately the NPR is entirely unclassified, although, naturally, some of the implementation directives will inevitably be secret. The Department of Energy's National Nuclear Security Administration (NNSA), responsible for designing and creating U.S. nuclear weapons, receives its guidance consistent with the NPR, as do the military departments responsible for raising, equipping, and training the military personnel.

It is clear to all, even those who believe that the elimination of nuclear weapons is infeasible or undesirable, that the 35,000 nuclear warheads the United States had at its peak in 1967 and the 45,000 Soviet nuclear warheads in 1982 are far in excess of a potential need for nuclear weaponry in the 21st century. These were built and maintained, together with their delivery means and the command and control system to make it absolutely clear that an attempt the Soviet Union to destroy the United States (or vice versa) would lead to the destruction of the attacking country itself.

Secretary of Defense Robert McNamara in 1964 established a criterion for "assured destruction", assumed to be the foundation of nuclear deterrence, as 400 equivalent 1-megaton warheads arriving at their targets in the Soviet Union. From that time on, the enormous numbers of strategic nuclear weapons were supported (largely after the fact) by the assumption of large-scale potential destruction of these weapons before they could be launched and by an assumed relatively effective ballistic missile defense system and air defense system that could intercept the nuclear weapons before they reached their targets. Despite secrecy about the numbers of warheads in every country, knowledgeable estimates were published that were not far from the mark.

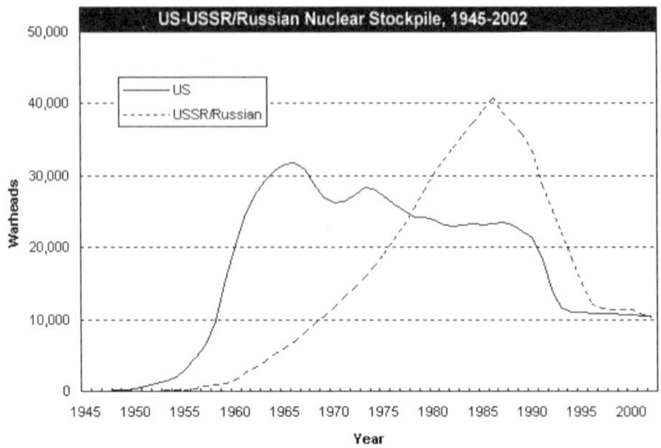

Fig. 5: Natural Resources Defense Council (NRDC) graph.

Today the threat of world-wide communism that could succeed only by the elimination of the United States is no longer credible, and with the enormous openness in the world in recent decades, together with the collapse of the Soviet Union, it is clear also that the United States is not planning the destruction of Russia or any other of the five nuclear states under the 1970 Nonproliferation Treaty (NPT).

But there are reasons other than excess cost to move to greatly reduce the numbers of nuclear weapons and to secure weapon usable-material. In May 2010, the United Nations in New York was held the Eighth Review Conference on the NPT, which in 1995 was converted from an agreement renewable for five year terms to one of indefinite duration. Much in discussion is the "grand bargain" of the NPT, by which almost all the nations of the world signed up as non-nuclear-weapon states—NNWS—while those that had detonated nuclear explosives by January 1, 1967 signed as nuclear weapon states with obligations to provide information and support for peaceful use of nuclear technology to the NNWS. In her speech at the NPT Conference May 3, Secretary of State Hillary Clinton presented official data on deployed U.S. nuclear warheads and those that are held in reserve. As of September 20, 2009, the United States possessed 5113 nuclear warheads and bombs of all types. An additional "some thousands" of warheads are not maintained ready for use and are scheduled to be dismantled. Figure 6 shows the progression of U.S. warhead numbers, as revealed by the U.S. government in May, 2010.

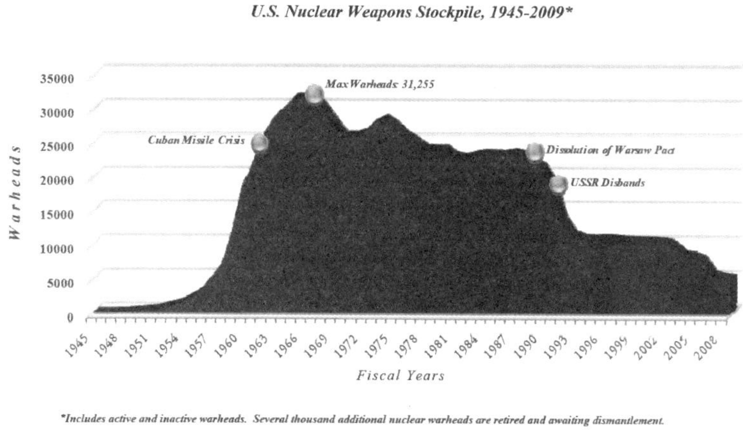

Fig. 6: U.S. Government Fact Sheet May 2010.

Do these thousands of nuclear weapons ensure world security, or even U.S. security? The 2010 NPR gives five U.S. objectives related to nuclear weapons:

1. Preventing nuclear proliferation and nuclear terrorism:
2. Reducing the role of nuclear weapons:
3. Maintaining strategic deterrence and stability at reduced nuclear force levels:
4. Strengthening regional deterrence and reassurance of U.S. allies and partners:
5. Sustaining a safe, secure, and effective nuclear arsenal:

First, there is the hazard of ***proliferation*** of nuclear weapons to additional states, as has occurred with North Korea and which, it seems to many, is the goal of Iran's nuclear program, despite protestations that Iran's enrichment of uranium from the natural abundance of 0.71% U-235 to 5% and more recently to almost 20% U-235 is for the production of nuclear-electric power in Iran and not for the acquisition of nuclear weapons. Iran concealed from the International Atomic Energy Agency (IAEA) for 18 years enrichment activities that under the 1970 nonproliferation treaty (NPT) it was committed to report in a timely fashion. Paradoxically, what indicates a bomb-oriented program is that the enrichment facilities are much too small to provide fuel for even the one large power reactor that has been built at Bushehr by Germany and completed and initial fuel supplied now by Russia.

An even greater threat, according to most analysts and world leaders, is the acquisition of nuclear weapons by terrorist groups such as Al Qaeda, which has vowed to use a nuclear weapon against the United States if they can acquire it.

The sheer magnitude of the problem is apparent from experience with the highly successful 1992 "megatons to megawatts" program under which the United States has been buying 500 tons of Russian highly enriched uranium (HEU) from their nuclear weapon stockpile. This HEU is blended in Russia with some of the depleted uranium left from the initial enrichment process to the 4% U-235 range and then fabricated in the United States into fuel rods for the 104 large U.S. nuclear power reactors; about half of the U.S. nuclear electricity is thus fueled by uranium from Soviet nuclear weapons.

An understanding of what 500 tons of HEU means in terms of fabricating simple nuclear weapons is available from the International Atomic Energy Agency (IAEA) long-time establishment of "significant quantity" (SQ) for HEU and for plutonium. One SQ is 25 kg of HEU or 10 kg of plutonium, although weapons can evidently be made with less, since the average in the U.S. weapon stockpile is about 4 kg of Pu per weapon.

The Hiroshima gun-type that did not compress the bomb fuel, so about 60 kg of HEU was required to produce some 13 kt of yield. The "implosion" method of assembly using high explosive is the only practical one for plutonium, for which an SQ is 10 kg. China's first nuclear weapons were implosion-assembled HEU, and they have formed the mainstay of Pakistan's nuclear forces. Here the SQ is 25 kg. Although the Hiroshima bomb weighed some 8000 lbs to produce a yield of 13 kt, later gun type weapons were much smaller and lighter—among them 6 such weapons secretly built by South Africa and later destroyed without having been tested. In 1957 the United States began the deployment of more than 1200 army artillery shells—the W33—that was tested in Nevada with a yield of 1 kt, but that had other selectable yields up to 40 kt (probably enhanced with fusion "boosting"). The W33 weighed 243 lbs, and was 8-inches in diameter and 37 in long.

The 500 tons of Russian HEU blended down to 4% U-235 for use in civil power reactors could, instead, have made 20,000 nuclear weapons. And there remains at least an equivalent amount of HEU still in Russia. Terrorists might steal or be given a nuclear weapon from the armory of Russia or Pakistan or even North Korea. Alternatively, they could probably more readily obtain HEU and fashion it into an Improvised Nuclear Explosive (INE) that could be detonated in a city.

Here I draw on my article on a terrorist nuclear detonation in a city that appeared in the June 2010 issue of *The Bridge*, a publication of the National Academy of Engineering. Because this is a detonation at ground level (presumably in a basement or a lower floor of a building, the consequences are quite different from a similar yield of 10 kt air-burst, as was the case in Hiroshima and Nagasaki. In particular, there may be fewer direct burns from thc fireball, which will be on the ground, but there will be a crater some 56-ft deep and 245 ft in diameter, which will mix with the radioactive fission products from the bomb and largely fall out in the local area and in a relatively narrow plume that will extend for ten miles or more from the point of detonation. Had the failed truck bomb in Times Square been an improvised nuclear explosive of 10-kt yield, some 300,000 people might have been killed, and a good many more depending on the prevailing winds and the track of the fallout. An example of a simplistic predicted fallout plume from an attack on Long Beach, California, is shown in Figure 7.

Figure A.7
Alternative Estimates of Fallout After First Day

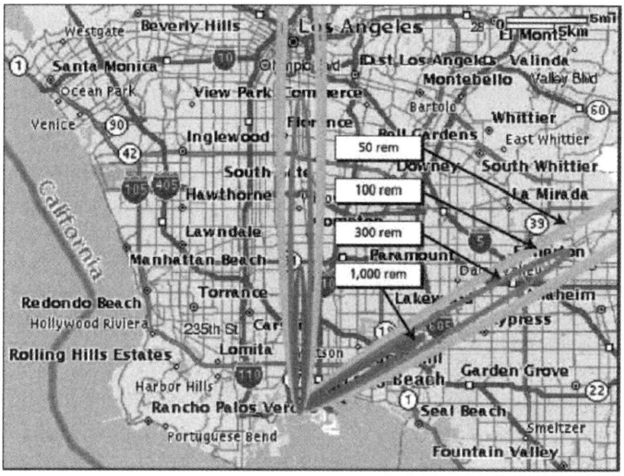

SOURCE: Map and content © 2006 by MapQuest, Inc. Used with permission.

Fig. 7: Charles Meade and Roger Molander (2006).

Naturally, the best approach to preventing such a terrorist nuclear explosion is to eliminate or to lock up all nuclear weapons and weapon-usable materials, and to enforce strict personnel security practices among those who have "legitimate" access to such items. Useful also would be to reduce the number of individuals who wish to carry out such an act. And other means are important to consider on the "supply side" of the terrorist nuclear explosive threat.

Since 1962, many U.S. nuclear weapons have been equipped with increasingly competent "permissive action links" (PAL) so that physical possession of a weapon does not convey the ability to obtain a nuclear explosion. In the United States, increasingly greater demands are being imposed on the security of nuclear weapons—with the ideal that in the wrong hands a weapon would be totally inert[2]. Many safety features on U.S. weapons also contribute to security—e.g., measures that ensure that a lightning strike cannot detonate the weapons, the requirement that the weapons be "one-point safe" so that it produces no damaging nuclear yield if its explosive (used to implode the plutonium "pit") is detonated at any one point, and the secure transport system that is used to move the nuclear weapons. Although attention is often focused on "intrinsic" surety features most readily provided in weapons of new design, much progress can be made with improvements in the weapon environment, and in procedures for handling, storing, and transporting nuclear weapons. I hope that presentations such as this will increase the demand among the public and the political leaders of those states possessing nuclear

[2] In a speech of January 31, 2008, U.S. STRATCOM Commander General Kevin P. Chilton expressed the need for a nuclear weapon so secure that in terrorist possession "it would be a doorstop."

weapons to ensure that these weapons are not only safe against accident but also secure against theft of a weapon and even secure against acts by those authorized to maintain nuclear weaponry.

But beyond that, public officials have the obligation to mitigate (i.e., reduce but not eliminate) the effects of terrorism or accident, and in that regard there are actions that can be taken at relatively low cost. It is unclear whether political leaders will accept the political/electoral cost of explaining this risk.

Education and communication are among these actions, and it should be possible to inform people that there is a remote hazard of a terrorist nuclear explosion in their city, corresponding to a significant possibility in a city in general, and that "if you see something, say something" in this regard as well as others. Not much can be done for medical care of people injured in the nuclear explosion, because the local resources for healthcare are microscopic compared with the need. It will be self-help, if anything, for the first day or so. In the United States, there is considerable discussion of the federal role in such an attack, in view of the fact that the rest of the country would be undamaged. If 300,000 people are lost, this is 0.1% of the total U.S. population, and a primary (strategic) federal role is to ensure that no single explosion, or even two could destroy the entire society as a consequence of undue concentration of records or talents in a particular fairly small region of a city. An important federal role is to analyze thoroughly the potential consequence of such a terrorist nuclear explosion and to assess the impact of the possible responses, sharing this with the public and local and state authorities. The federal government would need to work with cities to create and practice the management of the vast effort that would be required to bring external supplies and support to a stricken city.

What can be done in the immediate aftermath of a nuclear detonation is to reinforce discipline that people should get indoors and stay there until they are told that they can move. Within a couple of minutes it should be possible to obtain from the NARAC (National Atmospheric Release Advisory Center) a prediction of the fallout map, and within an hour, a local map of actual fallout could be obtained from aircraft at sufficient altitude to be flown even by pilots. At that time, people in the fallout pattern might be advised urgently to walk half a mile to either side, which could be the difference for them between life and death, although most will be safe by remaining indoors.

And if they don't move, they could, without measuring radiation levels themselves, do a good deal to reduce their hazard by moving to a different location within the building as exemplified by the shielding factors in Figure 8.

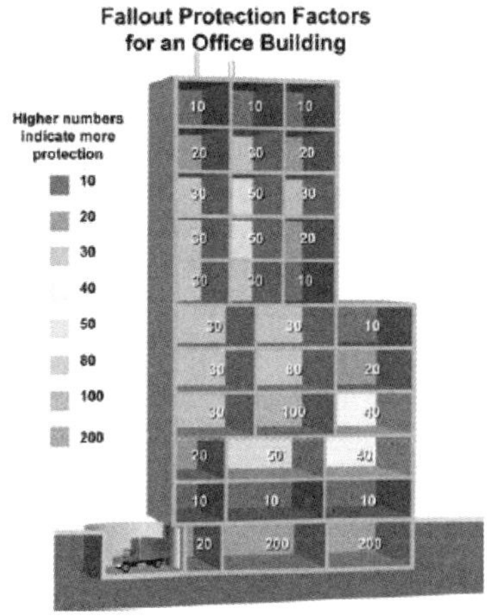

Fig. 8a: Building as shielding: numbers represent a dose reduction factor. A dose reduction factor of 10 indicates that a person in that area would receive 1/10th of the dose of a person in the open.

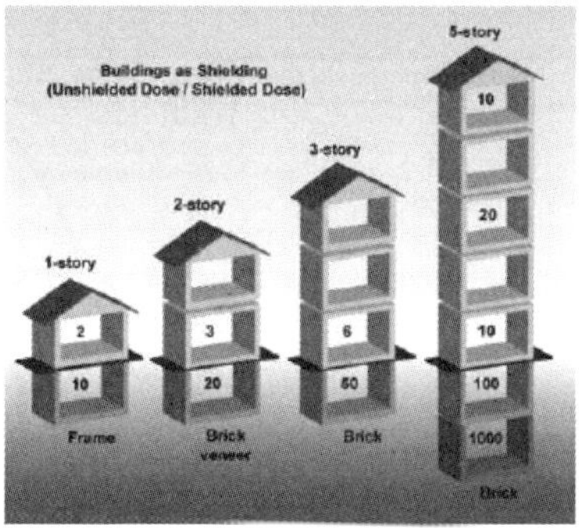

Fig. 8b: Building as shielding: numbers represent a dose reduction factor. A dose reduction factor of 200 indicates that a person in that area would receive 1/200th of the dose of a person out in the open.

With the new perception of danger from an improvised or stolen nuclear explosive detonated in a city, the U.S. Congress in 2007 required the Department of Homeland Security (DHS) to undertake a quantification of the hazard and what interventions might minimize that hazard.. I had addressed this in my presentation in April 2007 to a high-level meeting in Washington convened by the "Preventive Defense Project" of Ashton B. Carter and William J. Perry.

A window on the work that has been done under the mandate to DHS on urban nuclear explosions is an excellent paper by Brooke Buddemeier[3]. Here are a couple of illustrations from that paper.

Figure 9 is an example of the complex nature of the fallout, given actual winds aloft at various levels.

---

3 "Reducing the Consequences of a Nuclear Detonation: Recent Research," by Brooke Buddemeier, *The Bridge*, Summer 2010, pp. 28-38, to be found at http://www.nae.edu/File.aspx?id=19815.

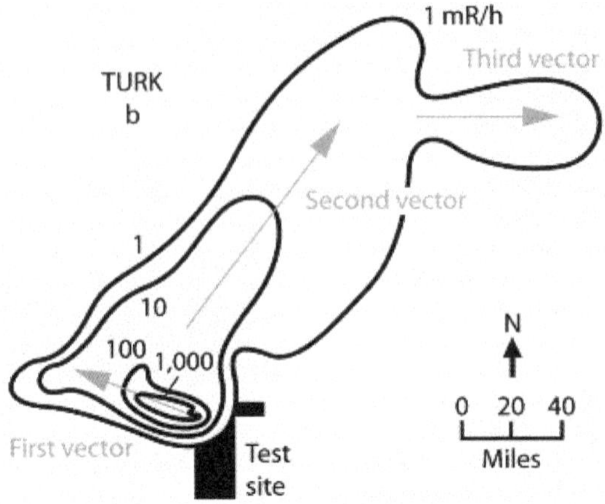

Fig. 9: Early fallout dose-rate contours from the TURK test at the Nevada Test Site. SOURCE: Lawrence Livermore National Laboratory. (from B. Buddemeier, June 2010, by permission.

Buddemeier provides estimates of the radiation dose from fallout as influenced by delay time to evacuation and makes the point that to stay for delayed evacuation in a portion of a building with a reasonable protection factor (PF) substantially reduces the dose and the potential damage from radiation exposure—a point we have been making in our Permanent Monitoring Panel in Erice for several years. Naturally, the evacuation route also influences the dose, not only because the fallout itself is non-uniform, but because the time spent along different evacuation routes varies with capacity and congestion.

Of course, choosing a route and an estimated speed is only a tiny part of what needs to be done in planning for the aftermath of a nuclear explosion, where police and military supervision need to be organized for a situation in which hundreds of thousands of people may already have died and the survivors are in great peril.

Paradoxically, the greatest psychological problems in working in this field arise from discussions of how best to save lives and preserve health among the survivors. Other interesting aspects of the problem, such as that illustrated in Figure 10 for the beneficial shadowing effects of buildings against the thermal radiation from the fireball are easier to contemplate.

Fig. 10: Integrated thermal flux from a 10-kiloton ground-level nuclear detonation in a small U.S. city. SOURCE: Lawrence Livermore National Laboratory. For more information, contact brooke2@llnl.gov. (From Buddemeier, June 2010, by permission)

Proliferation brings to mind North Korea but also Iran. The possibility that Iran will obtain nuclear weapons is encouraging similar thoughts in the Middle East and elsewhere, as Arab states face the possibility that a fellow Muslim (but not Arab) country acquires what has long been called "the ultimate weapon."

It may be that some of these countries are simply hedging their energy bets, believing that the era of oil is drawing near its end, and that they should begin to make a transition to nuclear power. In fact, just before the overthrow of the Shah of Iran of 1979, the United States government was proposing to Iran the acquisition of 20 nuclear power plants, and the full fuel cycle for supplying them. As you know from the news, Iran has had a very active program for mining uranium, converting in into UF_6 gas, and enriching that gas in U-235.

Both for Iran and for the rest of the world, it would be cheaper and safer for individual countries to buy enriched uranium fuel or enriched uranium material on an assured market rather than to build and operate enrichment plants themselves. If they don't trust the reliability of supply, they could buy fuel as long as ten years in advance, because the fuel cost for a nuclear reactor—unlike for a coal or gas-powered plant—is a tiny fraction of the overall cost of electrical energy.

Similarly, rather than directly entomb the highly radioactive spent fuel (or the separated fission products) within the country that generates it, as is the legal requirement now, nations ought to encourage the development of regional mined geologic repositories so that only the reactor itself would be on national territory.

CONCLUSION

Although the possibility of destruction of the United States by an onslaught of hundreds or thousands of Russian warheads is much reduced, there is a growing probability of the detonation of a stolen or improvised nuclear weapon in a large city—especially in a U.S. city-- which might kill directly 300,000 people—mostly within a few weeks by exposure to radioactive fallout within first hour or day after the explosion. Although the loss of 0.1% of the U.S. or European population need not destroy the entire society, without analysis and planning at great cost, it might do so. Hence the recognition that nuclear weapons anywhere are a great threat to our society.

I believe that such an urban nuclear explosion is probable over the next ten years, and that probability can be greatly influenced by a collaborative effort, world over, to protect stores of weapon-usable plutonium and HEU, to reduce holdings of nuclear weaponry and to improve greatly the security of nuclear weapons. Most important is to recognize that this is a problem on which progress can be made both in prevention and mitigation.

The final figure reminds us of the consequences of failure in this mission.

As a remedy to security problems, nuclear weapons have potentially lethal side effects.

Fig. 11: WHICH COUNTRY? WHICH CITY? OCTOBER 2010?

ENERGY AND THE ENVIRONMENT IN THE NEXT MILLENNIUM

RICHARD WILSON
Department of Physics, Harvard University
Cambridge, USA

GENERAL

In a sense, talking about the next millennium is an immediate planetary emergency. For, if we do not think about it carefully, we will lose any ability to control or even influence it. The result might be self destruction by disease or atomic bombs or a lapse into poverty. But the discussion will be very different from the discussions we have had in the last 10 years when we discuss situations which call for, some would say demand, immediate attention. I argue, even insist, that one must think positively. We must not be Luddites—the social movement of British textile workers in the nineteenth century who protested, often by destroying mechanized looms,—against the changes produced by the Industrial Revolution, which they felt was leaving them without work and changing their way of life. Luddites were unable to stop the industrial revolution but, in my naive view, they may have delayed the proper reaction of environmental scientists to it. The environmental movement in the USA and, hence, the world includes people who resist change and this subgroup are sometimes called neo-Luddites. I sometimes accuse them of exclaiming: "Stop the world. I want to get off".

Fig. 1: Ned Ludd, the leader of the Luddites.

My family on my father's side were weavers. Their main complaint was different from that of Ned Ludd. They were concerned that no one would buy woolen products. It was the 1897 McKinley tariff that the USA enacted which cut imports of British wool into the U.S. and caused major unemployment and drove my grandfather into a different job as a publican and then a baker.

I assume that one of the major tasks of human society is to advance civilization especially as written in the world convention of human rights.

SOME HISTORY

I first emphasize the importance of energy in the last few millennia.

4000 years ago, man could only use the energy performed by his own body—about 300 watts. He was also limited in fuels. Many, such as crude oil, he was unable to digest.

3000 years ago, man discovered fire: enabling him to do various tasks he was otherwise unable to do. Wood and charcoal were used for steel making.

2000 years ago, man domesticated animals, particularly the horse. Not only could a horse perform functions which demanded more energy use than man could directly assemble, but he could eat and digest fuels that a man could not.

1000 years or so ago, man learned how to use water for water wheels and mill—in a water mill. He also found out how to tame the wind in windmills.

Interestingly, the Dutch windmills of the 14th century AD were about 20% efficient in tapping the energy of the wind. We are only now getting to 40%.

The progression of fuel use in England is instructive. In about 1100 AD, people were bringing "sea-coales" from Newcastle to London. This increased dramatically over the centuries as England ran out of wood.

Although elementary steam engines had been used 2000 years ago (Hero's Engine) it is usual to mark the major use of coal in the steam engine of Newcomen in 1712, and improved and developed by James Watt a few years later. This started the Industrial Revolution and rapid increase of prosperity and population.

The next step was the realization that oil could fire the internal combustion engine, followed by nuclear energy in 1939-1945.

All these steps were made by an unfettered inventor, with some, often limited scientific training, but fundamental science concepts were quickly necessary. Indeed, the further down the list of technological steps one goes the more basic science was necessary.

KING COAL WINS FOR EIGHT CENTURIES

Contrary to a common present day view, the environment was not neglected although in retrospect firm action was delayed far too long. For example in 1207 King Edward I of England objected to coal because of its air pollution and so wrote to the Sheriff of Southwark (in latin of course). "...we instruct you to make public proclamation in the aforesaid village of Southwark that all those who wish to operate kilns … should bake them in the customary fashion from brushwood or charcoal and henceforth should make absolutely no use of sea-coales...". His successor King Edward II tortured people who "fouled the air". King Richard II was more moderate. He sought to contain it through taxation. Henry V took a different tack. He formed a royal commission. We all know that is the first step to ignoring the issue! For the next 0.5 millenia there was little but talk.

But as England ran out of wood, "Sea-coales", coal coming in by sea from the mines in Sunderland and Newcastle, took over. King James I (of England VI of Scotland) wrote a tract objecting to the developing habit of "tobacco drinking". The diarist John Evelyn in 1661 wrote about "this horrible smoake". 260 years ago William Blake in his famous poem "Jerusalem" wrote about the "dark satanic mills" with their air pollution. In the 1850s the churches got into the rhetoric with the prayer: "From Hell, Hull and Halifax, good Lord deliver us."

But action was not rapid enough. Not for 100 years after the major expansion of the industrial revolution, did England build chimney stacks 100 in the policy stated in 1920 by Hetherington: "the solution to pollution is dilution". This was a scientific slogan which was valid for the time but it sent pollution to Scandinavia. The first measurements were, perhaps, by a young physician, Dr Richard, later Sir Richard, Doll who measured benzo(a)pyrene outside St Bartholemew's Hospital in London. But it took the air pollution "incident" of December 1952 to inspire rapid action. Within 6 months the British parliament banned the burning of soft coal in major cities. This was accepted because we now had cheap Middle Eastern oil as a replacement.

This was certainly a correct decision but most details of the science were wrong. The science was slow to be done but it is ongoing. By 1973 it was clear that particulates were responsible not SO_2. We thought there was a threshold but we cannot find one. Not till this meeting do we hear that it is elemental carbon that matters.

But we are now more ambitious. We are trying to control the environment and climate. We do not want to wait for it to control us. This has been a major discussion item in Erice for the last 10 years. Should we try to influence the climate and even control for change due to natural events. I think that we can and should, and the issue is how and how much. This will need the best possible scientific imagination, thought and technological practice. I reject arguments such as by Nigel Lawson of the UK as neo-Luddism.

THE KARDASHEV SCALE

I argue that mankind needs energy and we need more of it to eradicate poverty. This was a national policy until 1970. During the presidencies of President Eisenhower, President Kennedy and President Johnson, the theme was to provide plentiful cheap energy. Admiral Strauss, Chairman of the U.S. Atomic Energy Commission, expressed this in a very over optimistic manner when referring to nuclear fusion, "Electricity will be too cheap to meter". Many others have argued the importance of having affordable energy including Benjamin Franklin over 200 years ago (on the Energy PMP website). In the USA and the developed world we can afford other incentives. But many people would oppose putting coping with global warming ahead of raising the living standards of the poor in China and India. But we should figure out how to do both.

An important correlation exists between the availability of energy and the world population. The rapid increase in population from 1750 to the present accompanied a rapid increase in fuel availability, and the energy intensity (energy per GNP) increased rapidly. Of course Malthus argued that only wars and pestilence would limit population. But as many parts of the world have developed a net reproduction rate below 2, this can now be questioned and possibly modified. `The early 1960s were a time of great optimism. Nino Zichichi and Dick Garwin measured g-2 for the muon. Contrary to a common belief that the communist government in the USSR was a complete disaster, they recovered from the terrible suffering of World War II with enthusiasm and optimism. When I first visited in 1959, 0.014 millennia after WWII, I saw the expansion of Moscow. As I was told at the time, "behind every window is a happy man or woman". These were the first apartments built since 1914, nearly 0.05 millennia before. Meaningful scientific contacts were established between the two large industrial powers, the USA and USSR resulting in real collaborations and real friendships. Of course we would now have to consider China as a major power. The future seemed limitless.

In this optimistic context a young Russian astrophysicist proposed that we look at energy use in a very positive light. He considered three levels of civilization. These three levels of the Kardashev Scale can be quantified in units of power (watts) and plotted on an increasing logarithmic scale. I discuss here a modified updated, definition:

- Type I—a civilization that is able to harness all of the power available on a single planet—has approximately 10^{16} or 10^{17} W available. Earth specifically has an available power of 1.74×10^{17} W (174 peta watts.
- Type II—a civilization that is able to harness all of the power available from a single star, approximately 4×10^{26} W. Our sun outputs approximately 3.86×10^{26} W.
- Type III—a civilization that is able to harness all of the power available from a single galaxy, approximately 4×10^{37} W. This figure is extremely variable, since galaxies vary widely in size; and the stated figure is the approximate power output of the Milky Way.

Two western scientists supported and developed the idea, Carl Sagan and Freeman Dyson. A formula was proposed to calculate intermediate numbers: For perspective it is useful to translate the units of power and energy into "practical" terms. The energy from exploding TNT, the largest nuclear weapon ever detonated, released an estimated 57 megaton yield; even a Type I civilization makes use of roughly 25 megatons of TNT equivalent a second. A Type II civilization would control 4×10^9 times more energy (4 billion hydrogen bombs per second), and a type III 10^{11} times more yet. That is a lot of energy and the questions arise:

(1) Can we in the next millennium learn how to control that much energy and fuel? and
(2) Do we want to do so?

I assume the answer to (2) is maybe, but only if we can control the environmental consequences.

CURRENT STATUS OF HUMAN CIVILIZATION

Human civilization is currently somewhere below Type I, as it is able to harness only a portion of the energy that is available on Earth. The current state of human civilization has thus been named Type 0. Kardashev's followers assigned an intermediate number by interpolating and extrapolating the values given above using the formula:

K = frac{\log_{10}{W}-6}

A recent table shows how the Kardashev number has changed over the last century. The entry for AD 1000 is my own and less well documented.

Year	Exjoules/year	Terrawatts	Quads/year	Mtoes/year	Kardashev rating
1000					0.49
1900	21	.067	20	500	0.58
1970	190	6.0	180	4500	0.67
1973	260	8.2	240	6200	0.69
1985	290	9.2	270	6900	0.69
1989	320	10	300	7600	0.70
1993	340	11	320	8100	0.70
1995	360	12	340	8700	0.70
2000	420	13	400	10000	0.71
2001	420	13	400	10000	0.71
2002	430	14	410	10400	0.71
2004	440	14	420	10600	0.71
2010	510	16	480	12100	0.72

The Kardashev rating of 0.72 today means that we are using approximately 0.16% of the total available planetary energy budget. A very poor energy efficiency!

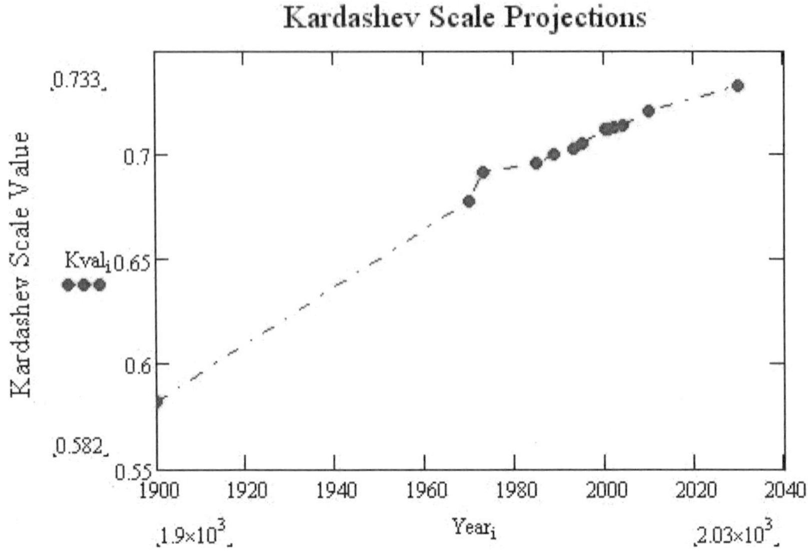

Fig. 3: Kardashev scale projections for human civilization ranging from years 1900 to 2030, based on data from the International Energy Agency World Energy Outlook.

METHODS BY WHICH A CIVILIZATION COULD FEASIBLY ADVANCE TO TYPE I

I now assume that we will try to reach a Type I civilization by AD 3000. I will make some comments on how:

1. Using more fossil fuels. All too often in the last couple of centuries someone has said we are running out of coal, of gas, of oil, or one or another fossil fuels. They were pessimistic as we found more deposits and learned how to use less concentrated deposits. But a simple calculation suggests that indeed we will do so within the next millennium. A slowing down in their use for environmental reasons will stretch this time. This, therefore, seems a good idea whether or not there is a problem with carbon-caused global warming.
2. Using more nuclear fission. If we find a way of economically extracting uranium from seawater, and of course extracting uranium from the particulate residues of coal burning, nuclear power at the present rate can be used in present style reactors for the next millennium. But this could be extended another factor of 100 by using breeder reactors. This might seem enough. But it still would not get us to the Type I civilization.
3. Large scale application of fusion power. In this we have to be far more optimistic than our colleagues Jeff Ongena and Miyahara. Type I implies the

conversion of about 5 kg of matter to energy per second. This can be achieved by fusing approximately 1,000 kg of hydrogen into helium each second, a rate roughly equivalent to 3×10^{10} kg/year. A cubic km of water contains about 10^{11} kg of hydrogen, and the Earth's oceans contain about 1.3×10^{9} cubic km of water, meaning that this rate of consumption can be sustained over geological time scales.

4. Solar energy—converting sunlight into electricity by either solar cells or indirectly through wind and hydroelectric power is very inefficient. Moreover to achieve Kardashev Type I by solar power alone would demand (almost by definition) 100% efficiency over the whole of the earth's surface. To completely coat the surface of the earth with the required man-made structures is presently not feasible. If we set 10% as the efficiency of conversion and cover 10% of the earth with such structures we would, however advance beyond 0.8. However, if a civilization constructed very large space- based power satellites, reradiating the energy at higher power density back to earth by focused microwaves, this could be increased. This was seriously suggested by NASA in 1985.
5. Antimatter. Any civilization with the technological ability to collect antimatter in large quantities cheaply would have a mechanism to produce power on a scale several factors above our current level of technology. In antimatter-matter collisions, the entire rest mass of the particles and not merely the mass defect is converted to kinetic energy. The energy per unit mass is about 10 orders of magnitude greater than chemical energy (compared to TNT), about 4 orders of magnitude greater than the energy that humans liberated today using nuclear fission, and about 2 orders of magnitude greater than the best possible from fusion.
6. Dyson spheres and storms. These are hypothetical mega structures originally described by Freeman Dyson to completely enclose a star and capture most or all of its energy output.

ADVANCING BEYOND TYPE I CIVILIZATION

Type II civilizations might use the same techniques employed by a Type I civilization, but applied to all of the planets in our solar system. They may also be able to tap into the energy released from the super massive black holes which are believed to exist at the center of most galaxies. But I suggest that we defer detailed consideration of going to a Type II society for another millennium (AD 3000-AD 4000).

SOCIAL AND ENVIRONMENTAL UPHEAVAL

Few historians have expressed the large-scale transitions, such as the Industrial Revolution in terms of the Kardashev scale. But because of the importance of energy and fuel use in society rapid advancing on the Kardashev scale could potentially describe periods of social upheaval, since they entail surpassing the hard limits of the resources available in a civilization's existing territory. Common speculation suggests that as we advance from K-0.72 to a Type I civilization there is a strong risk of self-destruction. In the most common scenario of self destruction nuclear fission gets out of control as the use thereof is expanded I remember clearly discussing this with Andrei Dmitreyvich Sakharov in 1979 at our first meeting. I asked, "Do you still support nuclear energy after the Three Mile Island accident?" The reply was unequivocal and repeated publicly several times since then, "We must not stop progress". Andrei was optimistic. He thought that mankind was bright enough to control its progress. In a real sense his work on human rights was a necessary part of mankind's struggle for progress.

But, there are more sophisticated disaster scenarios. If population expands as energy supply increases, as it did rapidly between 1800 and 2000, there would no longer be room for further expansion on the civilization's home planet. Excessive use of energy without adequate disposal of heat, an ultimate environmental problem, could plausibly make the sea temperatures rise above 35°C.

Professor Antonino's (Nino's) call for use of science in the next millennium inherently involves some optimism. This includes optimism about mankind's ability to control the environment.

A RETURN TO THE PRESENT CENTURY

All the above discussions seem very fanciful. But they illustrate the type of thinking of which the human race is (presently) capable. I admit to not being as optimistic as I should be. In 1939 when my brother's college friend Arthur C. Clarke talked and wrote about going to the moon, and (possibly) even beyond, I thought he was talking nonsense. But in the very short time interval of 0.03 millennia (1939 to 1969) mankind achieved it. I have written about the difficulty of going beyond—to the nearest galaxy. I note that the Gian Carlo Wick prize this year is going to my friend Tom Kinoshita. He has been working on g – 2 for the electron for 0.05 millennia! I call attention to his patience of which I, alas, am not capable. I first brought public attention of the possible carbon-caused climate change to the attention of the American Physical Society 0.04 millennia ago. It is still an unresolved issue.

The velocity of space travel would have to be great—v/c about 0.5—to go to the next galaxy within a human lifetime, and the huge adverse effects of the radiation as one ploughs through even the very dilute interstellar sea of hydrogen seem into be insuperable. But scientists and technologists will dream and some dreams will come true. But we must also follow the environmental consequences of our dreams rather faster than we did as carbon fuels were being burned. Are scientists up to the task? I think so. I hope so. It is essential.

As I see it, in all these considerations, expansion of fuel supply and use, control of the environment, limit to "creative enthusiasm" in increasing the population, a clear understanding of the basic sciences of all sorts is essential. Only when the scientific limitations are understood can the social scientists do their work. This, I suggest is a role for discussion at Erice.

ENVIRONMENTAL CONSIDERATIONS

I hope we will not have to await the 0.8 millennia before the first concerns about coal burning were fully addressed by data and scientific study of possible causes. Indeed, I think this is an essential role for science and scientists. We must find a way of discussing the issues seriously without the exaggerations, political shenanigans and insults that have characterized the global warming issue since about 1990. We must look at each claim calmly and seriously.

> NIMBY—Not In My Back Yard
> I offered my back yard for nuclear waste (at a nominal but lucrative cost) 20 years ago but this led to:
>
> NIMNBY—Not In My Neighbor's Back Yard
> I was not the only person rebuffed. In 1990 the town of Martindale, Illinois wanted low-level nuclear waste. It would pay their taxes. But the state would not let them have it. Senator Barbara Boxer objects to low-level waste in Ward Valley which is nowhere near where she lives or the people she represents.
>
> This leads at once to:
> BANANA—Build Absolutely Nothing Anywhere Near Anybody

We must also be wary of the undefined "Precautionary Principle". The only clear definition I have heard is: "If you don't understand what you are doing, don't do it." But that invites the question, "Who is to do the understanding?" Even the European Union has adopted scientific nonsense in one regulation. It has called for abolition of Depleted Uranium (DU) in warfare. Yet depleted uranium has less radioactivity than ordinary uranium and similar chemistry. This is an example of how decisions can be made by people with no understanding of the science and technology. Obviously one must look carefully at every one of these issues that arise. It must be done scientifically but with understanding of and patience with, those who do not understand the science.

Nuclear power advocates are apt to think the problems only apply to them. But in Massachusetts it is windmills that are now the problem. This was known historically in Europe 0.3 millennia ago. Cervantes wrote about Don Quixote who "tilted at windmills" as shown in the following figure. Those are perhaps suitable words to describe many of the present excesses in the environmental movement.

Fig. 4: Don Quixote.

Finally, I suggest that a Kardashev scale might be helpful in looking at catastrophes, whether man-made or natural. I think also that mankind can and even must in the next millennium exercise more control over the environment that in the previous millennium.

It may not matter whether an incipient hazard is man-made (anthropogenic) or natural.

In that sense, I urge that we use our imaginations to envisage future problems and the best possible science to study them. Let us be more careful than the dinosaurs that, according to this cartoon, argued about what to do and did nothing.

"All I'm saying is now is the time to develop the technology to deflect an asteroid."

AN INCOMPLETE SET OF REFERENCES

1. Kardashev, Nikolai (1964)
2. Sagan, Carl (October 2000) [1973]. Jerome Agel. ed. Cosmic Connection: An Extraterrestrial Perspective. Freeman J. Dyson, David Morrison. Cambridge Press.
3. World Energy Outlook. Paris, France: International Energy Agency.
4. "Key World Energy Statistics" (PDF). International Energy Agency. 2004. http://www.iea.org/dbtw-wpd/textbase/nppdf/free/2004/keyworld2004.pdf.
5. Quads: 1 quadrillion BT.
6. Mtoes: Million tonnes (metric tons) of oil equivalents.
7. Dyson, Freeman J. (1966). "The Search for Extraterrestrial Technology". Perspectives in Modern Physics (New York: John Wiley & Sons).
8. Kaku, Michio (2005). Parallel Worlds: The Science of Alternative Universes and Our Future in the Cosmos. New York: Doubleday.
9. Dyson, Freeman J. Energy in the Universe Article in September 1971 Scientific American magazine (Special September Issue on Energy).
10. http://en.wikipedia.org/wiki/International_Energy_Agency.

SCIENCE OF CHEMISTRY AND CHEMICAL SECURITY IN THE THIRD MILLENNIUM

VALERY P. KUKHAR
Ukrainian Academy of Sciences, Laboratory of Fine Organic Synthesis
Kiev, Ukraine

Chemistry is an important constituent of the scientific culture of our civilization. Wikipedia writes: "Chemistry is the science of matter that studies matter and the interactions among different types of matter. The science of matter is also addressed by physics, but while physics takes a more general and fundamental approach, chemistry is more specialized, being concerned with the composition, behavior, structure, and properties of matter, as well as the changes it undergoes during chemical reactions." Chemistry explains chemical transformations of various substances, their chemical properties and an ability to interact with each other. Chemical processes underlie biochemical transformations and in life.

The Earth's nature is a giant chemical plant that every second produces or converts organic and inorganic substances and materials of our planet. These are the natural processes during some billion years. Thanks to chemistry, we are surrounded by an oxygen atmosphere that is necessary for living organisms.

From the past, chemical processes are the fundamental technological principle to convert raw material into materials needed for human activity and life. No doubt, chemical processes and chemicals will be used by society in everyday practice in the future from purification of water to production of very sophisticated materials for high-tech and pharmaceuticals. If we look about more precisely, we can find that chemical processes dominate in industrial technology, e.g., combustion of organic fuels for energy production or metallurgical processes are also chemical processes—oxidation or reduction, correspondingly. Food processing is based on chemical or biochemical conversion carbohydrates, proteins, water and other components into final desired products.

The second half of the 20th century was especially successful for chemistry. New synthetic materials—polyethylene and synthetic rubber, silicones and fluorocarbons with unusual properties, special additives and synthetic surfactants, liquid crystals and many others chemicals as well as agrochemicals, fertilizers and herbicides, very fast became integral components of a modern life. Actively using physics and mathematics, new scientific instrumentations and high-precision devices, the chemistry has formulated correctly some fundamental "chemical lows", determined numerous fine peculiarities of numerous chemical reactions and a structure of very complex natural compounds. Chemistry proposed numerous new materials with excellent properties that meet the requirements of new technique. Powerful pharmaceuticals created in chemical laboratories help us to treat serious illnesses.

Indeed, chemistry and chemical technology gave new constructional materials for space missiles, high-clean metals and their derivatives for electronics, new powerful pharmaceuticals, so-called "synthetic metals". The special report of NAS, "Pimentel' report", published in the end of the 1980s, has predicted that chemistry in the 21st century would be oriented mainly on two general areas of activity—a development of new materials and fast convergence with biological sciences. In my vision, recent decades confirm total accuracy of this conclusion.

Determination of chemical composition and fine structure of proteins and enzyme enabled scientists to understand mechanisms of complex biochemical transformations and to create a basis in development of new target-directed inhibitors and bioregulators. Chemists discovered new organic reactions which are mimics of natural biochemical processes in living organisms. Modern organic synthesis can resolve very serious tasks, and the "total synthesis" of complicated natural compounds is realized in numerous laboratories. As one example, I'd like to demonstrate the chemical formula of a completely synthetic vaccine prepared by fragment condensation of a few blocks.

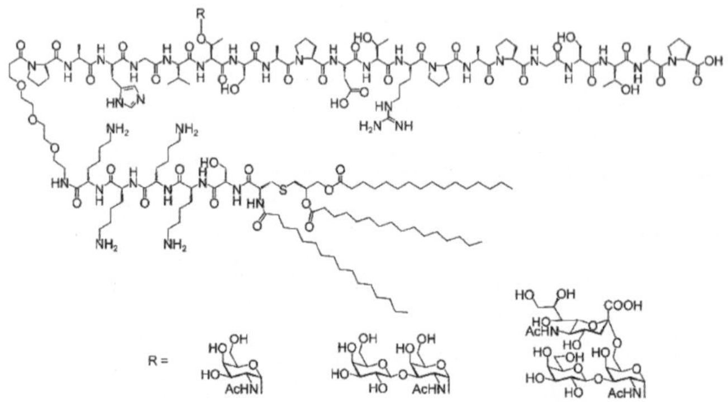

Fig. 1: Synthesis of MUC1 glycopeptide TLR2 agonist vaccines 9–11 by fragment condensation. H. Kunz et al. Angew. Chem. Int. *Ed. 2010, 49, 3688–3692.*

The outstanding achievement of chemical science is undoubtedly the result of Craig Venter and his team who have reported the creation of the first self-replicating artificial microorganism with completely synthetic genome (*Science 2010,* DOI: 10.1126/science.1190719).

"The M. mycoides JCVI-syn1.0 genome synthesized in the most recent study was 1.08 million base pairs (Mbp) long. This again was created from synthetic DNA oligonucleotides by combination of cassettes first from 10 kb fragments, then assembled to 100 kb intermediates, and finally by completing the full length genome in yeast (Figure 1). The synthetic genome was then transplanted into the recipient cell Mycoplasma capricolum. It was subsequently shown that the cells containing the synthetic genome were self-replicating (and hence, the synthetic DNA software built its own M. mycoides hardware within the M. capricolum surrogate host), showed logarithmic growth, and was a typical Mycoplasma phenotype." (Uwe T. Bornscheuer. Angew. Chem. Int. Ed. 2010, 49, 5228 – 5230)

Productivity of chemists may be demonstrated by some numbers. Special service, CAS, Chemical Abstracts Service, which registers new synthesized chemicals, contains more than 32 million identified molecular substances and nearly 12 million reactions. Chemical synthesis of new compounds is everyday "non-stop" process in numerous laboratories all over the world.

But unfortunately, chemistry has two sides—"white" and "black". The "black", negative side of chemicals is toxicity. For millennia these dangerous properties of chemicals have been used by man. Make a note, the first "chemical weapons" curare has been produced by extraction of toxic components from some plants. Numerous generations of alchemists in search of phlogiston and easy way from copper to gold simultaneously prepared poisons on requests of their "sponsors" and bosses. I used this generalization in order to underline that our civilization had and will have the problems with toxic chemical substances at all times.

One day our society discovered that our world was polluted by various chemicals —NO_x and SO_x, chlorofluorocarbons HFC, lead and mercury compounds. River and surface waters contain numerous pollutants and need fine cleaning. Agricultural lands are contaminated with chlorine-containing pesticides and nitrates. Industrial and municipal wastes containing numerous polymers and plastics polluted soil and collected in ocean waters formed real "plastic seas". To this long list it's necessary to add military depots of chemical weapons with highly toxic chemicals. Well-known heavy accidents in chemical plants, Bhopal, Seveso, oil spills, the last accident in the Gulf of Mexico on the BP platform, as well as numerous small-scale accidents in transportation or using chemicals (chlorine, phosphorus, acids or ammonia) clearly demonstrate a high risk of chemicals to life and environment. Sometimes chemical pollution or migration of chemicals doesn't "know" national borders, and toxic chemicals contaminate very rapidly water or atmosphere of neighboring states.

The **Seveso disaster** was an industrial accident that occurred around 12:37 pm July 10, 1976, in a small chemical manufacturing plant approximately 15 km north of Milan in the Lombardy region in Italy. It resulted in the highest known exposure to 2,3,7,8-tetrachlorodibenzo-p-dioxin (TCDD) in residential populations which standardized industrial safety regulations. The EU industrial safety regulations are known as the Seveso II Directive

The **Flixborough disaster** was an explosion at a chemical plant close to the village of Flixborough (North Lincolnshire), England, on 1 June 1974. It killed 28 people and seriously injured 36. The chemical plant produced caprolactam by the process involved oxidation of cyclohexane with air in a series of six reactors. Within a minute, about 40 tonnes of the plant's 400 ton store of cyclohexane leaked from the pipe and formed a vapour cloud 100–200 metres (320–650 feet) in diameter. The cloud, on coming in contact with an ignition source (probably a furnace at a nearby hydrogen production plant) exploded, completely destroying the plant. Around 1,800 buildings within a mile radius of the site were damaged. (Wikipedia. *Flixborough disaster*).

The 2005, March 23, disaster at UK oil major BP's Texas City refinery, in Texas, U.S., was considered the nation's worst industrial disaster in 15 years. A series of explosions occurred when a hydrocarbon isomerization unit was restarted and a distillation tower flooded with hydrocarbons. As a result, 15 were killed and another 180 were injured.

Nov. 13, 2005. A series of explosions rocked China-based **Jilin** Petrochemical's 70,000 tonne/year aniline complex in Northeast China, killing five and injuring 70. Benzene also leaked into the Songhua river and caused millions of people to go without drinking water, with many fleeing their homes. Water and electricity supplies had to be cut off as local residents reported tap water turning red or yellow. There were also concerns that water supplies to some Russian towns could be affected by the contamination of the river.

All chemicals are toxic generally and differ only by levels of toxicity. Between all chemical substances, which have been synthesized on the moment, comparatively very small group of compounds has been tested on toxicity to human and animals. Chemical synthesis of new compounds, predominantly—organics, continues actively in academic and industrial laboratories through all world. Observation of the last years showed an increase of the fine chemicals synthesis and production in so-called "BRIC" states (Brazil, Russia, India and China). Chemical industry sector are growing very fast in various developing countries. At the same time, new scientific discoveries may bring new risks, including the potential of new chemical compounds being abused as chemical weapons or toxic matter by terrorists. Even humane targets of pharmaceutical industry can result in some tragic consequences that it has been in the case of thalidomide use. This case has induced an application of new principles in the pharmaceutical industry for a process to search new drugs and an introduction of additional investigations of possible side-effects for new molecules.

Chemical science at the beginning of the 21st century is marked by activity in nano-chemistry and new chemical technology. Naturally, scientific results in these areas induced a numerous questions connected with chemical security. Have new nano-materials the new toxic properties? Is it possible to use nano-material as chemical weapons?

New chemical technology is concentrated in so-called "micro-reactors", which enable one to realize a chemical reaction in very small reactor space and which could be placed in a private house or a truck. Productivity of these types of new chemical reactors is comparable with a standard chemical installation or whole plants. To use micro-reactors, a "drop-by-drop" process could yield, per year, the hundreds tonnes of desired chemical, and it is not improbable, a toxic chemical.

In order to use chemistry, science and technology, to the benefit of society, to reduce adverse effects of chemicals and chemical accidents, we have to organize an appropriate system of chemical security. All above mentioned determines necessity to have strong control and corresponding regulation for chemical production as well as for usage of chemicals overall a life-cycle. The first very important result and excellent example of possibility to organize this complex system of chemical security, on my opinion, is the Chemical Weapons Convention.

Chemical weapons. After decades and long-term negotiations at last, international community achieved the prohibition to use chemicals in military purposes. 13 years ago the Chemical Weapons Convention (Convention on the Prohibition of the Development, Production, Stockpiling and Use of Chemical Weapons and on their Destruction) has entered into force.

The Convention is a unique international agreement, since most negotiations on disarmament were finished without comprehensive solutions of problems. The Chemical Weapons Convention proclaims the total rejection by State Parties (188 States on the beginning of 2010) to develop, to produce or to use chemicals as weapons, and each State Party undertakes to destroy its own chemical weapons. I like to underline that the Convention prohibits the military use of toxic chemicals both known and existed today and those which could be synthesized in future. The Article II of the Convention determines:

"1. "Chemical Weapons" means the following, together or separately:
(a) Toxic chemicals and their precursors, except where intended for purposes not prohibited under this Convention, as long as the types and quantities are consistent with such purposes; ...
2. "Toxic Chemical" means:
Any chemical which through its chemical action on life processes can cause death, temporary incapacitation or permanent harm to humans or animals. This includes all such chemicals, regardless of their origin or of their method of production, and regardless of whether they are produced in facilities, in munitions or elsewhere." (http://www.opcw.org/chemical-weapons-convention/articles/article-ii-definitions-and-criteria/)

For 13 years States Parties have destroyed 42,507, or 59.71%, of the world's declared stockpile of 71,194 metric tones of chemical agent, 45.56%, of the 8.67 million chemical munitions and containers have been destroyed. Other, to my vision, the very important peculiarity of the Convention, in contrast to other treaties, for example—BWC (Biological Weapons Convention), is the verification regime and inspection applied not only for military depots and arsenals. The chemical industry enterprises are also the object of control and inspection. The CWC implementation and verification regime now covers over 98% of the world's population, but what is more important, 98% of the world's chemical industry. 4,134 inspections have taken place at 195 chemical weapon-related and 1,103 industrial sites on the territory of 81 States Parties since April 1997. Worldwide, 4,919 industrial facilities are liable to inspection.

Chemical weapons also proliferated to several countries in the developing world and were used on the battlefield in the Yemen Civil War (1963-67) and the Iran-Iraq War (1980-88). During the latter conflict, Saddam Hussein first ordered the use of mustard gas in 1983 to counter Iran's numerical superiority and "human-wave" infantry tactics, which were overwhelming Iraqi positions. Towards the end of the war, Saddam Hussein used chemical weapons as an instrument of terror against the restive Kurdish population in northern Iraq. In a notorious attack on March 16-17, 1988, the Iraqi Air Force dropped bombs containing mustard gas and nerve agents on the Kurdish town of Halabja, killing an estimated 5,000 civilians, many of them women and children. Terrorist groups such as Aum Shinrikyo in Japan and al-Qaeda in Afghanistan have also attempted to acquire and use chemical weapons

In my opinion, it's possible to do the next step in the toxic chemical control and to find common position on "riot-control" and "non lethal" agents problem (production and usage), so called "mid-spectrum agents" etc., to think on prohibition at whole so-called "chemical wars", large-scale military operations with pesticides or contamination of water sources. I consider that the CWC regime should be easily expanded, perhaps with some different principles of control, firstly on natural and synthetic toxins and other above mentioned chemicals weapons, riot control agents. (For more information: J.B. Tucker. The Future of Chemical Weapons. *The New Atlantis. Fall 2009/Winter 2010. P.3*; J. Matousek. Chemical Weapons Convention and Role of Engineers and Scientists. Kemija u industriji. (2010) Vol. 59. No. 2. p. 69.; M. Balali-Mood, P.S. Steyn, L.K. Sydnes, R. Trapp. Impact of scientific developments on the CWC. *Pure Appl. Chem.* **80** (2008), 175-200.)

Chemical Weapons Convention is the important constituent of general strategy in prohibition of mass-destruction weapons and reduction of war threat. And at the same time, The Convention preserves State sovereignty and equality and does not limit conventional activity of chemical industry and production of "peaceful" chemicals. Thus, the Convention represents, per se, one of the first international treaties on a regulation of toxic chemical management. (J.D. Fry. Sovereign Equality under the Chemical Weapons Convention: Doughnuts over Holes. *Journal of Conflict & Security Law* (2010), Vol. 15 No. 1, 45–63).

The CWC is the first, but not the last, an international treaty on chemicals regulation.

Chemical Wastes and POP—Persistent Organic Pollutants. Toxic wastes being produced by various sectors of economy, from chemical industry to housing and municipal services, are a reason to introduce national legislation on waste management and corresponding systems of regulation and control. Transboundary transfers of toxic wastes, especially illegal shipping operations of these materials from develop countries to developing ones, became the real risk at the end of the last century. To prevent illegal transfer of toxic wastes from one place to other, the **Basel Convention** on the Control of Transboundary Movements of Hazardous Wastes and Their Disposal was proposed and adopted by world community (170 state-parties).

The next steps in international regulation of some chemical were **Rotterdam Convention** on the Prior Informed Consent Procedure for Certain Hazardous Chemicals and Pesticides in International Trade (September 1998. Entry into force: 24.02. 2004, Status: Parties 117) and **Stockholm Convention** on Persistent Organic Pollutants (May 2001. Entry into force: 17 May 2004, Status: Signatories: 152, Parties: 147). The objectives of the **Rotterdam Convention** are:

- To promote shared responsibility and cooperative efforts among Parties in the international trade of certain hazardous chemicals in order to protect human health and the environment from potential harm;
- To contribute to the environmentally sound use of those hazardous chemicals, by facilitating information exchange about their characteristics, by providing for a national decision-making process on their import and export and by disseminating these decisions to Parties".

The Convention covers pesticides and industrial chemicals that have been banned or severely restricted for health or environmental reasons by Parties and which have been notified by Parties for inclusion in the PIC procedure. There are 39 chemicals listed in Annex III of the Convention and subject to the PIC procedure, including 24 pesticides, 4 severely hazardous pesticide formulations and 11 industrial chemicals. Many more chemicals are expected to be added in the future. By this wording of the Convention, the today existing list of hazardous chemicals is only beginning and will be enlarge with years when new data on negative consequences of some industrial chemicals are obtained.

The Stockholm Convention on Persistent Organic Pollutants (POPs) was adopted in 2001 in response to the urgent need for global action to protect human health and environment from "POPs". These are chemicals that are highly toxic, persistent and bioaccumulated. In the list of banned chemical POPs there are aldrin, heptachlor, hexachlorobenzene and polychlorinated biphenyls. The Convention is directed on the elimination or restriction of production and use of all listed POPs (i.e., industrial chemicals and pesticides). It also seeks the continuing minimization and ultimate elimination of releases of unintentionally produced POPs such as dioxins and furans. Stockpiles of POPs have to be managed and disposed of in a safe, efficient and environmentally sound manner. These tasks of the Stockholm Convention are similar to those by Article 1 of Chemical Weapons Convention on destroying the chemical weapons. I suppose that the POPs list of the Convention will be also expanded in future years.

OBSOLETE PESTICIDES: A TICKING TIME BOMB AND WHY WE HAVE TO ACT NOW

Table 1. Estimated tonnes of OPs, by region

Region	Estimated tonnes	Estimated costs (€ million)*
ENP & Russian Federation	151,500	454.5
Central Asia	47,500	142.5
EU accession countries	36,000 to 41,5000	108 to 124.5
(New) EU member states	21,000 to 23,000	63 to 69
Total	**256,000 to 263,500**	**768 to 790.5**

* Assuming €3,000 per tonne destroyed or safely stored.

Source: IHPA data, POPs Convention NIPs (see Appendix 1 – small differences in Appendix 1 are due to rounding).

J. Vijgen. International HCH& Pesticides Association, IHPA. C. Egenhofer. Centre for European Policy Studies. CEPS. (www.ihpa.info)

As we can see, all above listed conventions have similar targets to establish regulatory regime for a certain number of chemicals that are dangerous for human health and environment. Three years ago, Conferences of the Parties of Basel, Rotterdam and Stockholm conventions proposed and established the **ad hoc joint working group** on enhancing cooperation and coordination. The mandate of the group is to prepare joint recommendations on enhanced cooperation and coordination among the three conventions for submission to the conferences of the Parties of all three conventions. Major objectives in cooperation of the three conventions are directed on strengthening implementation at the national, regional and international levels and to analyze possibilities to organize joint development of documentation, technical guidance and standards, training and capacity-building activities, shared use of regional offices, centers and facilities, etc.

UNEP Chemicals, the center for all chemicals-related activities of the UNEP, has organized an elaboration of the Strategic Approach to International Chemicals Management (SAICM), a policy framework for international action on chemical hazards. The Strategic Approach, endorsed by the Johannesburg World Summit on Sustainable Development in 2002 and the New York World Summit in September 2005, has been developed by Preparatory Committee, co-convened by UNEP, the Intergovernmental Forum on Chemical Safety (IFCS) and the Inter-Organization Programme for the Sound Management of Chemicals (IOMC). It was adopted by the International Conference on Chemicals Management (ICCM) on 6 February 2006 in Dubai. European states have introduced **REACH** initiative (REACH—Registration, Evaluation, Authorisation and restriction of Chemicals). On the other hand, chemical industry of member countries of ICCA (International Council of Chemical Association) is implementing the **Responsible Care**® program to improve product stewardship, workers' safety and health, p!ant safety, and transport safety. In the industry's own words: Responsible Care promotes co-operation with governments and organizations in the development and implementation of effective regulations and standards, and helps companies meet or exceed these requirements". These self-regulatory measures aid transparency, facilitate verification, and provide additional safeguards with respect to the security of chemical plants.

All above mentioned demonstrate 1.) necessity in chemical security regulation; 2.) possibility to reach a common agreement; and 3.) good basis to construct well-composed system of chemical security on international and national levels.

A brief comparative analysis of all conventions on chemicals shows some important and attractive claims of CWC in contrast to others. First, the Chemical Weapons Convention has the comprehensive nature of the prohibition in the CWC and covers both old and new possible chemicals as weapons. The number of known toxic compounds with potential for abuse will steadily increase. The advances in the life sciences will create new risks of the dual-use potential of many chemicals. States as well as non-state proliferators may alternatively opt for toxic industrial materials as less-effective chemical weapons. Therefore, non-proliferation efforts, including measures to control access to relevant chemicals, equipment and technologies remain and will remain important safeguards. The Convention contains a mechanism to amend the schedules; chemicals can be added to or deleted from the schedules subject to agreement by States Parties. They can also be moved from one schedule to another should the States Parties consider this beneficial for verification purposes. It is important that the CWC does not become "frozen in time" but that it adapts to the changes in science, technology and industrial practice.

Secondly, CWC contains well-developed mechanism of national and international control, verification and reporting. Use of modern analytical instruments on site and network of designated analytical laboratories as well as significant database of toxic chemicals allow organizing the desired control system and for other toxic chemicals.

Other above-mentioned conventions have no similar comprehensive control system. On today, OPCW has experience in inspections, verification and control, qualified specialists etc.

Now, I think, we have quite good basis to think on the integration of various chemical management regimes and, perhaps, OPCW should to be as basic model on international chemicals management. As example of similar approach, I would like to give an example the model of IAEA status and directions of activity. An independent international organization related to the United Nations system, IAEA transformed into the world´s center of cooperation in the nuclear field, dissemination of the best practice and safety standards in nuclear energy technology as well as creation of additional legal instruments. Three main areas of IAEA´s work are: Safety and Security; Science and Technology; and Safeguards and Verification. These are the same what are necessary for chemical security on national and international levels.

I'm convinced that peaceful and safe use of chemicals should be coordinated by international community in the unified center on high professional level, responsibility and international collaboration.

The very important element in the chemical security and safety is the education. Unfortunately, last years we can see shortening education programs on chemistry in schools and some deficiencies in its curricula. Simultaneously with basic knowledge in chemistry, chemical education has to explain the dual-use nature of advances in chemical science and technology as well as chemicals. University curricula should include appropriate references to the Conventions, which regulate a management with chemicals on today, their requirements, related information and ethical guidance. Ethic codes of conduct and professional codes are also important tools in chemical security.

Under sharp criticisms by the public, chemistry also tries to change a "face of chemical industry". Indeed, modern technological advances in chemical manufacturing are significant. It is obvious that numerous traditional chemical operations based on oil- or gas-based feedstock will use a renewable feedstock and bio-processing technologies. Process efficiency also can be significantly improved in order to minimize the environmental impact of manufacturing and to minimize by-products and wastes. Principles of "Green Chemistry", which have been recently formulated, make provision for the implementation of energy-efficient processes with (near)-100% atoms efficiency leading to drastically reduced generation of unwanted by-products, to increase selectivity and to improve the economy of reactions using standard equipment. Efficient optimizations of processes have been proposed and applied to produce medicines, pesticides, fine chemicals and other high-value products for market demands.

We are awaiting new, impressive results in the combination of catalysis and biocatalysis with nanotechnology that offers new possibilities to increase reactivity, selectivity and efficiency of catalysts and finally processes. Another trend relates to the emerging use of micro reactors in industrial production. One of the driving factors is safety: chemicals that are otherwise hazardous to manufacture, handle or store can be produced on-site when needed. Other important factors include the smaller amounts of reaction by-products, better kinetic data for reaction, lower capital costs and very precise control of the reaction scenario. Many chemical reactions have now been demonstrated to show improved reactivity, product yield and selectivity when performed in micro reactors compared to conventional laboratory practices. Furthermore, many of the traditional problems associated with scaling-up production from laboratory to industrial-scale volumes can be avoided by what is called “numbering-up”, i.e., running large numbers of micro-reactors in parallel.

Chemical industry has to take the best principle of modern nuclear energy industry—this is “**safety culture**”. But safety culture in use of chemicals should be also implemented in all spheres of human activity.

I’d like to finish with the words of Roald Hoffmann, Nobel Price Laureate**: “For chemistry, and chemists, there is no choice other than responsibility, personal and collective”** (Cornell University, 1990).

HOW TO AVOID HAVING RESEARCH FUNDING FASHION KILL THE GOOSE THAT LAYS THE GOLDEN EGGS

DR. CHRISTOPHER ESSEX
Department of Applied Mathematics, University of Western Ontario
London, Canada

WHAT THIS IS ABOUT

I've been asked to speak about a cause that Prof. Lee Smolin and I have been engaged in for a few years. Lee is at the Perimeter Institute for Theoretical Physics, and is known for his work in quantum gravity. We have both written books dealing with the troubles of science in society.[1] Together we are attempting to have research funding policy reflect lessons from the history of science.

This is not about journal impact factors, publication performance indicators, application success rates, or some strategic this or that. In fact it will suggest that such criteria can be distractions that set us back rather than move us forward. For a healthy scientific culture, governments must resist their urge to "manage" what no one truly comprehends: the process of scientific discovery. We recommend a humbler approach for superior results.

A BELIEF IN THE IGNORANCE OF EXPERTS

Scientific research is about the unknown of the natural world. History shows time and again how easily we fool ourselves about the truths of nature. So I usually define scientific thinking as the struggle to overcome our own human fallibility. But for this talk Richard Feynman's definition[2] is better:

> "Science is the belief in the ignorance of experts."

Nonetheless, modern funding schemes depend on the opinions of experts. Who gets how much funding to ponder Feynman's "ignorance of experts" is awarded by those very experts. If this is not a paradox, it does at least seem like a conflict of interest. Of course some might rightly respond by asking, what else can we do? Ask non-experts for advice? Good questions. I'll address them, but first I emphasize that much more is at stake than any individual's funding.

The history of science is riddled with stories of scientists with unfashionable thinking successfully challenging both the prevailing scientific conventions and the experts who defend them. Feynman's definition resonates in such stories. Naturally we remember better the high profile cases where the unlikely idea was proved right against the odds. Unlikely triumphs over bullies are popular stories, but only a fraction of them in science are widely known.

---

[1] L. Smolin, The Trouble With Physics: The Rise of String Theory, the Fall of a Science, and What Comes Next , Houghton Mifflin(2006); C. Essex and R. McKitrick, *Taken by Storm: the troubled science, policy and politics of global warming,* Key Porter (2002, 2007).

[2] Feynman, "What is Science?" The Physics Teacher, Vol. 7, issue 6 (1969).

The fact that many of the most important advances in science have this pattern should capture the imagination of people who set science policy. It should inspire respect for scientific nonconformity. But this isn't happening. Instead there's more contempt for unusual thinking than ever. Modern scientific mavericks have as hard a time as they have ever had. The best career strategy for such a maverick today is to become mainstream.

Thomas Kuhn is famed for explaining that our knowledge advances in a jerky way because of the permanent tension between consensus and revolution in science. The term "transformative" was later coined to reflect the revolutionary side, but the term has been co-opted and debased. Now nearly everything funded is described by someone as "transformative." So instead I will just call true transformative advances "golden eggs," which brings me to the goose in the title of this talk.

AESOP AND NESCIENCE

You may know Aesop's fable about the goose that lays the golden eggs. The goose was killed to get at its gold. However, to the surprise of the killers, there was no gold in the dead goose's body, despite what was greedily imagined. Not only did their intervention fail to deliver the extra gold they lusted for, but also as an additional unintended consequence there would be no more golden eggs either. Of course this fable is usually assigned morals about greed and acting before thinking, but there is a less common moral here too, a moral about hubris—hubris that the mystery of the golden eggs was understood well enough to justify using a knife at all.

For research funding, the issue is neither greed nor thoughtlessness. There's been plenty of thinking, if not over-thinking. Hubris is the issue. Some fancy that they can manage the unknown. But the unknown is beyond managing. The architects of research funding schemes nonetheless confidently forge ahead aided by experts to dissect the research goose with selective funding using all of the fashionable performance criteria.

Of course performance measures or other indices in themselves aren't hubristic. They are just tools, like knives. But we are talking about a knife used unwisely. The architects imagine that these made-up recipes for numbers actually weigh and measure the unknown. And being duly metered, management as a regular business follows. That is hubris.

Some might object that all managers routinely face unknowns. True, but research unknowns are more than knowns with error bars. They aren't like guessing next year's interest rates or market conditions. Research includes the unknown in the big sense of the word, where we don't know what the variables are or even what questions to ask. I like the word "nescience" for that kind of unknown.

This is a hard point to make, as it is an indulgence of every age to believe there's no nescience. 19th century physicists Lord Kelvin and Albert Michelson famously claimed that the unknown would vanish simply as a matter of precision, shortly *before* being soundly undone by the major revolutions in physics of the 20th century. But even that stunning object lesson did not put it to rest. A journalist revived the argument at the end of the 20th century by arguing that the unknown is a "substance", like oil, that we are running out of because the Universe is being researched out. This too dropped out of sight in the wake of subsequent discoveries.

The equation on the front of this podium, The Dirac equation, is a key component of that 20th century revolution. It is not the only or even the first golden egg of the 20th century, but it represents the culmination of a long line dating back into history that tells us that nescience will remain with us. Of course if one doesn't believe in nescience, then there are no more golden eggs to be laid—might as well cook the goose.

HOW "SEEMINGLY-WRONG" CAN TURN OUT RIGHT

Let's return to the archetypical story of a lone scientist standing against an established worldview. After much resistance from the experts the lone scientist is proven right against the odds and the establishment crumbles. Let's call this *the transformative scientist's story*. This story seems like something dreamed up by Hollywood, but it really happens over and over. It's not a string of isolated incidents; it's an inevitable part of how humans advance. We experience the arrival at new truth sequentially: there's a front end and a back end.

Let me explain. I've made one or two small discoveries in my career. Discovery feels like being in a bright and vibrant place after awakening from a dark fitful sleep. The front end is darkness and confusion before waking, while the back end is the light and simplicity afterward. You might recall something like it while learning mathematics in school. At first a mathematical idea may seem completely incomprehensible, then suddenly the light comes on and you get it. Ironically, it may seem obvious or even trivial afterward. We all make these jumps, but we forget the dark struggle before breaking out of the front end. So there is often a gulf between those on the front end and those on the back end.

Suppose now *you* are the first to discover something very important. You have found your very own golden egg. Congratulations! But there's a catch: while you have made it happily to the back end, everyone else in the world is at the front end. Your job is to convince other experts that it is worth their effort to wade through their confusion at the front end to meet you at the back end. Why should they divert from their own ideas to give your idea the effort that it needs? An unestablished discovery can simply seem wrong viewed from the front end and it might even seem like a threat to legacies as well. Just assuming it's rubbish and you are a lunatic is much easier.

If you have no track record or you are seen as a troublemaker, an outsider, or worse unfashionable, you will be on a certain trip down the rabbit hole. Good luck getting past referees to get published, let alone getting past still other referees to get funded. As George Orwell wrote, "an unfashionable idea rarely gets a fair hearing." This is the plight of those who take Feynman seriously by believing in the ignorance of experts. If you have never experienced this yourself, trust me, you don't want to.

THREE EXAMPLES

Here are three of many lesser-known examples:

Hannes Alfén discovered previously unknown waves in plasmas. He couldn't convince other experts. Eventually Fermi intervened causing experts to look more seriously. They concluded he was right after all. Alfén won the Nobel Prize for what are now known as Alfén waves. But how many years would we all be set back if there was no Fermi to intervene? Alfén never became fully accepted despite his Nobel Prize. He rubbed experts in the wrong way. I saw him speak when I was a student. He told us that space plasma physics was all wrong. We students enjoyed it as an object lesson in "believing in the ignorance of experts," but the space plasma experts didn't feel so warmly.

Another better-known example is Alfred Wegener. He proposed in 1912 that the solid ground under our feet should float like a cork and drift over the planet over thousands of millennia. What a crazy idea! Or at least that is how it seemed to many in the scientific community until after Wegener's death. But now, long after he is gone, we teach continental drift to children in school—front end to back end in half a century.

I think the example most immediate to everyone is the case of Ignaz Semmelweis. We all live daily with the back end of his discovery. We are even repelled by those who don't follow his thinking. Semmelweis discovered that thoroughly washing your hands between examining hospital patients reduced death rates. Imagine that! We are so immersed in the back end of this golden egg that washing just seems obvious. But in the 1840's it wasn't obvious. It was more than not obvious: the experts not only did not believe it; they were angry about it! Semmelweis was driven out of Vienna.

Don't conclude that the 1840's experts were fools while we are now safely sensible. We're all fools at the front end. That's the defining feature of the human experience with the unknown. That's why the transformative scientist's story will always be with us. Do I know examples of this from today? Of course I do! But I'll keep them to myself for now. You might think its rubbish and me a lunatic if I tell you where today's conventional thinking is wrong.

KILLING THE GOOSE WITH THE "EXCELLENCE" KNIFE

"Excellence" is a fashionable catchword of modern research funding bureaucrats: the more "excellent" the research, the more the funding. This kind of "excellence" is determined by a panel of experts and referees. As the experts only judge proposals about the unknown we must, in the spirit of Feynman, believe in the ignorance of this "excellence." This "excellence" is not excellent.

Unsurprisingly, research challenging orthodoxy, or even with the slightest whiff of peculiarity is inherently not excellent in this deficient system. The iconic lone scientific challenger is at an extreme disadvantage. For young scientists, being too challenging or adventurous can be a career ender—cooking their geese so to speak. And as sauce for the goose, the "excellence" definition has deal breaking criteria grafted on that are foreign to research: numbers of students, papers ranked by journal impact factors, citation numbers, anticipated business applicability to name but a few. Success in such a system comes from paying close attention to fashion and not upsetting referees, while going through motions that raise the superficial criteria profile.

BLAMING THE VICTIM

But I know many scientists who follow principle by keeping their eyes on the unknown, caring little for expert panels, fashion, or criteria. As always in history, their imaginative ideas are called strange or wrong because of the ignorance of experts sitting as the smug gatekeeper at the front end of new ideas. Consequently, papers have to be written many times over to have a chance to get them published at all. Criterial productivity drops. Soon they are unfunded. Many simply give in and do conventional things, or they quit. And if they do quit, they are blamed.

Iconic challenges to scientific orthodoxy have labels in the research policy field. It's "high risk" or it's deemed perniciously as "disruptive research". The result is a risk averse, authoritarian environment—the opposite of what history tells us good science requires.

KEEPING THE GOLDEN GOOSE ALIVE DOESN'T TAKE MUCH

So is there an alternative to using experts? Yes. Ironically, it's the business investment world offering one. They make money on the unknown all the time. These venture capitalists support many small long-shots minimally. That gives them "a positive exposure" to the typically high returns that just one of their small investments can pay. They know implicitly that any expert they might consult will be no more expert than the people actually doing the work. So an expert panel or referees are uninformative if not counterproductive.

Nassim Taleb, the author of *the Black Swan: the impact of the highly improbable*[3], recommends a "dumbbell strategy" for investment portfolios. 90%, say, should be in relatively safe choices, while a minority of funds should be put into the venture capital approach—minimal investment in many independent risky ventures. Safe research is just as valid as safe investment. But contemporary research funding schemes don't get the "positive exposure" to potential golden eggs. Wise science funding can and should account for the essential dichotomous impulses of consensus and revolution in science. Lee and I proposed a program[4] that could do a good job along these lines with 1% of the available funds. That's round off noise or chicken feed if you like. 99% would be available to continue as before.

Many people have trouble grasping how special rules near zero funding would help, but physicists understand implicitly about edge or boundary effects. It doesn't take much to keep scientific mavericks going or just encourage the inner maverick in all scientists. It only takes some chicken feed to keep the golden goose alive. But when the next golden egg comes along, you won't complain that the chicken feed cost too much.

---

3 p. 276, N. N. Taleb, The Black Swan: the Impact of the Highly Improbable, Random House, New York, (2007) pp 366.

4 http://www.apmaths.uwo.ca/~essex/NSERCSmallScience.pdf

THE INVESTIGATION OF HUMAN BRAIN FUNCTION BY PHYSICAL METHODS: A Central Frontier for Science and a Way to Encounter the Emergency of the Neurodegenerative Diseases

BRUNO MARAVIGLIA
Department of Physics, University of Rome "Sapienza"
Enrico Fermi Centre, Rome, Italy

Physicists, in general, investigate nature by studying simple systems, which are easier to conduct experimental tests and theoretical analysis. Living organisms are quite complex; in particular, the human brain is probably the most extraordinary frontier of research because of its vast amount of unknown mechanisms and functions, which are responsible of the human life.

The Physics Research on Human Brain Structure and Function will be both:

1. A way to solve some of the fundamental questions of knowledge concerning memory, logic, conscience, the definition of the border between life and death, etc.
2. A crucial contribution to understand the causes and the evolution of the neurodegenerative diseases, a dramatic emergency hitting several million human beings.

In the last few decades, some new physical techniques like Positron Emission Tomography (PET), Magnetic Resonance Imaging (MRI) and functional Magnetic Resonance Imaging (fMRI) together with Electro Encephalography (EEG) and Magneto Encephalography (MEG) have opened new investigative lines for studying the brain's structural and functional properties, without substantial perturbation.

My research for over a decade has exploited the great versatility of Nuclear Magnetic Resonance (NMR) to investigate the Human Brain Function. The power of NMR is much greater than that of other modalities, due to its capability of measuring several observables and their spatial dependence (imaging) like: spin density, the spin-lattice relaxation time T1, (related to molecular dynamics), the spin-spin relaxation time T2, the chemical shift (which allows the mapping of several metabolites), the J coupling, the diffusion tensor D (whose imaging provides the axonal cabling of different neuronal areas), flow, etc.

New physical methods related to NMR in which my group is strongly involved:

- Dynamics of metabolites concentration under brain stimulation,
- Combination of fMRI and EEG,
- Resting state fMRI
- Ultra-low field MRI-MEG,
- Neurocurrents imaging,etc.

Neuroenergetics: experimental lactate dynamics

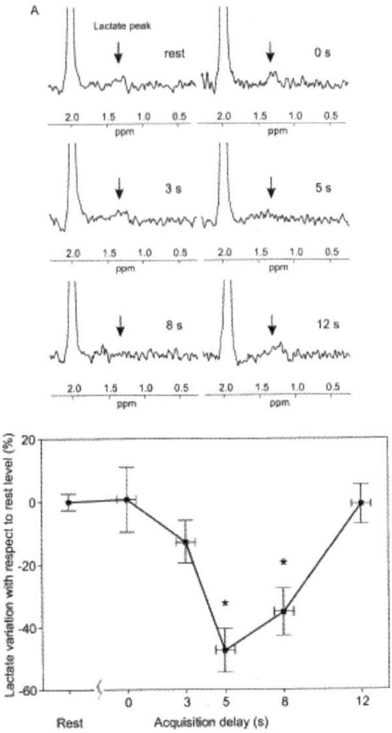

(*a*)

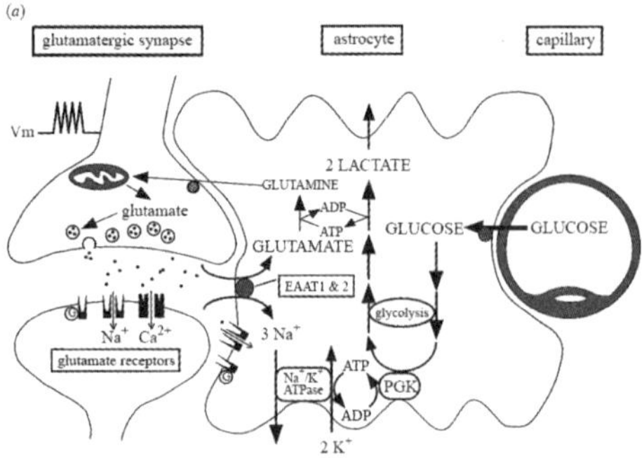

Neuroscience **118:**7-10 (2003).

Simultaneous fMRI and Evoked Potentials

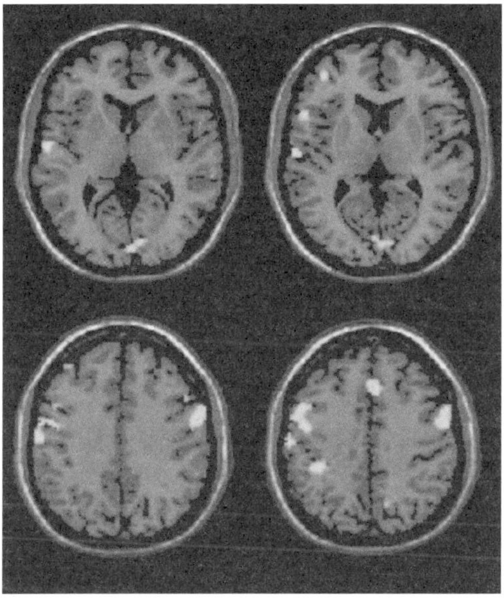

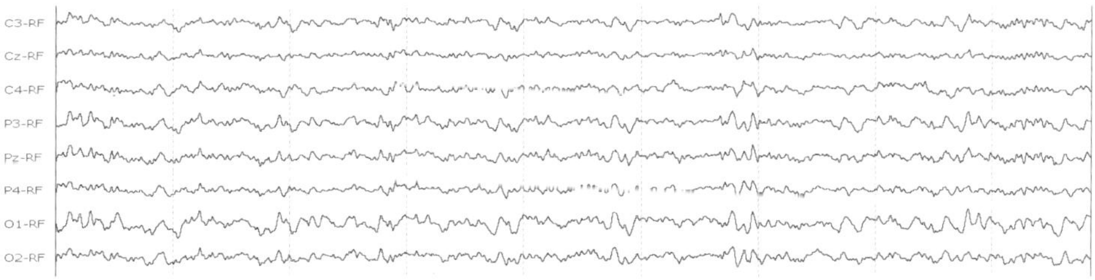

fMRI → EEG

Improved EEG source localization by constraining with fMRI clusters (spontaneous & evoked activity)

EEG → fMRI

Improved prediction of time fMRI signal changes in order to localize with increased sensitivity:

- unpredictable spontaneous activity (epilepsy, sleep, brain rhythms, drug effects)
- unpredictable changes in evoked activity

MRS&EEG
Metabolism ↔ long time: visual stimulus / rest
EEG Signal ↔ short time: alternate ON/OFF

Concentration
Glu+Gln (mmol)

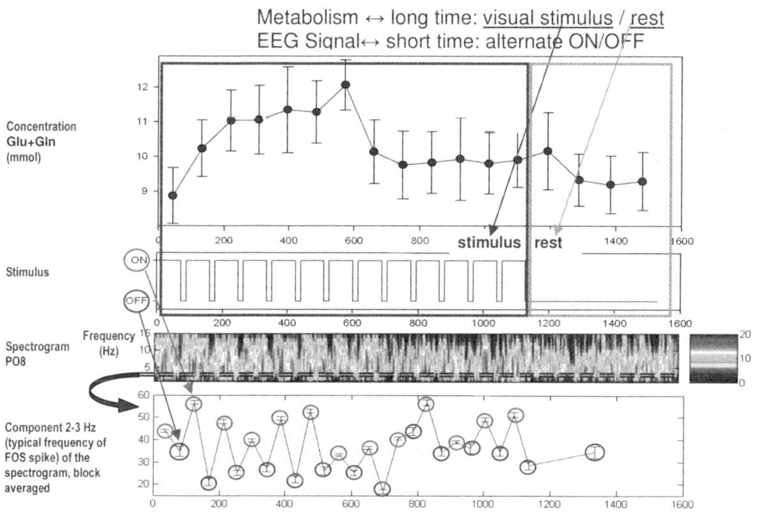

The brain accounts for 20% of the oxygen, and hence calories, consumed by the body. In the average adult human, the brain represents 2% of the body weight.

Only about 20% of brain energy consumption is due to its interaction with the external world (stimulated activity), all the rest (80%) is due to autonomous, unexplored activity.

Resting state
Low frequency intrinsic correlations during rest the default mode network(s)

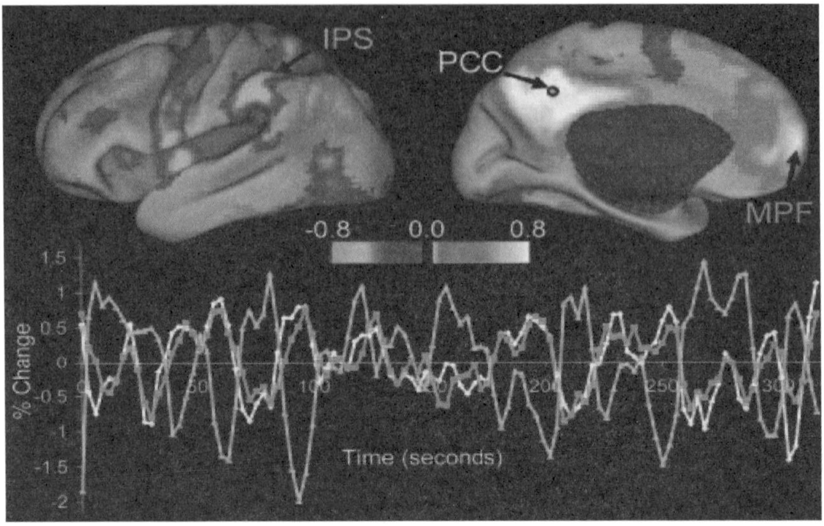

IPS—Intra Parietal Sulcus; **PCC**—Posterior Cingulate Cortex; **MPF**—Medial PreFrontal Cortex

After all the target of understanding, by physical means, the brain architecture and the way it operates is just a humble but Galilean new approach to:

NOSCE TE IPSUM

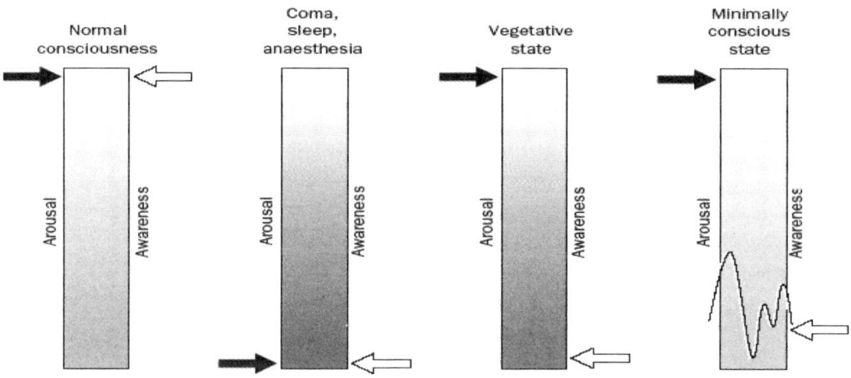

Conscience = *Arousal* + *awareness*
(vigil, awake) + (conscious of self and of the environment)

Task 1: "imagine playing tennis"
Task 2: "imagine walking around home"

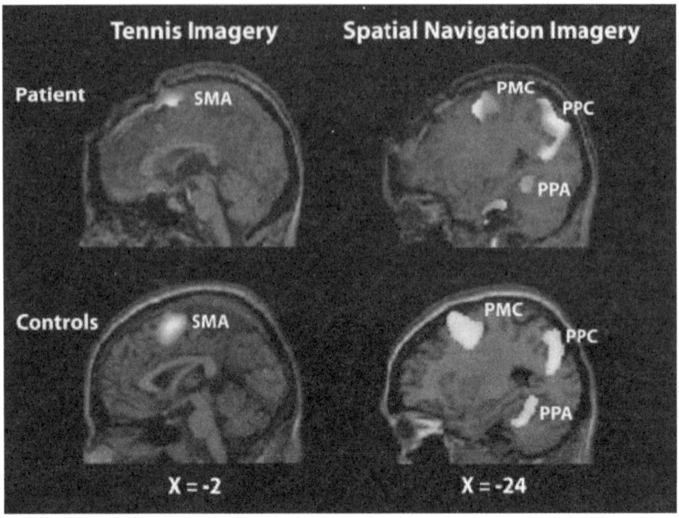

Fig. 1. We observed supplementary motor area (SMA) activity during tennis imagery in the patient and a group of 12 healthy volunteers (controls). We detected parahippocampal gyrus (PPA), posterior parietal-lobe (PPC), and lateral premotor cortex (PMC) activity while the patient and the same group of volunteers imagined moving around a house. All results are thesholded at $P < 0.05$ corrected for multiple comparisons. X values refer to distance in mm from the midline in stereotaxic space (SOM text).

New physical methods can be used, in vegetative states, to detect perception and even conscious awareness. This perspective will be relevant for diagnosis, medical decision making, besides for the fundamental questions about the nature of consciousness, thought and will.

Among the brain dysfunctions the neurodegenerative diseases cannot, in general, be objectively diagnosed; the mechanisms which generate them are unknown and medical treatments, in practice, are non-existing.

EMERGENCY

New insight, with the support of original physical methods, is dramatically needed in order to discover the causes of these modern plagues.

I thank you for your attention.

GLOBAL BRIGHTENING AND CLIMATE SENSITIVITY

CHRISTOPHER MONCKTON OF BRENCHLEY [1]
Carie, Rannoch, Scotland

ABSTRACT

The rapid global warming rate of 0.16 K/decade from 1975-2001, near-identical to those observed from 1860-1880 and 1910-1940 that pre-dated significant anthropogenic influence, may also have been chiefly natural. Based on simplified methods for determining climate sensitivity without resorting to general-circulation models, the natural global brightening of 2.6 W m^{-2} observed in the 18 years 1983-2001 should have caused a transient global warming of 0.6 K or, after adding anthropogenic warming, 0.8 K: yet <0.35 K warming was observed. Likewise, over the 56 years 1950-2005, assuming no net global brightening or dimming, anthropogenic transient warming of 1.45 K should have occurred: yet only 0.65 K was observed. Accordingly, current estimates may overstate climate sensitivity by a factor >2. Improved analysis of variations in cloud cover will be essential for the eventual determination of climate sensitivity.

INTRODUCTION

The radiative-forcing method of the Intergovernmental Panel on Climate Change (IPCC: Houghton et al. 2001, ch. 6.1) for determining climate sensitivity without resorting to atmosphere-ocean general-circulation models is deployed to determine the quantum of projected transient global warming from the global brightening of 1983-2001, as well as from anthropogenic influences over the same period and, separately, from 1950-2005. Results, compared with observations, shed light on the reliability of the IPCC's method.

THE DATA

The Hadley/CRU global mean surface temperature dataset (Jones et al. 1999, and Brohan et al. 2006, cited in Solomon et al. 2007) shows a rapid and sustained global warming from 1975-2001 at 0.16 K/decade, a rate identical, within measurement error, to those observed from 1860-1880 and again from 1910-1940. Warming in the two earlier periods preceded any possible significant anthropogenic influence on the climate. Was the third and most recent period of rapid warming also chiefly natural?

Satellites first measured global surface temperature, cloud cover, and radiant-energy flux leaving the top of the atmosphere in the early 1980s. Pinker et al. (2005), finding agreement between satellite and terrestrial records for the first time, reported a strong global brightening from 1983-2001, possibly caused by a decline in cloud cover with changes in water vapor and aerosols. Pinker, relying chiefly on the International Satellite Cloud Climatology Project (ISCCP: Rossow & Schiffer, 1991, 1999), applied linear and second-order least-squares fits to the satellite-derived time-series of globally-averaged short-wave anomalies in solar radiative flux at the Earth's surface from 1983-

[1] Carie, Rannoch, Scotland, PH17 2QJ: monckton@mail.com.

2001, after removing the annual cycle. The linear slope was positive at 0.16 W m^{-2} yr^{-1} (Figure 1), close to the 0.143 W m^{-2} yr^{-1} found in a recent reanalysis by Boston (Appendix).

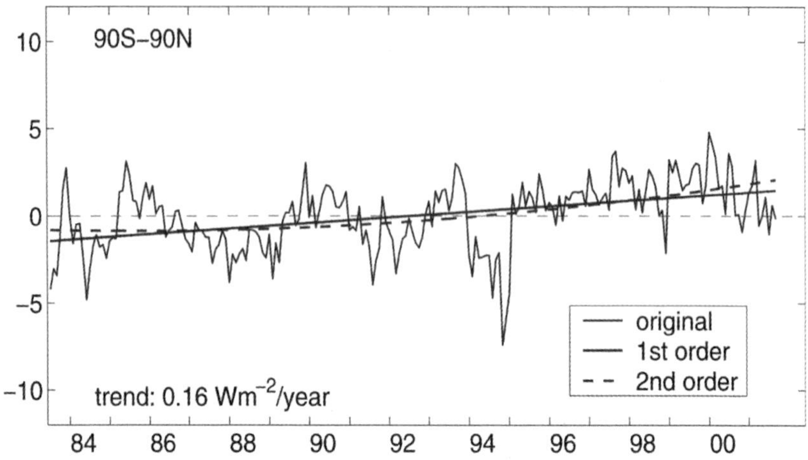

Fig. 1: A globally-averaged + 0.16 W m^{-2} yr^{-1} trend in the short-wave solar surface radiative flux anomaly, 1983-2001, after removal of the mean annual cycle. From Pinker.

The Earth Radiation Budget Experiment satellites (ERBE: Barkstrom, 1984) also detected a reduction in short-wave radiation reflected from clouds to space during the same period, with a corresponding increase in long-wave radiation as more short-wave radiation reached the Earth's surface and was Wien-displaced to the near-infrared, consistent with a reduction in global cloud cover, especially at low altitudes and latitudes, and particularly in the 1990s.

At the time when Pinker reported, the ERBE outgoing-radiation data presented in Wielicki et al. (2002a, b) had not been corrected to allow for orbital decay. After allowing for adjustments published by Wong et al. (2006), Boston (Appendix) used the ERBE data as a validity check for the ISCCP data, concluding that there was reasonably close agreement with them after allowance for differences in spatial and temporal coverage, and that the annual net surface flux trend for 1983-2001 was +0.143 W m^{-2}. This value is taken as the starting-point for the calculations in the present paper.

Other observations confirm the fact of the global brightening, while disagreeing as to its magnitude. Solomon et al. (2007, at Table 3.5) compare tropical (20°S-20°N) top-of-atmosphere long-wave and short-wave radiative-fluxes for 1994-1997 with the fluxes for 1985-1989, citing the ERBE satellite data, which showed outgoing long-wave radiation increasing by 0.7 W m^{-2} and outgoing short-wave radiation decreasing by 2.1 W m^{-2} over the period. Solomon also cites the ISCCP data as indicating that outgoing long-wave radiation increased by 0.5 W m^{-2}, while outgoing short-wave radiation decreased by 2.4 W m^{-2}. The AVHRR Pathfinder gave an opposite result: but, as

Solomon notes, "Calibration issues, conversion from narrow to broadband, and satellite orbit changes are thought to render the AVHRR record less reliable for decadal changes compared to ERBS."

Wild et al. (2005), focusing on changes in mean surface temperature over the land, concluded that the global dimming up to the 1980s was offset by the period of brightening (or at least the absence of dimming) thereafter, and that the rapid warming that followed the transition was attributable almost entirely to anthropogenic influences. However, Pinker (at Figure 5A) shows a very slight dimming over land only from the 1980s onward. Though Wild makes a passing reference to Pinker, he relied not upon satellite data but upon data from the Global Energy Balance archive and the Baseline Surface Radiation Network. Wild focused solely on land surfaces. The absence of solar dimming and the insignificant net brightening over land after 1980, deduced from surface-station measurements, is consistent with the analysis of land-only satellite data in Pinker.

Solomon (2007: ch. 3) concludes the sub-chapter on clouds as follows:

> "In summary, while there is some consistency between ISCCP, ERBS, SAGE II and surface observations of a reduction in high cloud cover during the 1990s relative to the 1980s, there are substantial uncertainties in decadal trends in all datasets and at present there is no clear consensus on changes in total cloudiness over decadal time-scales."

However, the data indicate that from 1983-2001 there was a significant decline in cloud cover generally, not merely in high cloud cover. It is the optically-dense clouds at low altitude and latitude that are most influential in global brightening or dimming.

ANALYSIS AND RESULTS

For clarity, this analysis is confined to central estimates. First, established methods are used to determine the surface pre-feedback warming ΔT_0 caused by global brightening. Next, simplified methods provided in Solomon are used to determine post-feedback surface climate sensitivity ΔT both to natural and to anthropogenic radiative forcings.

Boston's net global surface short-wave flux trend +0.143 W m^{-2} yr^{-1} represents a flux increment $\Delta F_{\text{nat}} \approx 2.586$ W m^{-2} over the 18.083 years 1983-2001 (Appendix: Figure A3). Warming before feedbacks is the product of ΔF_{nat} and the surface value of the Planck pre-feedback climate-sensitivity parameter κ_{sfc}, which, from data in Kiehl & Trenberth (2008) and methods in Kiehl & Trenberth (1997) and Kimoto (2009), is the first differential of the Stefan-Boltzmann equation where emissivity ε is unity, mean surface temperature T is 288 K, and radiative fluxes F_{sfc}, F_{eva}, F_{thm} from surface radiation, evapo-transpiration, and thermal convection are 390, 80, 17 W m^{-2} respectively:

$$\kappa_{\text{sfc}} = \Delta T \,/\, \Delta F_{\text{sfc}} = T \,/\, [4(F_{\text{sfc}} + F_{\text{eva}} + F_{\text{thm}})] \approx 0.1478 \text{ K W}^{-1} \text{ m}^2. \quad (1)$$

Accordingly, the naturally-occurring transient pre-feedbacks warming $\Delta T_{0,\text{nat}}$ that would be expected to have arisen from the global brightening mentioned in Pinker was:

$$\Delta T_{0,\text{nat}} = \Delta F_{\text{nat}}\, \kappa_{\text{sfc}} \approx 0.38 \text{ K}. \tag{2}$$

Allowance is then made for the amplifying influence of temperature feedbacks arising in response to the surface warming at Eq. (2). The feedback multiplier f, where b is the sum of all unamplified climate-relevant feedbacks and where, at the characteristic-emission altitude, the Planck parameter $\kappa_{\text{cel}} \approx 3.2^{-1} = 0.3125$ K W^{-1} m^2 (Solomon, ch. 10, p. 631, footnote), is given by the feedback-amplification function in Bode (1945):

$$f = (1 - b\kappa_{\text{cel}})^{-1}. \tag{3}$$

At a doubling of atmospheric concentration, where the multi-model mean projected climate sensitivity $\Delta T_{2x} \approx 3.26$ K (Solomon, ch. 10, p. 798, box 10.2) and the anthropogenic CO_2 radiative forcing $\Delta F_{2x} = 5.35 \ln 2 \approx 3.71$ W m^{-2} (Myhre et al. 1998, cited in Houghton, 2001, and in Solomon), the IPCC's implicit central estimate is:

$$f_{\text{man}} = \Delta T_{2x} \,/\, (\kappa_{\text{cel}}\, \Delta F_{2x}) \approx 2.813. \tag{4}$$

With this central estimate of f_{man}, Eq. (3) is rearranged to derive a central estimate of the sum b_{man} of all unamplified temperature feedbacks that is implicit in Solomon:

$$b_{\text{man}} = (f_{\text{man}} - 1) \,/\, (f_{\text{man}}\, \kappa_{\text{cel}}) \approx 2.062 \text{ W m}^{-2} \text{ K}^{-1}. \tag{5}$$

However, it is necessary to deduct from b_{man} the cloud feedback of 0.69 W m^{-2} K^{-1} and the surface albedo feedback of 0.26 W m^{-2} K^{-1} (Soden & Held, 2006, cited in Solomon), since the observed global brightening will have encompassed these feedbacks. Thus,

$$b_{\text{nat}} = b_{\text{man}} - 0.69 - 0.26 \approx 1.112 \text{ W m}^{-2} \text{ K}^{-1}. \tag{6}$$

Then

$$f_{\text{nat}} = (1 - b_{\text{nat}}\kappa_{\text{cel}})^{-1} \approx 1.533, \tag{7}$$

and

$$\begin{aligned} \Delta T_{\text{nat}} &= \Delta T_{0,\text{nat}} f_{\text{nat}} \\ &= \Delta F_{\text{nat}}\, \kappa_{\text{sfc}} f_{\text{nat}} = \Delta F_{\text{nat}}\, \kappa_{\text{sfc}}\, (1 - b_{\text{nat}}\, \kappa_{\text{cel}})^{-1} \\ &\approx 0.59 \text{ K}. \end{aligned} \tag{8}$$

Accordingly, Solomon's radiative-forcing method implies that, in response to a natural global brightening of the magnitude determined by Boston, pre-feedback and post-feedback warmings would be 0.38 K (Eq. 2) and 0.59 K (Eq. 8) respectively.

In addition, anthropogenic warming would be expected to have occurred over the period. Table 1, again relying upon simplified methods described in Solomon, gives central estimates of the radiative forcings arising from observed increases in the concentrations of the major greenhouse gases from 1983-2001.

Table 1.

Greenhouse gas	1983	2001	Radiative forcing
Carbon dioxide	342 ppmv	370 ppmv	0.421 W m^{-2}
CFC-12	343 pptv	541 pptv	0.065 W m^{-2}
Methane	1630 ppbv	1775 ppbv	0.054 W m^{-2}
Ozone	Estimate	Estimate	0.048 W m^{-2}
Nitrous oxide	304 ppbv	312 ppbv	0.024 W m^{-2}
CFC-11	183 pptv	256 pptv	0.018 W m^{-2}
		Total forcing	ΔF_{man} = 0.630 W m^{-2}

Radiative forcings arising from changes in the atmospheric concentrations of key greenhouse gases from 1983-2001, determined from functions given in Myhre et al. (1998) and cited in Houghton et al. (2001) and in Solomon. CFC concentrations are from Hartley et al. 1996. The ozone forcing is an estimate derived from relative magnitudes of forcings in Solomon. The negative forcings from aerosol effects are omitted, since the measured surface radiative flux anomaly occasioned by the global brightening of 1983-2001 implicitly takes them into account. Forcings from minor halocarbons, land-use changes, and aircraft contrails are omitted as *de minimis.*

Since the forcings in Table 1 represent changes in net (down minus up) radiative flux at the tropopause rather than at the surface, the surface value κ_{sfc} of the Planck parameter is inapplicable. Instead, the substantially greater value $\kappa_{cel} \approx 3.2^{-1} \approx 0.3125$ K W^{-1} m^2 at the characteristic-emission altitude applies. However, the feedback factor f_{nat} remains appropriate thanks to the dominance of the global brightening over the period.

The warming ΔT_{nat} is *transient*, since—in the sufficiently short term, at any rate—global brightenings and dimmings come and go. However, in Solomon's methodology, the temperature change ΔT_{man} that is projected to arise in consequence of greenhouse-gas forcings is *equilibrium* warming: i.e., the global-temperature anomaly that would arise only after the climate had settled to a new equilibrium after the perturbation of the climate arising from the forcing. Thus, at equilibrium:

$$\Delta T_{man} = \Delta F_{man}\, \kappa_{cel} f_{nat} = \Delta F_{man}\, \kappa_{cel}\, (1 - b_{nat}\kappa_{cel})^{-1} = 0.30 \text{ K} \quad (9)$$

To determine transient temperature change, we introduce an additional term r_y to represent the ratio of transient to equilibrium warming. On the A2 scenario, Solomon projected 0.2 K/decade *transient* warming in response to CO_2 forcing for 2000-2010, all other anthropogenic influences being treated as broadly self-canceling, but the observed increase in CO_2 concentration was 20 ppmv, implying *equilibrium* warming of 4.7 ln(388/368) $\approx$ 0.25 K, and hence a decadal transience ratio $r_{10} \approx 0.2/0.25 \approx 0.80$.

Over all the 21st century, on Solomon's A2 emissions scenario, with predicted transient warming of 3.4 K and CO_2 concentration rising from a measured 368 to a projected 836 ppmv, the centennial transience ratio r_{100} is 3.4 / [4.7 ln (836/368)] $\approx$ 0.88. Therefore, for periods $10 \leq y \leq 200$ years, the following simplifying approximation r is effective:

$$r = r_y \approx 0.79 + y/1000 \quad (10)$$

Thus, for instance, $r_{10} \approx 0.8$; $r_{18} \approx 0.808$ over the 18 years 1983-2001; $r_{56} \approx 0.846$ over the 56 years 1950-2005; and $r_{100} \approx 0.89$. Taking $r_{18} \approx 0.808$:

$$\Delta T_{\text{man}} = r_{18}\, \Delta F_{\text{man}}\, \kappa_{\text{cel}} f_{\text{nat}} \approx 0.24 \text{ K}. \tag{11}$$

Then

$$\Delta T = \Delta T_{\text{nat}} + \Delta T_{\text{man}} \approx 0.83 \text{ K}. \tag{12}$$

Of this 0.83 K warming (Eq. 12) over the 18-year period, >70% would be attributable to the naturally-occurring global brightening. However, the linear regression trend on the HadCRUt monthly global land and sea surface temperature anomalies shows observed transient warming ΔT_{obs} = 0.34 K. Solomon's methods appear to lead to an overstatement of climate sensitivity by a factor 2.4.

Even if the flux anomaly occasioned by the global brightening of 1983-2001 had been only one-third of that which Pinker had reported, it would have caused warming approximately equivalent to the observed warming, consequently leaving little room for any contribution from anthropogenic influences. Alternatively, even if temperature feedbacks were taken as net-zero rather than strongly positive, the warming arising from the global brightening over the period would be at least equal to the global warming that was actually observed, again leaving little room for any anthropogenic component.

It is possible that the global brightening from 1983-2001 was less than Pinker or Boston had found, or that the reported rise in radiative flux arising from the global brightening was to some extent offset by unreported natural influences. It is also possible that over so short a period the influence of long-acting feedbacks may not have made itself felt, though this possibility is to some extent accounted for in the transience ratio *r*. Besides, if the initial warming were small, long-acting feedbacks would also be small.

In the sufficiently long term even a significant transient global brightening such as that of 1983-2001 might be canceled out. Solomon finds, with 90% confidence, that more than half of the global warming from 1950-2005 was of anthropogenic origin. On the assumption that all of it was anthropogenic, and that in particular no substantial net global brightening or dimming occurred from 1950-2005, radiative forcings from 1950-2005 were determined using the functions in Solomon, and are listed in Table 2.

To determine the warming that would be expected to have arisen over the 56-year period, it is again appropriate to omit Solomon's strongly-negative very-long-term forcing from anthropogenic particulate pollution, since no global brightening or dimming is assumed to have occurred over the period. However, over so long a period as half a century, the positive forcings from the cloud and surface-albedo feedbacks posited by Solomon would be expected to have operated.

Table 2.

Greenhouse gas	**1950**	**2005**	**Radiative forcing**
Carbon dioxide	308 ppmv	378 ppmv	1.095 W m^{-2}
Methane	1100 ppbv	1775 ppbv	0.281 W m^{-2}
Ozone	Estimate	Estimate	0.237 W m^{-2}
CFC-12	20 pptv	541 pptv	0.172 W m^{-2}
Nitrous oxide	287 ppbv	319 ppbv	0.102 W m^{-2}
CFC-11	10 pptv	248 pptv	0.060 W m^{-2}
		Total forcing	ΔF_{man} = 1.948 W m^{-2}

Radiative forcings arising from changes in the atmospheric concentrations of greenhouse gases from 1950-2005, determined on the same basis as Table 1.

Where $\Delta F_{man} \approx 1.948$ W m^{-2} (Table 2), where $r = r_{56} \sim 0.846$ to reflect the 56 years 1950-2005, and where the feedback multiplier $f_{man} \approx 2.813$ (Eqs. 3, 4) encompasses cloud and surface-albedo feedbacks, Solomon's methods (Eq. 11) would indicate transient warming of 1.45 K from 1950-2005 in the absence of any global brightening or dimming. However, observed warming ΔT_{obs}, taken as the least-squares linear-regression trend on the HadCRUt global mean surface temperature anomalies from 1950-2005, was only 0.65 K, again suggesting that use of Solomon's methodology may lead to overstatement of climate sensitivity, this time by a factor 2.2.

Since 2001, linear regression on the global surface-temperature data for 2001-2008 shows the trend to be $dT/dt = -0.012$ K yr^{-1}, yielding a total cooling $\Delta T \approx -0.1$ K for the eight years: an indication, perhaps, that cloud-cover has stabilized since 2001. (See also Appendix, Figure A1). A reversal of this magnitude may represent the early years of a sustained long-term cooling trend (Keenlyside et al. 2008) comparable to that of 1877-1910, a 34-year period when the trend also approached -0.01 K yr^{-1}.

DISCUSSION

The contribution of the naturally-occurring global brightening from 1983-2001 to warming over the period that was determined using Solomon's methods substantially exceeded observed warming. Even if the true global brightening had been only half of that reported by Pinker, its magnitude would have been sufficient to account for all of the observed warming over the period, leaving little room for any anthropogenic contribution. The surface brightening from 1983-2001 appears to have been real, substantial, and of natural origin. CO_2 concentration has continued to rise near-monotonically until the present, but the monotonicity of its increase, set against the stochasticity of the fluctuations in global brightening and dimming, implies an absence of correlation and hence of causation between the former and the latter, at least on decadal-to-subdecadal timescales. By contrast, at least from 1983-2001, there is some agreement between the global brightening and the observed warming, suggesting a perhaps causative and certainly far from counter-intuitive correlation between the two. The question arises whether Solomon was correct in listing direct and indirect aerosol forcings as being very strongly negative. If not, then the true magnitude of net anthropogenic forcings may have been considerably greater than the values given in Solomon. If so, Solomon's method may tend appreciably to overstate climate sensitivity.

The anthropogenic warming from 1950-2005 that was determined using Solomon's methods again substantially exceeded observation, raising the question whether there were any negative natural changes in net surface flux during that period. The natural cycles of brightening and dimming that may contribute to the quasi-periodicity of the global temperature record appear to correspond to some extent with the warming and cooling phases of the Pacific Decadal Oscillation, but they may not necessarily cancel one another out even over long periods. However, in the absence of adequate satellite instrumentation before the early 1980s, that question cannot be definitively answered.

In future the deployment of standardized, automated, surface-mounted thermometers and pyranometers at locations all over the planet, reporting by satellite much as the Argo bathythermographs do for ocean temperature and salinity today, may provide better measurement of changes in cloud cover. Pyranometers deployed in Japan for >100 years show a remarkably close, and possibly causative, correlation between changes in surface solar flux (expressed as hours of sunlight) and changes in surface temperature in the region (W.W.-H. Soon, 2009, personal communication). However, data sources from many other regions do not show a similar correlation.

It is not likely that the net global brightening from 1983-2001 was caused in part by a systemic decline in particulate aerosols resulting from environmental measures in Western nations to improve the quality of the atmosphere: particulate aerosols continue to be emitted in increasing quantities by emerging nations such as China. If there had indeed been an enduring global clearing of the air sufficient to influence global temperatures, the warming from 1983-2001 might have been expected to continue thereafter: however, there has been a small cooling trend since late 2001.

A note on the Planck parameter

The Planck parameter κ is of particular significance in the determination of climate sensitivity because not only pre-feedback temperature change ΔT_0 but also the feedback factor f are dependent upon it, so that post-feedback temperature change ΔT is dependent upon κ^2:

$$
\begin{aligned}
& \Delta T && = (\Delta F + b\Delta T)\kappa \\
\Rightarrow\quad & \Delta T(1 - b\kappa) && = \Delta F\,\kappa \\
\Rightarrow\quad & \Delta T && = \Delta F\,\kappa(1 - b\kappa)^{-1} \\
\Rightarrow\quad & \Delta T/\Delta F && = \kappa(1 - b\kappa)^{-1} \\
& && = \kappa f \\
\Rightarrow\quad & f && = (1 - b\kappa)^{-1} \qquad (13)
\end{aligned}
$$

Equivalently, expressing the feedback loop as the sum of an infinite series,

$$
\begin{aligned}
\Delta T &= \Delta F\,\kappa + \Delta F\,\kappa^2 b + \Delta F\,\kappa^3 b^2 + \ldots \\
&= \Delta F\,\kappa(1 + \kappa\, b + \kappa^2 b^2 + \ldots) \\
&= \Delta F\,\kappa(1 - \kappa\, b)^{-1} \\
&= \Delta F\,\kappa f \qquad (14)
\end{aligned}
$$

From Eqs. (14, 15) it is evident that it is inappropriate to take different values of κ when determining pre-feedback warming T_0 and the feedback factor f. Yet that is what occurred in Eqs. (7, 8), where $T_{0,nat}$ was taken as the product of the naturally-occurring surface brightening ΔF_{nat} and the surface value κ_{sfc} of the Planck parameter, but $f_{nat} = (1 - b\kappa_{cel})^{-1}$ was determined using the characteristic-emission-altitude value κ_{cel}.

We acted thus because in Eq. (7) it was surely appropriate to convert a surface radiative-flux change ΔF_{nat} to pre-feedback surface temperature change $T_{0,nat}$ using the surface value κ_{sfc} of the Planck parameter, and because in Eq. (8) it was appropriate to adopt the characteristic-emission-altitude value κ_{cel}, which Solomon, following Houghton, uses to determine f in response to the pre-feedback surface temperature change T_0.

The dilemma thus posed by the IPCC's treatment of the Planck parameter has received surprisingly little attention in the literature. However, Kimoto (2009) addresses it in some detail, and concludes that the surface value κ_{sfc} is preferable even for determination of climate sensitivity to radiative forcings at the tropopause. In that event, and on that ground alone, all of the IPCC's climate-sensitivity values would be overstated by $(\kappa_{cel}/\kappa_{sfc})^2$, approximately a factor 4.4.

If we had used κ_{sfc} not only for the initial surface warming caused by the net global brightening of 1983-2001 but also for the consequent feedbacks, on the IPCC's methodology the consequent natural warming ΔT_{nat} shown in Eq. (8) would have been 0.46 K rather than 0.59 K, and projected ΔT including anthropogenic influences would have been 0.7 K, greater than the observed 0.34 K by a factor 2.1. On the other hand, if we had used κ_{cel} throughout Eqs. (7, 8), projected ΔT_{nat} would have been 1.24 K and projected ΔT 1.48 K over the 18-year period, above observed warming by a factor 4.4.

A note on the transience ratio r_y

Hitherto we have determined the transience ratio r_y under the assumption that anthropogenic forcings other than CO_2 are broadly self-canceling, as Solomon's table of radiative forcings shows them to have been since 1750. However, we might test the skill of Solomon's methods differently, deriving what Eq. (16) shows to be an implausibly low value $r = r_{56} \approx 0.378$ from observed warming over the 56 years 1950-2005:

$$r = r_{56} = \Delta T_{obs} / (\Delta F_{man}\, \kappa_{cel} f_{man}) \approx 0.378 \tag{16}$$

A note on forcing from particulate aerosols

Solomon finds that from 1750-2005 the forcing from particulate aerosols was strongly negative at –1.2 W m^{-2}, reflecting the theory that anthropogenic particulates cause a global dimming that cools the Earth's surface. However, during the 56-year period of study, most Western countries greatly reduced their emission of particulates through clean-air legislation, while Asian countries increased particulate emissions. Accordingly, no net forcing from particulates is assumed here.

A negative particulate-aerosol forcing would have the paradoxical effect of increasing climate sensitivity to greenhouse-gas forcing.

CONCLUSIONS

If the global brightening of 1983-2001 was as significant as Pinker reported and Boston broadly confirms, the fact that application of Solomon's method to the brightening flux implies a pre-feedback warming $\Delta T_0 \approx 0.38$ K from 1983-2001, when only 0.34 K global warming was observed, raises the question whether Solomon was correct in regarding short-acting feedbacks as strongly net-positive. Use of Solomon's method implies that the naturally-occurring global brightening of 1983-2001 should have caused a transient warming of 0.83 K, more than thrice the 0.24 K anthropogenic warming over the period inferred by use of the same method, and more than twice the 0.34 K observed transient warming of those 18 years.

Likewise, even if no net global brightening were assumed to have occurred over the 56 years 1950-2005, use of Solomon's method would lead us to expect a warming of 1.45 K, which is more than twice the observed transient anthropogenic warming of 0.65 K. These discrepancies are too great to be attributable merely to the failure to use atmosphere-ocean general-circulation models when determining climate sensitivity. In any event, the IPCC's simplified methods used here were derived from the models.

If the data are in substance correct, the positive anthropogenic forcing over the period may have been offset to some extent by unidentified negative forcings or temperature feedbacks. For instance, a temporary reduction in relative humidity associated with a possibly cyclical reduction in cloud cover (in approximate alignment with the warming phases of the Pacific Decadal Oscillation) may have transiently altered the sign of the water-vapor feedback. The cloud feedback may also be strongly negative rather than strongly positive (Spencer & Braswell, 2007). Temperature feedbacks in general may be somewhat net-negative (Lindzen & Choi, 2009, 2010 [submitted]), rather than strongly net-positive, as Solomon, citing Soden & Held (2006), finds them.

It is also possible that, in today's atmosphere, true climate sensitivity not only to atmospheric CO_2 enrichment but also to other anthropogenic and natural radiative forcings is less by a factor 2-3 than the estimates in Solomon.

It has been determined theoretically (e.g. Lindzen, 2007; Schwartz, 2007; Monckton of Brenchley, 2008) and confirmed empirically by direct measurement of outgoing radiation from the Earth's characteristic-emission level (e.g. Covey, 1995; Chen et al. 2002; Cess & Udelhofen, 2003; Hatzidimitriou et al. 2004; Clement & Soden, 2005; and Lindzen and Choi, 2009, 2010 [submitted]), and by direct measurement of ocean temperatures in the mixed layer (Lyman et al. 2006 as amended, Gouretski & Koltermann, 2007; Willis, 2008; Willis et al. 2009; and Loehle, 2009), that the IPCC's current estimates of climate sensitivity to atmospheric CO_2 enrichment may be excessive.

It was perhaps no mere coincidence that the global brightening of 1983-2001—a naturally-occurring event—coincided with the greater part of the only supra-decadal period of sustained and rapid warming observed since 1950.

ACKNOWLEDGEMENTS

I am most grateful to Dr. Joseph Boston for having kindly re-evaluated (in the Appendix) the satellite data examined by Pinker, and for having made many other vital contributions to this work; to Dr. David Evans for having provided a helpful discussion of the Planck parameter; to Professor Larry Gould for having eliminated some errors; to Dr. Susan Solomon for some constructive comments; to Professor Richard Lindzen for having patiently answered many questions; to Dr. Willie Wei-Hock Soon for having drawn my attention to the literature on the global brightening of 1983-2001, and for having supplied much other useful material; and to Professor Antonino Zichichi and my fellow-participants in the 43rd annual Seminar of the World Federation of Scientists on Planetary Emergencies at the Ettore Majorana Center for Scientific Culture, Erice, Sicily, in August 2010 for several important observations on climate sensitivity. Any errors that remain are mine alone.

APPENDIX

REANALYSIS OF SATELLITE RADIATIVE-FLUX DATA

Joseph Boston[2]

PERIOD OF STUDY

The 18 years 1 month from mid-1983 to just after mid-2001 were selected as the period of study (Figure A1) for three reasons: first, satellite radiative-flux datasets began in the early 1980s; secondly, global surface temperature rose rapidly until the end of the period, and global temperatures have exhibited no trend since then; and thirdly, the period of study is near-coincident with that of Pinker (2005).

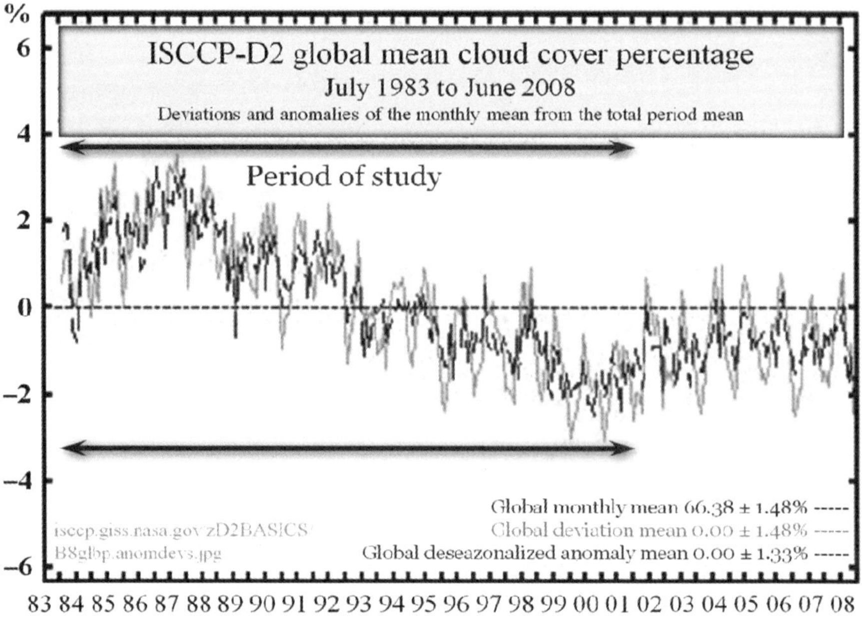

Fig. A1: International Satellite Cloud Climatology Project (ISCCP-D2) global mean percentage cloud cover, July 1983 to June 2008, showing deviations (green) and deseasonalized anomalies (blue) of the monthly mean from the total period mean.

[2] Dr. Joseph Boston: j.boston@earthlink.net

ISCCP FD AND ERBE SATELLITE RADIATIVE-FLUX DATA SOURCES

Two sources of satellite-based radiation data were examined. The International Satellite Cloud Climatology Project (ISCCP) was established in 1982 to collect and analyze satellite radiance measurements.

Data were collected from a suite of weather satellites operated by several nations and processed by groups in several government agencies, laboratories and universities. Measurements were obtained every three hours over the whole globe on a 280 km equal-area global grid (equivalent to 2.5 degrees latitude-longitude at the equator) from July 1983 to December 2007.

Results include global averages of upwelling and downwelling short-wave and long-wave radiative fluxes at top of atmosphere, in the atmosphere, and at the surface, for full, clear, and cloudy-sky conditions.

These results were compiled in the ISCCP FD datasets, which are available in specified latitude groupings. For this study separate FD datasets were obtained for five latitude groupings, all covering the entire 1983-2001 period. The principal groupings analyzed in the study were 20N-20S, 60N-60S and 90N-90S.

The Earth Radiation Budget Experiment (ERBE) datasets are based upon non-scanner wide-field-of-view data obtained from three NASA satellites (ERBS, NOAA-9 and NOAA-10) from November 1984 to September 1999.

The instrument package included five active-cavity radiometers, two medium-field-of-view, two wide-field-of-view and a solar monitor, with a combined ability to measure incoming and earth-reflected solar irradiances in the short-wave (0.2-5 micron) spectral region, and earth-emitted long-wave radiances up to 100 microns.

The wide-field-of-view instruments observed irradiances from the entire earth disc enclosed by a 2-angular-degree space ring. The medium-field-of-view instruments viewed earth irradiances from regions with earth-centered angles of 10 degrees.

The raw data were processed by the data management team in the Atmospheric Science Division of NASA to develop temporally and spatially averaged short-wave and long-wave fluxes at the top of the atmosphere, which the team compiled into a number of different data products.

The team discovered two sources of bias, which were quantified, and for which corrections were subsequently introduced. One of the biases was caused by slight satellite orbital decay, the other by a slow drift in the short-wave sensor.

The two datasets used in this study are edition 3, revision 1 of the 72-day mean datasets for the tropics (20N-20S) and near-global (60N-60S), both of which incorporate corrections for the two sources of bias. The datasets encompass the 14.5 years for which ERBE data (Barkstrom, 1984) are available, from 1985 to mid-1999, approximately centered on the 18.1-year period of the ISCCP FD datasets.

DETERMINATION OF SHORT-WAVE FLUX TRENDS

A central objective of this work was to examine the impact on the global warming trend of increasing solar flux to the Earth's surface and into the atmosphere. For this purpose it was essential to determine the underlying trends in the radiative fluxes.

Since the eruption of Mt. Pinatubo in mid-1991 caused a significant short-term distortion in the underlying trend, for both the ISCCP FD and ERBE datasets all of the flux data were pre-processed by removing the Pinatubo-affected data over 18 months from mid-1991 to end 1992. Remaining data were deseasonalized by determining the average value for a given calendar month over the whole period and subtracting that value from the individual yearly values for that calendar month. This procedure was repeated for each of the 12 months in turn. The Pinatubo-adjusted, deseasonalized monthly data were then regressed to obtain least-squares linear trends in W m^{-2} yr^{-1}.

The key item of satellite data for this work was the net mean global surface short-wave flux trend from 1983-2001. The ERBE satellites did not produce data beyond 60N and 60S; they covered only a 14.5-year subset of the period of interest; and the data did not explicitly include the net short-wave surface flux. Therefore the ISCCP 90N-90S dataset was used as the basis for the study, and the ERBE data were used as a validity check with the ISCCP 60N-60S data shown in Figure A2.

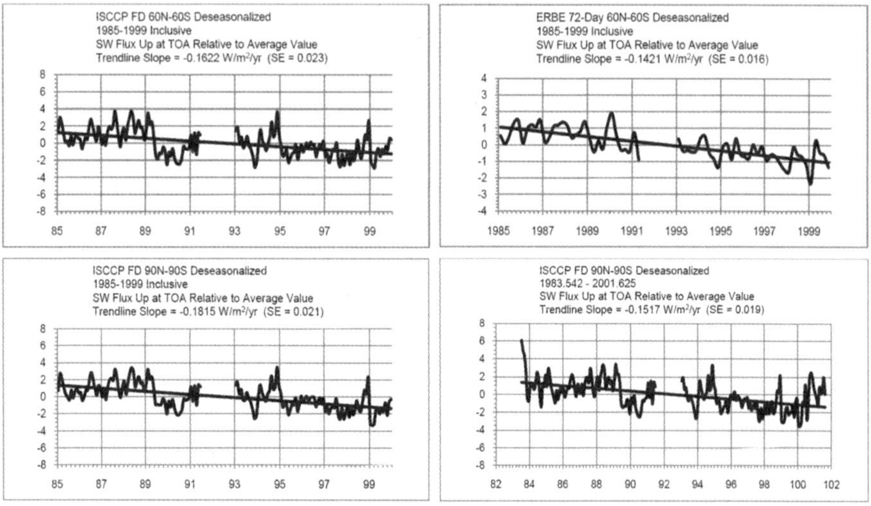

Fig. A2: Deseasonalized east squares linear-regression trends on data from ISCCP FD 60N-60S, 1985-1999 (top left); ERBE 60N-60S, 1985-1999 (top right); ISCCP FD 90N-90S, 1985-1999 (bottom left); and ISCCP FD 90N-90S, 1983.542-2001.625 (bottom right, for comparison with Pinker).

The ERBE data, covering 1985-1999, are divided into five 72-day periods for each year, a total of 75 periods. Owing to orbital characteristics, periods that are near-exact multiples of 36 days are essential for accurate analysis of the ERBE data. For seven of those periods (three each in 1993 and 1998 and one in 1999) the data are missing.

The analysis examined the possibility that the missing data might have a significant impact on deseasonalization. A quadratic regression was carried out over 1985-1999 for each 72-day period in turn, whereupon the five second-order fits were used to provide estimates of the missing values. There were two missing values for period 1, one missing value for each of periods 2-4, and two missing values for period 5. The deseasonalization and the overall regression were then repeated to obtain the least-squares linear-regression trend-line for upwelling short-wave flux at the top of the atmosphere. Agreement between the 60N-60S ISCCP and ERBE trends, and between the ISCCP results and those in Pinker (Table 3) was found to be close.

Table 3. Deseasonalized least-squares linear-regression trends for upwelling short-wave radiative flux at the top of the atmosphere, 1985-1999 (ISCCP-FD and ERBE), and 1983.542-2001.625 (ISCCP-FD and Pinker).

Flux (W m^{-2} yr^{-1})	Latitudes	ISCCP	ERBE	Pinker	Variance
1985 1999	60N-60S	–0.162	–0.142		–12%
1985-1999	90N-90S	–0.182			
1983.542-2001.625	90N-90S	–0.152		–0.160	+5%

Figure A3 is a short-wave energy budget diagram showing how the average fluxes and flux trends are distributed among the significant flux components, including the surface, cloud, and aerosol albedo components. Effects that are *de minimis* are not shown. Net surface flux is found to be 0.143 W m^{-2}.

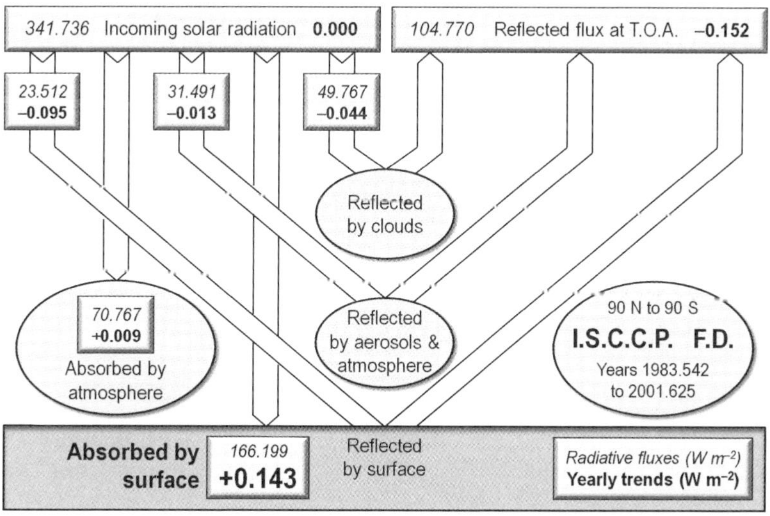

Fig. A3: Simple radiation budget showing the net increase in surface solar radiance, the net brightening flux, for 1983.542-2001.625. Total brightening flux, the sum of the components 0.143 W m^{-2} absorbed by the surface and 0.009 W m^{-2} absorbed by the atmosphere, is 0.152 W m^{-2}.

REFERENCES

1. Barkstrom, B.R. (1984) "The Earth Radiation Budget Experiment (ERBE)." *Bull. Am. Met. Soc.* **65**:1170.
2. Bode, H.W. (1945) *Network analysis and feedback amplifier design.* 551pp. Van Nostrand, New York.
3. Brohan, P., J.J. Kennedy, I. Harris, S.F.B. Tett, and P.D. Jones (2006) "Uncertainty estimates in regional and global observed temperature changes: a new dataset from 1850." *J. Geophys. Res.* **111**:D12106, doi:10.1029/2005JD006548
4. Cess, R.D., and P.M. Udelhofen (2003) "Climate change during 1985–1999: Cloud interactions determined from satellite measurements." *Geophysical Researh Letters* **30,** 1: 1019, doi:10.1029/2002GL016128.
5. Chen, J., B.E. Carlson, and A.D. Del Genio (2002) "Evidence for strengthening of the tropical general circulation in the 1990s." *Science* **295:**838-841.
6. Clement, A.C., and B. Soden (2005) "The sensitivity of the tropical-mean radiation budget." *J. Clim.* **18:**3189-3203.
7. Covey, C. (1995) Correlation between outgoing long-wave radiation and surface temperature in the tropical Pacific: a model interpretation. Lawrence Livermore National Laboratory, Livermore, CA 94551, November. UCRL-ID-122565.
8. Gouretski, V., and K.P. Koltermann (2007) "How much is the ocean really warming?" *Geophysical Research Letters* **34:**10.1029/2006GL027834.
9. Hartley, D.E. T. Kindler, D.E. Cunnold, and R.G. Prinn (1996) "Evaluating chemical transport models: Comparison of effects of different CFC-11 emission scenarios." *J. Geophys. Res.* 101:14381-14385.
10. Hatzidimitriou, D., I. Vardavas, K. G. Pavlakis, N. Hatzianastassiou, C. Matsoukas, and E. Drakakis (2004) "On the decadal increase in the tropical mean outgoing longwave radiation for the period 1984–2000." *Atmos. Chem. Phys.* **4:**1419–1425.
11. Houghton, J.T., and Coauthors (eds.), 2001: *Climate Change 2001: The Scientific Basis. Contribution of Working Group I to the Third Assessment Report of the Inter-governmental Panel on Climate Change.* Cambridge University Press, Cambridge, United Kingdom and New York, NY, USA, 881pp.
12. Jones, P.D., M. New, D.E. Parker, S. Martin, and I.G. Rigor (1999) "Surface air temperature and its variations over the last 150 years." *Reviews of Geophys.* **37**:173-199.
13. Keenlyside, N.S., M. Latif, J. Jungclaus, L. Kornblueh and E. Roeckner (2008) "Advancing decadal-scale climate prediction in the North Atlantic sector." *Nature* **453:**84-88, doi:10.1038/nature06921.
14. Kiehl, J.T., and K.E. Trenberth (1997) "The Earth's Radiation Budget." *Bull. Am. Met. Soc.* **78:**197.
15. Kiehl, J.T., and K.E. Trenberth (2009) "Earth's global energy budget." *Bull. Am. Met. Soc.*, doi 10.1175/2008BAMS2634.1.
16. Kimoto, K. (2009) "On the confusion of Planck feedback parameters." *Energy & Environment* **20:**1057-1066.

17. Lindzen, R.S. (2007) "Taking greenhouse warming seriously." *Energy & Environment* **18:**7-8, 937-950.
18. Lindzen, R.S., and Y.-S. Choi (2009, 2010) [submitted]. "On the observational determination of climate sensitivity and its implications." *J. Geophys. Res.*
19. Lyman, J.M., J.K. Willis, and G.C. Johnson (2006) "Recent cooling of the upper ocean." *Geophysical Research Letters* **33:**L18604, doi:10.1029/2006GL027033.
20. Masarie, K.A., and P.P. Tans (1995) "Extension and integration of atmospheric carbon dioxide data into a globally consistent measurement record." *J. Geophys. Res.* **100:**11593-11610.
21. Monckton of Brenchley, C.W. (2008) "Climate sensitivity reconsidered." *Physics & Society* **37:**3.
22. Myhre, G., E.J. Highwood, K.P. Shine, and F. Stordal (1998) "New estimates of radiative forcing due to well-mixed greenhouse gases." *Geophys. Res. Lett.* **25:**(14), 2715-2718.
23. Pinker, R.T., B. Zhang, and E.G. Dutton (2005) "Do satellites detect trends in surface solar radiation?" *Science*, **308**:850–854.
24. Rossow, W.B., and R. A. Schiffer, (1991) "ISCCP cloud data products." *Bull. Am. Met. Soc.* **72**:2.
25. Rossow, W. B., and R. A. Schiffer (1999) "Advances in understanding clouds from ISCCP." *Bull. Am. Met. Soc.* **80**:2261.
26. Schwartz, S. (2007) "Heat capacity, time constant, and sensitivity of Earth's climate system." *J. Geophys. R.*
27. Soden, B.J., and I.M. Held (2006) "An assessment of climate feedbacks in coupled ocean-atmosphere models." *J. Cli.* **19:**3354–3360.
28. Solomon, S., and Coauthors (eds.), 2007: *Climate Change 2007: The Physical Science Basis. Contribution of Working Group I to the Fourth Assessment Report of the Intergovernmental Panel on Climate Change.* Cambridge University Press, Cambridge, United Kingdom and New York, NY, USA.
29. Spencer, R.W., W.D. Braswell, J.R. Christy, and J. Hnilo (2007) "Cloud and radiation budget changes associated with tropical intraseasonal oscillations." *Geophys. Res. Lett.* **34:**L15707, doi:10.1029/2007GL029698.
30. Wielicki, B.A., and Coauthors (2002a) "Evidence for large decadal variability in the tropical mean radiative energy budget." *Science* **295:**841–844.
31. Wielicki, B.A., and Coauthors (2002b) "Changes in tropical clouds and radiation: Response." *Science* **296:**2095a.
32. Wild, M.A., and Coauthors (2005) "From dimming to brightening: Decadal changes in solar radiation at Earth's surface." *Science* **308**:847–850.
33. Willis, J. K. (2008) "Is it me, or did the oceans cool?" *U.S. Cli. Var.* **6:**2.
34. Willis, J.K., J.M. Lyman, G.C. Johnson and J. Gilson (2009) "In-situ data biases and recent oean heat content variability." *J. Atmos. & Oceanic Technology* **26:**846-852.
35. Wong, T., B.A. Wielicki, and R.B. Lee III (2006) "Reexamination of the observed decadal variability of the Earth radiation budget using altitude-corrected ERBE/ERBS nonscanner WFOV Data." *J. Cli.* **19:**4028–4040.

SECTION 2

SESSION 2

ENERGY

FOCUS: PASSIVE SAFETY IN
NUCLEAR POWER PLANTS

OVERVIEW OF PASSIVE SAFETY FEATURES OF ADVANCED LIGHT WATER REACTORS

ROBERT J. BUDNITZ
Lawrence Berkeley National Laboratory
University of California, Berkeley, USA

INTRODUCTION

This paper summarizes a verbal presentation that the author made in Erice on 20 August 2010. Like that presentation, this paper is not a comprehensive discussion of the topic; rather, it is an overview summary intended to introduce the subject to an audience of technically trained experts in other areas who may or may not have been introduced to this topic before.

For a nuclear power reactor, a *"passive" safety feature* is a design feature that performs one of the vital safety functions, such as reactivity control, inventory control, heat removal, or containment of the radioactive inventory, without the use of active equipment or the active intervention of the operating crew. These features are particularly useful during off-normal or upset conditions when improper performance of the design or the operating crew could lead to an accidental release of radioactivity, but they can also be beneficial in improving the reliability of normal plant operation.

In addition, there are *other passive features* that are not specifically safety features. These features mostly provide functions that help the power reactor to run and perform its desired function, namely producing electricity for the grid.

From the beginning, all power reactors have had a few such features, some useful in normal operation and others useful during off-normal events. However, advances in design and operation over the years have made more and more such passive features available and/or have made the existing features increasingly effective. Almost all of the advanced light water power reactors (LWRs) now being offered for sale by reactor vendors in today's worldwide market have some advanced passive features (either safety features or non-safety features), and a few of them have significantly more and better such features than the others.

However, it is important to note that starting with the earliest power reactors and continuing right up to those running today, many of the most important safety issues are coped with using active features.

BACKGROUND

As background, the general approach to both achieving and enhancing safe and reliable operation, given the existence of a reactor design that is already sufficiently robust that it will perform the basic functions, *begins with performing an analysis* of all "off normal" or "upset" scenarios that, were they to occur, would force the reactor to respond, to go into an off-normal operating mode, or in unusual circumstances to lead to a radiological accident. If a concern arises that involves one of the scenarios that are identified, the traditional design approach, which of course is entirely sensible, would be to *design an engineering "fix"* of one kind or another. There are three broad categories of such "fixes":

- Fixes that reduce the probability of the initiating event;
- Fixes that reduce the likelihood that the rest of the scenario will evolve;
- Fixes that reduce the consequences of the scenario.

For the large power reactors (here I concentrate on the LWRs), the historical fact is that, for engineering reasons, *most of the early fixes involved "active" features.* This was because such active features were usually the least expensive and most effective available engineering solutions. This was sensible at the time, and in fact criticizing those decisions in the light of hindsight is inappropriate.

The decisions were sensible because the imperative of the time was to design, build, and operate the reactors on time scales (a few years total) that pushed for rapid solutions to problems. There was also, and appropriately so, diligent attention to safety that took precedence over attention to costs, and a preference to use existing technologies wherever feasible. And finally, the national regulatory agencies, like the U.S. Nuclear Regulatory Commission in my own country, were generally in alignment with the approach of using active safety systems as the best approach for solving the problems that arose—or at the least they were not insisting that less active and more passive approaches be evaluated too. (These agencies generally do not have the authority to impose a particular technological approach to solving a safety problem. They were and are generally in the mode of simply reviewing the adequacy of the design solutions presented to them, except in rare compelling circumstances.)

Also, given the consensus for the need for large safety margins, it was generally judged best to design and deploy systems and functions with demonstrably large margins, using very conservative criteria and design rules. And it was generally easier to carry out those demonstrations, in the early days, for active features.

LIMITATIONS

Crucially, a major limitation was that for many of the most important off-normal scenarios, whether they be scenarios affecting public safety or scenarios that only threaten the economic viability of the plant, *it was not possible to do a realistic analysis of the phenomena involved.* Given this limitation, various conservative analyses were done, generally following rules prescribed by the regulatory agency. Often, a design solution involving a passive system, if considered, was not adopted because to implement it would require an analysis that needed to be more realistic than was feasible. (In fact, even today this is a major barrier to the penetration of some passive-features into the designs.)

Another major limitation was that *the relative likelihood of the most important scenarios was not known very well.* Therefore, it was difficult to know whether an active-feature or a passive-feature design solution was truly better—the transparent benefits we now ascertain of certain passive features (in decreasing the likelihood or the consequences of a scenario of concern) could not be understood well enough to allow a designer to take advantage of this knowledge.

As mentioned, the designers' response was generally to design and introduce "fixes" for problems that relied on "conservative" (often unrealistic) analysis, and then to introduce the simplest design solution consistent with budget, schedule, and the ability to obtain regulatory approvals.

WHAT CAUSED A CHANGE, AND WHY, AND WHEN?

In retrospect, a watershed event in the evolution of the designers' approach to how reactor safety and reliability are achieved was the 1973-1975 "Reactor Safety Study", the so-called "Rasmussen Report" performed by an American team under the guidance of N. Rasmussen and S. Levine, and supported by the U.S. government. It was the first large-scale "PSA" (probabilistic safety assessment) on a large technical system, and it single-handedly revolutionized the way reactor safety engineers, and later all reactor engineers, think about their work. The PSA that this team performed enabled an engineer, or a member of the larger public, to understand the following:

- The probabilities and consequences of major accident scenarios;
- The probabilities and consequences of upset conditions without safety implications;
- The efficacy of various proposed "fixes" that might reduce either the probability or the consequences of a specific scenario of concern.

A PSA analysis enabled not only the designers but the whole reactor community to put into a probabilistic or a risk context the benefits of various available design options. Some of the benefits were safety benefits, but a larger number were benefits in terms of more reliable operation.

It is tempting to say that with the advent of PSA methods and insights, "everything changed". It was not that simple—it took a decade or more for these methods to become widely accepted even in the countries which were most amenable or receptive, and in other countries, such as in the United States, it took 20 or even 25 years for PSA methods and insights to become widely used and the benefits realized.

However, major progress did occur, some of it very quickly. The safety and operational results have been dramatic, albeit it took two or more decades, and the progress is still underway in terms of improving the safety and operability of our existing LWRs.

A mark of the progress is captured in a couple of summary items, where I am using U.S. plant data here as my example, but the same results occurred worldwide:

- The typical core damage frequency (CDF) for the U.S. LWRs has been reduced from around 10^{-4} to around 10^{-5} per year;
- The typical capacity factor has improved from around 60% to around 90%.

While this major progress was not entirely due, of course, to the advent of using PSA methods to analyze safety and operations, this analysis technology has been a major force. Other major forces have been the broad advances in our ability to analyze the scenarios of concern more realistically, and the careful attention to operations issues (control room design and simulator training, widespread emphasis on reliance on procedures, learning from operational experience, a safety-culture change to emphasize broad sharing of lessons across the entire industry, and a few others).

Still, most of the improvements in LWR reactor safety and operational effectiveness in the current plants have *not* involved a switch from active features to passive features. This is because a majority of the passive features that we now consider to be most beneficial are difficult or impossible to introduce into the existing LWR designs. [More on this just below.]

A way of saying this is that, although huge progress has been achieved in improving reactor safety and operability, a broad limit is being reached. Specifically, *the current LWR designs have some intrinsic limitations that prevent introduction of certain of the passive features* that represent one of the major new ways to get to the next level of improvements in terms of both safety and operability. To say it in a succinct way, backfits to introduce many of the most beneficial passive features to existing plants are sometimes simply not feasible.

WHAT IS NEXT?

What is next is the advent of the new LWR designs. (Here I emphasize the new large LWR designs.) The availability to the designer of a "clean sheet of paper" brings with it the possibility to introduce a set of passive features that can overcome some of the major limitations faced in the current designs, in terms of bringing down the probabilities and the consequences of the major scenarios of concern.

And the widespread use of such passive features is among the most important differences that characterize the new LWR designs now being built for the first time (or about to be built around the world in the next few years.)

Attention has been given by the designers of these new LWR designs to the following issues, all of which use the logic of PSA and the availability of better analysis tools to supplement the more traditional engineering tools:

- Design attention to reduce the probability of all of the important initiating events, or in some cases to entirely eliminating the possibility of the initiator;
- Design attention to reduce the likelihood that the follow-on events in each important scenario will evolve to reach an undesired endpoint;
- Design attention to reduce the consequences of the scenarios of concern.

Often, a major barrier to earlier use of some of these features, which was that one could not perform realistic analysis of how the feature would perform, has been overcome *by assuring that analyzability is a design criterion.* That is, *assuring that one can analyze the performance of a proposed design feature has become a major aspect of the design approach.*

It turns out that often, the designers have chosen passive features as the best way to accomplish the above.

The bottom line is that analysis shows that this broad set of approaches has worked very well:

- The CDF frequencies for the new LWR designs, according to the PSA analyses, are generally in the range around 10^{-6} per year or even lower.
- Control is easier, human errors are shown by analysis to be less likely, there is more time to respond, and operability is generally improved a lot.
- The reliability of key equipment has been emphasized in the design, including assuring that ease of maintenance is a design criterion.

Crucially, the way this has been achieved involves many more "passive features".

EXAMPLES OF PASSIVE FEATURES

This paper is not a place where a comprehensive list of passive features will be developed. Other papers during this same session at Erice cover this topic. But a few examples will be given to give a flavor for what is involved.

- Some vital safety systems rely on gravity instead of electric pumping.
- Response to some major scenarios avoids requiring operator intervention.
- Some major sequences rely on natural circulation for heat flow and removal.
- Some designs rely heavily on passive containment cooling.

All of these have the characteristic that they are simple to design, simple to understand in a qualitative sense, and simple to analyze quantitatively. These three attributes are a major reason why there is great confidence that the new designs will in fact perform very well in terms of both safety and operability.

PASSIVE SAFETY FEATURES FOR SMALL MODULAR REACTORS

DANIEL T. INGERSOLL
Oak Ridge National Laboratory
Oak Ridge, USA

ABSTRACT

The rapid growth in the size and complexity of commercial nuclear power plants in the 1970s spawned an interest in smaller, simpler designs that are inherently or intrinsically safe through the use of passive design features. Several designs were developed, but none were ever built, although some of their passive safety features were incorporated into large commercial plant designs that are being planned or built today. In recent years, several reactor vendors are actively redeveloping small modular reactor (SMR) designs with even greater use of passive features. Several designs incorporate the ultimate in passive safety—they completely eliminate specific accident initiators from the design. Other design features help to reduce the likelihood of an accident or help to mitigate the accident's consequences, should one occur. While some passive safety features are common to most SMR designs, irrespective of the coolant technology, other features are specific to water, gas, or liquid-metal cooled SMR designs. The extensive use of passive safety features in SMRs promise to make these plants highly robust, protecting both the general public and the owner/investor. Once demonstrated, these plants should allow nuclear power to be used confidently for a broader range of customers and applications than will be possible with large plants alone.

INTRODUCTION

The first commercial nuclear power plants to be constructed in the United States had relatively low power capacity, generally less than 200 MWe, and were based on the successful operation of even smaller designs used for marine propulsion by the U.S. Navy. Between 1970 and 1980, plant sizes grew rapidly to plants having capacities exceeding 1000 MWe. In many cases, the detailed designs for the large plants were substantially completed before their smaller predecessors had been operated for any significant time. Problems encountered during the operation of the smaller plants were then accommodated as last-minute engineering patches to the larger plant designs. Also, an increasing concern for potential accidents that could have severe consequences drove the designers to add an increasing number of active backup systems to the plants. For example, the concern for the extreme consequences of a double-ended large pipe break in the primary coolant circuit of a pressurized water reactor (PWR), caused designers to add systems such as high-pressure water injection systems to ensure that the reactor core would remain covered with water in the unlikely occurrence of this extreme accident. By the end of the 1970s, punctuated by the core-damaging accident at the Three Mile Island plant [Kemeny, 1979] on March 28, 1979, plant designs had become encumbered by a complex layering of engineering patches and active safety systems, which drove up the capital and operating costs of the plants while reducing their availability.

The accident at Three Mile Island catalyzed a number of additional design and operational changes to nuclear power plants and also stimulated a lot of soul-searching by industry leaders who were uncomfortable with the established trend toward larger, more complex plants. Alvin Weinberg, a pioneer in the nuclear energy community, led a 1983 study [Weinberg, 1985] by the Institute for Energy Analysis to evaluate the viability of designing nuclear reactors that are "inherently safe," i.e., they require no active engineered systems or operator intervention to mitigate the consequences of an accident. Weinberg's team evaluated several new concepts and designs that were being developed world-wide and concluded that it should be technically feasible to develop inherently safe designs. In particular, they favored two emerging designs: the water-cooled Process Inherent Ultimately Safe (PIUS) design and the helium-cooled Modular High-Temperature Gas-cooled Reactor (MHTGR) design. Their primary concern with these designs was that they were necessarily constrained to low power capacities and might not be economically viable due to the same economy-of-scale factors that previously drove the industry to very large plants.

Additional studies and new government-sponsored programs were initiated in the 1980s, most of which focused on smaller, simpler reactor designs with inherent or "passive" safety features. Although none of these designs were built, several of the more promising features have been incorporated into the newest plant designs that are just now beginning to be ordered and constructed. Several of the designs from the 80s and 90s were initially small output designs, but were subsequently enlarged due to economy-of-scale considerations. An example of this is Westinghouse's AP-600 design developed within the Advanced Light-Water Reactor program. Although originally certified by the U.S. Nuclear Regulatory Commission (NRC) at 600 MWe, the design was later upsized to 1140 MWe for commercialization.

Despite a continued trend toward large plants, there are an increasing number of nuclear plant vendors that are responding to Weinberg's recommendations and are developing designs that are small and simple and consequently very robust. In response to Weinberg's concerns about economic competitiveness, these designs further introduce "modularity" at the reactor level, i.e., they replace economy-of-scale with economy-of-replication. These small modular reactor (SMR) designs can be fully fabricated in a factory setting and easily transported and installed at the plant site. Additionally, these designs take maximum advantage of passive safety features in order to keep the design sufficiently simple to remain economically competitive with large plants. The following sections will discuss examples of the types of passive safety features used by the range of SMRs currently under development.

PASSIVE SAFETY FEATURES

Like their bigger siblings, SMRs are being developed that use water coolant (light water or heavy water), gas coolant (typically helium), liquid-metal coolant (sodium, lead, or lead-bismuth), or liquid-salt coolant (typically fluoride salt). While each of these classes has unique safety features, many of the same passive safety principles apply to all reactor classes. This paper emphasized these cross-cutting features, although some specific examples are also cited.

Many SMR designers share a common set of principles [IAEA, 2009] to improve the safety of their designs: (1) eliminate from the design as many features as possible that have the potential to initiate a serious accident, (2) for those features that can not be eliminated, reduce the probability that an accident will be initiated, and (3) design the system to substantially mitigate the consequences of remaining potential accidents. As an example the developers of the International Reactor Innovative and Secure (IRIS) design claim that of the eight Class IV design basis accidents for a large loop-type PWR, three are completely eliminated from their design, two have reduced probability of occurrence, and two have reduced consequences. [Carelli, 2004] According to the IRIS developers, only the refueling accident remains unchanged from large plant designs. Of the three fundamental design principles, the first approach is clearly the most elegant solution and is a key feature of many SMRs being developed, as discussed in the following section.

Elimination of Accident Vulnerabilities

Very early in the development of large commercial power plants, the most challenging accident scenario was defined as the double-ended break of the largest diameter coolant pipe, which (for PWRs) was typically the pipe connecting the primary reactor vessel to the steam generator vessel. This postulated break could very quickly remove coolant water from the primary vessel and result in an uncovering of the reactor core, which in turn would quickly cause major damage to the fuel and release a significant fraction of the radioactive materials contained in the fuel. To mitigate the severe consequences of such an accident, complex water injection systems were added to the plant designs. These safety systems, which are never expected to be used for a successfully operated plant, add significant cost, both for initial construction and routine testing and maintenance. Eliminating these safety systems requires that the vulnerability be eliminated, i.e., all large primary coolant pipes must be eliminated from the design. The answer to this issue is the integral primary system reactor (IPSR).

An IPSR, sometimes referred to as in integral pressurized water reactor (iPWR), incorporates the reactor core, steam generator(s), and pressurizer into a single common pressure vessel. Figure 1 shows a comparison of a traditional loop-type PWR design compared to an IPSR design. The first (and only) commercial IPSR was the NS Otto Hahn merchant ship, which was commissioned in 1968 and powered by a 35 MW integral reactor [Nuclear Ship, 1968]. In the late 1980s, the Safe Integral Reactor (SIR) was developed by a consortium including the U.K. Atomic Energy Agency, Combustion Engineering, Stone and Webster, and Rolls Royce [Matzie, 1992]. The 320 MWe SIR design was developed specifically to respond to the safety challenges encountered by the early large loop PWRs and represents a precursor to many of the SMR designs that have emerged in the past few years, including the Westinghouse IRIS design referenced earlier.

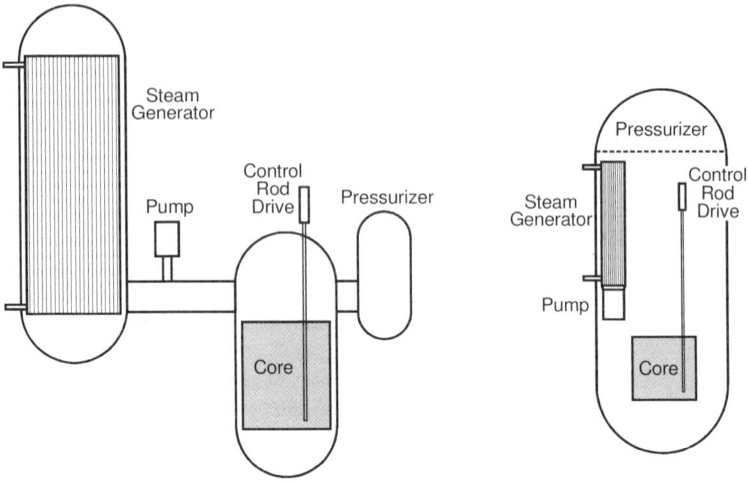

Fig. 1: Simplified schematic of loop (left) and integral (right) designs for pressurized water reactors.

Packing all of the primary system components into a single vessel has a number of safety-related advantages:

- All large coolant pipes are eliminated. Only small (5-7 cm diameter) feed water and steam outlet pipes penetrate the primary vessel wall compared to large (80-90 cm diameter) pipes for loop-type PWRs.
- The total inventory of primary coolant contained within the primary vessel is much larger than for an external loop PWR, which increases the heat capacity and thermal inertia of the system. Hence the primary system exhibits a much slower response to core temperature transients.
- The presence of the steam generators within the primary vessel provides an effective heat sink for decay heat removal in a loss-of-flow accident.
- Typically the heat exchangers are placed above the core creating a relatively tall system that facilitates more effective natural circulation of the primary coolant in the case of a coolant pump failure. Some designs have sufficient natural circulation flow rates to eliminate the primary coolant pumps and rely only on natural circulation for normal operation.
- The vessel accommodates a relatively large pressurizer volume that provides better control of under/overpressure transients. For example, the IRIS pressurizer volume is designed to be nearly 5 times larger per unit power than for a large plant.
- The extended riser area above the core provides the possibility for internal placement of the control rod drive mechanisms (CRDM), thus avoiding another potentially serious accident scenario: the rod ejection accident. Also, the reduced number of penetrations in the reactor vessel head simplifies

opening of the vessel and eliminated the near-accident situation that occurred at the Davis-Besse plant in which boric acid in the coolant water leaked through a control rod penetration seal and corroded the vessel base material. [Davis-Besse, 2008] Only a thin stainless steel inner liner prevented a major breech in the vessel pressure boundary. Some IPSR designs take advantage of this feature while others use conventional external CRDMs.

- The larger downcomer region between the core and the primary vessel provides extra shielding and results in a much lower fluence to the reactor vessel. This reduces the activation level of the vessel material and reduces another major safety concern: pressurized thermal shock, which results from radiation-induced embrittlement of the reactor vessel.

The major down side of the integral design is that it constrains the reactor to relatively small power levels, i.e., it forces the reactor to be "deliberately small." [Ingersoll, 2009] Beyond some practical limit, probably in the 300-400 MWe range, the size of the vessel would become prohibitively large to manufacture and transport. To compensate for its inability to scale dimensionally, IPSR designers depend instead on scaling by replication, hence the interest in small modular reactors. An example of this is the NuScale design, [Lorenzini, 2010] which uses twelve 45-MWe reactor IPSR modules to comprise their reference 540 MWe plant design and the Babcock and Wilcox mPower design, [Halfinger, 2010] which uses four 125-MWe modules to comprise a 500 MWe plant.

Reduced Source Term

Fundamentally, reactor safety comes down to ensuring that a hazardous amount of radiation is not released to operations personnel or the general public. The total quantity of radionuclides in a reactor core is referred to as the reactor "source term." Although the source term depends on a number of factors, including fuel composition and operational history, it is roughly proportional to the power level of the reactor, i.e., twice the power level yields twice the source term. Hence, small output reactors will have proportionately smaller source terms. The benefits of a lower source term can be manifested in a number of ways, including: reduced shielding, reduced site exclusion boundary, reduced emergency planning zone (EPZ), increased plant safety margins, or some combination of these options. Of course if the total plant is comprised of several reactor modules, the total plant inventory may be comparable to a single-core large plant. However, even in this case, it is more difficult to imagine an accident scenario that results in fuel failure in all reactor modules simultaneously, hence the probability of a large radiation release is reduced.

Improved Decay Heat Removal

In the event of a loss of forced circulation of the primary coolant, mechanisms must be provided for removal of the residual power that continues to generate heat in the reactor core from the decay of radionuclides. The decay heat can be quite significant—approximately 6% of the full operating power immediately after shutdown. This decay power drops by an order of magnitude within the first day after shutdown, so the primary concern is the short-term heat removal. Conduction of heat through the core material and natural convection of the primary coolant help to move the heat from the fuel to the reactor vessel boundary. But at some point, additional external cooling of the vessel is needed to transfer the heat to the ultimate heat sink (typically the ground or atmosphere).

Small plant designs are able to accommodate this heat transfer through passive measures better than large plants for a number of reasons. Firstly, a lower core operating power of the SMR will result in lower decay power because, as with the source term, the total decay power is proportional to the operating power. This advantage is further enhanced by using a lower power density in the fuel, which many SMR designers choose to do in order to lengthen the fuel cycle duration.

Secondly, the smaller core volume enables more effective conduction of the decay power to the reactor vessel. This is a result of the smaller core radius, which results in a shorter conduction path from the core centerline (the hottest part of the core) to the reactor vessel.

Thirdly, the removal of heat from the external surface area of the vessel is more effective for smaller systems. This feature results from the fact that core power, and hence the decay power, is proportional to the volume of the core, which varies as the cube of the effective core radius. On the other hand, heat removal from the exterior surface of the vessel is proportional to the vessel surface area, which varies roughly as the square of the core radius. As the power level is reduced, the core volume decreases faster than the surface area of the vessel, or conversely, the surface-to-volume ratio increases, hence increasing the relative efficiency of external heat removal. The 1/r relationship between surface area and volume is shown graphically in Figure 2. [Lorenzini, 2010] As a result, most small reactor designs are able to easily accommodate removal of the decay heat using fully passive, natural convection air or water circulation systems. Also shown in Figure 2 is the fact that for a given vessel thickness, its ability to withstand internal pressure also varies as 1/r. Consequently, the smaller diameter vessel is able to contain larger pressure transients than a large vessel.

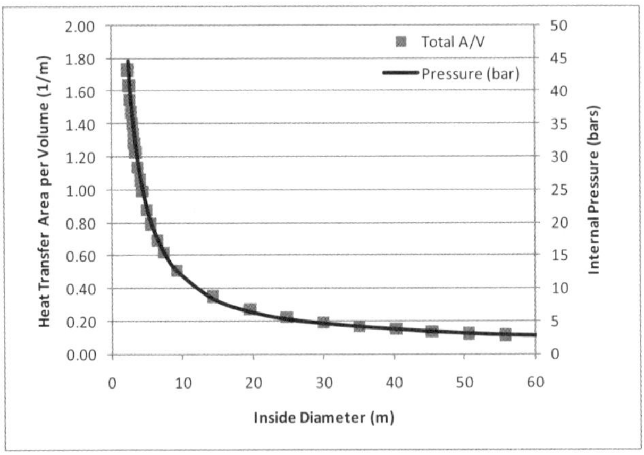

Fig. 2: Relation between heat transfer area (left) or internal pressure (right) to reactor vessel radius. [Lorenzini, 2010]

Core Geometry

The ability to passively remove decay heat has become a central part of the enhanced safety case for SMRs. In some cases, merely reducing the size of the core (and power) may not be sufficient to provide full passive removal of the decay heat. Such is the case with high-temperature gas-cooled reactors (HTGR). The initial power level of General Atomic's HTGR design was 2240 MWt but was later downsized to 600 MWt to improve its safety case. [Labar, 2002] Even at this power level, however, it was difficult to ensure adequate decay heat removal using passive systems, owing largely to the poor thermal properties of the helium coolant. During a loss-of-flow transient, natural circulation of the coolant can remove some of the decay heat, but the limited radial conduction to the vessel surface caused unacceptably high temperatures at the core centerline. As a result, General Atomics adopted an annular core design, in which the central portion of the core is replaced by a graphite reflector. The annular-core design and the large thermal inertia provided by the graphite moderator in the modular high-temperature reactor (MHR) provides a similar favorable transient response as the water-cooled IPSR.

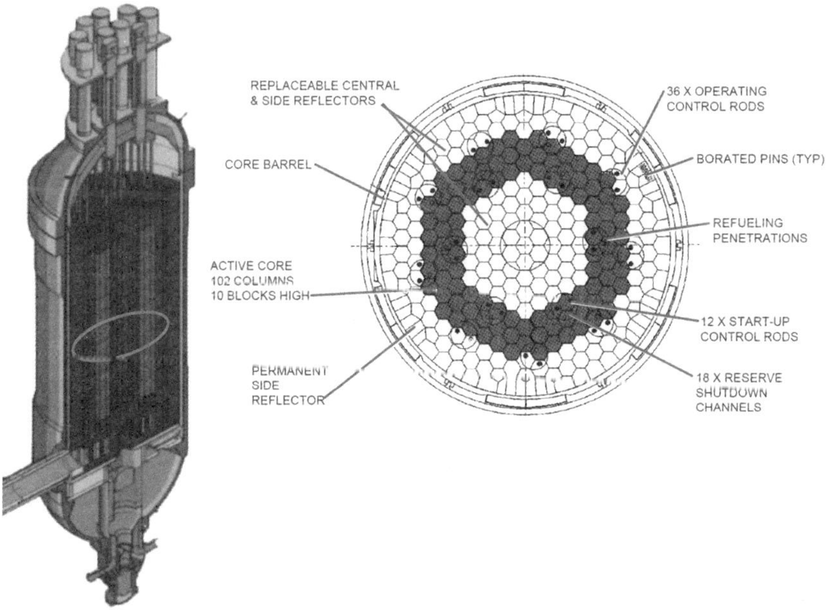

Fig. 3: Cutaway of GT-MHR reactor vessel (left) and cross-section of core region (right). [Labar, 2002]

OTHER PASSIVE SAFETY DESIGN FEATURES

Some SMR designs such as IRIS and NuScale utilize a helical coil steam generator rather than the more traditional straight tube steam generator. During normal heat-up of the tubes, the helical coil configuration allows the tube bundle to expand radially outward in response to thermal expansion of the tubes rather than longitudinally, which reduces mechanical stresses on the tubes at the upper and lower headers. Furthermore, the higher pressure primary coolant is placed on the shell side of the steam generator, which places the tubes in compression and reduces the likelihood of a tube rupture. These design choices are expected to reduce the anticipated failure rate of the steam generators, and hence reduce radiological exposures due to repair and replacement activities.

In addition to immersing the containment vessel in a large pool of water to facilitate decay heat removal, the NuScale design has a vacuum region between the reactor vessel and the containment vessel. This feature increases the steam condensation rate in the case of a steam release from the reactor vessel. The steam condensate collects at the bottom of the containment vessel and provides an effective conduction pathway from the reactor vessel to the containment pool. During normal operation, the vacuum region acts like a thermos bottle and obviates the need for vessel insulation, which creates a sump plugging challenge for traditional plants.

Some water-cooled SMR designs choose to remove soluble boron from the primary coolant as an added safety feature. Eliminating the soluble boron, which is present in traditional large plants to compensate for fission product buildup and fissile depletion, must be balanced by another form of reactivity control such as burnable poisons or additional control rods. However, eliminating the soluble boron provides a strongly negative moderator feedback mechanism during coolant voiding transients, and avoids boric acid corrosion issues such as encountered at the Davis-Besse plant cited earlier.

CONCLUSIONS

The concept of small modular reactors evolved from lessons learned during the rapid escalation of nuclear plant size during the 1970s as a way to achieve plant safety without the cost and complexity of the large plants. There continues to be interest in SMRs today for the same reasons and for a range of additional reasons such as affordability, site flexibility, operational flexibility and reduction in fossil power emissions. Clever design and engineering, such as the integral primary system reactor configuration, can result in designs that appear to offer the same or better safety performance while offering enhanced plant robustness and resiliency to perturbations, thus protecting the considerable capital investment. The trade-off is that these designs are constrained to be of smaller capacity, and the commercial viability of SMRs continues to be a dominant and unresolved concern. Several SMR vendors believe that the cost savings from economy-of-replication will be sufficient to overcome the economy-of-scale factor, and they are moving forward aggressively to bring new designs to market. With a desperate global need for safe, clean and abundant energy, the time has come to prove them right.

REFERENCES

1. [Carelli, 2004] M.D. Carelli, et al. "The design and safety features of the IRIS reactor," *Nuclear Engineering and Design*, **230**, p. 151-167 (2004).
2. [Davis-Besse, 2008] "Davis-Besse Reactor Pressure Vessel Head Degradation: Overview, Lessons Learned, and NRC Actions Based on Lessons Learned," U.S. Nuclear Regulatory Commission, NUREG/BR-0353, Rev. 1 (2008).
3. [Matzie, 1992] R. Matzie, et al. "Design of the Safe Integral Reactor," *Nuclear Engineering and Design*, **136**, p.73-83 (1992).
4. [Halfinger, 2010] J. Halfinger, "A practical, scalable, modular ALWR," presentation at the International Conference on Advances in Nuclear Power Plants (ICAPP) 2010, San Diego, CA, June 13-17, 2010 (2010).
5. [IAEA, 2009] "Design Features to Achieve Defense in Depth in Small and Medium Reactors," International Atomic Energy Agency, NP-T-2.2 (2009).
6. [Ingersoll, 2009] D.T. Ingersoll, "Deliberately small reactors and the second nuclear era," *Progress in Nuclear Energy*, **51**, p. 589-603 (2009).
7. [Kemeny, 1979] J. Kemeny, "Report of the President's Commission on the Accident at Three Mile Island," U.S. Government Printing Office, Washington, D.C. (1979).
8. [LaBar, 2002] M.P. LaBar, "The Gas Turbine–Modular Helium Reactor: A Promising Option for Near Term Deployment," presented at the International Congress on Advanced Nuclear Power Plants, Embedded Topical American Nuclear Society 2002 Annual Meeting, Hollywood, FL, June 9–13, 2002, GA-A23952 (2002).
9. [Lorenzini, 2010] P. Lorenzini, "NuScale Power: Capturing the 'Economies of Small'," presentation at the International Conference on Advances in Nuclear Power Plants (ICAPP) 2010, San Diego, CA, June 13-17, 2010 (2010).
10. [Nuclear Ship, 1968] "Nuclear Ship 'Otto Hahn'," *Atomwirk. Atomtech.* 13, p294-330 (1968).
11. [Weinberg, 1985] A.M. Weinberg, et al. *The Second Nuclear Era*, Praeger Publishers (1985).

SODIUM-COOLED FAST REACTOR (SFR) SAFETY

TANJU SOFU
Engineering Analysis Department, Nuclear Engineering Division
Argonne National Laboratory, Illinois, USA

INTRODUCTION

Current renewed interest in nuclear energy is largely driven by the need to develop carbon free energy sources, demographics and rapid development of emerging economics, and security of supply concerns. The pace at which the nuclear energy option is embraced seems to be accelerating worldwide, with the energy availability concerns causing especially the emerging economies to give it a serious consideration.

Although nuclear energy can help meet the growing energy demand without contributing to global warming, the current path relying heavily on light water reactors (LWRs) and "once-through" fuel cycle may not be sustainable. Conventional LWRs extract energy from only a small fraction of the natural uranium in the fuel: used LWR fuel still contains more than 95% of the original uranium, as well as heavier elements that are usable as fuel. The once-through fuel cycle with no provision for recycling or conditioning of used fuel also puts heavy burdens on the spent fuel repository storage capacity.

Next generation nuclear energy systems currently under consideration aim for significant advances over existing and evolutionary LWRs in the areas of sustainability, economics, safety, reliability, and non-proliferation. Development of these systems is an international effort, involving collaborations under the framework of the OECD's Generation-IV International Forum (GIF) and IAEA's International Project on Innovative Nuclear Reactors and Fuel Cycles (INPRO).

Recent studies under these programs highlight the importance of closed fuel cycle systems using fast-neutron reactors to meet especially the sustainability goals through efficient resource utilization. In comparison to LWRs, fast reactors can extract about two orders of magnitude more energy from the same amount of fuel. Moreover, nearly all long-lived heavy elements (transuranic waste that remains radioactive for a long time) can be consumed in a fast reactor with a closed fuel cycle, greatly reducing the amount of repository space needed for waste isolation.

Although reconsidered as part of the next generation nuclear reactors, the fast-spectrum systems, particularly the liquid sodium cooled reactors (SFRs), are not new concepts. Since the 1950s, SFR technologies have been pursued and demonstrated worldwide, leading to the construction and operation of several experimental and prototype fast reactors in the U.S. (EBR-I and -II, FERMI, and FFTF), Soviet Union (BR-10, BOR-60, BN-350 and -600), U.K. (DFR and PFR), France (RAPSODIE, Phénix, and Superphénix), Germany (KNK and SNR-300) and Japan (JOYO and Monju). These fast reactors have achieved well over 300 reactor-years of accumulated operation experience. After declining interest in nuclear energy during the past several decades, there is again a renewed interest in SFRs with the restart of Monju in Japan and ongoing construction of CEFR in China, PFBR in India, and BN-800 in Russia.

SFR DESIGN AND SAFETY

Schematic of a typical pool type SFR design is shown in Figure 1. Although they have many similarities in comparison to LWRs, many of the SFR concepts pose new design challenges. Historically, the most frequently considered SFR safety issues are related to the fact that the fuel in an SFR core, with an inventory of many times the critical mass, is not in the most reactive configuration. As a result, relocation of fuel in a hypothetical core disruptive accident has the potential to significantly increase reactivity, leading to power excursions. However, through proper design and taking advantage of the superb heat removal and heat transport characteristics of liquid sodium metal coolant and inherent negative reactivity feedback, it is possible to achieve a benign termination to a wide range of accident initiators. Such enhanced safety of SFRs has been demonstrated in test programs in the United States and the rest of the world.

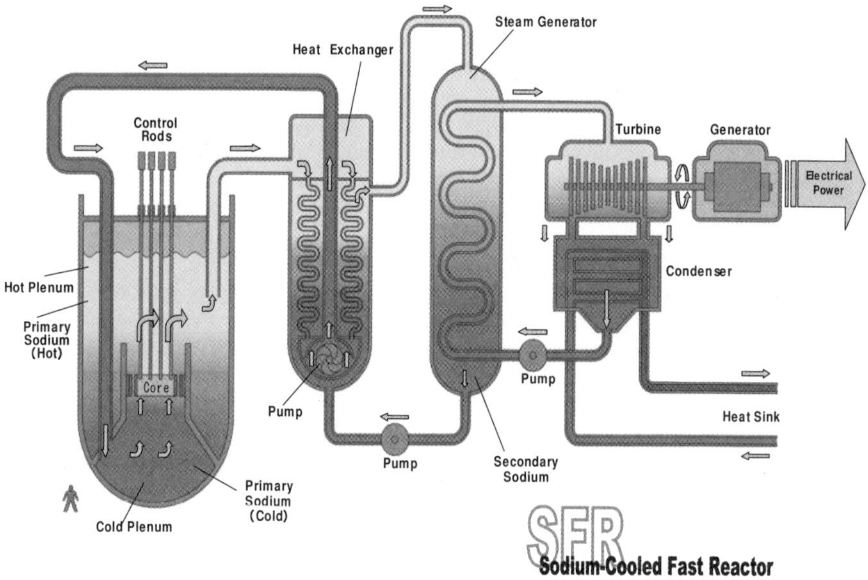

Fig. 1: Simplified diagram of a typical pool-type sodium-cooled fast reactor.

Similar to an LWR, the SFR safety is first based on utilization of multiple redundant engineered protection systems (independent scram systems, multiple coolant pumps, auxiliary decay heat removal systems, and multiple barriers to release of radioactive materials) to lower the probability of accident occurrence and to limit its consequences. But the typical SFR safety goes beyond dependence on these standard preventive and mitigative systems. Relying on fundamental phenomena such as thermal expansion, buoyancy-driven flow and gravity, the safety design features that enhance inherent negative reactivity feedback as well as passive decay heat removal provide additional measures to protect the reactor during very low probability accidents even when the engineered protection systems fail.

The focus of SFR safety is to avoid large uncontrolled increases in core power, insufficient cooling of the reactor core, and rearrangement of fuel that would lead to energetic recriticalities. Therefore, inherent safety uses three basic principles: favorable reactivity feedback (through core physics, structural design, and selection of materials), sufficient natural circulation cooling and passive decay heat removal systems, and appropriate selection of fuel and cladding materials, all contributing to reducing the probability of severe consequences even during the extremely unlikely accidents.

The fast neutron spectrum in SFRs leads to long neutron path lengths, enhancing the neutron leakage from the core and causing the reactivity feedback effects impact the reactor as a whole, not just locally. High neutron leakage also implies that in an SFR the reactivity is sensitive to minor geometric changes: as the core temperature increases and the materials expand, a net negative reactivity feedback is inherently introduced into the core in response to the accident initiators. The elements of negative total reactivity feedback in an SFR include the Doppler feedback, axial fuel expansion, radial core expansion, and control rod driveline expansion. With proper design choices, these contributions are more than sufficient to offset the coolant density and void worth.

Low system pressure in an SFR also offers significant advantages in terms of safety: The minimal pressure loading on the coolant boundary reduces the concern for coolant pipe breaks and resulting loss-of-coolant accidents. Since the coolant leaks are unlikely to propagate to a large-scale failure, there is no need for high-pressure injection cooling like those employed in LWRs. The liquid sodium coolant in SFRs provides a large margin to coolant boiling (about 400°C as opposed to ~15°C in a PWR).

DEMONSTRATION OF SFR PASSIVE SAFETY

In the U.S., past and present SFR R&D programs have focused on development and demonstration by testing of the concepts to enhance safety. The potential for a sodium-cooled fast reactor to survive severe accident initiators with no core damage has been extensively demonstrated during the landmark EBR-II tests. Testing in EBR-II spanned several decades, but a significant, whole plant testing period took place between 1984 and 1987. During this time period, the shutdown heat removal and plant inherent control tests were performed to assess whole-plant safety performance during a wide variety of transients and load conditions.

The inherent safety measures that do not require the functioning of any active system to provide sufficient margins for safe shutdown even during the beyond design basis accidents were also demonstrated in the EBR-II test program. Such beyond design basis accidents considered during the EBR-II testing program included the unprotected (un-scrammed) loss-of-flow, unprotected loss of main heat sink, and unprotected inadvertent withdrawal of reactor control rods resulting in a transient overpower condition. In Figures 2-5, the results for such an unprotected test are presented.

This unprotected loss of flow test simulates an accident condition assumed to be initiated by a total loss of offsite power, causing the electrical power to be lost to all primary pumps, intermediate loop pumps, and feed-water pumps. There is also a total failure to scram the reactor so that the reactor power changes only in response to the reactivity feedbacks. The primary and intermediate pumps coast down according to their internal characteristics only, as shown in Figure 2 for the primary pumps. The loss of the feed-water pumps also causes the loss of heat rejection to water side at the steam generators, but the decay heat removal system continues to operate at their rated capacity.

The pump coast-down without scram causes reactor core temperatures to initially rise. As seen in Figure 3, the peak coolant temperature reaches to 533°C in 70 seconds while the coolant saturation temperature is greater than 900°C, leaving more than 360°C margin to coolant boiling. Although the pumps stop, the rising coolant temperature establish the natural circulation of the sodium coolant through the core, stabilizing the flow rate at about 5% of the nominal flow observed during the normal operation as shown in Figure 4.

As shown in Figure 5, the change in core flow is accompanied by a change in power, with the flow dropping faster than the power. The power-to-flow ratio reaches 2.0 at 12 seconds after the start of the transient, and peaks at 2.3 at about 30 seconds. This mismatch between power and flow rate results in a proportional increase in the core temperatures, with the coolant outlet temperature reaching its peak at around 70 seconds as seen in Figure 3. The rising reactor core and coolant temperatures eventually introduce a net negative reactivity and reduce the reactor power to decay heat level.

The analysis of this test shows that, the initial reactivity feedback is dominated by the positive coolant density contribution such that the net reactivity is positive for very short time (first few seconds) after the start of the transient. Then, the negative contributions from core radial expansion, Doppler, and to the lesser degree, core axial expansion and control rod driveline expansion reverse the reactivity trend such that the net reactivity becomes negative and remains negative for the remainder of the transient. All reactivity feedbacks tend to return to zero as power-to-flow ratio returns to 1.0 with slightly negative net reactivity. The temperatures stabilize at about nominal core-outlet coolant temperatures in the long term.

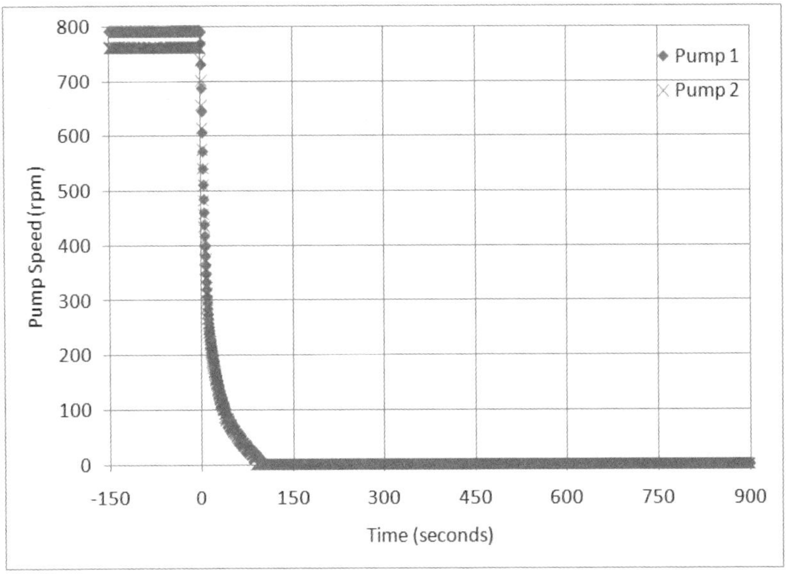

Fig. 2: The measured pump coast down curves during an unprotected EBR-II test.

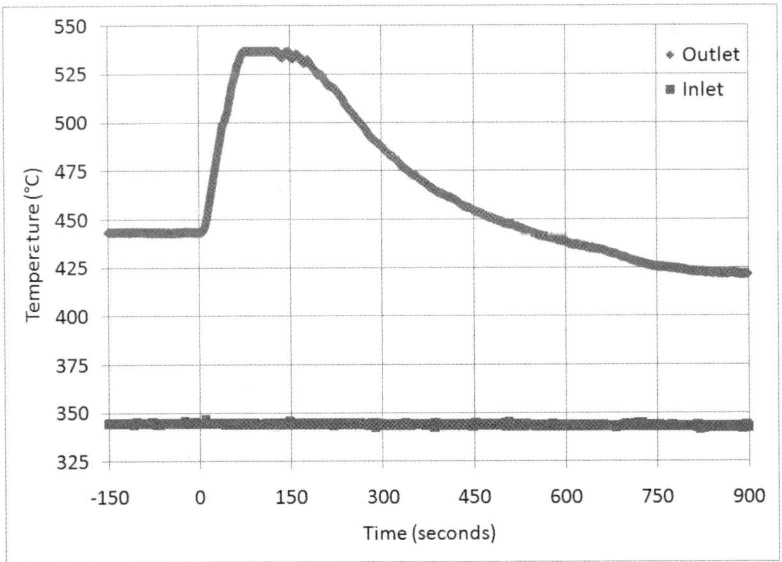

Fig. 3: The measured coolant inlet and outlet temperatures during an unprotected EBR-II test.

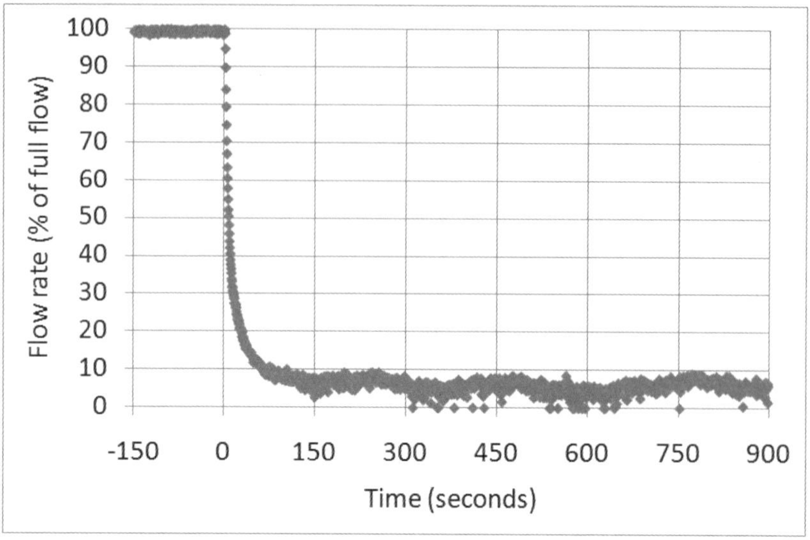

Fig. 4: The measured core flow rate during an unprotected EBR-II test.

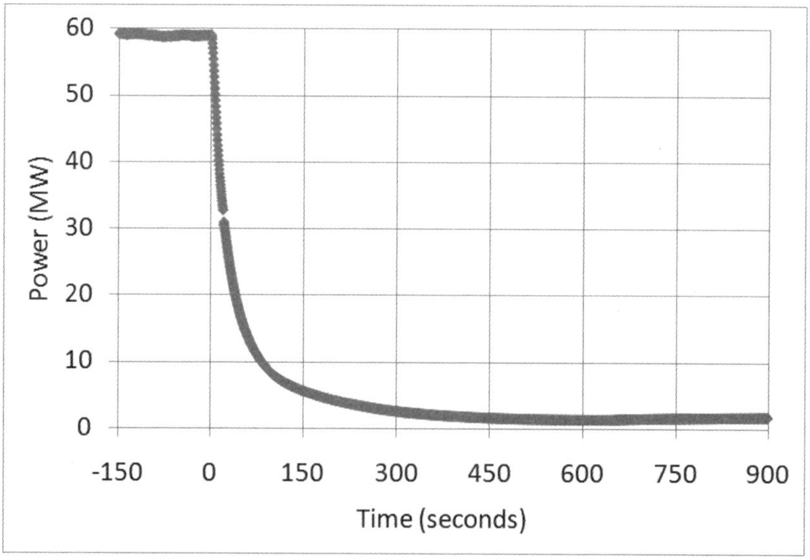

Fig. 5: The measured core power during an unprotected EBR-II test.

SUMMARY AND CONCLUSIONS

In summary, the landmark EBR-II passive safety tests show that, with safety principles of defense-in-depth with multiple redundant systems and barriers to release of radioactivity and the added inherent safety phenomena to prevent severe consequences from unprotected accidents, SFRs are capable of inherent self-protection. Through favorable reactivity feedback and natural circulation cooling, and considering only mechanistic, physically-realizable accident conditions to be relevant for safety (guided by probabilistic risk assessments), it may be possible to virtually eliminate accident-related large radioactive releases from an SFR design.

SECTION 3

SESSION 3

ENERGY & POLLUTION

FOCUS: OIL SPILLS AND DEEP WATER DRILLING

THE GULF OF MEXICO OIL SPILL AND DEEPWATER OIL DRILLING: INTRODUCTION

CARMEN DIFIGLIO, PH.D.
Office of Policy and International Affairs
U.S. Department of Energy, Washington, USA

DEEPWATER HORIZON: MAJOR EVENTS 2010

- 4/20: Explosion and fire; oil and gas flow up riser to ocean surface
- 4/22: Rig sinks; oil and gas continue flowing into ocean from sunken riser on sea floor
- 5/8: Initial containment system fails; icy methane hydrates foil hydrocarbon recovery
- 5/16: Riser Insertion Tube Tool installed to recover oil from riser to surface ship
- 6/3: Top Hat #4 placed on top of Lower Marine Riser Package
- 6/4: First oil and gas successfully routed to surface from Top Hat #4
- 7/13: Well integrity testing and shut-in operations initiated
- 7/15: Shut-in operations completed and flow of oil ceased
- 8/3: 'Static Kill' procedure initiated; mud and cement channeled into blown-out well
- 8/8: Testing reveals well casing successfully plugged by 'Static Kill' procedure
- Going Forward. Response efforts are now concentrated on relief well drilling ('Bottom Kill') to permanently seal the Macondo well

U.S. DEPARTMENT OF ENERGY SECRETARY CHU'S LAB TEAM

The U.S. Department of Energy was tasked to begin evaluating ways to plug the Macondo well. In response to the President's direction, Secretary Chu assembled a team of top scientists to help BP resolve key technical issues and evaluate the safety of proposed procedures to cap the spill. Secretary Chu regularly visits the command center in Houston, helping BP to design and implement spill response strategies.

The U.S. Department of Energy's scientific team includes, besides Secretary Chu, the following experts:

- Dr. Richard Garwin
- Dr. Tom Hunter (Director, Sandia National Laboratories)
- Dr. George A. Cooper
- Dr. Alexander H. Slocum

DEPARTMENT OF INTERIOR RESPONSES

The U.S. Department of Interior immediately called for a 30-day Safety Review:

- Temporarily suspended applications for permits to drill in offshore areas (pending the outcome of the 30-day safety review);
- Ordered immediate inspections of all deepwater operations in the Gulf of Mexico; two Incidents of Non-Compliance were noted and corrected;
- Issued a safety notice to all offshore operators, citing need to adhere to U.S. regulations and to initiate thorough tests of equipment/installations;
- Established the Outer Continental Shelf Safety Oversight Board to improve operations, regulation, and oversight;
- Launched a joint agency investigation with the U.S. Coast Guard to determine key factors leading to the Macondo well blowout.

MAJOR OFFSHORE POLICY DECISIONS

- The *National Commission on the BP Deepwater Horizon Oil Spill and Offshore Drilling* is focusing on key facts and root causes of the Gulf of Mexico spill.
- The U.S. Department of Interior's 30-day Safety Review report made recommendations to enhance enforcement efforts and deepwater technology/systems reliability.
- A Deepwater Moratorium was instituted and is now in place until 11/30/10 (may end sooner), disallowing new permits to drill new deepwater wells.
- Suspended action on 33 Gulf of Mexico deepwater exploratory wells and cancelled plans for exploratory drilling and new lease sales for Alaska and Virginia.
- The Obama Administration issued a new order in response to June Federal Court decision in Louisiana to block the 6-month moratorium.
- The U.S. Congress considering a series of legislative proposals to reform oil spill liability limitations, offshore leasing, and royalty policy provisions.

U.S. DEEPWATER PRODUCTION: SIGNIFICANT AND INCREASING

Figure 2. Gulf of Mexico Federal Offshore Oil Production

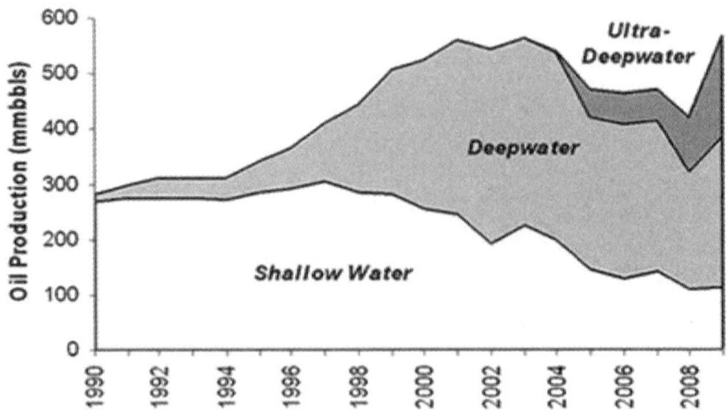

Sources: MMS, EIA Office of Oil and Gas; includes lease condensate

MORATORIUM PRODUCTION IMPACTS

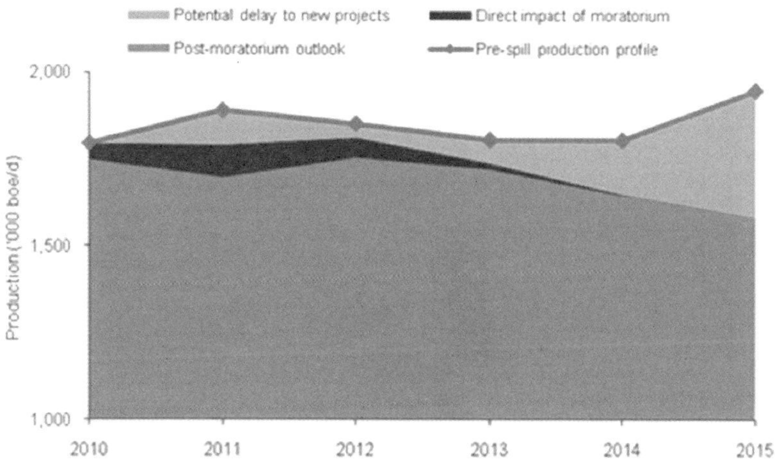

Source Wood Mackenzie

ESTIMATED ECONOMIC IMPACTS OF THE DRILLING MORATORIUM

Deep Water Moratorium:

- Will suspend operation on 33 wells
- Estimates indicate 6,000 to 9,000 rig workers will be temporarily displaced

Deferred Production Estimates.

- Ranging from 70,000 and 200,000 barrels per day (bpd) in 2011
- Majority of estimates are on the lower end of this 70-200k bpd range (This amount of oil would increase U.S. oil imports expense $1–$3 billion)
- For the six- month period, assumes an oil price of $75 per barrel
- Represents roughly 0.5 percent to 1.0 percent of U.S. oil imports

OVERVIEW OF INDUSTRY INITIATIVE

Chevron, ConocoPhillips, ExxonMobil and Shell have committed $1 billion to develop new rapid-response deepwater technologies and capabilities. The industry partners are planning to use modular containment systems to be kept on standby in the event of a future Gulf of Mexico deepwater well blowout. They also plan enhancements such as pre-engineered, constructed and tested response technologies and systems. These systems are intended to be responsive to different deepwater scenarios, spill flow rates, and weather conditions. The primary objective, stated by the industry partners, is to fully contain oil, without flow to the open sea, in the event of a future blowout.

Chevron ConocoPhillips ExxonMobil

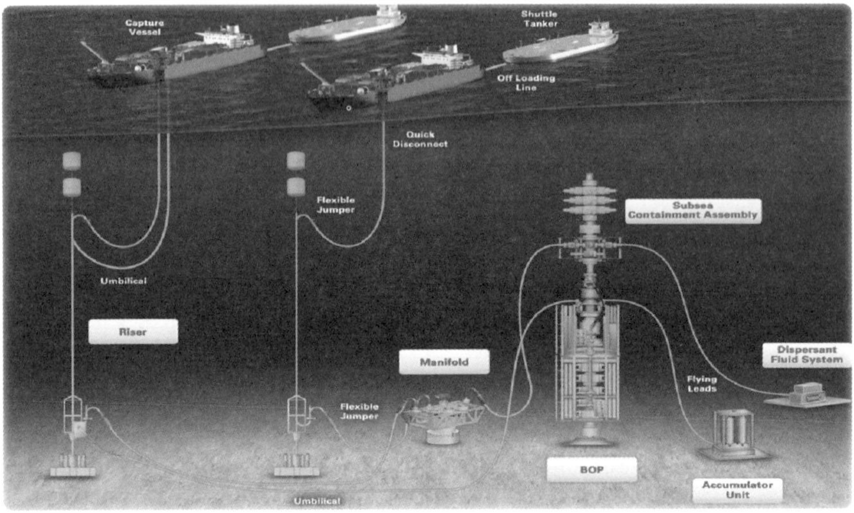

U.S. DEPARTMENT OF ENERGY RESEARCH ON DEEPWATER CHALLENGES

- The U.S. Department of Energy is elevating scientific expertise in fluid flow, imaging, fire science, and ultra-deepwater.
- U.S. Department of Energy's research portfolio integrates across academia, national labs, and industry.
- Unique capabilities now being developed to contain high pressure fluids/gases in deepwater under extreme conditions:
  - *prediction of materials behavior under extreme conditions*
  - *dynamic monitoring for containment*
- The U.S. Department of Energy's research and technology development role is separate from the Federal government's regulatory role.

REMAINING SPEAKERS

Approaches and Solutions to Taming a Wild Deep-Sea Oil Well
Dr. Richard Garwin, Member, U.S. Secretary of Energy's Science Team for the Gulf Oil Spill

Consequences of the Gulf Oil Spill on Future Offshore Oil Developments
Mr. Ivan Sandrea, Statoil

Environmental Consequences of the Gulf Oil Spill: Habitat and Ecosystem Destruction
Professor John W. Day, Jr., Department of Oceanography and Coastal Sciences, Louisiana State University, Baton Rouge, USA

Environmental Consequences of the Gulf Oil Spill: Human Toxicology
Professor LuAnn White, Tulane University, New Orleans, USA

Thanks also to Professor Lorne Everett, Pollution PMP Chair, for his role in organizing this session.

APPROACHES AND SOLUTIONS TO TAMING A WILD DEEP-SEA OIL WELL

RICHARD L. GARWIN
Thomas J. Watson Research Center, IBM Research Division
Yorktown Heights, USA

Thanks to Carmen Difiglio for his fine introduction of this session. Thousands of oil and gas wells have been drilled in the ocean floor ranging from the North Sea to the South Atlantic and beyond. Of most of these you never hear, as they contribute to the gas and oil supply of world commerce. Many more are being drilled for exploration and production, as you will hear from Ivan Sandrea.

Figure 1 is a schematic of a Kuwait oil well, from my paper[1] with Henry Kendall of November 1991, reporting on the world's success in stopping the flow of 640 wells that had been destroyed and set aflame in March 1991 by explosives by Saddam Hussein's military engineers.

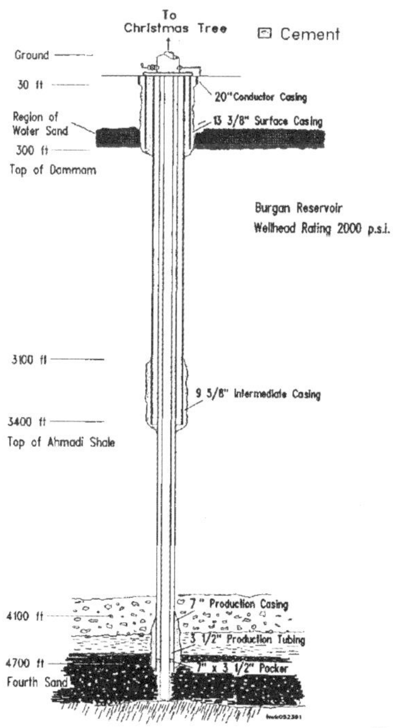

FIG. 1 A typical oil well in the greater Burgan field, south of Kuwait city, not to scale.

---

1 "Quenching the wild wells of Kuwait," by R.L. Garwin and H.W. Kendall in *Nature*Vol. 354, No. 6348 (pp. 11-14), Nov. 7, 1991 (with brief comment by Peter Aldhous, p. 5). http://preview.tinyurl.com/2fv6ko4

A seafloor well is not much different, except that it is drilled by a long string of drill pipe hanging as much as 3000 meters below a drill ship or drilling platform. In the case of the Deepwater Horizon (DH) well owned by BP in the Gulf of Mexico, the seafloor is at 5000 ft (1500 meters) and the well was completed for exploration with a blow-out preventer (BOP) and on top of that a Lower Marine Riser Package (LMRP) to connect to the mile-long 21-inch diameter steel tube (riser) that extended to the surface. The riser is buoyant, in order that it need not support the dead weight of a mile of pipe. Inside the riser is run the drill pipe, of various diameters, including that for drilling the hole for the 18-in casing. The riser allows control of the well environment, the circulation of drilling mud from tanks on ships or platforms at the sea surface, etc.

In reality, the ocean floor is drilled first with a "conductor casing," 36-inches diameter, and there are various cementing operations that go on as longer casings of smaller diameters are installed. Deepwater Horizon was drilled to a depth of 13,293 ft (4052 m) below the mud line (seafloor) or 18,360 ft (5596 m) below the sea surface.

DH was reportedly filled with a modest 50 bbl of cement at depth, and seawater in the 9-inch casing instead of mud to keep it safe while the drill platform was moved away, and another sea surface platform was to come in to complete the well for production.

A WELL FULL OF MUD OR CEMENT OF DENSITY COMPARABLE WITH THE SURROUNDING ROCK IS SAFE BECAUSE FLUID FROM THE RESERVOIR AT ITS FOOT CANNOT ENTER THE WELL

Drilling platforms are of various kinds, with many of them these days dynamically positioned via GPS to whatever accuracy is needed. Alongside the riser pipe are attached two smaller lines (4-inches in diameter), the Choke and Kill lines that have valves that enter the BOP among the various "rams" that can hold and seal (or shear) various diameters of pipes in a redundant fashion. The DH well, aka "Macondo 252" is more complex[2] than the wells in Kuwait. Figure 2 shows the diameters and depths of the various steel "casings" of DH, beginning at the "mudline" at 5067-ft depth below sea surface (BSS), to a total depth of 18,360 ft below the sea surface.

---

[2] http://www.energy.gov/open/documents/3.1_Item_2_Macondo_Well_07_Jun_1900.pdf

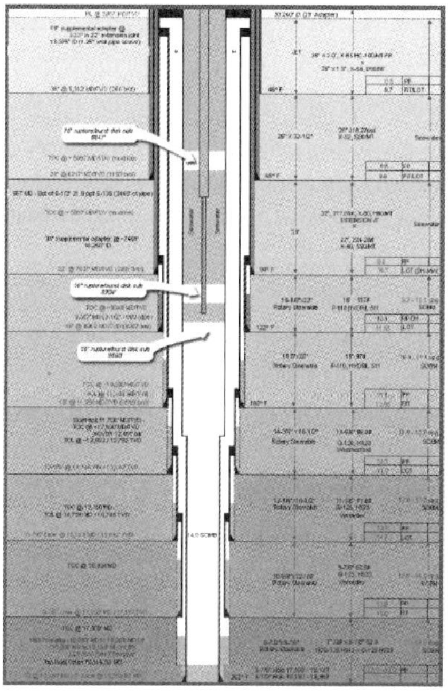

Fig. 2:

On April 20, with no warning, there was a whoosh, and a stream of solids and fluids shot up the pipe. The gas (methane) caught fire, and the drilling platform soon sank, with the loss of 11 lives and a rescue of 115. My own involvement with DH began on May 11, when I was called by Secretary of Energy Steven Chu to participate as a small team of outside scientists, together with many from government laboratories and the Department of Energy national laboratories (Livermore, Los Alamos, and Sandia), in order to provide scientific support and analyses for government direction of BP activities in stopping the flow of oil into the Gulf or collecting as much oil at the source as possible. In his introduction, Carmen Difiglio has given a timeline of the well's activity.

Our team had no mandate to investigate the accident itself, despite its relevance to the well configuration and what could be done in order to stem the flow. Initially BP had not characterized the flow quantitatively except to indicate that it was small. Later the government and BP[3] characterized the flow from the well as "1000 barrels of oil per day" (bopd). In fact, the science team was able to derive an estimate of the flow by the use of simple scaling laws, resulting in an official statement of August 2 that the initial flow into the Gulf had been 62,000 bopd, decreasing to 53,000 bopd before the well was capped on July 15. The decline was due largely to partial depletion of the contents of the sandstone reservoir.

---

[3] The Wikipedia article is substantial and largely but not entirely correct: http://en.wikipedia.org/wiki/Deepwater_Horizon_oil_spill

The 1500-m-long riser to the DH drilling platform broke free at its top when the platform sank, and fell along the seafloor, still attached to the top of the LMRP, in turn attached to the top of the BOP. The LMRP and BOP together weigh several hundred tons and extend about 30 m above the seafloor. The 21-in-diameter steel riser kinked a meter or so above its attachment flange, and first videos by remotely operated vehicles (ROV) showed oil and gas emerging from several tears in the heavy steel riser at the kink (Figure 3)

Fig. 3: Diamond-wire saw in position to attempt to cut the riser. Video[4] shows the oil and gas billowing from several tears in the "kink" of the 21-inch-diameter riser.

Most of the hydrocarbons emerged from the end of the fallen riser on the seafloor 1500 m away, with the gas emerging from the upper half of the horizontal pipe cross-section and the oil from the lower. At the seafloor ambient pressure of about 150 bar (15 megapascal—MPa) gas is more buoyant than oil, but oil itself with a density of about 0.6 is highly buoyant in seawater of density 1.02, accounting for the great difference in local environment between the deepwater wild well and a wild well on land.

The BOP and LMRP were damaged prior to or during the catastrophic accident and did not perform their function of stopping the flow of oil and gas, despite the fact that the hydraulic actuators on the various valves and rams were, in fact, "closed."

Attempts were made to capture some of the oil emerging from the broken end of the riser, but only small amounts were "contained" in this way until the broken riser was cut, not without difficulty, by enormous shears.

Here is a still from an ROV video of the gas/oil plume emerging from the clean-cut 21-inch-diameter riser (Figure 4).

---

[4] This clip and others are for public use from www.BP.com

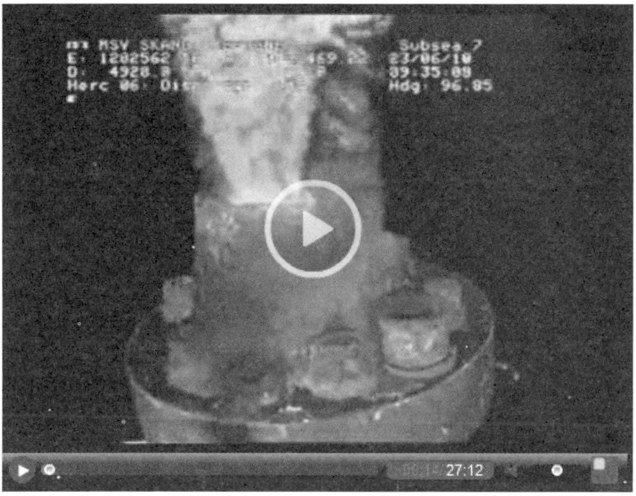

Fig. 4: A still from an ROV video of the gas/oil plume emerging from the clean-cut 21-inch-diameter riser. The oil is orange in color.

A covering dome or bell attached to a riser was used to take oil and gas to the surface, where it was separated and the oil collected while the gas (and some of the oil) was "flared" (burned). Mixing the oil with seawater before burning prevents the usual smoky flame that one would get from flaring crude oil. Up to 25,000 bopd was collected in this way, as depicted in Figure 5.

Fig. 5: Schematic of the TopHat 4 for capturing as much of 50% of the Deepwater Horizon flow.

The massive block rubber seal affixed to the TopHat was broken in handling, so substantial amounts of oil still emerged from below the skirt of the TopHat, as shown in Figures 5a and 5b.

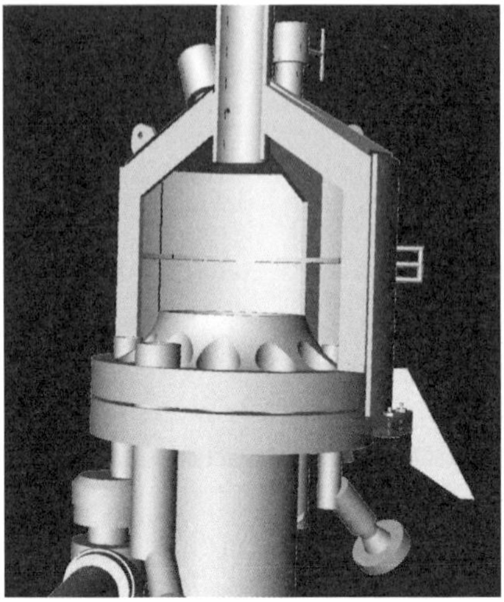

Fig. 5a: The TopHat4 as conceived mounted on the seafloor structure (yellow). The block rubber gasket is shown in red.

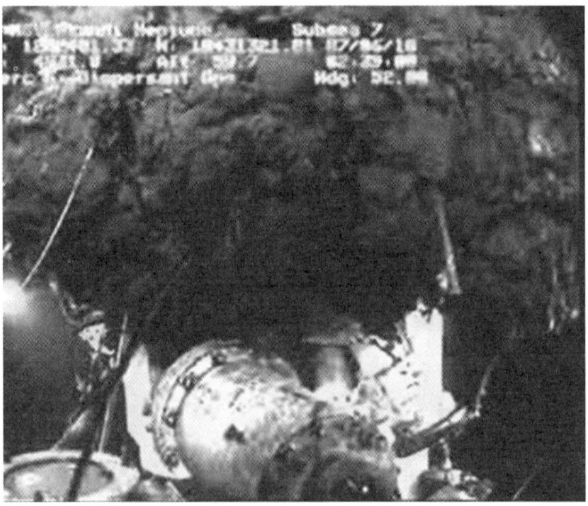

Fig. 5b: Oil billowing from below the skirt of the TopHat4.

Nevertheless, even with the small pressure difference[5] of the order of 0.1 bar between the skirt and the top of the TopHat, when one of the ports on the TopHat was opened to the sea it was possible to calculate the flow through that port and thus to estimate the total flow from the well. The derived flow agrees reasonably well with the later calculation by the government team of 50-66,000 bopd for the flow. At a world price of $75/bbl, the 50,000 bopd number corresponds to a price for that oil on the market of $3.75 million per day or $1.3 billion per year.

The wells in Kuwait suffered explosive cutting of their superstructure, and many were producing at a rate that when the oil or gas was flared to the atmosphere provided a fluid velocity that allowed entrained sand to act as an abrasive cutter. As detailed in our 1991 paper, the wells were quenched unprecedentedly rapidly because the primary approach was not the traditional one of putting out the flame and *then* dealing with the oil and gas, but rather to *stop the flow* in various ways, as with a massive steel needle that could be put into the cylindrical well bore. If the tapered needle does not suffice, one can have an inflatable "packer" on the needle that expands against the well bore and stops the flow. After that, a pit can be dug around the casing, and conventional large-diameter tubing and valves attached to the now-quiet well in order to provide a normal "Christmas tree" to conduct the effluent into gathering lines and pipelines in the normal fashion.

For DH, the damaged wellhead was 1500 m below the sea surface, so that gas expansion and resulting erosion velocity was limited. But one needs to rely on highly capable ROVs that can do construction work as an extension of the human operators on the mother ships, observed in real time at BP's headquarters in Houston and eventually put live onto the internet for public viewing.

The option of quenching the flow into the Gulf by installation of a valve was the first to be considered, and if the 9-in casing of the well were intact and sealed at its top with its rubber gasket, that would have been an excellent approach. But the oil reservoir tapped by the well had an initial pressure of more than 11,000 psia (pounds per square inch—absolute), and the pressure that could be confidently tolerated by the BOP is no more than 9000 psia. Because the mixture of gas and oil was thought to have a density of only 0.3, the hydrostatic head of the fluid column between seafloor and reservoir was only 3000 psia, so that the BOP would barely tolerate a valve that closed the flow, which would eliminate the non-hydrostatic friction loss of the fluid on its way from the reservoir to free egress to the seafloor.

But if the 9-in casing hanger were damaged, then the wellhead pressure could communicate with the "16-in" diameter casing, which has (as can be seen by the diagram on the doe.gov site) six "burst disks" at various levels in the pipe, (and six "collapse" disks). Although the burst disks are small, it was calculated that they would indeed burst and then pressurize the "18-in" casing that would allow high-pressure hydrocarbons into the rock formation at levels where the rock could fracture indefinitely as a result of transfer of hydrocarbons from the deep reservoir to the intermediate rock.

Closing the well was thus not an option unless one could determine that the 9-in casing hanger was intact.

There are striking differences between a wild well at 1500 m depth and one on the surface. The Devil's Cigarette Lighter comes to mind (Algerian desert, 1961) and other

[5] Resulting from the relative buoyancy of gas/oil mixture and seawater.

wells. The Devil's Cigarette Lighter was an enormous torch of gas from Hell, whereas the Lakeview Gusher of 1910 never did catch fire during its 17 months of flow but produced large lakes of oil in the scrub land near Bakersfield, California. It spewed 380 million gallons (9 million barrels of oil, as compared with the 4.9 million barrels from the DH.

Of the 690 wells beheaded by Saddam Hussein in 1991, 640 were on fire. In many cases the well did form enormous lakes of crude in the desert, extending into the Persian Gulf.

In contrast, as can be seen in the slides, at the seafloor there are no humans present to observe, and the ROVs do not have to be protected from the blistering heat of gas and oil fires, but there is another hazard at depth, and that is the formation of methane hydrate—a rock-hard compound of methane and water that is stable[6] in deep water on the seafloor the world over, within the methane-hydrate stability diagram as shown in Figure 6.

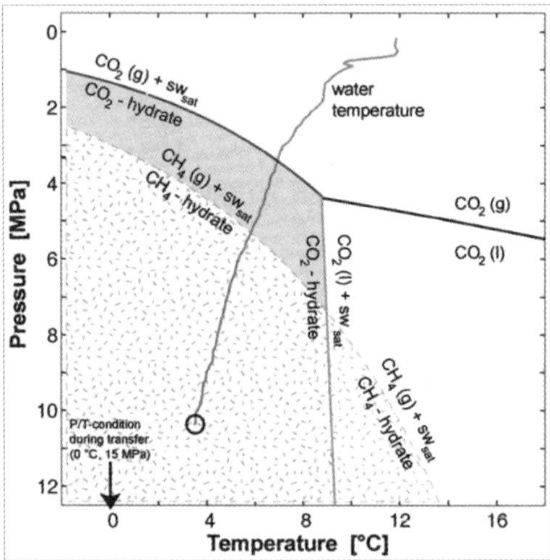

Fig. 6: Methane-hydrate stability diagram. Please ignore the CO_2 hydrate curves.
(Ignore the notation "P/T-condition during transfer (0 C, 15 MPa)"; the diagram is from another document that is concerned with CO_2 hydrate as well)

In fact, the first well-intentioned "containment dome" that had worked in shallow waters failed with Deepwater Horizon because the copious flow of methane from the well combined with seawater at seafloor ambient pressure of 150 bar and about 5 C to form methane hydrate that plugged a riser to the surface so that the dome was useless—a result predictable by simple modeling of the proposed operation.

[6] Methane hydrate slowly dissolves in seawater.

Later efforts to capture the effluent after the riser was cut had more success, because the stream from the well is hot, and despite a potential small admixture of seawater, is above the stability limit of methane hydrate at seafloor ambient pressure. I have shown a slide of the setup in which about half of the flow from the well was being taken up the riser to be flared or collected on surface ships.

From May 26-29 an effort was made to implement a "dynamic top kill" in the flowing well. So far as I know this had never been tried before, and was doomed to failure. I don't have time here to go into the details, but the idea was to inject heavy mud, in this case water-based, of density such that a mud column from the reservoir to the surface would provide sufficient hydrostatic pressure that gas or oil could not emerge from the reservoir into the casings of the well. The reason for the adjective "dynamic" is that the well was flowing at a rate that was later determined to be 50-66,000 bopd, so that if the mud were to fulfill its intended purpose of displacing the hydrocarbon in the casing, it would need either to move in "plug flow" down the casing or globs of mud would need to sink faster than the upwelling hydrocarbon.

Plug flow would be possible if the pressure at the wellhead (within the failed BOP) could be raised above the "shut-in pressure" either by shut-in or by pumping mud at such a rate that the pressure at the wellhead exceeds the shut-in pressure; this was not to be permitted and was not achieved.

As for globs of mud falling down the well cylinder against the flow of hydrocarbon, given the scaling laws for speeds and size, it is clear that a glob of mud almost the size of the casing—perhaps 10-cm in diameter—would fall at greater than the necessary speed, but such an interpenetrating fluid situation is unstable to what is called Kelvin-Helmholtz instability. The Bernoulli pressure at the bounding surface of the droplet and the upwelling fluid ($0.5\ \rho V^2$) far exceeds the surface tension of the globule, for globules of a size that they would fall by gravity against the upward current. In the future, greater attention to the rheology of the mud could preserve globules against the K-H instability; indeed a mechanism that would encase the mud in sausage casing (literally) could ensure that it would fall through the rising oil column. Details of its fate at the bottom of the well also need to be analyzed.

When as a test, mud was injected at a few barrels of mud per minute above the closed "rams" in the failed BOP, copious mudflow was evident through the cracks by the kink of the riser; when the primary mudflow aggregating 30,000 barrels over a few hours at a rate of 80 barrels per minute was injected into the well, there was only the usual gas and oil flow apparent at the riser kink.

I was watching these ROV videos via the internet, as could any member of the public, so my own view is that the mud was going down the so-called "9-in casing," or via some other path into the reservoir. In all, 30,000 bbl of mud was pumped into the well, the volume of which was less than 3000 bbl. Hence, at least 27,000 bbl of heavy mud either went into the reservoir by hydro-fracturing or through a convoluted path out the end of the broken riser onto the seafloor.

When the top kill ceased, gas and oil flow resumed within seconds, with no apparent residual effect of the Topkill. An alternative hypothesis was that there was no seal at the 9-in casing hanger, and that hydrocarbons were coming up from the "16-in annulus" (the annular space between the 9-in casing and the 16-in casing) while the

Topkill was being attempted, with the mud going down the 9-in casing at the same time as the hydrocarbons were flowing unimpeded up the annulus.

More complicated possibilities exist, with the 9-in casing being blocked at its base, but that casing being broken at the "crossover" between 9-in and 7-in casing diameters. In any case, after the failed dynamic top kill, operations resumed to collecting as much gas and oil as possible through TopHat 4, and bringing in more ship capacity on the surface to do the same.

A lesson for the future is that the industry should master seafloor storage and separation of mixed oil and gas on various scales—ranging from a buffer of a few seconds, on to days or weeks of product storage.

But hurricane season was approaching, during which time the ships could not be there on the surface, and a more durable solution was sought (which to my mind should have been pursued in parallel over the previous months and by the industry over the previous years).

Two "relief wells" had been initiated early on, as is conventional for wild wells on land or in the ocean. A relief well sometimes has the task of intercepting the bore of the wild well, and providing an alternative open path for collection of the hydrocarbons, to enable the valves at the surface of the wild well to be closed without excessive pressure at the wellhead. The relief well first to intercept, DD3, could intercept the DH well within days, and DD2, coming from a different azimuth, is a couple of weeks behind.

But before the relief well was completed, BP, with the approval of the USG, affixed a three-ram Capping Stack to the top of the failed blowout preventer. This was done by using the ROVs to unscrew the six massive bolts in a large flange visible in the slide, and bolt the Capping Stack to the top of the BOP system. The valves were cautiously closed, and the pressure at the wellhead rose to a still acceptable level against the seafloor ambient. Fortunately, even the damaged BOP sustained this pressure, and when the valve was closed, the pressure remained essentially constant, following some transient due to the cooling of a gas bubble that had been adiabatically heated by the increase in pressure. It was determined that there was no large leak of reservoir fluids from the well, and on July 15 the well was shut-in with no further leakage into the Gulf.

At this point, with hydrocarbons no longer flowing at the well, BP proposed and was authorized to do a static top kill, in which heavy synthetic oil based mud (SOBM) was introduced (thus without danger of hydrate formation) through the open test ram in the failed BOP. About 3000 bbl of SOBM was thus introduced, with a leader and a follower of base oil, in order to reduce the hydrostatic pressure at the wellhead to near-seafloor ambient. BP carried out the injection of mud on August 2. On August 5, BP completed displacement of much of the mud by a total of 500 bbl of cement, but with the probability that the well annulus was not filled with mud or cement.

In this case, the relief well will have the task of injecting heavy mud into the wild well, Deepwater Horizon, so that the mud will flow up the annulus path to the surface, once the "capping stack" valve is opened and communication from the top of the annulus ensured, for instance by perforation of the 9-inch casing near its top. Any fluid in the annulus must evidently be displaced by mud and eventually by cement from the relief well.

The simplest approach to allowing mud and cement from the relief well to displaced the current contents of the annulus is to open a valve on the capping stack, which, under some circumstances might allow 1000 bbl of oil to flow into the Gulf—less than 30 minutes of the unconstrained flow through the failed BOP that persisted almost 3 months. An alternative would be to master the technology of having even a relatively small buffer tank at seafloor that could hold the gas-oil mixture in the amount of 1000 bbl (about 150 cubic meters) above seawater, until it could be taken away.

The cementing was completed August 5 and appears to have been successful, in that the pressure at the wellhead has been steady for the last 10 days or so, with no evidence of leaks into the rock formation or up around the outside of the casing of the well. But it is not at all clear whether the cement went down to the reservoir through the 9-in central casing of the well or crossed over and went down through the 16-in casing of the well. The operation is well covered by a briefing[7] by BP's Kent Wells, of which I provide here a static Figure 7.

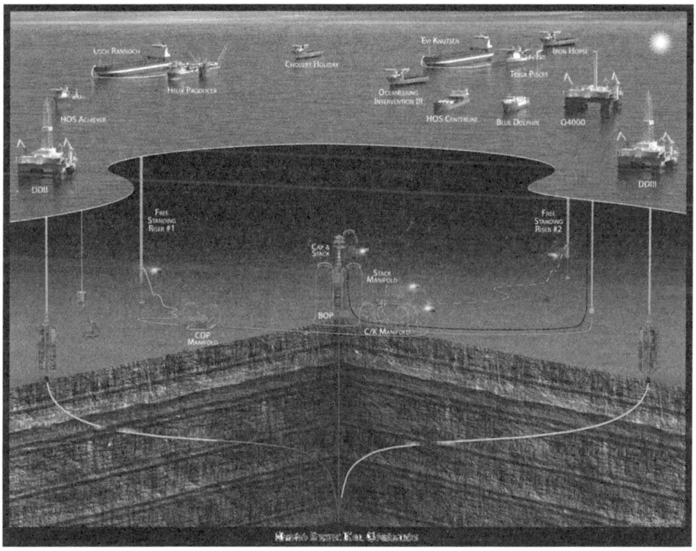

Fig. 7: From an animated BP video briefing by BP's Kent Wells, covering Static Kill and relief well, DD3.

In an interview on August 18, National Incident Commander, Admiral Thad Allen, noted the constraint that when DD3 makes the intercept with heavy mud, and if the annulus of Deepwater Horizon is filled with light hydrocarbons, the wellhead must not be over-pressured.

---

[7] http://www.bp.com/genericarticle.do?categoryId=9034436&contentId=7064379

Clearly the solutions include to open valves on the wellhead, and either to take the hydrocarbons into a TopHat filled with methanol or base oil, or into a storage tank on the seafloor, initially filled with base oil to avoid formation of methane hydrates.

BP has announced that following its success in opening of the wellhead to ambient seawater, the failed BOP will be replaced with the BOP that has been mounted on the DD2 relief well, and that the DD3 relief well operation and cementing will be delayed until September.

With due attention to every detail of this activity, the mudding and cementing of the annulus of the Deepwater Horizon well should be completed safely and the well transferred to the routine "P&A"—Plug and Abandonment.

ECOLOGICAL IMPACTS OF THE GULF OIL SPILL: FROM WETLANDS AND ANIMALS TO ECOSYSTEM SERVICES AND NET ENERGY

JOHN W. DAY, JR.
Dept. of Oceanography and Coastal Sciences, Louisiana State University, Baton Rouge, USA

MATHEW MOERSBAECHER
Dept. of Oceanography and Coastal Sciences, School of Renewable Natural Resources, Louisiana State University, Baton Rouge, USA

The BP oil spill is one of the largest marine oil spills in history. It is becoming clear that the spill was the result of a lack of emphasis on safety relative to the pressure to produce oil. Drilling in ultra-deep water is due to the growing global demand for oil at the same time as land and shallow water discoveries have declined rapidly and oil reservoirs are being exhausted.

In addition, the net energy of new oil discoveries, as measured by energy return on investment (EROI) has been declining for decades as a result of a decrease in discovery reservoir size and number. EROI analysis is a tool used to measure the net energy of an energy supply process (Cleveland et al. 1984). The net energy of an energy source is the amount of energy output minus the energy input in the extraction process. Domestic oil is becoming scarcer and more difficult to produce due to reservoir depletion and a decrease in the number of large discoveries (Hofmeister, 2008; Hall and Day, 2009). The net energy of domestic oil in the U.S. has been in decline for several decades (Cleveland et al. 1984; Cleveland, 2005). In 1930, the average domestic oil discovery yielded about 100 units of energy equivalent output production for every unit of input (Cleveland et al. 1984). Today the average net energy measured by EROI of domestic oil production has declined to 20:1 or 20 units of output for every unit of input (Cleveland, 2005). The EROI of ultra-deep oil in the Gulf of Mexico is in the range of 4-18 (Moerschbaecher et al. 2010). Thus, the trend to oil deposits in ultra deep water and other difficult areas is combined with a decreasing net yield.

Over five million barrels of oil were spilled over the several months before the well was sealed. The ecological consequences of the spilled oil are still being determined. After oil is spilled, some components degrade rapidly while others persist for longer periods of time. Low molecular weight alphatic and aromatic hydrocarbons degrade in days to weeks while high molecular weight alphatic and aromatic hydrocarbons take months to years to degrade. A number of authors have reported that many fractions of crude oil are broken down by bacteria (Head et al. 2006, Margesin and Schinner, 2001, Venosa and Holder, 2007, Zahed et al. 2010).

The oil spill affected both animal and plant communities. The toxic effects of oil on vegetation include disruption of plant water relationships, direct impacts on plant metabolism, toxicity to living cells, and reduced oxygen exchange between soil and air. Ko and Day (2004) reviewed the impacts of oil on vegetation in coastal Louisiana (Table 1). Impacts included reductions in photosynthesis, aboveground biomass, belowground biomass, and plant cover. Most of the marshes recovered within five years.

Table 1. Effects of oil exposure on plants in Louisiana coastal marshes from Ko and Day (2004). References cited in the table are in Ko and Day.

Species	Oil type	Exposure	Duration	Research type	Effects	Reference
Distichlis spicata	Crude oil	0.28 l/m^2	3 months	Field	64% decrease in live cover in mixed species assemblage	[13]
Distichlis spicata	Crude oil	0.28 l/m^2	5 years	Field	Overall recovery of vegetation completed in 5 years	[16]
Juncus roemerianus	Crude oil	2 l/m^2	5 weeks	Laboratory	6–30% decrease in photosynthesis for first 4 weeks and partially improved after that	[14]
Phragmites australis	Crude oil		1 year	Field	Significant reduction in stem density, plant shoot height, and aboveground biomass	[10]
Sagittaria lancifolia	Crude oil	24 l/m^2	3 months	Greenhouse	No significant changes in photosynthesis; increased aboveground biomass	[7]
Spartina alterniflora	Crude oil	Up to 8 l/m^2	4 months	Greenhouse	No significant difference in aboveground biomass	[12]
Spartina alterniflora	Crude oil	2 l/m^2	5 weeks	Laboratory	6–30% decrease in photosynthesis for first 4 weeks and partially improved after that	[14]
Spartina alterniflora	Crude oil	0.28 l/m^2	3 months	Field	64% decrease in live cover in mixed species assemblage	[13]
Spartina alterniflora	Crude oil	24 l/m^2	3 months	Greenhouse	50% reduction in photosynthesis	[7]
Spartina alterniflora	Crude oil	16–32 l/m^2	16 months	Greenhouse	No new shoots after first harvest	[12]
Spartina alterniflora	Crude oil	0.28 l/m^2	5 years	Field	Overall recovery of vegetation completed in 5 years	[16]
Spartina alterniflora	No. 2	29 mg/g dry soil	3 months	Greenhouse	Significant decrease in belowground biomass	[11]
Spartina alterniflora	No. 2	228 mg/g dry soil	3 months	Greenhouse	Constrained plant growth and microbial activities	[11]
Spartina patens	Crude oil	0.28 l/m^2	3 months	Field	64% decrease in live cover in mixed species assemblage	[13]
Spartina patens	Crude oil	24 l/m^2	3 months	Greenhouse	75% decrease in photosynthesis; 80% reduction in aboveground biomass	[7]
Spartina patens	Crude oil	0.28 l/m^2	5 years	Field	Overall recovery of vegetation completed in 5 years	[16]
Taxodium distichum	Crude and refined oils	Chronic exposure	70 years	Field	Higher Pb uptake by tree than ambient environment	[20]

DeLaune et al. (2003) reported that oiling of the salt marsh plants *Spartina alterniflora* and *Spartina patens* reduced photosynthetic CO_2 uptake for up to 16 weeks. Hoff (1995) reported on factors that affected the time of recovery of oiled marshes (Table 2). Marshes that took longer than five years to recover were in cooler climates, were in more sheltered locations, experienced heavy oiling, and were affected by spills of fuel oil (bunker C or number 2 fuel).

Table 2. Factors affecting the recovery time of oiled marshes. Adapted from Hoff (1995)

Characteristic	Recovery in less than five years	Recovery in greater than five years
Location	warm climate	north or south temperate (cold) environments sheltered location
Amount of oil	light to moderate oiling	heavy oiling
Type of oil spill	usually spills of light-to medium crude oil	spills of fuel oils (bunker C or no. 2 fuel)
Cleanup methods	variety of cleanup methods used often no cleanup resulted in fastest recovery time	in some cases intensive cleanup methods delayed recovery

Oiled marshes that recovered in less than five years tended to be in warmer climates and had light to moderate oiling with light to medium crude oil. Hoff also reported that intensive cleanup methods often delayed recovery and often no cleanup had the fastest recovery times (Table 3).

Table 3. Cleanup techniques used in marshes and their advantages and disadvantages (from Hoff 1995).

Advantages	Disadvantages
No response	
minimal impact (if oil degrades quickly)	potential oiling of birds or wildlife
no physical impact	oil may impact adjacent areas
	heavy oils may degrade slowly or form asphalt
Vacuum/pumping	
can remove large quantities of oil	access /deployment of equipment
	physical impacts
Low pressure flushing	
assists in removal by herding oil	requires careful monitoring
lifts oil off sediment surface	pressure must be controlled
	physical impacts
Burning	
potential to remove oil quickly	potential damage to plant roots and rhizomes
can minimize impacts from trampling	little known about impacts due to season, inundation of marsh, species composition,
	air pollution, regulatory concerns
Sediment removal	
may be only remediation possible for heavily oiled sediments	"destroy marsh to save it"
	increased erosion potential
	elevation changes may impede regrowth of plants
	replanting necessary
Vegetation cutting	
leaves most of plant intact	may kill plant
prevents oiling of birds	potential for increased erosion
	must be carefully monitored
Bioremediation	
great theoretical potential	few case studies available
low impact	potential for nutrient enrichment
	oxygen may be limiting

Animal populations are also impacted by oil. Ko and Day (2004) reviewed the impact of oil on benthic communities in the Louisiana coastal zone (Table 4). Impacts included decreases in population numbers, reduction in grazing on microphytobenthos which stimulated primary production, shifts in the importance of different groups, changes in nitrogen limitation, lowered diversity, and accumulation of hydrocarbons.

Table 4. Effects of oil spill and cleanup activities on the benthic communities in Louisiana wetlands (Ko and Day 2004). References in the table can be found in Ko and Day.

Species	Oil type	Exposure	Effects	Reference
Macrofauna	Crude oil	Constant exposure	Significant reduction in macrofaunal populations	[44]
Macrofauna	Crude oil	2 l/m^2	No immediate initial die-back, generally low densities in oiled area	[24]
Macrofauna	Crude oil	2 l/m^2	Significant population decrease	[24]
Macrofauna	Crude oil	2 l/m^2	Highly significant population decrease	[24]
Meiofauna	Crude oil	2 l/m^2	Stimulated population increase after 5–60 days	[24]
Meiofauna	Crude oil	2 l/m^2	Population increase after 30 days	[24]
Microphytobenthos	PAH	Chronic contamination	Reduced benthic grazers and increased microphytobenthos	[48]
Darter Gobies	PAH	Chronic contamination	No acclimation, no adaptation	[56]
Microbial, meiofaunal	PAH	Chronic contamination	Possible adaptation of microbial/meiofaunal community to PAH	[42]
Microalgae, mieofauna	Diesel	Chronic contamination	Competition between copepod and nematode over microalgae, increased copepod mortality	[46]
Meiofauna		Chronic contamination	Reductions in abundance and grazing activity of crustaceans, enhanced algal biomass	[55]
Meiofauna	Diesel	Chronic contamination	Initial nitrogen limitation by reduced grazer pressure and increased microalgae growth, enhanced NH_4^+ in a longer term	[49]
Barnacle	Crude oil	Chronic contamination	Higher recruitment in open-coast sites	[50]
Grass shrimp	PAH	Chronic contamination	No catches in contaminated area during summer	[45]
Macrofauna	PAH	Chronic contamination	Low diversity and low abundance	[45]
Meiofauna	PAH	Chronic contamination	Nematodes, oligochaetes, rotifers found dominant	[45]
Meifouna	PAH	Chronic contamination	No feeding habit changes, higher mortality risk	[52]
Epibenthic species	PAH	Chronic contamination	Bioaccumulation of PAHs depending on characteristics of site and PAHs, and exposure time	[54]
Macrofauna	Crude oil	Chronic contamination	Sparse population, lower biodiversity	[53]
Meiofauna	Crude oil	Chronic contamination	Deposit feeders were more sensitive to metal contamination than particle feeders.	[51]
Macrobenthic	Crude oil	Chronic contamination	Lower populations, contaminated detritus-based food web	[44]

Oberdorster et al. (1993), for example, reported decreased populations of macrobenthos and meiobenthos in heavily oiled areas (Figures 2 and 3, from Oberdorster et al. 1993).

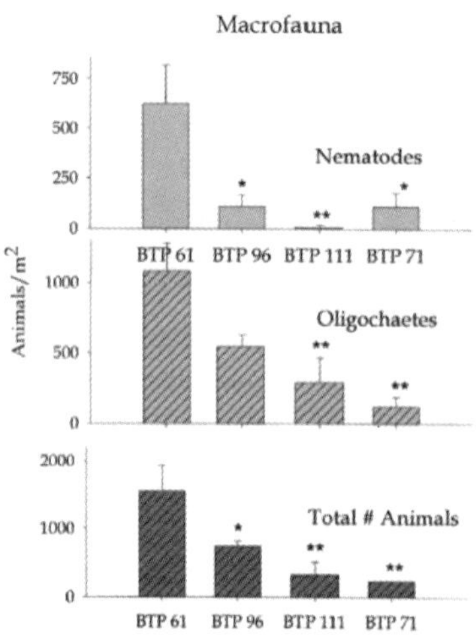

Fig. 2. Macrofaunal density along the contamination gradient in Bayou Trepagnier. There is a significant decrease in total number of animals at the most contaminated sites (BTP 96, 111, and 71); n = 3, bars are SEM; *p < 0.05; **p < 0.01

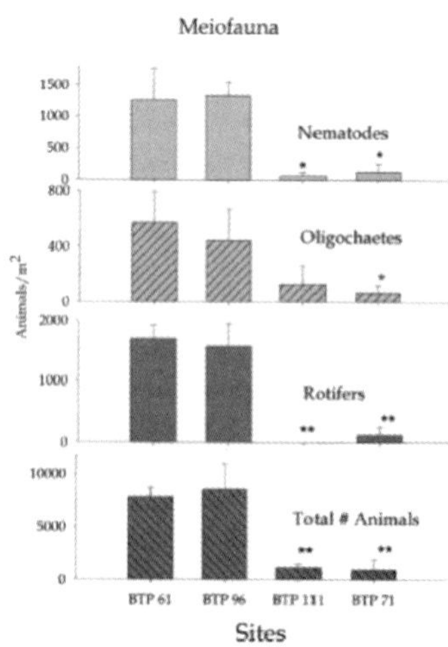

Fig. 3. Meiofaunal density along the contamination gradient in Bayou Trepagnier. There is a significant decrease in total number of animals at the most contaminated sites (BTP 96, 111, and 71); n = 3, bars are SEM; *p < 0.05; **p < 0.01

Figs 2 and 3.

We want to put the Deep Water Horizon oil spill into a larger context of the broader energy picture and the value of the natural assets of the Mississippi delta. It is estimated that the Macondo field contained 50 to 100 million barrels of oil while the current estimates of total ultra-deepwater oil in the northern Gulf of Mexico are 4 to 6 billion barrels (EIA 2010). Five of the top 15 largest oil fields in the U.S. are in federal waters of the northern Gulf. At current rates of usage, the Macondo field would supply less than two days of domestic consumption and the ultra deep reserves, based on current estimates of reserves size, of the northern Gulf would supply less than 70 days of domestic consumption.

As stated in the introduction, the EROI of domestic oil production (the 100:1 number is for discoveries in 1930) has been declining for decades. In the 1930s, the EROI for domestic discoveries was about 100:1 while EROI domestic production has since declined to between 10:1 and 20:1 (Cleveland, 2005). We calculated the EROI of ultra-deepwater oil in the northern Gulf of Mexico in 2009 and for the Macondo well (Moerschbaecher and Day, 2010). Production and operational expenditure data is used to estimate a second order standard EROI for 2009 ultra-deepwater production and the potential for production of Macondo well reserves. EROI estimates for 2009 ultra-deepwater oil and natural gas production range from 7-22:1. Estimates of Macondo well oil reserves range from 4-16:1. Estimates are directly related to the energy intensity ratio used to quantify costs. These EROI estimates suggest that ultra-deepwater energy reserves in the Gulf of Mexico come at increasing energetic and economic costs to society.

It is interesting to compare the value of ecosystem services of the Mississippi delta to that of oil in the northern Gulf of Mexico. Batker et al. (2010) analyzed the value of ecosystem services of the delta. They concluded that the Mississippi River delta ecosystems provide at least $12-47 billion in benefits to people every year. If this natural capital were treated as an economic asset, the delta's minimum asset value would be $330 billion to $1.3 trillion using a 3.5% discount rate. This is significantly more than the total value of remaining oil reserves in the northern Gulf. The study is the most comprehensive measure of the economic value of Mississippi River Delta natural systems to date. Marine waters, wetlands, swamps, agricultural lands, and forests provide natural goods and services. The goods and ecosystem services valued in the study include hurricane and flood protection, water supply, water quality, recreation, and fisheries. The Mississippi River delta is a vast natural asset, a basis for national employment and economic productivity. It was built by with sediments, freshwater, and the energy of the Mississippi River.

Yet, this vast national economic asset is being lost at a tremendous cost. While the oil spill in the Gulf of Mexico focused the country's attention on the value and vulnerability of coastal wetlands, decades of wetland loss remains the greater threat to the long-term ecological and economic health of the Gulf. The Mississippi Delta lost over 1.2 million acres of land in the last 80 years. In some areas, the coastline has retreated by as much as 30 miles. The lower Mississippi River has been constricted by levees since the 1930s, resulting in billions of tons of valuable sediment and trillions of gallons of valuable freshwater being channeled into deep water off the edge of the continental shelf. The Mississippi's energy to move vast amounts of sediment and water could have built additional land and provided hurricane protection and other economic benefits at no significant cost.

In summary, the Deep Water Horizon spill released large volumes of oil into the Gulf and bays and wetlands of the Mississippi delta. This has impacted and will continue to impact both deepwater and shallow water environments. This giant spill is the result of a relentless search for oil in deeper and more difficult areas that pose significant technical challenges and dangers. The value of the remaining oil and the damages due to the spill should be balanced against the ecological values of the Mississippi delta. The natural asset value of the delta is worth more than the remaining oil reserves in the northern Gulf. In the near future, the remaining oil reserves will be exhausted. If at the same time, the deterioration of the delta continues, Louisiana will be left with neither fossil fuels nor the ecosystem service values of the delta.

REFERENCES

1. Batker, D., I. de al Torre, R. Costanza, P. Swedeen, J. Day, R. Boumans, and K. Bagstad (2010) Gaining Ground: Wetlands, Hurricanes and the Economy: The Value of Restoring the Mississippi River Delta. Earth Economics. 98 p. Available at:http://www.eartheconomics.org/FileLibrary/file/Reports/Louisiana/Earth_Economics_Report_on_the_Mississippi_River_Delta_compressed.pdf.
2. Cleveland, C.J. (2005) "Net energy from oil and gas extraction in the United States, 1954–1997." *Energy*. 30: 769–782.
3. Cleveland C.J., Costanza R., Hall C.A.S. and Kauffmann R. (1984) "Energy and the U.S. Economy: A Biophysical Perspective." *Science* 225: 890-97.
4. EIA 2010. United States Energy Information Agency. Crude oil proved reserves, reserves changes, and production. EIA website: http://www.eia.gov/dnav/pet/PET_CRD_PRES_A_EPC0_R01_MMBBL_A.htm Accessed 2 November 2010.
5. Hall, C. and J. Day. (2009) "Revisiting limits to growth after peak oil." *American Scientist.* May-June. 92: 230-237.
6. Head, I., D. Jones, and W. Roling (2006) "Marine microorganisms make a meal of oil." *Nature Reviews Microbiology*. 4: 173-182.
7. DeLaune, R.D., Pezeshki S.R., Jugsujinda A., and Lindau C.W. (2003) "Sensitivity of U.S. Gulf of Mexico coastal marsh vegetation to crude oil: Comparison of greenhouse and field responses." *Aquatic Ecology* 37**:** 351–360, 2003.
8. Hoff, R.Z. (1995) Responding to oil spills in coastal marshes: The fine line between help and hindrance. HAZMAT Report 96-1. December 1995. Hazardous Materials Response and Assessments Division. National Oceanic and Atmospheric Administration. Seattle, Washington USA.
9. Hofmeister, J. (2008) Statement before the House Select Committee on Energy Independence and Global Warming, Washington, D.C., USA, April 1, 2008.
10. Ko, J., and J. Day (2004) "A review of ecological impacts of oil and gas development on coastal ecosystems in the Mississippi delta." *Ocean and Coastal Management*. 47: 671-691.

11. Margesin, R. and F. Schinner (2001) "Biodegradation and bioremediation of hydrocarbons in extreme environments." *Applied Microbiology and Biotechnology*. 56: 650-663.
12. Moerschbaecher, M., J. Day, and C. Hall (2010) "Energy return on investment for ultra-deep water oil in the Gulf of Mexico, including a case study of the Macondo well." Submitted to the journal *Sustainability*.
13. Oberdorster, E., Martin M., Ide, C.F. and McLachlan J.A. (1999) "Benthic Community Structure and Biomarker Induction in Grass Shrimp in and Estuarine System." *Arch. Environ. Contam. Toxicol.* 37, 512–518
14. Venosa, A.D. and E.L. Holder (2007) "Biodegradability of dispersed crude oil at two different temperatures." *Marine Poll. Bulletin* 54: 545-553.
15. Zahed, M.A., Aziz, H.A., Isa, M.H., and Mohajeri, L. (2010) "Effect of initial oil concentration and dispersant on crude oil biodegradation in contaminated seawater." *Bull Environ Contam Toxicol* 84:438–442.

TOXICOLOGY AND PUBLIC HEALTH FACTORS OF THE USE OF DISPERSANTS IN THE GULF OIL SPILL

LUANN E. WHITE, PHD, DABT
Tulane Center for Applied Environmental Health, Tulane University
New Orleans, USA

INTRODUCTION

On April 20, 2010, the Deepwater Horizon drilling rig exploded, engulfing the rig with fire. After two days, the rig sank into the Gulf of Mexico. Eleven of the 125 workers lost their lives and 114 were rescued. The well is located in deep water 1 mile (5000 ft 1500 m) below the surface of the Gulf of Mexico; the well extends down another 13,000 ft into the Mercando Prospect (Mississippi Canyon Block 252). The well is located approximately 50 miles from the mouth of the Mississippi River off the Louisiana coastline and further distances from the coast of other Gulf States including Mississippi, Alabama and Florida. The winds, currents and tides moved the oil around the Gulf for a few weeks before impacting the shore. The oil spread out as a light sheen with ribbons of red emulsified oil. The depth of the water and its distance from the shore are factors in the impact of the oil spill.

The initial reports indicated minimal leakage, but it became apparent that large volumes of oil were leaking into the Gulf. The volume was estimated at approximately 53,000 per day for a total of 4.9 million gallons by time the leak was plugged. Crude oil contains thousands of hydrocarbons and the composition varies with the type and source of crude oil. Most of the Gulf of Mexico oil is Louisiana sweet light crude oil. Light crude oil has a greater percentage of smaller molecular weight compounds that include natural gas and volatile organic compounds (VOC); correspondingly, there are a lower percentage of the high molecular weight compounds that have denser characteristics. Sweet crude oil indicates low amounts of sulfur in the crude oil.

Every oil spill is different and must be evaluated with respect to the type of oil, the location and geography. The characteristics of the oil and the incident greatly influence how the oil impacts the environment and people. Initially the Gulf of Mexico Oil spill was compared to the Exxon Valdez oil spill, however this comparison is misleading and the differences became apparent. The Exxon Valdez was a tanker spill with heavier crude oil that drenched the rocky coast of Prince William Sound in Alaska with a cold climate. This differs from the Gulf Oil Spill in distance from the shoreline, the depth of the leaking well, the type of crude oil and the hot climate. The oil spills that occurred after Hurricane Katrina in Louisiana provide some insight into the fate of the oil and the influence of the hot climate on the natural breakdown of the oil.

COMPONENTS OF LOUISIANA LIGHT CRUDE OIL

Crude oil consists of thousands of hydrocarbons with different physical, chemical and toxicological characteristics. To assess the toxicology of crude oil, the various classes of compounds need to be assessed independently with respect to the innate toxicity of the class of compounds and the pathways by which those compounds might come into contact with people. Compounds in crude oil range from a single carbon (methane) to compounds with more than 45 carbons. Various classes include the following:

C_1 to C_4: Gases

This is the component of crude that produces the natural gas and is responsible for the explosive characteristics associated with the well blow outs. If not contained or burned at the well site, the gases dissipate into the air and are ultimately broken down. The greatest hazard of the gases is explosion and the direct toxic effects are minimal. In confined spaces, these gases can displace oxygen leading to asphyxiation due to lack of oxygen.

C_5 to $_{10\text{-}14}$: Volatile Organic Compounds (VOCs)

This is the liquid fraction of crude oil from which gasoline, kerosene, diesel fuel and other fuels are extracted. This fraction contains both aliphatic and aromatic compounds. From a toxicological perspective, the aliphatic compounds have low degrees of toxicity. However, the volatile aromatic compounds include benzene, toluene, ethylbenzene and xylene (BTEX) which are moderately toxic effects including neurotoxic effects. Benzene is also a proven human carcinogen. BTEX are compounds of concern and benzene is especially a concern because of its carcinogenic effects. A percentage of these compounds are removed from the oil in the mile long trip from the leak site at the bottom of the Gulf to the surface of the water. On the surface, remaining VOCs rapidly evaporate from the crude in short periods of time in the hot weather of summertime in Louisiana.

$C_{14} – C_{24}$: Oils

This fraction contains the lubricating oils, mineral oils and other denser liquids. These compounds have a low degree of volatility and remain in the liquid state. The toxic effects are low with dermal irritation in sensitive individuals. This fraction of the crude oil becomes the thick gooey liquid that has reached the coast line and estuaries in Louisiana. However, this fraction also contains the polycyclic aromatic hydrocarbons (PAH) that have carcinogenic characteristics. The PAHs are compounds of concern in toxicological assessments. The PAHs do not bioaccumulate in organisms or humans because they are readily metabolized and excreted.

C_{25} – above

The heavier solid components of crude oil including the paraffins and asphaltenes are present in this fraction. These are usually in a solid form that produce the tar patties and tar balls that wash up on the coast line. They have little toxicity and in some sensitive individuals may cause dermal irritation.

Weathered Oil

The characteristic of the oil change and begin to degrade as the oil moves from the bottom of the gulf to the surface, and while on the surface of the water. The gas readily dissipates. The oil becomes emulsified with seawater during its mile long trek to the surface and changes to a thick reddish consistency. On the surface, any remaining VOCs evaporate and photo-oxidation begins. This process is called weathering of the oil and it greatly changes its composition and physical characteristics. The oil takes on a thicker gooey consistency largely composed of the heavier hydrocarbons. Weathering is an important factor in assessing the impact of the crude oil that reaches the coast. Analysis of the weathered oil shows the higher (C_{10}-C_{28} and C_{29}-C_{39}) aliphatic hydrocarbons and some PAHs. However, as the oil further weathers in the heat and sunlight, additional photo-oxidation occurs and PAHs are not detected in analysis of the mousse and tarballs. Table 1 lists the summary analyses of weathered oil.

Table 1: Analysis of Weathered Oil

	Weathered oil	Mousse	Tar ball	Water
Total Petro Hydrocarbons C^6-C^{10}	ND	ND	ND	ND
Total Petro Hydrocarbons C^{10}-C^{28}	YES	YES	YES	YES
Total Petro Hydrocarbons $>C^{28}$-C^{35}	YES	YES	YES	YES
PAH	YES	ND	ND	ND
BTEX	ND	ND	ND	ND

Spilled crude oil will eventually breakdown in the environment. Components of the oil break down naturally through evaporation, dissolution in water, emulsification with water, photo-oxidation and biodegradation by microorganisms in the water and on shore. Microbial degradation is a major factor as the micro-organisms in the gulf are primed from natural seepage and oil spills to breakdown oil. The issue is how long will the biodegradation take given the massive amount of oil spilled.

DISPERSANTS

A large controversy has arisen on the use of dispersants to fight the oil spill in the Gulf of Mexico. Dispersants mix with the oil breaking it into small droplets that move into the water column. Dispersants are composed of solvents that allow the dispersant to mix with oil and surfactants which make the oil miscible with water. The controversy arose immediately upon their use. People questioned adding more chemicals to the Gulf with the massive amount of oil leaking into it every day. Further, the components of the dispersant used, Corexit 9500 (and some 9527 early on) were propriety and the company would not release its formulation. The assumption was that the components are very toxic and that BP and others wanted to make the oil disappear from sight. The news media questioned the use of dispersants repeatedly raising fear among the general public. Further, the dispersants had never been used in deep sea conditions and little was known of their characteristics at the depths and temperatures a mile below the surface.

The use of dispersants was effective in reducing the amount of oil on the surface of the water. Dispersants do not affect the amount of oil, but change where it is located. The dispersants break up the oil into small droplets that have more surface area and are more readily biodegraded by microorganisms. In the Gulf Oil Spill, the dispersants are being administered at the site of the lead near the floor of the gulf. This is dispersing oil at greater depths than ever before and has raised concerns of underwater plumes of oil. Monitoring has detected oil in the part per billion range 5-10 miles from the site of the spill. Little is known about its fate at that depth and how it may distribute within the water column.

Table 2: Composition of Corexit 9500 as published by EPA.

Item	CAS Number	Chemical Name
1	57-55-6	1,2-Propanediol
2	577-11-7	Butanedioic acid, 2-sulfo-, 1,4-bis(2-ethylhexyl) ester, sodium salt (1:1)
3	1338-43-8	Sorbitan, mono-(9Z)-9-ctadecenoate
4	9005-65-6	Sorbitan, mono-(9Z)-9-octadecenoate, poly(oxy-1,2-ethanediyl) derivs.
5	9005-70-3	Sorbitan, tri-(9Z)-9-octadecenoate, poly(oxy-1,2-ethanediyl) derivs
6	29911-28-2	2-Propanol, 1-(2-butoxy-1-methylethoxy)-
7	64742-47-8	Distillates (petroleum), hydrotreated light

List of chemicals developed and distributed by EPA
Note: Corexit 9527A contains solvent 2-butoxyethanol

The EPA conducted aquatic toxicity test of 8 dispersants; all were found to have low toxicity and Corexit 9500 was midrange and had low toxicity. In another set of testing that combined the oil plus dispersant, toxicity test showed that the combination of oil and dispersants had the same toxicity as oil alone. This indicated that the toxic effects were due to the oil, and not the dispersants. In spite of these tests, fear still persists.

EXPOSURE ASSESSMENT AND HUMAN HEALTH

To evaluate the impact of oil on health, the toxic components must be identified and pathways of exposure to humans determined. The major pathways for the oil include: 1) Inhalation of the volatile compounds, particularly the BTEX; 2) ingestion through seafood is a possible pathway, particularly for the PAH which have been shown to temporarily build up on oysters because oysters only slowly metabolize PAH. When oil with the PAH is removed, the PAH in the oysters decreases. 3) Dermal irritation when some hydrocarbons come into contact with the skin.

Inhalation

To assess inhalation exposure, extensive air monitoring was conducted along the coast line of Louisiana, Mississippi, Alabama and the panhandle of Florida. Air samples taken directly above the well site detected only low levels of BTEX. This provided further evidence that volatiles were extracted from the oil in the mile long trip from the bottom of the gulf to the surface. Several types of tests were conducted including direct read monitoring in vans that drove along the coastline, specific sampling for BTEX, and monitoring from particulates. The direct read monitoring detected background levels of BTEX associated with the use of petroleum products. Particulates were detected periodically following *in situ* burning of the oil offshore (a means to capture the oil) and the wind was blowing toward shore. The weathered oil, mousse and tar balls were also analyzed for BTEX. None were detected in these samples confirming the dissolution or volatilization of the BTEX compounds from the oil occurred.

Seafood Monitoring

Because of the amount of oil spilled, there was concern of seafood becoming contaminated with oil. Louisiana had a substantial seafood monitoring program prior to the oil spill and this was greatly enhanced to look for components of the oil. As soon as oil was observed in fishing areas, they were closed to fishing. To reopen, the areas must fulfill all aspects of the seafood monitoring program that includes three aspects:

1. No visible oil within the designated monitoring areas
2. Sensory analysis: The smelling of seafood has long been a method for detecting tainted seafood. Further, hydrocarbons have a distinctive smell that the human nose can detect at levels far below those which cause health effects. The sensory analysis requires special training of the "sniffers." There are several levels sniffers based on experience and training. The seafood monitoring program requires that seafood taken from a closed area be smelled at the location and then by a special panel of ten highly experienced sniffers.
3. Chemical analysis: if the seafood passes the sniff test, it is then chemically analyzed for aliphatic hydrocarbons and PAHs.

To date, the chemical analyses have not detected contamination in the seafood samples. As additional evidence, water testing in the areas also does not detect components of the oil or dispersants.

Human Health Surveillance

The Louisiana Department of Health and Hospitals has tracked reports of oil related illness through hospitals, emergency departments, urgent care facilities, clinics and the poison control centers. To date, there have been 361 reported complaints; of these, 277 are in workers and 84 in the general public,

The major symptoms are headache, nausea, dizziness, vomiting, weakness and fatigue and irritation associated with odors. Heat stress in workers (which presents with many of the same symptoms) was associated with 133 of these complaints.
The use of Tyvek suits as protective gear appears to be contributing to the heat stress.

The acute effects of the oil spill on human health have been minimal. Workers are most at risk. Few effects have been observed in residents along the coast. This can be attributed to the oil spill being located 50 miles from the coastline and the weathering of the oil that has removed many of the VOCs and the breakdown of other compounds of concern. Long term effects are being studied in workers and other issues may emerge. Dispersants remain a source of concern and fear for many. Even with testing and a body of literature showing short half lives, concern remains.

SECTION 4

SESSION 4

ENERGY

FOCUS: SCIENCE FOR SUSTAINABLE ENERGY SYSTEMS

SCIENCE FOR ENERGY TECHNOLOGY: DISCOVERING SILVER BULLETS

WILLIAM FULKERSON
Institute for a Secure and Sustainable Environment, University of Tennessee, Knoxville, USA

ABSTRACT

Achieving sustainable energy systems for the world is a massive challenge that will require much better energy technologies than are currently commercially available. Inventing these better technologies is underway in many countries. By sustainable is meant technologies that endure, that do no harm, and that leave no permanent negative change in the environment. The evolution toward this desirable state is illustrated, for example, by the efforts in the United States over the past decade to analyze basic research needs for achieving sustainable energy technologies and to pursue them systematically. Needless to say the objective has not been attained although substantial progress is being made. Silver bullets are rare. Much larger research investments, public and private are needed, and they must be sustained for decades. Can the political will be motivated? Sustainable energy will require not only better technology but also better policies to incentivize invention and to encourage deployment.

INTRODUCTION

This session of International Seminars on Planetary Emergency we call "Science for Sustainable Energy." All the authors are Energy PMP members. We devised this session to be in harmony with the theme of the Seminars this year, i.e., "The Role of Science for the 3rd Millennium." This talk is an introduction to the Session.

For me, "Science for Sustainable Energy" means basic and applied R&D that can lead to substantial improvement in energy technologies, possibly to transformative or even disruptive technologies. Sustainable means technology that is enduring—it lasts a long time (and is expandable and secure), also it does no harm and finally that it leaves no or very little change in the environment. This is the definition proposed by George Crabtree and John Sarrao in their paper "The Road to Sustainability" (Crabtree and Sarrao, 2009).[1] In fact, much of this talk is derived from the thinking and writing of George Crabtree.

There is one recurring conclusion from most of the comprehensive studies of the energy situation of the world, and that is that there is no silver bullet. There is no perfect technology that meets the sustainability criteria and is also cost-competitive (with the possible exception of some efficiency technologies). The only way to find a silver bullet is through research, development and demonstration (RD&D) applied comprehensively to all potentially important energy sources and services.

---

1 Crabtree, George and John Sarrao (2009) "The Road to Sustainability," *Physics World*, October 2009, p. 24-30.

So I call my introductory talk "Science for Energy Technology: Discovering Silver Bullets." I will outline some relevant recent energy history, and I apologize for focusing too much on the United States. I will give an example of a potential silver bullet that is sustainable and end by talking about the RD&D funding dilemma.

Recent Energy History

Since the second world war there has been an American faith expressed by the report commissioned by President Roosevelt called "Science the Endless Frontier" and written by Vannevar Bush. It was published in 1945. The faith is that no matter the problem in energy, in defense, in economics, in health, in agriculture, etc., science (and resulting advanced technology) can solve it, and, in particular, the government should be involved at least to encourage it. In the United States this faith derived, I believe, from the development of the nuclear bomb, and much later to man's landing on the moon. *[I wondered if this faith might be dampened by the Gulf oil blowout, but the session yesterday, put together by Carmen Difiglio of the Energy PMP and by Lorne Everett and Frank Parker of the Pollution PMP suggested that just the opposite should happen. The technologies of deep-water production are amazing.]*

Actually this American faith is in the methods of Galileo Galilei.

Of course, the bomb itself resulted from the work of a bunch of brilliant physicists who discovered fission before the War and projected its potential consequences from the bomb to the power reactor. Thus, science led to nuclear power for peaceful purposes as well as to the most destructive forces in human experience.

It was relatively simple to make a controlled fission reactor as Enrico Fermi, the Italian Navigator, demonstrated so brilliantly in Chicago in 1942. Fission power continues to evolve today as illustrated by the session yesterday on passive safety that Professor Zichichi asked Dick Wilson, Dick Garwin and me to organize which we did with the help of Bob Budnitz. You will hear more in this session from Dick Garwin about fast reactors around the world.

Of course, it has proved much more difficult to get peaceful power out of fusion, and Jef Ongena will summarize the progress.

The U.S. government support of the nuclear enterprise included fundamental research in materials, nuclear physics, chemistry, biology etc., and the creation of the National Laboratory System that continues to this day.

The Arab Oil Embargo of 1973-74 changed everything. In 1975 the Executive Branch of the U.S. government was reorganized to bring most energy research into a single agency, the Energy Research and Development Administration (ERDA), later to be transformed into the Department of Energy (DOE) and including the National Labs. Gone was the Atomic Energy Commission but added was the Nuclear Regulatory Commission. Since then, the U.S. government has supported work on all energy sources and services. This included not only basic research but also applied research including development and occasionally demonstration. Before the War most basic research was conducted in universities or in private sector labs such as Bell Labs. Increasingly after the Embargo the federal government supported most basic energy research, but it is done in national labs, in universities and in the private sector.

Through the seventies and eighties the problem was oil and energy security. It is still a problem.

But another major concern emerged that was articulated in 1962 by Rachel Carson in her book "Silent Spring"[2]. The concern was damage to the environment, and it led to the environmental movement that grows ever stronger today. Attention was brought to what man was doing to himself and the environment including pesticides, smog, acid rain, lose of species, and climate change, many of these related to energy. Science has contributed greatly to our understanding of these problems.

In 1987 the Bruntland Commission report[3]: "Our Common Future" made sustainable development the focus of the environmental movement. Today that includes the realization that we must move toward sustainable well-being as John Holdren, President Obama's Science Advisor, has put it[4], and not just sustainable development.

Where from there?

For the past decade or so the United States has been pursuing our faith that science can reinvent the energy systems of the world to be sustainable. It is summarized, for example, in 14 reports from DOE covering such topics as materials under extreme environments, catalysts for energy, electrical energy storage, geosciences facilitating 21st century energy systems, solid state lighting, superconductivity, solar energy, hydrogen and advanced nuclear energy systems.

See Table 1. All reports are on the web. They give a clear and logical way to think about finding silver bullets.

Table 1. U.S. DOE Reports on Science Needs for Energy Technologies (http://www.sc.doe.gov/bes/reports/list.html)

1.	Science for Energy Technology: Strengthening the Link between Basic Research and Industry
2.	New Science for a Secure and Sustainable Energy Future
3.	Directing Matter and Energy: Five Challenges for Science and the Imagination
4.	Basic Research Needs for Materials under Extreme Environments
5.	Basic Research Needs: Catalysis for Energy
6.	Basic Research Needs for Electrical Energy Storage
7.	Basic Research Needs for Geosciences: Facilitating 21st Century Energy Systems
8.	Basic Research Needs for Clean and Efficient Combustion of 21st Century Transportation Fuels
9.	Basic Research Needs for Advanced Nuclear Energy Systems
10.	Basic Research Needs for Solid-State Lighting
11.	Basic Research Needs for Superconductivity
12.	Basic Research Needs for Solar Energy Utilization
13.	Basic Research Needs for the Hydrogen Economy
14.	Basic Research Needs To Assure A Secure Energy Future

2 Rachel L. Carson, (1962) Silent Spring, Houghton Mifflin Co. New York.

3 "Our Common Future", Report of the World Commission on Environment and Development, "(1987).

4 Holdren, John P. (2008) *Science Mag.* , 319, p 424-434.

Japan is doing similar things as Akira Miyahara told us yesterday at the Energy PMP meeting. You will also hear from Xiliang Zhang later in this session about China's science in renewables and efficiency. No country is moving faster than China.

The tough problems identified in these reports are being worked by the continuing R&D programs in DOE and across the federal and state systems. These ongoing efforts were boosted by about $29B over several years by the 2009 American Recovery and Reinvestment Act, the economic stimulus package. As part of this stimulus three innovative management ideas were initiated. One is called Energy Frontier Research Centers (EFRC's) that gather together the right people from across the nation and the right facilities to focus research on specific problems ($100M/y and 46 awards in FY 09) e.g., examining the fate of CO_2 in various geologic environments. The second management initiative is called Energy Innovation HUBs modeled after Bell Labs that are being set up as multidisciplinary teams under one organization to look at specific opportunities such as:

- Fuels from sunlight,
- Modeling and simulation tools for advanced nuclear reactors, and
- Improving building systems design.

Finally, the Advanced Research Projects Agency-Energy (ARPA-E) is focused on transformational and translational research that tends to be high risk/high payoff. Some 37 grants of up to a few million dollars per year each were awarded from the first round of solicitations. Many involve private sector partnerships. ARPA-E is modeled after the Defense Advanced Research Projects Agency or DARPA.

Steven Chu, the Secretary of Energy and Nobel Laureate, has been very aggressive in pursuing these cross cutting R&D management innovations.

Finding a silver bullet, an example

Let's look at solar electricity and in particular at photovoltaics as a possible example of finding a silver bullet that is sustainable. I have said for the past 25 years or so that all we need to solve world's energy problems are a high efficiency, low cost PV cell and a good inexpensive battery with long life that can be deep discharged many times, and its critical components can be effectively recycled. Neither of these technologies is ready to fulfill this promise, but we are getting much closer.

For PV's there has been enormous "learning by doing" progress incorporating many technology improvements. The cost has come down from $70/peak W in 1978 to less than $3.5 today. The efficiencies have gone up from 6% in the 1970's to near 40% for multiple junction devices under solar concentration. Can we bring the cost down by another order of magnitude to about $0.4/peak watt where PV's can compete with fossil electricity and nuclear? No we can't, not without a breakthrough.

One step is to improve the efficiency to greater than 50% and many ideas are being tried including[5]:

[5] Report on the Basic Energy Sciences Workshop on Solar Energy Utilization, April 18-21, 2005, Lewis, Nathan S., and Crabtree, George, chair and co-chair respectively, Office of Science, U.S. Department of Energy, p 91.

- Optical spectrum shifting,
- Multiple excitation generation using microdots,
- Multiple energy level cells,
- Hot carrier cells,
- And trying many inorganic new photocell materials and organic PV's that are less expensive to manufacture if the efficiency can be improved dramatically.

Maybe this is a bridge too far, but PV sales keep booming and for some of the poorest parts of the world PV brings lighting and even parts of the electronic communications world.

Bill Barletta will mention some of these ideas in his talk about photon science.

Batteries have improved enormously as well. Automotive companies are predicting plug-in hybrid vehicles within the next few years and EV's soon after that. I don't think the batteries are good enough, but we will see.

+ Systems: Hymotion™ L5 PCM

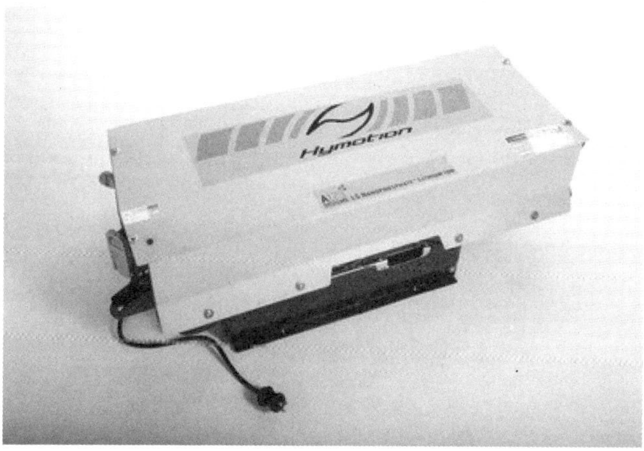

Our Hymotion L5 Plug-in Conversion Module turns a Prius into a plug-in hybrid capable of achieving 100+ miles per gallon. Plug in now and get:

Up to 60% reduction in fuel consumption and greenhouse gas emissions
Up to 100+ MPG for the first 30 - 40 miles of electrically assisted driving*
Automatically reverts to standard hybrid operation until next charge - driving range not limited by battery charge
Charges from standard household outlet

Certified Hymotion Installer Partners (Green CHIPs) are located around the country and are ready to convert your Prius!

One of the advances supported by DOE is the so-called A123 battery that depends on a number of innovations; prominent among them is the use of nanoscale deposits of lithium phosphate olivine on the cathode. This leads to higher energy densities. Energy density is a major stumbling block for PHEV and EV vehicles.

In fact, we find that nanoscience is important to both advanced batteries and revolutionary PV cells.

This is a picture from the A123 website showing a Hymotion battery pack that can be put in a new Prius to make it a PHEV. You can buy this to supplement the Ni metal hydride battery already in your car. The electricity assisted range is 30-40 miles and a projected 100 miles/gallon of gasoline energy use in hybrid mode while reducing CO_2 emissions substantially.

Of course, to make electrified vehicles and distributed PV power supplies more effective having a smart grid will really help, and Hisham Khatib will speak about that possibility.

Shorting funding for RD&D won't likely solve getting to a sustainable PV/battery system, or any other sustainable energy system. As I have mentioned U.S. funding for energy RD&D went up dramatically over the last year because of the economic stimulus package. A similar bump up occurred in a number of other countries, but will it be continued for several decades as needed? A recent IEA report[6] estimated that RD&D funding (public and private) worldwide is at least a factor of 3 too low for meeting the needs of accomplishing the IEA Blue Map Scenario designed to hold global warming to less than two degrees. This estimate, agrees qualitatively with a variety of other studies made over the past decade and a half e.g., Bob Schock et al.[7] for the U.S.

I am optimistic, however. I believe the nations of the world are beginning to pull together toward solutions for climate change, energy insecurity and energy poverty based on faith in science and technology.

Are there ways we can make RD&D expenditures more productive? The RD&D roadmaps for various clean energy technologies have been laid out. Can more, smart people around the Globe be brought to the party at less cost? It's an international problem, but can it be made an international opportunity and enterprise?

Bruce Stram will examine this topic in his talk.

---

6 "Global Gaps in Clean Energy Research, Development and Demonstration," IEA 2009, and also, Gaps (2010).

7 Shock, R.N., Fulkerson, W., Brown, M.L., San Martin, R.L., Greene, D.L., and Edmonds, J., (1999), " How much Is Energy Research and Development Worth as Insurance," Annual Review of Energy and Environment, 24: 487-512, http://arjournals.annualreviews.org/doi/pdf/10.1146/annurev.energy.24.1.487.

PHOTON SCIENCE, A DISRUPTIVE WILDCARD

WILLIAM A. BARLETTA[1]
U.S. Particle Accelerator School, Department of Physics
Massachusetts Institute of Technology, Cambridge, USA

ABSTRACT

Physics, chemistry and biology using synchrotron light sources and free electron lasers—now commonly called photon science—can be instrumental in providing the scientific basis for a sustainable energy future. As an example, it considers the use of sunlight to generate electricity and transportable fuels. Even in this limited arena this report makes no attempt to be comprehensive about the many ways in which photon science can be a disruptive agent in the research agenda for sustainable energy. Rather it is a call to action to use the several dozens of accelerator-based science facilities that are available worldwide.

INTRODUCTION: PHOTON SCIENCE FOR ENERGY RESEARCH

The late Professor Kai M.B. Siegbahn, Nobel laureate, President of the WFS from 1990 to 1998 and my friend and mentor, was a visionary proponent of the utility of the instruments of photon science—electron storage rings and free electron lasers—to advance fundamental science. Indeed for the theme of the 500th volume of Nuclear Instruments and Methods in Physics Research—the international journal of which he was the founding editor—Kai chose the development of new tools for photon science. In the editorial introduction to that issue Kai wrote:

> "...the previously harmful synchrotron radiation from accelerators has been found possible to "tame" in long time storage rings. With additional new concepts and further technical developments in X-ray optics these storage rings have been supplied with "wigglers" and "undulators" which greatly have enhanced the importance of synchrotron radiation as a tuneable X-ray light source for research in physics, chemistry and biology...Two new schemes related to the existing synchrotron radiation field have been under development. The first is the free electron laser (FEL). This scheme has already been realized and operates effectively in the infrared (IR) region. The second scheme is the so-called X-ray laser and is by now close to its realization. The X-ray laser is a more complex scheme, which has required several new concepts for its realization. The basic expected performance has already been tested successfully. The X-ray laser generates very short X-ray pulses in the femtosecond region at extremely high intensities, many orders of magnitude higher than can be obtained from the third generation of

[1] The author is the director of the United States Particle Accelerator School and co-editor and contributor to the book *Averting Disaster: Science for Peace in a Perilous Age*, a retrospective of the International Seminars on Nuclear War and Planetary Emergencies.

synchrotron radiation sources. There are good hopes that this new scheme will become a very powerful tool in physics, chemistry and biology."

With nearly seventy synchrotron light sources, i.e., electron storage rings, operating worldwide, with the remarkable first flash X-ray images of biological samples from the FLASH free electron laser at DESY and with the spectacular success of the Linac Coherent Light Source (LCLS) at SLAC—the world's first hard X-ray-free electron laser—the scientific vision that Kai expressed in 2002 has become an operational reality.

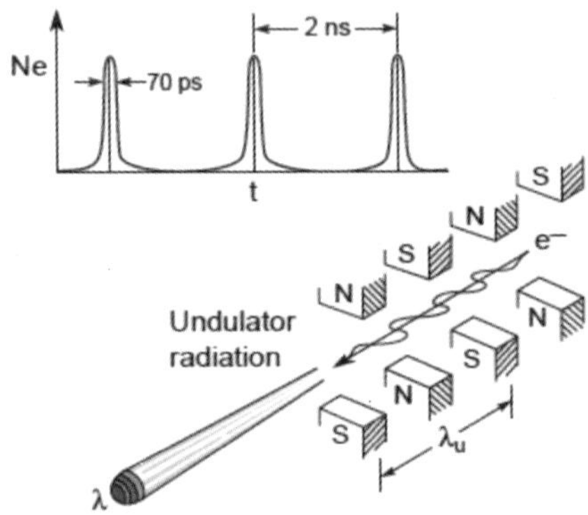

Fig. 1: Synchrotron radiation from an undulator.

While the storage ring light source produces incoherent synchrotron radiation from its stored electron beam at a level directly proportional to the circulating current, the free electron laser produces coherent self-amplification—gain—of the synchrotron radiation as the electron beam traverses a long periodic array of dipole magnets in an up-down configuration that is called an undulator. If the electron beam has sufficiently high energy, the result is a burst of X-rays that has a spectral brilliance—measured in photons/s/mm^2/mrad^2/0.1%bandwidth—that is billions of times greater than that generated by storage ring sources. Moreover, the X-ray bursts can be thousands of times shorter in duration than those produced with storage rings.

Over the past few decades of operation of three generations of storage ring sources, physicists have developed three broad categories of experimental techniques of photon science:

1. Microscopy
2. Diffraction and scattering
3. Various spectroscopies

These categories can be further sub-divided into time average and time-resolved techniques. Measurements can be energy-resolved, and angular correlations can be measured. The latter techniques, especially if combined, require very large numbers of photons (~10^{18}) to be incident on the sample.

One can visualize a photon science experiment in a schematic fashion. Photons of known and specified energy strike the sample to be studied. The experimenter may record elastically scattered or diffracted photons, photons emitted due to inelastic processes, photoelectrons emitted from the sample or ions ejected from the sample. Each measurement technique probes a different type of characteristic of the sample. The measurements with storage rings can reveal structural features and dynamics that happen on a time scale greater than tens of picoseconds. In contrast, ultra-short pulses from X-ray free electron lasers can reveal electron dynamics that take place on the femtosecond time scale. Moreover because of their extraordinary peak brilliance, the X-ray bursts can image tiny objects with diffracted photons on a time scale so fast (tens of femtoseconds) that the sample object cannot expand or explode due to the energy deposited by the far more numerous photons that are absorbed by and heat the object.

SUSTAINABLE ENERGY

The complementary use of storage ring sources and free electron lasers will enable finding economically attractive means of harnessing sunlight into a sustainable energy solution. What do we mean by sustainable?

Adopting the criteria of George Crabtree[2] in his keynote presentation to a study[3] conducted by the Basic Energy Science Advisory Committee (BESAC) of the U.S. Department of Energy (DOE), we might define sustainable energy in a pragmatic fashion as having the following characteristics:

1. It lasts a long time, at least from a contemporary perspective.
   Examples include oil in 1900, coal today, and nuclear fission
2. It does little-to-no harm to the environment.
   Examples: Energy conservation; nuclear electricity produces no CO_2
3. It leaves environment unchanged.
   An example would be energy from closed chemical cycles.

To Crabtree's three criteria one might add two others related to the scalability of the technology to have a global impact:

4. It is technologically scalable to global implementation.
   Does solar electricity meet this criterion?
   Does conversion of sunlight to fuel?
5. It is economically viable or practical on a global scale.
   Does this apply to solar electricity?
   Does it apply to converting sunlight to fuel?

---

[2] See G. Crabtree, "The Sustainable Energy Challenge," http://www.uic.edu/index.html/Chancellor/energylecture.shtml.

[3] This study held January 18-21, 2010 at the Rockville Hilton near Washington, D.C., is published at http://www.sc.doe.gov/bes/reports/list.html.

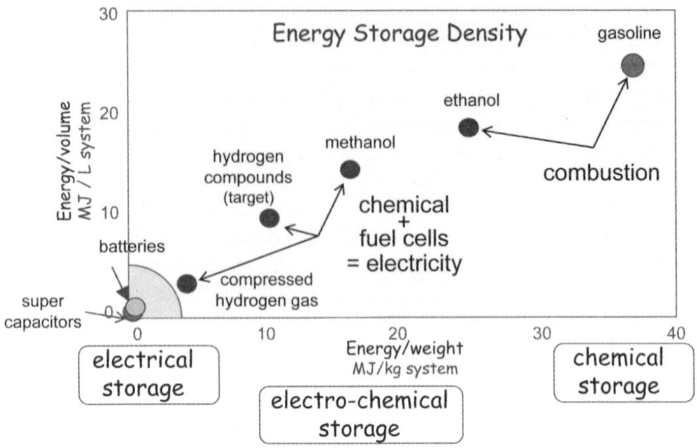

Fig. 2: Energy density by volume and by weight of gasoline compared with other forms of energy storage.

In other words, the certification of the sustainability of the full exploitation of solar- generated electricity for base load use requires the scalable, efficient and economic long distance transmission of electricity. Such efficient transmission from desert areas where sunlight is abundant with high availability to population centers thousands of kilometers away likely implies the availability of affordable high T_c superconductors and other DC transmission technologies that can be manufactured in an environmentally benign fashion from readily obtainable materials.

Perhaps even more challenging will be the development of scalable, adequate, and economic high energy density storage that also satisfies the sustainability criteria. The magnitude of the challenge of this latter requirement can be seen in Figure 2.[4] Especially with respect to transportable energy the challenge is great indeed. Gasoline is a remarkably high density chemical means of storing the energy from ancient sunlight.

Quantifying the number and magnitude of scientific and technological challenges is not meant to discourage research or to denigrate proponents. It is meant both to defuse exaggeration and to avoid raising false hope for quick solutions. However, with respect to setting both public policy and scientific research priorities, the whole truth must be articulated and understood. The citizens of all countries deserve no less.

THE PROMISE: SUNLIGHT TO COMPETITIVE, CONSUMABLE ENERGY

Realizing the full promise of solar energy will require the development and integration of a wide range of constituent technologies that span from the nanoscale at which photon science can reveal fundamental processes to millimeter-scale component cells to the

[4] From G. Crabtree, private communication, January, 2010.

multi-kilometer scale at which any technology must carpet the desert. Each scale will require specifically tailored interconnect, integration and control technologies. An example of these considerations for the U.S. DOE initiative to convert sunlight to fuel technology can be found in the presentation of Prof. N. Lewis to U.S. BESAC.[5]

Some simple arithmetic sets the scale of the problem. The sun delivers to Earth in one hour the amount of energy that humans consume globally in one year at present; that is, the sun supplies ~10^4 more energy to earth than humans consume. Consequently, an area of solar collectors ~200 km x 200 km could power the U.S. That does not sound too large for a large country with vast tracts of desert such as the U.S. Nonetheless, solar electricity supplies only 1.5% of the world's demand. The reasons include:

1. Higher cost relative to fossil fuels,
2. Non-economic energy storage to deal with <50% duty factor,
3. Inefficient energy transmission to deal with the non-locality of supply.

Point 1 is an obvious place where photon science can make a contribution. Widespread use of solar electricity for base-load application will require low cost, high-yield, and high-efficiency solar cells. Developing efficient photocells from plentiful materials rather than scarce materials such as tellurium will require major advances both in chemistry and in condensed matter physics.

For example, making more efficient, multi-crystalline, metal cluster solar cells requires controlling the metal impurities that degrade the efficiency of the cells. *X-ray emission (fluorescence) and absorption spectroscopy* using storage rings can yield the distribution, chemical state and electron–hole recombination activity of metal clusters before and after processing of the cells. The information learned can indicate ways to manipulate the size, distribution and chemical binding of impurities within clusters to reduce impact on efficiency. This process is sometimes called defect engineering.

An alternative photovoltaic technology, thin-film solar cells, has the promise to achieve conversion efficiencies of ~20% in next generation devices if researchers can acquire a sufficiently deep understanding of energies associated with electronic states, band gaps, excitation behavior and element- specific electronic wave functions. In this case one of the relevant approaches of photon science is to use undulator radiation plus a high-transmission x-ray spectrograph to acquire *resonant inelastic soft X-ray scattering* (RIXS) data. The greater the time-average spectral brilliance of a narrow bandwidth source, the more rapidly the data can be collected.

Assuming the science is positive, one returns to the practicalities of the 200 x 200 km^2 tract in the U.S. Certainly one can find huge tracts of suitable land, but 40,000 km^2 of solar collectors will require rapid manufacture and installation that is dirt-cheap and can be installed much faster than roads. To be quantitative, a cost goal of 5M$/$km^2$ implies a cost of ~$0.0002/gm of material. Installing ~10 km^2/day worldwide would imply a 30-year placement time. And if the cost and installation rate are acceptable, what about endangered species in this vast wilderness?

---

[5] Nate Lewis, "Joint Center for Artificial Photosynthesis 'Fuels from Sunlight'—Energy Innovation Hub," http://www.er.doe.gov/bes/besac/Meetings.html#0925.

Fig. 3: The Nevada Desert.

Once the photovoltaics are installed, how do we get power to the population?

Electricity could be transmitted at efficiencies much larger than in present practice and over much longer distances via high temperature superconducting transmission lines when those become practical and economically viable. Presently what are the barriers to commercial utility high temperature superconductors? Or even of BSSCO and YBCO?

1. The critical temperature is too low (less that 77 K).
2. The piece lengths that can be purchased are too short; production needs extrapolation to >100 km lengths.
3. The processing to form the superconductor is too cumbersome.
4. The materials are brittle and must be mechanically protected from the weather.

Fundamental scientific understanding of high T_c superconductors is essential to design new superconducting cable with electrical and mechanical properties suitable for long-distance electrical transmission. At this most fundamental level there is considerable hope that *time-resolved photoemission spectroscopy and pump-probe experiments with soft X-rays* will permit precision study of the dynamics of the superconducting phase transition. For example, one can measure the time evolution of the Fermi surface of the materials at the phase transition using time-resolved photoelectron spectroscopy.[6] The new free electron lasers at DESY, Stanford, Spring-8 in Japan and Elettra in Italy have great promise to provide incisive experimental insights into this hot scientific topic.

[6] The spectrographs used are a legacy of Kai Siegbahn's pioneering work.

But sun does not shine on any place for 24 hours every day, all year?

In his talk to BESAC Professor Lewis phrased the issue in this way: "He who does not store will snore after four." Some technology must provide energy storage for intermittent energy sources. Batteries are one possibility discussed. To be practical on the large scale they must provide far more energy/kg, recharge faster and survive far more duty cycles without degradation that present lithium ion batteries. (See Figure 2 again.) Again photon science may provide the scientific key, this time via *small angle X-ray scattering (SAXS)* that can measure rapidly changing electrode morphology on the nanoscale that is critical to Li^+ transport.

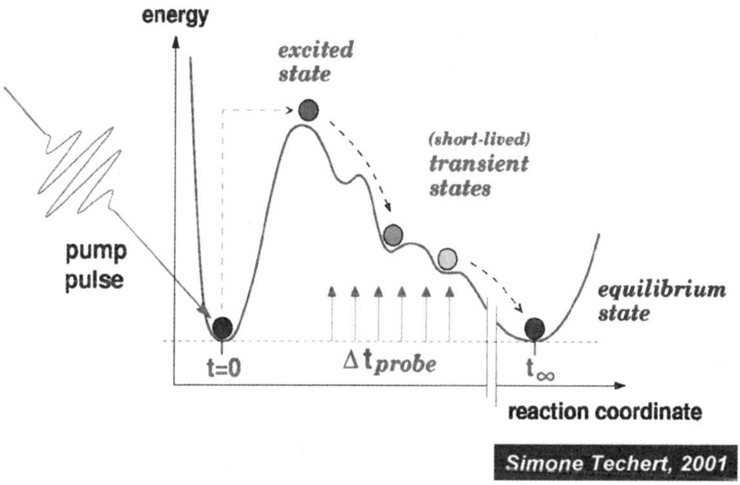

Fig. 4.

The promise and what to do now

Whether one is considering generating electricity from sunlight or making transportable fuels, a great scientific and technological endeavor is needed. The scientific pole of that endeavor must integrate bulk condensed matter physics, surface science and electro-chemistry with the photon science techniques of X-ray spectroscopy and high throughput experimentation. Much of this research can be done with electron storage rings. With respect to the relevant dynamics on femtosecond time scales, free electron lasers will enable X-ray movies of charge-transfer and of the breaking and formation of bonds with chemical and atomic specificity typically by exploiting pump-probe techniques[7].

The scientific road will be long, and many of the studies will be difficult and time-consuming. New instruments and experimental modalities will likely need to be invented. As the research proceeds there are crucial actions to take now:

[7] The illustration is from a presentation by Simone Techert in 2001.

1. Conserve energy;
2. Extend the lifetime and efficiency of installed energy generators.
3. Immediately plan means of adapting to climate change.
4. Extend the science basis worldwide for efficient, cost-effective, non-fossil energy sources.

Accelerators for photon science are a valuable way (though certainly not the only way) to accomplish item 4. A wonderful aspect of photon science is that it is not restricted to a small group of industrialized nations. Storage ring sources in Brazil, Thailand and soon in Jordan show that the scientific enterprise can find a productive home in developing nations. Novel approaches to building compact, relatively low-cost light sources abound. The benefits to developing nations and indeed to the entire world are many:

1. Create a world-class interdisciplinary research laboratory;
2. Promote basic and applied research & technology;
3. Address regional biomedical & environmental issues/concerns;
4. Provide an environment for collaborations and individual development;
5. Train graduate students who will no longer have to go abroad;
6. Attract scientists working abroad to return (*reversing the brain drain*);
7. Promote international scientific collaborations; and
8. Promote development of high-tech industry (*capacity building*).

Scientific creativity knows no borders North, South, East or West. In the end the human intellect is the ultimate disruptive wildcard, especially when the most penetrating scientific tools are at hand.

ACKNOWLEDGEMENT

I appreciate the opportunity provided by Professor Antonino Zichichi to present these remarks at the International Seminars in 2010 devoted to "Science for the 21st century." Many colleagues at the Advanced Light Source in Berkeley and at Sincrotrone Trieste provided valuable insights. At Sincrotrone Trieste particular thanks are due to Professor Fulvio Parmigiani for extensive discussions and to the Laboratory Director Alfonso Franciosi for his hospitality on many occasions.

SCIENCE ADVANCES RELATIVE TO ENERGY EFFICIENCY AND RENEWABLE ENERGY: SOME OBSERVATIONS FROM CHINA

XILIANG ZHANG
Institute of Energy, Environment & Economy, Tsinghua University
Beijing, P.R. China

ENERGY EFFICIENCY

The energy consumption per unit gross domestic product (GDP) in China decreased from 1.23 tons of coal equivalent (tce) per 10,000 Yuan in 2004 to 1.11 tce per 10,000 Yuan in 2008 while the energy consumption elasticity declined from 1.6 in 2004 to 0.41 in 2008 (Figures 1 and 2), leading to approximately 300 million tce of energy saving during 2005-2008. In 2008, as a sector of the economy industry consumed approximately 72% of China's energy whereas contributed 43% of China's GDP. Increasing the energy efficiency of the industry sector has been listed as a top priority in China's national energy programs. An empirical study conducted by Tsinghua University showed that more than 70% of the energy savings achieved in China during 2005-2008 could be attributed to the deployment of energy efficiency technologies in different industries (Figure 2), indicating the significance of science and technology advances. The study also demonstrated the declines in energy consumption per unit product, a major indicator of the physical energy use efficiency of an industry, had taken place in the major energy intensive industries of China since 2000 (Figure 3), giving further evidences of the contribution of science and technology changes in China's energy efficiency improvement.

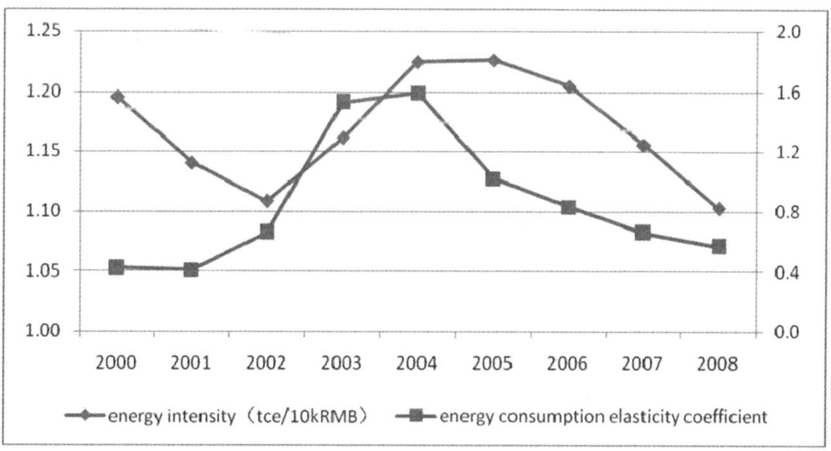

Fig. 1: Energy Intensity and Elasticity of Economy in China.

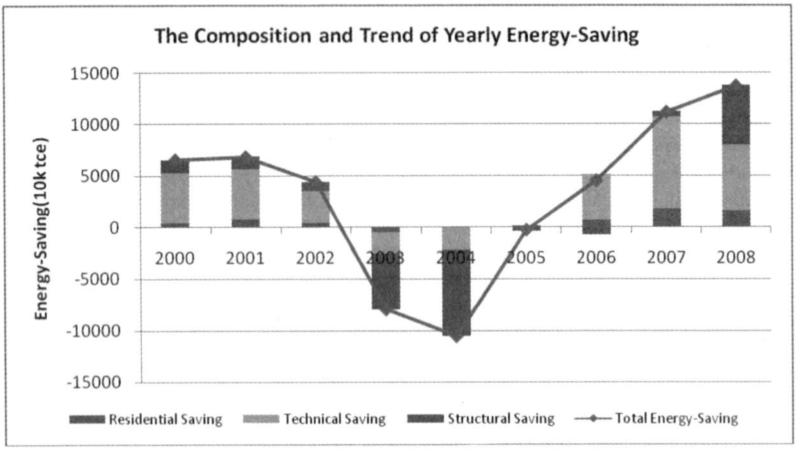

Fig. 2: Contributors to Energy Savings in China.

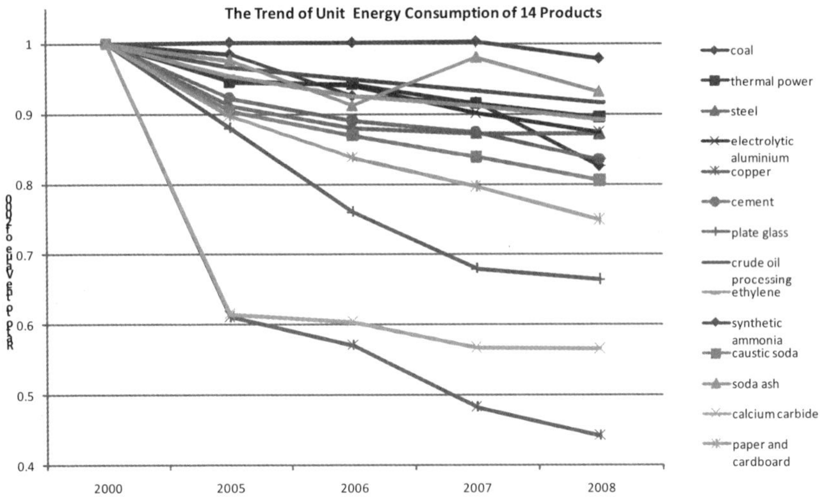

Fig. 3: Indices of Energy Consumption per Unit Product of the 14 Products.

Currently coal accounts for approximately 70% of China's primary energy supply. Approximately 65% of the coal is used for electricity generation in China. The coal consumption per kWh electricity generation decreased from 392 tce in 2003 to 349 tce in 2008 and the efficiency of coal-fired power plants in China has already surpassed that in the United States, indicating a substantial increase of power generation efficiency happened in China (Figure 4). China's improvement in electricity generation is largely attributed to the deployment of the high efficient super critical and ultra super critical coal generating technologies. Since 2005 the Government has requested all the new installations of coal-fired power plants should use either super critical or ultra super critical generating technologies, the efficiency of which is higher than that of sub critical generating technologies which dominated the new installations in the early 2000s (Table 1).

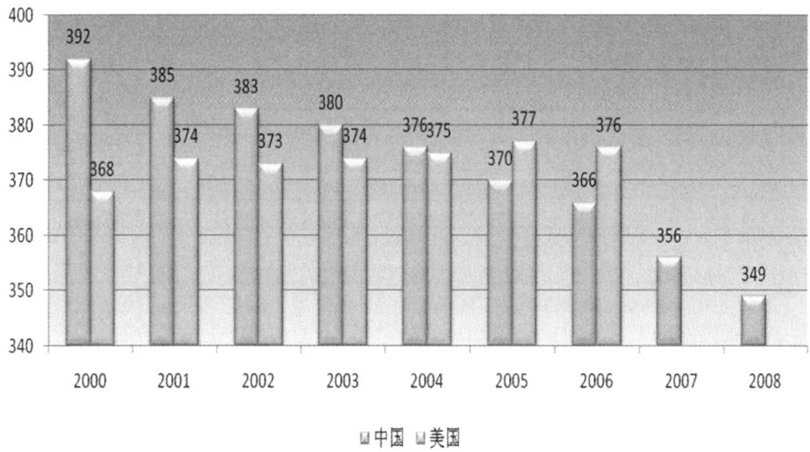

Fig. 4: Improvements in Efficiency of Power Generation in China.

Table 1. Technical Characterizes of Coal-fired Power Generating Technologies.

	Main steam pressure, MPa	Main steam temperature, °C	Reheat steam temperature, °C	Efficiency, %, net, HHV basis
Subcritical	<22.1	Up to 565	Up to 565	33-39
Supercritical	22.1-25	540-580	540-580	38-42
Ultrasupercritical	>25	>580	>580	>42

Please note that both super critical and ultra super critical generating technologies do not origin in China but are transfer from USA, Japan, and EU into China (Table 2). The development of super critical and ultra super critical generating technologies largely benefits from advances in material science and thermal physics in the world for many years. China has made a great progress in terms of manufacturing and reducing the costs of these technologies. The Chinese Government has made great R&D efforts in localization of super critical and ultra super critical generating technologies (Figure 5).

The cost of super critical and ultra super critical generating technologies made in China is approximately half of that in developed countries, making it possible for China to deploy super critical and ultra super critical generating technologies on a large scale. Further, China has already exported these technologies to India and other developing countries.

Table 2. Sources of SC/USC Technologies and Technology Transfer Pathways.

		Production capacity (MW)	Technology source	Transfer approach
Shanghai Electric Group	Boiler	4500	Alstom	Licensing
	Turbine	36000	Siemens	Joint-venture
	Generator		Siemens	Joint-venture
Harbin Electric Corporation	Boiler	53000	SC: Mitsui-Babcock; USC: Mitsuibishi	Licensing
	Turbine	12000	600 MW: Mitsuibish; 1000MW: Toshiba	Licensing
	Generator		Toshiba	Licensing
Dongfang Electric Corporation	Boiler	25000	BHK	Joint-venture
	Turbine	20560	Hitachi	License

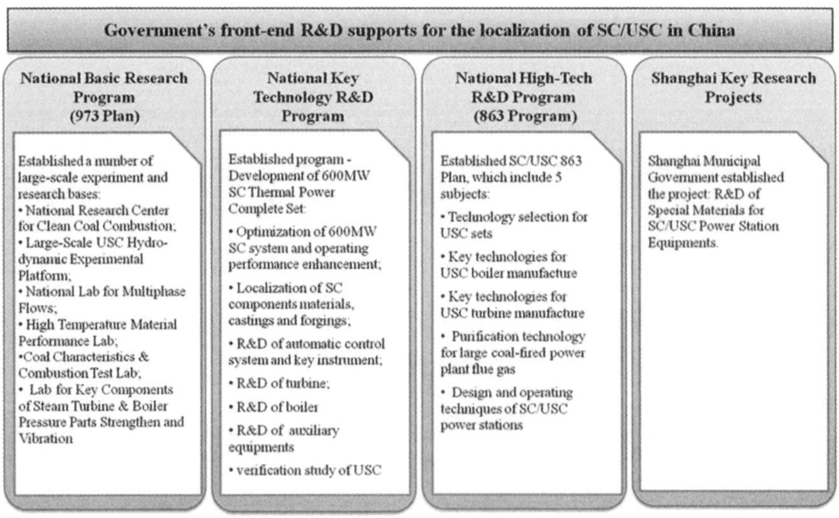

Fig. 5: R&D supports from the Chinese Government.

RENEWABLE ENERGY

China's *Renewable Energy Law* was passed by the Congress on February 28, 2005, and took effect on January 1, 2006. The Law recognizes the strategic role of renewable energies in optimizing China's energy supply mix, mitigating environmental pollution, improving energy supply security, and promoting rural social development. More importantly, the *Law* largely shapes an integrated renewable energy policy framework by providing a set of directives encouraging renewable energies, including national renewable energy targets, a feed-in tariff, a special fiscal fund, tax relief, and public R&D support as well as education and training. Thanks for the implementation of the *Law* and the science and technology advances in the field of wind and solar energy, China has made substantial progress in the development and deployment of wind and solar photovoltaic generation technologies (Figures 6 and 7).

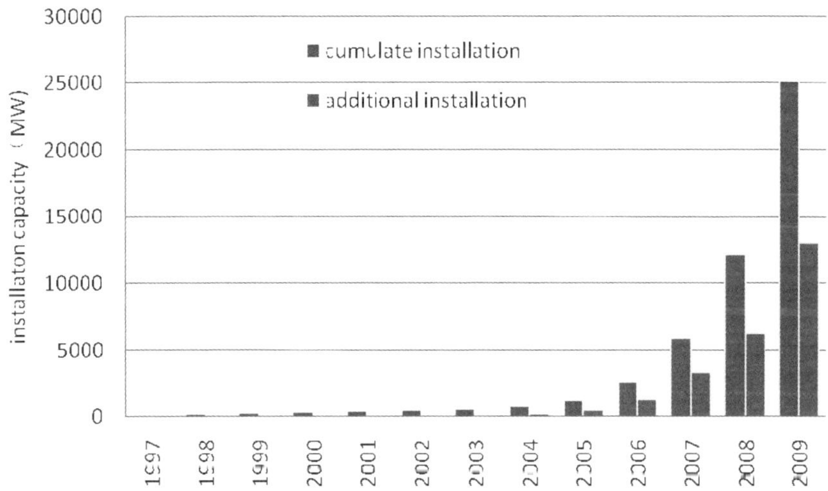

Fig. 6: Installations of Wind Turbines in China.

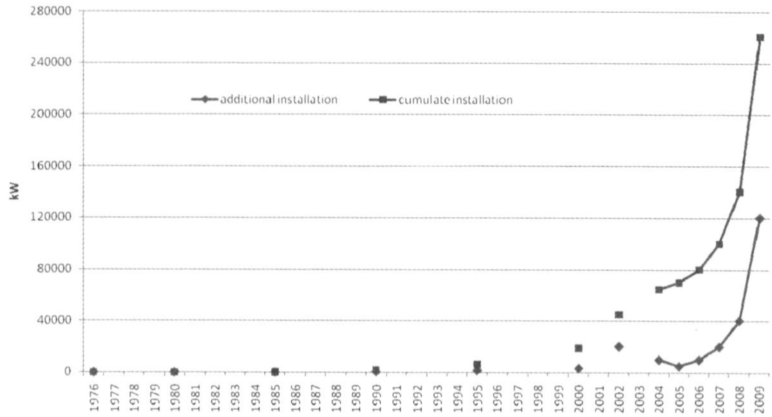

Fig. 7: Installations of Solar PV Systems in China.

China's wind energy resource is abundant. The technical development potential of onshore wind resources is as much as 297 GW at a height of 10 m, with annual electricity generation being more than 594 TWh, which is more than 18% of the total electricity production of China in 2007. Since 2004, China's wind power industry has been growing at an unprecedentedly high rate (Figure 6). The cumulative installed capacity of wind turbines went from 550 MW in 2003 to 25 GW in 2009. Most of Chinas' wind turbine technologies are transferred from EU countries like Germany and the Netherlands. The Chinese Government and companies have made great R& D efforts in the localization of wind turbines. Further, China has already had the capacity of manufacturing 3 MW wind turbines. The cost of wind turbines declined from 7000yuan/kW in 2005 to 4000 Yuan/kW in 2010. Such a wind development was also largely attributed to advances in material science, electronics, and aerodynamics in the world.

The development and deployment of solar PV systems followed a similar path to that of wind turbines. First, the installations of solar PV systems have been growing at an unprecedentedly high rate (Figure 7) since 2005 in China. Most of Chinas' solar PV technologies are transferred from EU countries like Germany. The Chinese Government and companies, however, have made great R&D efforts in the localization of solar PV technologies. China has become a major manufacturer of solar PV technologies in the world (Table 3). In 2009, nine of the top 15 solar cell manufacturers in the world were of China Mainland. Furthermore, the cost of solar PV panels declined substantially since 2005 in China. There are also evidences showing that China's development and deployment of solar PV technologies were closed related to advances in material science, electronics, and photovoltaic science in the world.

Table 3. Sources of SC/USC Technologies and Technology Transfer Pathways.

Rank	2008	产量(MW)	Rank	2009	产量(MW)
1	Q-cell (DE)	570.4	1	First Solar (US,DE,ML)	1100.0
2	First Solar (US,DE,ML)	504.0	2	Suntech Power (CH) 尚德	704.0
3	Suntech Power (CH) 尚德	497.5	3	Sharp (JP)	595.0
4	Sharp (JP)	473.0	4	Q-cell (DE)	586.0
5	Kyocera (JP)	290.0	5	Baoding Yingli (CH) 英利	525.0
6	Baoding Yingli (CH) 英利	281.5	6	JA Solar (CH) 晶澳	520.0
7	JA Solar (CH) 晶澳	277.0	7	Kyocera 日本京瓷	400.0
8	Motech (TW)	272.0	8	Trina Solar (CH) 天合	399.0
9	Sun Power (PH)	236.9	9	SunPower 美国	397.0
10	Solar World (US,DE)	221.0	10	Gintech 台湾昱晶	368.0
11	Sanyo (JP)	210.0	11	Canadian Solar (CH) 阿特斯	300.0
12	Trina Solar (CH) 天合	209.0	12	Solarfun (CH) 林洋	220.0
13	Solarfun (CH) 林洋	189.0	13	China Sunergy (CH) 中电	194.0
14	Mitsubishi Electric (JP,+HEL)	188.0	14	Changzhou Yijing (CH) 亿晶	135.0
15	GinTech (TW)台湾昱晶	180.0	15	Ningbo Solar (CH) 宁波	140.0
others		3300.7	others		2757.0
Total	**2008**	**7900**	**Total**	**2009**	**9340**

CONCLUDING REMARKS

China's energy efficiency improvement and renewable energy development has been closely associated with science advances in different disciplines in the world. Following concluding remarks regarding science advances relative to energy efficiency and renewable energy in China might be made:

1. Energy efficiency improvement and renewable energy development largely come from science advance in different disciplines;
2. China's energy efficiency improvement and renewable energy development benefit from science advance in the world;
3. China is also a contributor particularly for translating science advance into applications; and
4. Public policy and international cooperation could play an important role in science advance and technology innovation.

FUSION FOR THE THIRD MILLENNIUM

JEF ONGENA
Plasmaphysics Laboratory, Ecole Royale Militaire
Brussels, Belgium

WILLIAM BARLETTA
U.S. Particle Accelerator School, Department of Physics
Massachusetts Institute of Technology, Cambridge, USA

AKIRA MIYAHARA
Formerly at National Institute for Fusion Science
Tokyo, Japan

FRANÇOIS WAELBROECK
Formerly at Institut für Plasmaphysik, Forschungszentrum Jülich,
Jülich, Germany

Perhaps the greatest challenge facing the world is to develop the technology of affordable, clean and sustainable energy production. Currently more than 88% of the primary energy production in the world is based on fossil fuels. The disadvantages are well known: risk of irreversible changes to the climate system, limited reserves, dependency in supply. The successful development of fusion energy holds the promise of a means for producing large-scale base-load electricity, with extensive, well-distributed resources that is reasonably priced, safe, sustainable, carbon-free and with limited radioactive waste.

PRACTICAL REALISATIONS: MAGNETIC AND INERTIAL FUSION

Magnetic fusion

The first approach to realise fusion on earth for peaceful applications is by magnetic confinement. This line of research into controlled fusion is being funded in a large number of countries around the globe. Strong magnetic fields are used to keep the hot particles away from the walls of the confinement device. This is possible because of the property of charged particles to follow a helical path around magnetic field lines caused by the Lorentz force. In this way they literally 'stick' to the magnetic field lines, and possible movements perpendicular to the field are thereby highly restricted.

The most obvious way to realise this approach is a purely cylindrical configuration, which, however, suffers from plasma losses at both ends. These losses can be avoided by bending the magnetic field lines onto themselves, creating a doughnut-shaped or toroidal magnetic field configuration. A toroidal magnetic field, however, is not homogeneous. Its curvature destroys most of the nice confinement properties of a purely cylindrical configuration. The magnetic field is largest near the axis of the torus, and decreases towards the outboard side. As a result, ions and electrons move in opposite vertical directions, giving rise to an electric field which, when combined with the toroidal magnetic field, expels all particles to the outside, irrespective of their charge. Without special measures, it is thus impossible to confine a toroidal plasma in a stable way. However, if the toroidal magnetic field can be given an extra (poloidal) twist, vertical drifts can be cancelled and a suitable trap can be made for fusion plasmas. This extra twist can be produced by means of electrical currents flowing inside or outside of the plasma ring. The tokamak concept relies on currents produced inside the plasma, whereas devices from the stellarator family (heliotron, torsatron, heliac) rely on currents external to the plasma.

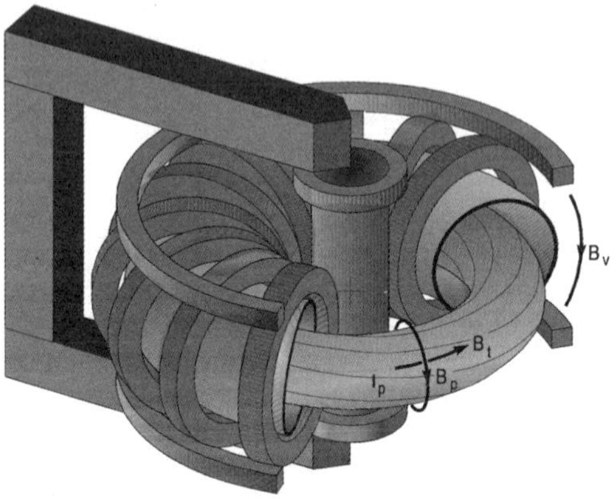

Fig. 1: Schematic representation of a tokamak. The main components are: (i) transformer (yokes and primary windings around the central column) to induce the plasma current I_p; (ii) set of coils around the plasma vessel for the toroidal field B_t; (iii) planar coils for the vertical field B_v. The poloidal field B_p associated with the plasma current adds to the toroidal field B_t resulting in a helical field, needed for stability. The vertical field B_v is needed to counteract the hoop force inherent to a current carrying ring.

In a tokamak (Figure 1), a set of coils placed around the doughnut-shaped plasma chamber produces the main toroidal magnetic field. The conducting plasma ring itself serves as the sole secondary winding of an enormous transformer.

A current pulse in the primary winding induces a large current in the secondary, i.e., in the plasma ring itself. This induced plasma current ranges from a few kA for small tokamaks up to 7MA in JET, currently the largest tokamak in the world (see below and Table I). This current generates a poloidal magnetic field and the sum of this poloidal field with the main toroidal field results in a helical magnetic field. The tokamak concept was proposed in 1951 by Andrei Sacharov and Igor Tamm of the Kurchatov Institute in Moscow[1], and the subsequent worldwide success of this device started with the announcement by Lev Artsimovich at the IAEA Novosibirsk Fusion Conference in 1968[2], of much higher temperatures compared to what was reached in other magnetic configurations at that time.

The tokamak is a pulsed device, since the transformer that induces the plasma current needs a steadily increasing current. For practical applications, continuous operation of such a device would clearly be a great advantage and it is an object of present research. To obtain continuous operation of a fusion device one has to avoid the need for the (pulsed) plasma current. Devices with a helical confinement geometry exploit this idea: currents outside the plasma, in a set of specially shaped coils generate the necessary helical magnetic field configuration. The largest helical devices are the stellarator Wendelstein 7-X (W7-X)[3], currently under construction at the Max Planck Institute for Plasmaphysik in Greifswald, Germany and the Large Helical Device (LHD), a heliotron/torsatron type device, located at the National Institute for Fusion Science in Toki-city, close to Nagoya, Japan.[4] The coils in both systems are superconducting. The magnet system in W7-X consists of 50 non-planar and 20 planar superconducting magnetic coils (Figure 2), delivering a field up to 3T on axis. The diameter of this toroidal device is about 16 m. First operation is foreseen for 2014. The coil system in LHD consists of two helical windings and six poloidal coils for the vertical magnetic field (Figures 3a and b), producing a magnetic field of up to 3T on axis. The diameter of the device is about 10m. LHD is in operation with full superconducting coils since 1998 and has obtained a wealth of very interesting results. Excellent plasma parameters have been obtained[5] and the longest plasma operation in the world (pulses up to 30 min) with relevant plasma parameters has been demonstrated on this device already in 2005.[6]

Several methods are used to bring the plasma in magnetic confinement devices to the required temperatures.

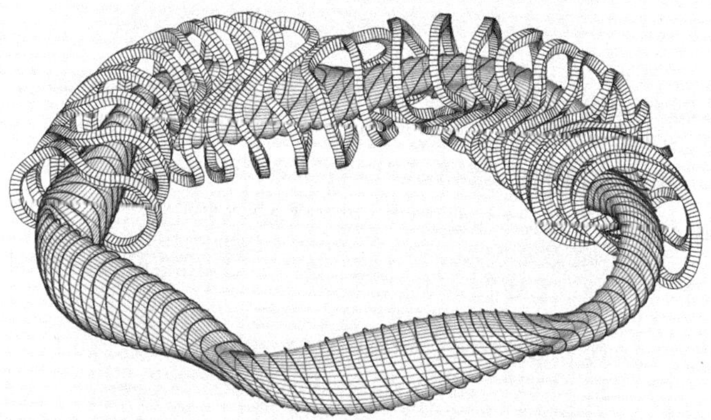

Fig. 2: Illustration of the complex coil system needed in a stellarator. The twisted plasma shape is also shown.

In the case of the tokamak a first method consists in making use of the ohmic heating caused by the plasma current needed anyhow for stability as explained above, but it is limited in possibilities. Indeed, the plasma resistance decreases with increasing plasma temperature, and at a temperature of about 1 keV (appr. 10 million of degrees), the efficiency of ohmic heating becomes too low[7] to be useful for further heating. For a stellarator type device ohmic heating is evidently not available, as explained above. Additional heating methods are therefore a necessity to reach the temperatures required for fusion.

A first additional heating method consists of injecting energetic beams of neutral hydrogen or its isotopes into the plasma. Neutral beam injection is possible because neutral particles can penetrate the magnetic fields needed to confine the plasma. To this end, ions are created in a plasma source, accelerated with voltages up to 150 kV and sent through a cloud of neutral gas.

Fig. 3a: The coil system in LHD: two twisted coils for the helical magnetic field and six planar coils for the vertical field.

Fig. 3b: Illustration of the plasma shape in LHD.

The accelerated ions 'steal' electrons from atoms in this neutral gas cloud, and become energetic neutral particles which enter the hot tokamak plasma ring without impediment from the confining magnetic fields of the fusion device. Once in the hot plasma, the neutral particles are almost immediately ionised again and deposit their energy via collisions with the rest of the plasma particles, thus raising the overall temperature. Powers of up to several MW per neutral injector can be delivered in this way.

Fig. 4: View of the Joint European Torus, the largest tokamak to date, located in Culham, close to Oxford, in the United Kingdom.

Yet another heating method injects electromagnetic waves into the plasma. The electromagnetic energy is delivered to the plasma by antennas or wave guides at the plasma edge. The principle underlying this method is similar to that of a microwave oven: the energy from the waves is most easily absorbed if the frequency used is equal to a 'natural' frequency of the particles to be heated. The cyclotron frequency, with which the charged plasma particles gyrate around the magnetic field lines, is most often used in fusion plasmas. If the waves have the right wavelength, their energy can be transferred into certain particles, which then transfer the energy through collisions with others. Thus, two variants of such heating systems exist: Ion Cyclotron Resonance Heating (ICRH) and Electron Cyclotron Resonance Heating (ECRH), depending on whether ions or electrons are heated. Ion cyclotron frequencies are in the MHz range (20 MHz and upwards), while electron cyclotron frequencies are approximately a 1000 times higher (up to 200GHz), due to the smaller mass of the electrons. Heating powers for high frequency systems range from 100 kW to several tens of MW.

The performance of fusion plasmas can be characterised by the product of "plasma density × temperature × energy confinement time", the so-called 'fusion triple product' ($n_i T_i \tau_E$). The energy confinement time τ_E is the characteristic time during which the energy remains confined in the plasma, in spite of losses due to the thermal conductivity of the plasma, particle transport and radiation. This confinement time is a measure of the thermal insulation provided by the magnetic field: the higher this value, the better the confinement of the plasma.

Another performance indicator is the power amplification factor Q, defined as the ratio of the power produced by fusion reactions to the total heating power. Two important milestones for the value of Q are customarily used in fusion research.

The first, breakeven, is reached when the heating power is equal to the power produced from fusion reactions, corresponding to a Q value of one. The second, ignition, is reached when the additional heating systems can be switched off and the heat of the fusion reactions alone is sufficient to maintain the high temperatures needed for fusion. This corresponds to an infinite value for Q. However, ignition is not needed for an economic fusion reactor: a commercial fusion reactor with a capacity of 3GW corresponds to Q ~ 50. The remaining small amount of additional heating would provide an additional means for control.

The fantastic progress obtained in fusion research in the past three decades is clearly visible from the successively obtained values for the fusion product. Compared to the first fusion experiments, this value has increased by a factor of more than 10 000! The results obtained on JET (Joint European Torus, Figure 4), the largest tokamak in the world, have contributed significantly to this progress, and to date, values very close to breakeven have been obtained in D-T experiments on this device. Already in 1989, temperatures of more than 250 million degrees were reached on JET for more than a second. Another milestone was reached in 1991 with the first production of large amounts of energy from controlled fusion reactions, as tritium was used for the first time as fuel in a tokamak.[8] These experiments, obtained with a mixture of 90% D and 10%T, generated fusion powers in the megawatt range for nearly 2 seconds, with a maximum of about 1.7 million watts (1.7 MW), corresponding to a Q value of about 0.15.

Further successes were obtained in early 1994 in the American tokamak TFTR (Tokamak Fusion Test Reactor) at the Plasma Physics Laboratory of Princeton University. (Unfortunately, this very interesting tokamak—and the only one outside Europe able to run the D-T reaction—was dismantled in 1997). In plasmas consisting of a mixture of 50% deuterium and 50% tritium, multi-megawatt level fusion powers were generated for about one second, with a maximum of 6.3 MW. Plasma temperatures in excess of 300 million degrees were reached in the plasma centre, 20 to 30 times hotter than in the centre of the sun! In November 1994, fusion powers of more than 10 MW were generated, corresponding to Q values of about 0.27.[9]

Other record values were reached in 1996 on the Japanese Tokamak JT-60U (Japan Torus 60 Upgrade), of the Naka Fusion Research Establishment, located 150 km north of Tokyo, a division of the Japanese Atomic Energy Research Institute (JAERI). This machine demonstrated temperatures in excess of 520 million degrees and a record value for the fusion triple product in pure deuterium plasmas. If the same conditions had been realised in deuterium-tritium plasmas, it would have resulted in a Q value of about 1.25, i.e., conditions better than break even.[10] Currently an upgrade of this device is planned as part of the ITER Broader Approach, as explained below.

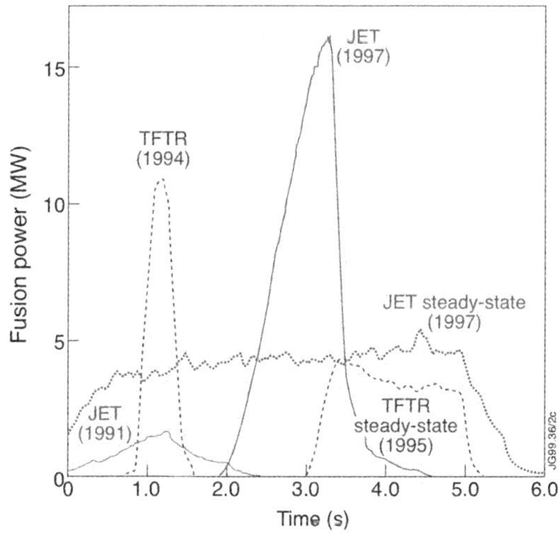

Fig. 5: Temporal evolution of the fusion power produced in high performance D-T plasmas in JET and TFTR. 16MW of fusion power from D-T plasmas was demonstrated in JET already in 1997.

The most impressive results in fusion research up to now were obtained in the JET, in October and November 1997. Experiments in 50% D - 50% T plasmas resulted in over 16MW of fusion power during about 1 second, with Q values in excess of 0.9.[11] These are the highest fusion powers and Q values ever reached, thereby effectively resulting in the first demonstration of breakeven in reactor grade D-T fusion plasmas. A quasi steady-state generation of fusion power has also been demonstrated: over 4MW of fusion power were produced for time intervals of more than 5 seconds, a duration only limited by the actual technical constraints of JET. With these results JET has effectively proven for what it was built: the scientific feasibility of magnetic fusion, with the practical demonstration of pulses very close to breakeven. A summary of the different high performance D-T pulses obtained on JET and TFTR, as discussed above, is presented in Figure 5.

Table 1: Main parameters of large tokamaks around the world: DIII-D (General Atomics, San Diego, USA), JT-60U (Japan Torus 60 Upgrade, Naka Fusion Institute, Ibaraki Prefecture, Japan), TFTR (Tokamak Fusion Test Reactor, Princeton University, Princeton, USA), JET (Joint European Torus, Culham, UK), ITER (International Thermonuclear Experimental Reactor, Cadarache, France). Except for TFTR (that was unfortunately prematurely closed down in 1997) all tokamaks in this list are in operation, being upgraded or under construction.

Tokamak name	**DIII-D**	**JT-60U**	**TFTR**	**JET**	**ITER**
Land / Organisation	USA	Japan	USA	UK/EURATOM	International
Plasma shape	elliptical (D)	elliptical(D)	circular	elliptical (D)	elliptical (D)
Minor radius a (m)	0.67(hor) 1.74(vert)	1.1(hor) 1.4(vert)	0.96	1.25(hor) 2.1(vert)	2.0 (hor) 3.9(vert)
Major Radius R (m)	1.67	3.4	3.1	2.96	6.2
Toroidal Magnetic field B_t (T)	2.2	4.2	6.0	4.0	5.2
Plasma Current I_p (MA)	3.0	5.0	3.0	7.0	15(17)
Pulse length (s)	10	10	7	60	400 - 3000
Power Amplification Q (in D-T discharges)	N.A.	N.A.	0.25	0.7	10
Injection of neutral particle beams (MW)	20	40	40	24	73 in total
Injection of electromagnetic waves (MW)	8	15	12	42	

Magnetic fusion research is now arrived at the point where large amounts of fusion energy (16MW in JET pulses of a few seconds) can be produced in a controlled way. The next step is to maintain a steady power output from fusion reactions in long pulses. To make this possible, superconducting coils for the magnetic fields will have to be used, because they are not subject to the huge resistive losses of classical copper coils and allow long pulse operation. The use of superconducting coils has been very successfully demonstrated e.g., in Tore Supra (CEA-Cadarache, France) and LHD in Japan. Both machines have record length pulses at relevant plasma parameters in hydrogen and deuterium plasmas: 6 minutes and 30 seconds on Tore Supra in 2004[12] and 30 minutes on LHD in 2005, as already mentioned above.[6] Long pulses in D-T plasmas are planned for ITER, the next step fusion machine designed to deliver 500 MW fusion power in D-T pulses of 400s and longer, with a Q value of 10.

The number of fusion reactions and the fusion power generated is determined by the energy confinement time τ_E, which is the characteristic time during which the energy remains confined in the plasma, in spite of losses due to the thermal conductivity of the plasma, particle transport and radiation. To generate 500 MW of fusion power scaling laws for the energy confinement time show that the plasma current should be about 15 MA, the toroidal magnetic field 5.3 T, the heating power 50 MW and the plasma volume about 1000 m^3.

Larger machines will also allow plasmas where the heating is dominantly provided by the alpha particles from fusion reactions, as it will be in future fusion reactors. This condition has been possible only to a very limited extent on existing tokamaks. The contribution f_α to the total plasma heating by the helium nuclei from the fusion reactions can be written as $f_\alpha = Q / (Q + 5)$. For the Q=10 in ITER we find $f_\alpha \sim 75\%$, close to the value for a fusion power reactor ($f_\alpha \geq 90\%$), and about 5 times larger than the best value achieved in JET ($Q \sim 1$, $f_\alpha \sim 16\%$).

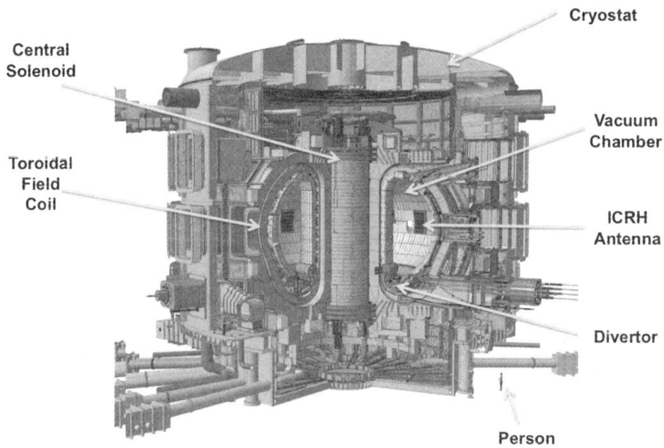

Fig. 6: Cross section of the ITER device, showing its main components.

ITER will still be an experimental machine. The results that ITER will provide are needed for the detailed design of a demonstration fusion reactor, DEMO. Besides scientific issues such as the further optimization of operational scenarios, investigating the stability of plasmas in which a large fraction (fast) He nuclei are present, etc., the emphasis in ITER will primarily be on testing the necessary technology: superconducting coils in a nuclear environment, remote maintenance techniques, systems to extract both energy and impurities from the plasma), in-situ production of tritium, the integrity of structural components for a future fusion reactor, etc. The construction of ITER in Cadarache (close to Aix-en-Provence in France) began in August 2010 shortly after the endorsement by the ITER Council of the final plans for the machine (end of July 2010). First plasmas in ITER are currently foreseen for 2019-2020. The duration of the experimental phase of ITER is currently planned for 10 years. Seven large nations are participating in the project: Europe, Russia, India, China, Japan, South Korea and the USA. Note that there is also a very important participation of emerging economies in the Far East. A schematic representation of the machine with the main components is shown in Figure 6 and the main parameters of ITER are summarized in Table 1.

In February 2007, EURATOM and Japan signed the "ITER Broader Approach" agreement. This aims to complement the ITER Project and to accelerate the realization of fusion energy by carrying out R&D and developing some advanced technologies for future demonstration fusion power reactors (DEMO). Within the Broader Approach, three main projects are being implemented:

The first project will complete the detailed and fully integrated engineering design of the International Fusion Materials Irradiation Facility (IFMIF). Fusion as a major energy source will require materials which maintain their essential physical properties and which do not remain highly radioactive for extended periods of time after exposure to the harsh thermal and irradiation conditions inside a fusion reactor. IFMIF is a joint project of EU, Japan, Russian Federation and the USA and consists of an accelerator-based, D-Li neutron source to produce high energy neutrons at sufficient intensity and irradiation volume and will allow testing and qualification of advanced materials in an environment similar to that of a future fusion power plant.

The second project is the International Fusion Energy Research Centre (IFERC). The missions of the centre include the co-ordination of DEMO Design and R&D activities, large scale simulation activities of fusion plasmas by super-computer and remote experimentation activities to facilitate a broad participation of scientists into ITER experiments.

The third project is the Japan-EU Satellite Tokamak Programme. During ITER construction, major experimental facilities will be required to develop operating scenarios and address key physics issues for an efficient start up of ITER experimentation and for research towards DEMO. This means in practice EU participation in (i) upgrading the largest Japanese tokamak JT-60U to JT-60SA (Super Advanced)[13], an advanced superconducting tokamak and (ii) in its exploitation: it will be used by Europe and Japan as a "satellite" facility to ITER.

Except for JT-60SA, the ITER Broader Approach activities are located in Rokkasho-mura in the Aomori Prefecture, close to Hokkaido. At present all buildings for the Broader Approach are constructed, and the activities can start very soon. Plans for the upgrade of JT-60U are finalized and construction is underway. First operation at JT-60SA is expected for 2015.

INERTIAL FUSION

This technique is the second approach to realize controlled fusion on earth. It involves compressing a small capsule (diameter of about 1mm, Figure 7) filled with a ~50/50 mixture of deuterium and tritium (density of about 0.1 g/cm^3). The inertia of the imploding fuel keeps it confined (typically about 1 ns). To achieve sufficient fusion reactions in this short time requires a density of about 1 kg/cm^3, a 10000-fold increase over the original density and about 100 times the density of lead or 1000 times that of water[14]. This approach is being investigated in far fewer countries than is magnetic fusion. Moreover, because of the relation of target physics and target design to military programs, some aspects of inertial fusion research remain classified in several countries.

The fuel capsule is both compressed and heated with very intense lasers, ion beams, or X-rays. The outermost layers of the capsule consist of material that can ablate easily. The intense radiation vaporizes these layers nearly immediately, generating an inward-propagating spherical shock wave, which leads to a very high compression of the central zone of the capsule.

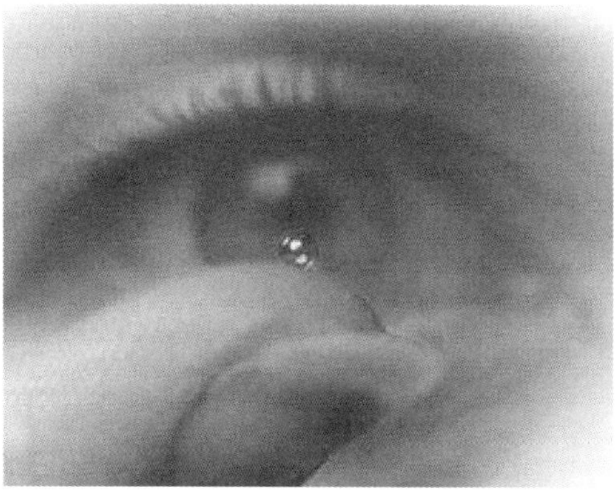

Fig. 7: Illustration of the size of a fuel capsule in use in inertial fusion research.

Two techniques are being investigated to obtain a symmetrical irradiation of the fuel capsule: direct-drive and indirect-drive. In direct-drive one tries to obtain uniform irradiation of the fuel capsule using a large number of laser or ion beams directly striking a spherical capsule. In indirect drive the fuel capsule is contained within a small, high Z, metal cylinder (so-called hohlraum, Figure 8a) with open ends that is irradiated by laser or ion beams focused on the inner surface of the cylinder rather than on the fuel capsule itself (Figure 8b). The intense beams transform the metal cylinder into superhot plasma that radiates mostly X-rays. Those X-rays are subsequently absorbed by the surface of the fuel capsule, imploding it in the same way as if it had been hit directly with lasers or particle beams. In this second method, the radiation striking the fuel capsule is far more isotropic, ensuring a spherically symmetrical implosion which is much more difficult to achieve with direct drive.

Inertial fusion experiments are characterized by the energy gain G, this is the ration of fusion energy released from the capsule to the energy used to heat and compress it. Ignition in this context is defined as $G \geq 1$. Note that the definition of energy gain in inertial fusion is different from the definition of the Q value in magnetic fusion: the first is a ratio of energies, the second a ratio of powers. Also the meaning of ignition in inertial fusion is fundamentally different than in magnetic fusion where it is defined as $Q=\infty$, corresponding to a self-sustained reaction with no external heating input.

The amount of laser energy needed to effectively compress the capsules to ignition conditions using indirect drive is in the MJ range. The use of "external match" however, should strongly reduce the total energy needed for ignition by decoupling the heating and compression phases of the implosion. A first "external match" is the so-called "fast ignition" technique.[14] The capsule is first compressed to hundreds of times solid density as described above. When the implosion reaches maximum density a second, ultra-short pulse (~10 ps) from a PW (10^{15} Watt) laser (~70 kJ, 4 PW) is delivered on one side of the core. The light from this pulse interacts with the fuel, generating a shower of high-energy (several MeV) relativistic electrons that are driven into the fuel thereby heating a spot on one side of the dense core. If this heating is sufficiently localized, it is expected to drive the dense center of the compressed capsule well beyond ignition temperatures.

Fig. 8a: Illustration of the size of a hohlraum as used in inertial fusion research.

How to get the laser light onto the small compressed spot (with a size of about 30 μm) is a subject of further research. Two fast ignition approaches are in under study: (i) the "plasma bore-through" method and (ii) the "cone-in-shell" method. In the first approach, successfully tested at the OMEGA device (see below), the ultra short pulse PW laser burns a hole through the compressed plasma heating the dense fuel inside. The second approach, tested successfully on the GEKKO XII laser in Japan (see below), uses a small gold cone that cuts through a small area of the capsule shell; on heating no plasma is created in this area, leaving a hole that can be aimed into by shining the laser into the inner surface of the cone.

More recently also "shock ignition" is proposed. This is an intermediate route between central ignition and fast ignition. The fuel is first irradiated symmetrically, but driven at lower velocity than for central ignition. The temperature of the hot spot generated at the end of the implosion is below the ignition threshold. At an appropriate time, toward the end of the implosion, the fuel capsule is irradiated by intense (quasi-symmetric) pulses, which drive a strong converging shock wave, with initial pressure of about 300 Mbar. This pressure is amplified by convergence, and further amplified as the converging shock collides with the outgoing shock bouncing from the center. The hot spot then undergoes additional heating and ignites. One could understand shock ignition as shock-assisted central ignition. The duration of the shock is between 200-500 ps with a energy of ~80 kJ corresponding to a power of 150-400 TW.[15] This new concept has been experimentally studied at the OMEGA laser facility[16], demonstrating the potential of a properly timed final shock to significantly enhance the neutron yield. Shock ignition has the advantage over fast ignition, in that it does not require any cone-in-a-shell target nor extreme high power multipetawatt laser systems. A comprehensive overview of the physics and current status of laser fusion research can be found in Atzeni, 2009.[17]

The National Ignition Facility (NIF) at the Lawrence Livermore National Laboratory in Livermore, California, USA is the largest and most powerful, laser-based inertial confinement fusion (ICF) device in the world, and the first that is expected to reach ignition. NIF aims to create a single 500 TW flash of light that reaches the capsule from numerous directions at virtually the same time, within a few ps. As designed, NIF will deposit 1.8MJ in the capsule using 192 individual beamlets, which are amplified in 48 beam lines that contain 16 laser amplifiers per line, each one amplifying four of the beamlets. NIF is designed primarily to use the indirect drive method of operation; it was completed in 2009. Ignition (G > 1) is expected soon (2010-2011).

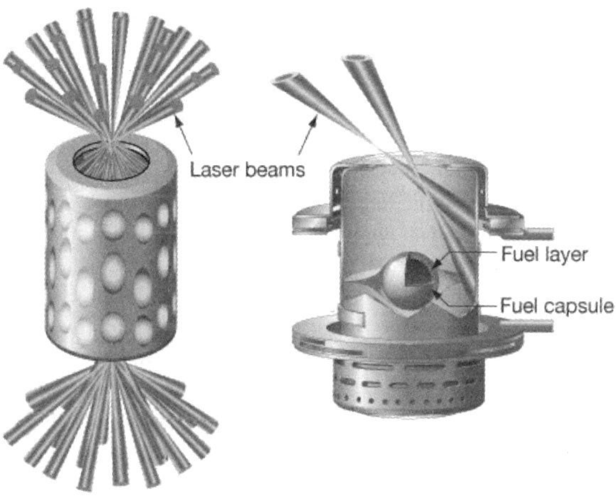

Fig. 8b: Principle of indirect drive using a hohlraum. The fuel capsule is located inside the hohlraum. The laser beams irradiate the inner side of the hohlraum (and not the fuel capsule). The X-rays generated from the superhot hohlraum material ablates the outer fuel layers of the capsule resulting in a shock wave that compresses the fuel capsule.

The European counterpart to NIF is the Laser Megajoule (LMJ), an experimental ICF device being built near Bordeaux, France by CEA, the French nuclear science directorate. Laser Mégajoule plans to deliver about 1.8 MJ of laser energy to its capsules, making it about as powerful as NIF. Laser Mégajoule is the largest ICF experiment being built outside the U.S. and uses a series of 240 laser beam lines, grouped into eight groups of 30 beams. Like NIF, LMJ intends to use the indirect drive approach. Construction on the Laser Mégajoule started with a single set of eight beam lines known as the Ligne d'Integration Laser (Laser Integration Line or LIL), which was completed in 2002. The entire LMJ system is expected to be completed in 2012.

GEKKO XII is a high-power 12-beam laser system at the Institute for Laser Engineering in Osaka, Japan in use since 1983. The 12 beams of the GEKKO laser are capable of delivering about 10 kJ per 1-2 ns pulse (i.e., 10-20 TW). Since 1997 the GEKKO system has also been equipped with a ~0.4 kJ, 0.5 PW ultra-short pulse laser used to investigate fast ignition. GEKKO, using the PW beam for fast ignition along with the original beams for compression, demonstrated a fusion yield enhancement of three orders of magnitude when the PW beam was used. GEKKO is currently being upgraded with the addition of a second "side-by-side" laser, the LFEX (Laser for Fast Ignition Experiment), part of the FIREX-1 program, in order to deliver a 10 kJ pulse of energy to a capsule in 10 ps, further exploring the fast ignition regime.

The OMEGA laser at the Laboratory for Laser Energetics (LLE) is part of the University of Rochester and is in use since 1995. It is a 60-beam laser, which is capable of delivering 30 kJ at up to 60 TW onto the capsule. OMEGA EP (extended performance) consists of four beam lines. Two of the OMEGA EP beam lines have the capability to operate in a pulse-width range of 1 to 100 ps. For widths of 10 ps or greater, the on-target energy can be up to 2.6 kJ. At the 1-ps pulse width, the system can produce 1 PW of optical power on target. All four OMEGA EP beam lines have a long-pulse capability with durations ranging from 1.0 to 10 ns. The long-pulse mode can deliver up to 2.5 kJ per beam on target for 1 ns pulses and 6.5 kJ per beam for 10 ns pulses. OMEGA EP is also equipped with a new target chamber and was completed in 2008.

The European project HiPER[18] intends to study the fast ignition concept. The compression provided by the driver is "good", but not nearly as good as that created by larger devices such as NIF. Specifically, HiPER's driver is about 200 kJ and produces densities of about 300 g/cm^3– about one-third that of NIF. However, the overall efficiency of this approach should be many times that of the conventional approach. In the case of NIF, the laser generates about 4 MJ to create ignition that should release about 20 MJ of energy. This output corresponds to an energy gain of about 5. If one uses the baseline assumptions for the current HiPER design, the two lasers (driver and heater) will produce about 270 kJ in total, yet generate 25 to 30 MJ, corresponding to an energy gain G ~ 100. Considering a variety of losses, the actual gain is predicted to be ~30% less. Recently also shock ignition schemes were also proposed.[19] Not only should HiPER outperform NIF in fusion gain by a wide margin, the smaller lasers are much less expensive to build as well. The HiPER project concluded its preparatory phase in 2008, and detailed designs are expected in 2011-2012.

A far different approach uses a plasma z-pinch. A z-pinch is the radial implosion of a cylindrical or annular plasma under the influence of a strong magnetic field produced by a current flowing down the length of the plasma. One of the principal z-pinch systems in the world is the Z machine located at the Sandia National Laboratory, Albuquerque, New Mexico, USA.[20] The Z machine sends an extremely large electrical current (up to 18 MA) during about 100 ns into an array consisting of several hundred thin (diameter of ~ 10μm), parallel tungsten or steel wires called a liner (see Figure 9).

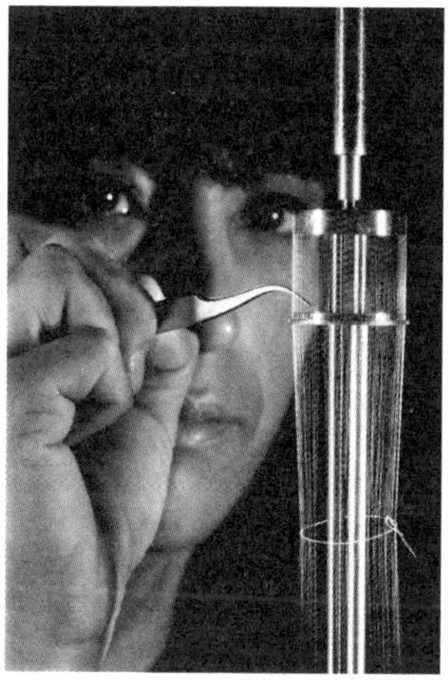

Fig. 9: A liner consisting of 300 tungsten wires, in use in Z-pinch research.

The high electrical current vaporizes the wires, which are transformed into a cylindrical plasma curtain that implodes under the force of its own radial magnetic field. The imploding plasma generates a high temperature and an X-ray pulse which creates a shock wave in a target structure located on the axis of the cylindrical configuration. The shock both compresses and heats the target. Fusion of deuterium using this system was observed in 2003. A refurbishment of the Z machine to increase its power by 50% was completed in 2007. This newer Z machine (called the ZR machine) can now produce up to 27 MA in 95 ns. The radiated power has been raised to 350 TW and the X-ray energy output to 2.7 MJ. Sandia's roadmap includes another Z machine version called ZN (Z Neutron) to test higher yields of fusion power and automation systems. ZN is planned to produce between 20 and 30 MJ of fusion power at a rate of one shot per hour. The next step planned would be the Z-IFE (Z-inertial fusion energy) test facility, the first true z-pinch driven prototype fusion power plant. Sandia has recently proposed a conceptual 1 PW Z-pinch power plant, where the electric discharge would reach 70 MA. An overview of the current status in Z-pinch research can be found in Cuneo et al. 2006.[21]

Yet another approach to compressing and igniting fusion capsules was proposed in 1973 by Al Maschke of Brookhaven National Laboratory.[22] In this approach accelerators would focus many high energy (GeV class) beams of heavy ions (such as lead or uranium) on the target. While heavy-ion fusion (HIF) has received a much smaller investment than laser-driven ICF, it is based on mature accelerator technologies that have the demonstrated capability to run at the 5 to 10 Hz required in a fusion power plant. At present the principal proponents of HIF is a three-laboratory collaboration (Lawrence Berkeley, Lawrence Livermore, and Princeton Plasma Physics Laboratories). The principal difference between an accelerator for inertial fusion and more conventional large accelerators is that the device must operate in a heavily space-charge dominated regime. The investigation of the beam dynamics in this regime and their implications for machine size and cost are the principal focus of the three-laboratory collaboration. A significant future step would be the construction of a multi-beam, Integrated Beam Experiment that would raise the temperature of a fuel capsule to 100 to 200 eV, thereby making close connection with the ignition experiments on NIF. An overview of the main parameters for several ICF devices / large laser systems in the world is given in Table 2. Recent information on heavy ion ICF research can be found in Logan et al. 2008.[23]

Table 2: Main parameters of existing and planned ICF laser systems in the world. HiPER is a European facility under consideration. LMJ in France is planned to be ready in 2012. The power for the separate extreme high power ultrafast fast ignition laser beam (f.i), if available, is listed separately.

Device Name	NIF	LMJ	GEKKO-XII	ISKRA-5	OMEGA OMEGA-EP	HiPER
Land/Organisation	USA	France	Japan	Russian Federation	USA	EU
Energy on target	1.8 MJ	1.8 MJ	10-12 kJ 0.4 kJ (f.i)	30kJ	30 kJ 2.6 kJ (f.i.)	250 kJ 70 kJ (f.i)
Power	500 TW	500 TW	10-20 TW 0.5 TW (f.i)	100 TW	60 TW 1PW (f.i.)	50-100 TW 4 EW (f.i)
Wavelength (nm)	350	350	532	1315	351, 1053	350
Pulse length (ns)	1-2 ps	1-2 ps	1-2 ns few ps (f.i)	250 ps	2 ns@351nm 1-100ps @1053nm(f.i)	~5 ns ~15 ps (f.i.)

Future plans in high power short pulse laser systems are ELI (Extreme Light Infrastructure) in Europe and GEKKO EXA in Japan. Both systems are intended to deliver 0.2 ExaWatt (2×10^{17} W) in femto (10^{-15}) or atto (10^{-18}) second pulses. The European system plans a pulse repetition rate of 1 minute, while the system in Japan will have a repetition rate of 2 hours; however at a 10 fold reduced cost.

The path to a ICF based power reactor is long: for a 1 GWe power station one foresees a laser delivering pulses of a few MJ, at a frequency of a few to 10 Hz, and burning targets with energy gain $\approx$ 100. Beyond the development of devices for proof-of-principle ignition, much long-term development is required to demonstrate: 1) high repetition rate operation at an affordable scale, 2) suitable scalable fusion chamber design, 3) sufficiently robust materials to withstand both the radiation flux and the mechanical shock of repeated explosive fusion events, and 4) fabrication of economically viable fuel capsules.

Since the first years of controlled fusion research, remarkable progress has been made. This is both illustrated by the increase in the fusion triple product in magnetic fusion, and the fusion gain in inertial fusion. Both in inertial and magnetic fusion we are witnessing a very interesting period to come. Several new devices are being built or will be in operation very soon. Values larger than break-even are expected in the next 10-15 years, and ignition seems to be in reach for inertial fusion. An overview of important milestones in fusion research is summarized in Table 3.

For the first time in history, mankind has sufficient knowledge and technical capability to heat and confine matter at the extreme temperatures needed for fusion of hydrogen isotopes. Many seemingly impossible difficulties have been solved, and as fusion research progresses, next step questions are now being looked into. A critical class of questions in the preparation of a demonstration reactor concerns material science. There should be no illusions about the time required bringing even the D-T reaction to a commercially viable system. However, there are at the moment no indications to cast doubt on the practical possibility of fusion as an energy source. History has repeatedly proven that major technological projects (not hampered by scientific limits) have finally reached a breakthrough. Given the potential advantages of nuclear fusion compared to the risks and impediments of all other alternatives for base load electricity generation and given the potential contribution of fusion to long-term, sustainable world development, it is our duty to future generations to continue the vigorous development of fusion without delay and with full commitment. The challenge is huge, but the rewards are enormous!

Table 3: Important milestones in the history of controlled fusion research.

1951 – Sakharov and Tamm (Kurchatov Institute, Moscow): Propose Tokamak
1955 – 1st UN conference - Bhabha Statement: "in 20 years fusion; water as fuel"
1956 – Lawson criterion: performance of fusion systems
1958 – 2nd UN conference: stellarator, pinch, tokamak presented
1961 – 1st IAEA conference: first plasmas at 1-10 million degrees (ZETA, pinch)
1965 – Failure of pinches: temperature not as high as expected (large thermal losses)
1968 – Novosibirsk IAEA Conference: Succes of tokamak
1971 – Tokamak fever; Conceptual design of reactor, Erice school (Brunelli, Martellucci)
1971 – First presentation of Inertial fusion systems (Frascati)
1972 – JET (EU), TFTR(USA), JT60 (Japan) proposed.
1973 – Start of INTOR, later ITER
1984 – JET, TFTR, JT60 operation start
1985 – Gorbachev, Reagan summit: ITER proposed
1991 – JET first D-T experiment in tokamak worldwide: 2MW fusion power
1994 – 10MW in TFTR: Q = 0.25
1997 – 16MW in JET: Q = 0.7
2005 – International agreement on ITER siting: Cadarache, France
2010 – ITER construction start

REFERENCES

1. А.Д. Сахаров "Теориа магнитного термоядерного реактора" Pt. II, in "Физика плазми и проблема управляемих термоядерних реакций", Vol. 1 (Moscow: Изд. АН СССР, 1958) p. 20; И.Е.Тамм, ibid. pp. 3, 31. English translation: A.D. Sacharov "Theory of the magnetic fusion reactor", Pt.II, in "Plasma Physics and the Problem of Controlled Thermonuclear Reactions" Responsible editor M.A Leontovich Vol.1 (London: Pergamon Press Ltd, 1961) p.21; I.E.Tamm, ibid., pp. 1, 35.
2. L.A. Artsimovitch et al. in *Plasma Phys. and Control. Nucl. Fusion Res.* (Proc. 3rd Int. Conf., Novosibirsk, 1968) Vol. 1 (Vienna: IAEA, 1969) p. 17. English translation: CCFE Culham (UK) Report CTO-536 (1968).
3. G. Grieger, et al. Plasma Physics and Controlled Nuclear Fusion Research, (Proc. 13th Int. Conf., Washington, 1990) Vol. 3, p 525, (Vienna: IAEA) (1991); G. Grieger et al. *Fusion Technology* 21, 1767 (1992).

4. A. Iiyoshi et al. *Nucl. Fusion.* **39**:1245 (1999).
5. H. Yamada et al. *Fusion Sci. Technol.* **58**:12 (2010).
6. R. Kumazawa et al. *Fusion Sci. Technol.* **58**:524 (2010).
7. R. Koch, Proc. 9th Carolus Magnus Summer School on Plasma and Fusion Energy Physics, *Fusion Sci.Technol.* **57**, No. 2T, 185 (2010).
8. JET Team, *Nucl. Fusion* **32** (1992) 187.
9. R.J. Hawryluk, H. Adler, P. Alling et al. in Plasma Physics and Controlled Nuclear Fusion Research, (Proc. 15th Int. Conf. Seville, Spain, 1994), Vol. I, p 11, (Vienna: IAEA, 1995); K.M. McGuire, H. Adler, P. Alling et al. *Phys. Plasmas* **2**:2176 (1995).
10. JT-60 Team in Plasma Physics and. Controlled Nuclear Fusion Research 1988 (Proc. 12th Int. Conf., Nice, France, 1988) Vol. 1, p 53 (Vienna: IAEA, 1989)
11. J. Jacquinot and the JET team, *Plasma Physics Control Fusion*, **41** (1999) A13.
12. J. Bucalossi et al. 32nd EPS Conference on Plasma Phys. Tarragona, 27 June–1 July 2005 ECA Vol. **29C**, O-4.005 (2005); G. Giruzzi et al. *Plasma Phys. Contr. Fusion*, **47**, 12B, B93 (2005).
13. S. Ishida, P. Barabaschi et al. "Status and prospect of the JT-60SA project", *Fusion Eng. Des.* (2010), doi:10.1016/j.fusengdes.2010.07.027, in print.
14. M. Tabak et al. *Phys. Plasmas* **1**:626 (1994).
15. Betti, R. et al. *Physical Review Letters* 98:155001 (2007).
16. W. Theobald et al. *Phys. Plasmas* **15**:056306 (2008); W. Theobald et al. *Plasma Phys. Control. Fusion* **51**:124052 (2009).
17. S. Atzeni, *Plasma Phys. Control. Fusion* **51:**124029 (2009).
18. M. Dunne, *Nature Physics* **2**:2 (2006).
19. X. Ribeyre et al. *Plasma Phys. Control. Fusion* **51:**124030 (2009).
20. D.D. Ryutov, M.S. Derzon and M.K. Matzen, *Rev. Mod. Phys.* **72**, Issue 1, pp. 167–223 (2000).
21. M.E. Cuneo et al. *Plasma Phys. Control. Fusion* **48** R1 (2006).
22. A.W. Maschke, IEEE Trans. on Nucl Scl., **NS-22**, No. 3, p. 1825 (1975).
23. B.G. Logan, L.J. Perkins and J.J. Barnard, *Phys. Plasmas* **15**:072701 (2008).

SMART GRIDS

HISHAM KHATIB
World Energy Council
Amman, Jordan

All through the twentieth century electricity revolutionized our life and economies. It brought "affordable" lighting and comforts to our homes and power to our industries and energy to our public transport. It simply has transformed our lives to the better, probably more than any other technology, which humanity knew, since steam power.

However, since its inception, progress in electrical power technologies has been slow and the inertia, to continue with business-as-usual, is high. The power business with its slow technological progress has been contrasted with the electronics and telecommunication business where rapid technological progress is taking place everyday.

The electrical power sector suffered from two main draw backs: the first is its low efficiency. Until now, average electricity generation efficiency world wide is less than 35%, and when system losses are taken into account only 30% of primary energy input to the sector is in final utilization. There has been progress in this regard, but it was slow and painful.

The second aspect is the unchanging hierarchy of the electricity system (Figure 1). Large inefficient central power stations, feeding into high voltage grids, transfer electric power into users through a myriad of distribution networks. Consumers have little, practically no choices and live communication between users and producers is almost non-existent.

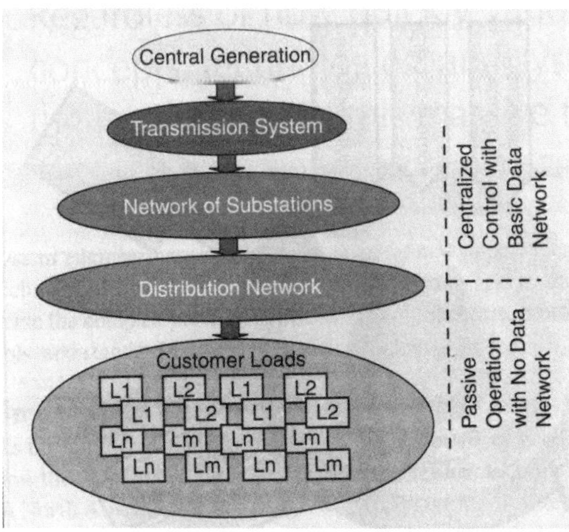

Fig. 1: The Existing Grid.

Therefore, the existing electricity grid is inefficient. Only 30% of the energy content of primary fuel input reaches consumers. Practically a third of the grid facilities are used for only few hours each year. Almost 20-25% of the installed capacity is idle most of the time and is only needed for emergencies which may or may not occur for a long time. Also interruptions are not uncommon. The system hierarchy can lead, in certain circumstances (like loss of major generation supply or breakdown of a transmission line), into major supply interruptions or blackouts that affect large numbers of customers. Therefore the existing electricity grid is not optimum. Over the last several decades it has well served consumers requirements but at relatively low efficiency and high cost and with lot of emissions.

Now with advancement in electronics and communications, as well as environmental awareness, things are changing at a relatively significant pace compared to past stagnation. We are now slowly entering the age of smart grids which are more efficient, interactive, consumers and environment friendly. This is going to gradually transform traditional grids into "smart grids". The transformation is going to be gradual and sometimes slow, but it is a significant departure from the past stagnation. Electric power is not the electronics and communication business; due to its capital intensive nature, it is a slowly evolving technology, but the results can mature over time.

Smart grids, in the minds of many power engineers and planners, are only associated with distribution. In this presentation we will present a broader view to cover electricity production (generation) as well.

Let us first of all start by defining the smart grid. There are many definitions, but I chose this one: "the collection of all technologies, concepts, topologies, and approaches that allow the silo hierarchies of generation, transmission, and distribution to be replaced with an end-to-end, organically intelligent, fully integrated environment where the business processes, objectives, and needs of all stakeholders are supported by the efficiency exchange of data, service, and transactions."[1]

A smart grid is, therefore, a grid that accommodates a wide variety of generation options and forms. It allows consumers to interact with the energy management system to optimize their use. It is a self healing system that predicts looming failures and mitigates problems. It relies on IT to optimise its economical performance. Therefore a smart grid incorporates four new activities which are not familiar in the existing electrical power grid.

1. It empowers consumers to interact with the energy management system in order to manage their demand and reduce their bills
2. It utilizes information technology and telecommunications to optimize the use of its capital assets, particularly generating facilities, in order to minimize operational cost.
3. It is a "self healing system", that can predict looming failures and take preventive action to limit interruptions and mitigate serious system failures
4. It accommodates a wide variety of generation options which, beside traditional generation, incorporate new and renewable sources including distributed generation and consumer premises, intermittent new sources like solar, wind, etc and also future possible mobile sources (like electric cars).

Most importantly, the smart grid aims at efficiency and also environmental sustainability. Therefore I shall start with this, then move to the other major features of the smart grid: metering, storage and intelligent micro-grids.

Smart Generation: The Issue of Efficiency

I am going to incorporate efficient generation as part of the smart grid. Although this is not usual however, it is most important. Post second-world war large coal firing power stations had an efficiency of around 25%. Such vintage units, some of them still operating, required almost half a ton of black coal for each MWh sent out of the power station, emitting tons CO_2 per MWh sent out.

During the last three decades with escalating fuel prices we moved towards more efficient generation which, beside nuclear and hydro, had much more efficient thermal generation with fewer emissions. Three major developments in electricity generation took place.

The first was new steam power stations firing coal and utilizing critical and super critical pressure. These already attained an efficiency of 45% with an efficiency of 55% in sight. Such new sets are replacing the old huge polluting plants and emitting only half the amount of CO_2 per unit of electricity sent out.

The second development is the introduction of combined cycle gas turbines plant (CCGT) firing clean natural gas and already attaining an efficiency of 60% with 70% efficiency in sight. Emissions per MWh are only one third of large coal firing plant.

The third major development is the increasing introduction of renewable new generation; particularly wind energy, but increasingly solar forms of clean zero emitting generation and also small distributed generation.

There are two approaches to improve thermal power generation efficiency and reduce its CO_2 emission, one is through increasing live steam parameters (pressure and temperature) to develop supercritical (SC) and ultra-supercritical (USC) technologies; another is by system integration. A typical example is Integrated Gasification Combined Cycle (IGCC). In the next 10 years, SC and USC will be built in significant numbers, for electricity production from coal, and these new plants are likely to remain in use until 2050 because of their flexibility and general advantages of lower cost, reliability, availability, maintainability, and operability. IGCC currently looks promising in its ability to produce deep CO_2 reductions at least cost.

New pulverized coal power plants—utilizing SC and USC—operate at increasingly higher temperatures and pressures and therefore achieve higher efficiencies than conventional units. Supercritical power generation has become the dominant technology for new plants in industrialized countries. Now considerable efforts are underway in United States, Europe, and Japan to develop 700°C-class Advanced Ultra-Supercritical (A-USC) steam turbines. If successful, this will raise the efficiency of the A-USC units to about 55% by 2020.[2]

Therefore future smart grids will be connected to smart generation systems which have more than 50% higher generation efficiency over past plant and emitting carbon of less than one half of past generating systems.

Metering

A central component in the smart grid is metering. Until now, the most widely used of electricity metering worldwide is the electromechanical meters. These have been in existence since the inception of electricity distribution. They proved to be resilient, relatively accurate and served the purpose of the industry relatively well. However they required frequent reading and manual billing. Correspondingly the introduction of the new smart meters remotely measuring energy consumption provided a direct communications link between energy suppliers and their consumers. That removes the need for meter readings at the premises and ensures accurate bills based on actual, not estimated, consumption. But this is one way communication from the consumer to the utility. It is termed one-way-automated meter reading (AMR). What is more important is to move from this into two-way-automated metering infrastructure (AMI). In this domestic customers will be able to see the information about their energy use through an integrated in-home display. More important than the real-time billing information, that will also give consumers more control over their energy use and carbon emissions, and create new opportunities for energy retail services, infra-structure management and renewable energy generation. Such AMI smart metering is crucial in the effort to make future energy saving and to tackle climate change.

It is a major step forward in the development of the smart grid as well as in overcoming bottlenecks in the operation of the generating system through load management with available generating capacity, and allowing pervasive control.

The European Union EC has now mandated that, by 2020, all existing electromechanical meters at consumers' premises should be replaced by smart meters.[3] This is the Automated Meter Reading (AMR) step. As explained above it is only a first step in the long path to the smart grid. The next step is the installation of the two way Automated Metering Infrastructures (AMI). It will allow consumers to communicate with the grid to obtain live information on costs and tariffs and to make choices that will not only reduce their costs but also help to optimize the grid operation.

Storage Issues

Electrical energy storage is the biggest problem that has besieged the electricity supply industry since its inception. Of course limited storage in batteries and similar facilities are available, but here we are talking about the availability of massive storage of energy in the form of GWhs. The only available large facility is the well known pumped hydro storage schemes, which have been available for decades. They are however capital extensive, with an efficiency of only 70%–80% and require certain topographical and geological formations, which does not necessarily exist everywhere.

Viable large storage is important for the functioning of the smart grid; correspondingly interest in developing such storage technologies is growing. There are few ideas being investigated / developed.

The first major development is the likely introduction of large batteries. Now a Japanese firm NGK sells a large room sized, sodium-sulfur battery. Each large NGK battery can provide one megawatt for six hours. When clustered in grid requirement standards they are still small and not cheap ($2500 per kilowatt). However it's an encouraging start.[4]

The second development is that of the provision of efficient lithium-ion batteries for plug-in hybrid electric vehicles (PHEV). They have the ability to store vast amounts of cheap off-peak electricity for use during the past day during which there is peak demand, with the possibility of releasing it to the grid during system shortages.

The third possibility is the large thermal storage. In the long term, particularly after 2030, concentrating solar power plants (CSP) provide means of indirect electricity storage by storing heat. CSP is able to store heat during the day and convert it to electricity at night, thus providing flexible round-the-clock electricity from solar power. Figure 2 demonstrates this, in which excess heat collected in the solar field is sent to the heat exchanger and warms the molten salts going from the cold tank to the hot tank. When needed, the heat from the hot tank can be returned to the heat transfer fluid and sent to the steam generator.

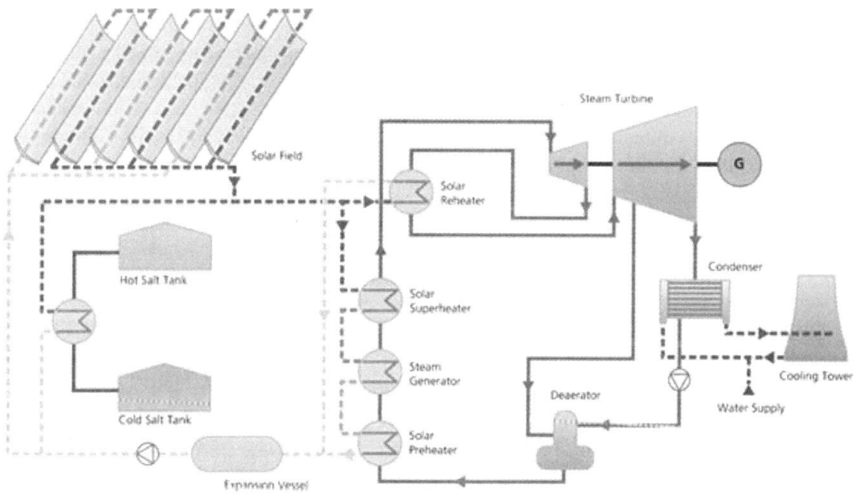

Fig. 2: Source—Layout of the Solana (Abengon Solar).

In view of the IEA, between 2030 and 2050, the researchers expect CSP to expand beyond pure power generation and into solar-assisted natural gas reforming. This is a process in which CSP plants use solar energy to produce hydrogen that can be mixed into gas as well as fuel products. 'Solar-generated hydrogen can help decarbonise the transport and other end-user sectors by mixing hydrogen with natural gas in pipelines and distribution grids, and by producing cleaner liquid fuels,' the IEA says.[5]

Another of these storage facilities is the concept of the isentropic pumped heat electricity storage system (PHES). According to its developers, the Isentropic PHES system uses a highly reversible heat engine/heat pump to transfer heat between two insulated storage vessels containing particulate mineral (in its simplest form, gravel). A gas circulates through the machine and, to store energy, is first compressed, which raises its temperature to 500ºC. It is then passed through one of the stores in which it heats the mineral by direct contact, which also cools the gas to close to the original temperature. It is then expanded back to its original pressure, which cools it to around -160ºC, and then passed through the other store where it cools the mineral by direct contact and is warmed back to close to its original temperature. This process requires energy that can be supplied electrically via a motor, and is thus an energy storage process. Discharge is the opposite of the charge process and releases energy, in which situation the machine drives an electrical generator. This results in an electricity-in to electricity-out (round trip) efficiency in the range of 72 to 85%. This compares favorably with a typical pumped-hydro value of 74%.[6]

There have been never ending attempts to store energy in pressurizing air into salt caves or similar facilities, and also in the form of inertia in flywheels, etc. However one must admit that success, till now has been modest. There is no economical solution, yet, in sight.

However, the smart grid is providing alternatives to the problem of storage. Storage is needed to store cheaply produced energy (say off-peak) and release it during shortages and peak demand. The smart grid can provide the same results, but in a different manner. Many appliances, and also batteries of PHEV, can be made to utilize cheap off-peak electricity. Such supplies will be interrupted during peak and shortages, for a prescribed time, without causing users undue inconvenience; thus "restricting demand". Such facilities can only be widely made available through the smart grid, correspondingly the smart grid will provide "demand management" alternative that can allow optimal operating and economical supplies without the need for large bulk energy storage.

EVOLUTION OF THE SMART GRID

It is not expected that the new smart grid will soon replace the existing electricity grid; it would rather slowly complement it, improving its performance and adding to its capabilities. This necessitates a topology for the smart grid with compatibility of the existing traditional grid, but allowing for incorporation of innovative technologies and organic growth.

This can happen through the integration of "intelligent micro grids". Such micro grids are defined as "interconnected networks of distributed energy systems (loads and resources) that can function whether they are connected to or separate from the electricity grid".

This micro grid will incorporate micro generation mainly renewables, local and distributed storage capability and load management facilities, all of which are essential to smooth the intermittent performance of renewables. Therefore smart meters, as described above, capable of remote controlling of the load, are an integral part of the smart grid. An intelligent core communication infrastructure, capable of exchanging information and undertaking micro-control, is also a basic requirement of the micro-grid.

Many observers think that we are now in transition to widely distributed electricity systems, often labeled as "micro-power" in which the customer own and operate their small (micro) generation facilities. Their devices are also connected to the distribution grid for exchange of power. The availability of storage facilities (electrical or thermal) will greatly enhance the prospects of micro-power facilities.

Many observers think that we are new in transition to widely distributed electricity systems, often labeled as "micro-power" in which the customer own and operate their small (micro) generation facilities. Their devices are also connected to the distribution grid for exchange of power. The availability of storage facilities (electrical or thermal) will greatly enhance the prospects of micro-power facilities.

Such micro grids will be incorporated to a smart grid through dedicated highways for command, data and power exchange. The smart grid will coexist beside the already existing traditional grid, gradually expanding and eventually, over the long term, replacing a lot of the infrastructure and functions of the already existing traditional grid (Figure 3).

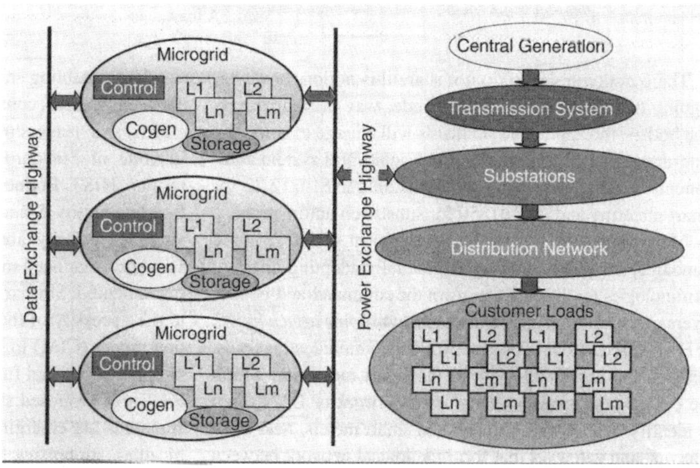

Fig. 3: Incorporating the smart grid into the existing grid (Copied from Ref. 1).

SUMMARY

The technology of the electricity supply industry which suffered from low efficiency and stagnation in the last century is now in the process of evolution through the introduction of more efficient generation and smart grids. Smart grids rely on two-way-automated metering infrastructure that will give consumers more control over their energy use and create new opportunities for energy retail services, infrastructure management and renewable energy generation.

Beside improved technologies for energy storage, the smart grid will provide an alternative through demand management of consumer's appliances and possible PHEV energy exchange with the grid.

In the core of the smart grid lies the development of the intelligent micro-grid. Such micro-grid will enhance distributed generation and will be incorporated to a smart grid through dedicated highways for command data and power exchange. The ultimate results will hopefully be more efficient supply that better serve consumer needs and lower costs and help to mitigate undue harmful emissions

REFERENCES

1. Hassan Farhangi: "The Path of the Smart Grid". *IEEE Power and Energy*, Vol. 8, No 1, Jan/Feb/2010.
2. H. Lui and G.S. Gallagher "Catalysing Strategic transformation to a low-carbon economy: A CCS roadmap for China". *Energy Policy Journal*, Col. 38, Jan 2010, pp 59-74.
3. M. Venables "Smart Policy?" *E&T*. Vol 4, Issue 18, Oct-Nov 2009.
4. Al Gore: "Our Choice—A Plan to solve The Climate Crises", Rodale 2009.
5. "The IEA Sees the Light", *European Energy Review*, June 4, 2010.
6. *Engineering & Technology* (E&T), Vol 5, issue 7, pp 6, 8 May 2010.

AMBITIOUS TARGETS FOR ENERGY RESEARCH AND DEVELOPMENT

DR. BRUCE STRAM
Element Markets
Houston, USA

INTRODUCTION

Almost 20 years ago, the nations of the world organized themselves to remediate what was perceived by many scientists and others as a planetary emergency: namely global warming which was thought to result from human emissions of GHG almost entirely the result of burning fossil fuels for energy. This problem was sharply focused by reports from the Intergovernmental Panel on Climate Change that had been established by the UN and World Meteorological Organization to specifically study this problem. Several reports from this effort concluded there was a high likelihood that, indeed, the indisputable increases in GHG concentration would lead to substantial and perhaps catastrophic climate warming. Further it was obvious that no individual nation acting alone could adequately address the problem. Hence was convened the UN Conference on Environment and Development in Rio de Janeiro in 1992. Virtually all nations sent delegations (though the U.S. was a close call). The principle result of this effort was the creation of a UN Framework Convention on Climate Change. That Convention set GHG emission reduction targets for virtually all signatories. Virtually none of those have been met. This was not entirely unexpected since the Convention was viewed as a starting point. However, it is a fact that in the almost 20 years since, and several more such meetings, GHG emissions have continued to rise. Further, the most recent meeting in Copenhagen made little progress and was a great disappointment to those highly concerned about the problem.

This outcome is not a surprise to many who predicted that serious flaws in the approach and divisions among nations as to costs and benefits would forestall any such meaningful agreement.[1]

It is reasonable to ask therefore, whether there are alternative courses of action which might achieve GHG reductions.

THE CURRENT REDUCTION PATH

The IEA (and others) have produced "scenarios" of GHG reductions which outline how the world might alter its various economies so as to hold global warming to minimally "acceptable" levels (2°C) for 2050. The analysis behind the scenarios is thorough and well documented and indicates very substantial changes that could (must?) occur in carbon related sectors of all major economies.

---

[1] Skepticism was contemporaneously expressed by myself, receiving little notice, and others. See "A Carbon Tax Strategy for Global Climate Change" in "Shaping National Responses to Climate Change", edited by Henry Lee, 1995 Island Press and "The Fix: A Global Warming Policy Practitioner's Handbook" Bruce N. Stram. Related Publication. May 2001, published on the AEI-Brookings Joint Center for Regulatory Studies web site (now housed at AEI).

These reductions must be made and coordinated among a number of sectors of the world economy. These reductions are usefully summarized in what has become known as "wedges". The following graphic is from the IEA report.

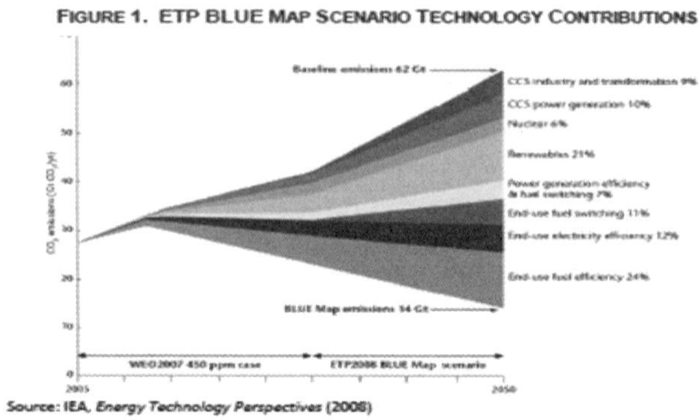

Fig. 1: ETP Blue Map Scenario Technology Contributions.

Causing these changes requires very substantial and costly government intervention in all major economies. Some of these changes are achieved through aggressive energy efficiency policies many of which are thought to more than pay for the cost of implementing the changes ("No Regrets Policy"). Further the policies inducing these changes must be negotiated and coordinated among these countries so that the burden is "fairly" shared and the goal is achieved. The policy outcomes per carbon are interrelated: for example a massive switch to electrification of transport systems will only have the projected effect if the power generation system is substantially decarbonized, which requires widespread implementation of CCS, and large increases in renewable generation. None of these are directly fomented by favorable market incentives, i.e., without aggressive governmental intervention, these efforts will not be profitable. The intervention must be in the form of command and control regulation or some version of cap and trade, or more likely, a combination of the two.

Needless to say these efforts will be costly (See the table below), and therefore will be extremely difficult to implement. While energy efficiency measures may achieve savings in the long run, the immediate impact will also be a burden of increased cost. Even among those who agree with the goal of GHG reduction (or purport to), there is the question of sharing the burden. Each affected sector can be foreseen to mount arguments and exert pressure to avoid or reduce impact. (This is already occurring.) At a minimum this effect will greatly slow down and perhaps ultimately forestall efforts.

Table 14.1 Annual investment by sector in Baseline and BLUE, 2015, 2030, 2050

	Baseline			BLUE		
USD billion	2010-2015	2015-2030	2030-2050	2010-2015	2015-2030	2030-2050
Power Geneneration	254	345	608	287	409	919
Transmission & Distribution	181	219	221	202	179	280
Industry	92	223	270	96	243	323
Transport	3975	4482	7386	4209	4673	8746
Total	3184	4329	6605	3204	4323	7670

Note: Industry includes iron and steel, chemicals, cement, pulp and paper and aluminum

Source: IEA analysis

Here is a short list of policies (not complete) that must be implemented to achieve the results indicated.

- Impose CCS requirements for power and industrial coal combustion or strongly incentivize them.
- Impose aggressive energy efficiency requirements so as to double the rate of decline in energy intensity.
- Create aggressive incentives or requirements for renewable energy sources.
- Impose aggressive requirements for carbon free vehicles.
- Require fuel switching to reduce carbon.

Obviously all this represents a very high degree of difficulty, and requires a singleness of purpose among nations to achieve a goal unseen in human history except perhaps in times of world war. And of course a war by definition has at least two sides. Further, achieving the goal has differential impacts among nations: some may benefit at least for a time, from warming.

THE POSTURE OF "SCIENCE"

The role of much climate science and climate scientists in this difficult situation is one of a messenger of bad news. At a minimum that surely leads to efforts to discredit and defame the messenger. And this is exacerbated by the fact that by its nature, climate science cannot make certain predictions. There are those scientists who with good reason disagree with what are perceived to be the consensus opinions.

The temptation among messengers bearing important bad news that "must" be acted upon is to slip over into the role of advocates. Any such proselytizing is a can of worms. First it is not in the nature of scientists in general to be very good at it. Second, advocacy can lead to biased analysis. And finally, and perhaps most important, it can lead to perceptions of bias among the body politic that must ultimately agree to very unpleasant measures.

Again while many of these difficulties could have been foreseen (and were by some), the litany here is more one of reportage rather than prediction. These things are already manifest.

A BETTER GOAL: IMPROVE HUMAN WELFARE

Perhaps achieving GHG reductions by widespread coercion is the wrong goal.

Human welfare has been fostered by dramatic improvements in technology and capital investment in substantial part induced by new technology. In particular, the reduction in cost for primary sources of energy, and reduction in the cost of related, end use services has been a vital key to these gains. (Some call energy the master resource.)

A goal of achieving even greater reductions in the cost of energy supply and use offers the promise of further gains in welfare. These gains include the benefits of lower costs for everyone but these could be particularly meaningful to those billion or more who are energy poor and therefore materially poor.

Fossil fuels have contributed substantially to past gains. They have been effectively exploited to reduce the cost of primary energy and secondary energy.

However, such things have their limits. Further cost reductions from fossil fuels seem to be reaching limits: primary sources, even if ultimate exhaustion is a long way off, clearly must be obtained under more challenging conditions; and efficiency in power production has thermodynamic limits as does internal combustion for transportation.

Thus a quest for lower cost energy must seemingly require alternatives to fossil fuels. Much lower cost nuclear and renewables come readily to mind. But such gains may not be on today's drawing boards and are not (and reasonably so) part of scenarios of GHG reduction. If such primary energy sources were available at lower cost than today's sources and coupled with low cost energy storage, human welfare could be advanced by displacing fossil fuels. Global warming would ultimately be mitigated by default (but should not be ignored). Other benefits would accrue: substantial fossil fuel externalities would be eliminated as well as substantial energy security costs related to fossil fuels.

Such goals could be stated as: expanding, on an internationally coordinated basis, energy RD&D targeted toward achieving lower cost energy supplies, and cost of energy services. While still very ambitious, such an effort could be similar to the Manhattan project (30 bil over 4 years, or .4% of U.S. GDP, equivalent to \$50 bil today, and \$240 bil worldwide) rather than a massive intervention in the world economies. This effort, coupled with "no regrets" policies are much more in the realm of feasibility than the massive interventions contemplated in the energy scenarios cited above and would ultimately generate net benefits, not costs. (Tying "no regrets" with an aggressive technology effort makes sense. The benefits of "no regrets" reduced energy survive at least in part the success of such technology whereas CCS, for example, does not. Secondly, it is in accord with the goal of increasing overall benefits. And third it acts as a partial backstop against a lack of success.)

A BETTER ROLE FOR SCIENCE?

Unequivocally creating lower cost energy supplies will be a great boon to mankind, differentially creating benefits for the poorest. Further, scientists and related engineers who have been given a very ambitious mission are often successful. The Manhattan Project is a case in point. Scientists can seek to initiate an effort to achieve this goal.

Obviously organizing and funding such an effort is a daunting undertaking. But the relevant question is not its degree of difficulty, but rather how that difficulty compares to the alternative path of mitigation outlined above. Here is a brief summary that captures the essence of such an enterprise even if not fully thought through:

Organize Multiple Manhattan Projects

- Senior scientists (i.e., Einstein's and Szilard's) and others as needed are gathered to select technology areas to drive toward lower cost energy target.
  - First order of business: are there such technologies that want for funding and focus?
  - Technology areas identified
- Zero to six government "Manhattan" Projects: China, EU, India, Japan, Russia and the U.S. are agreed to
- Each government and its Academy of Sciences (or other as appropriate) is given charge of organizing and funding one effort
- Periodically (every 3 years?), scientists reconvene to assess progress and prospects
- "No regrets" implementation is part of the deal
- Successful technologies are shared with an advantage to the developer

This author is really not qualified to determine that such technologies exist, and others are much more qualified to opine on how this might be approached, but several advantages compared to current mitigation efforts should b e apparent. First, $240 billion is not small change, but it much less costly than the burdens imposed by mitigation. Less cost means less political resistance. Second, the differential cost of mitigation tends to divide, whereas such an effort will tend to unite: everyone stands to gain rather than lose. Third, incentives to shirk, which are inherent in mitigation, are reversed: those who succeed first and most will get greater gain (even though the technologies should be substantially shared). Fourth, there would be less government and less international bureaucracy: mitigation seemingly requires creation of substantial multilateral administrations with real power over people's day to day well being. Surely few would accept this except in case of dire need with no alternative.

Finally, such a role should be much more copasetic with science skills. Scientists are problem solvers. Because of the financial and political stakes associated with current mitigation proposals, they are cast in the role of nags for which they are ill suited.

RISKS

Without question such a path is not risk free. First the research and development efforts may run up against insuperable barriers dictated by physical laws, or the technologies may be beyond our reach. Surely these goals are more of a stretch than even the Manhattan project itself. And of course any effort to reach agreement may founder.

But such risks cannot be evaluated in isolation. For those very concerned about global warming, these must be compared to the risks associated with alternative paths. The most germane comparison is to mitigation whose probability, at least in this author's view, for reaching agreement is much lower. And some critical technologies anticipated may fail: CCS and large gains in renewables as we now know them. For others there is the risk that the benefits of this effort don't exceed the costs even though they are relatively low in the grand scheme of things.

Such risks can be mitigated. As above, pursuing consistent "no regrets" policies will result in significant GHG reductions. Geoengineering can also be utilized to mitigate risk. And of course, mitigation versus aggressive pursuit of technology, presented here as either or paths, may be jointly entertained, mixing the most cost effective of either.

One risk, potentially present in both paths, is that scientists will be tempted to organize the selection of research and development to further their own goals rather than serve the greater goal. Here it is assumed that "Senior Scientists" who are especially immune to such incentives but have appropriate expertise can be persuaded to participate to set fundamental guidelines. Oversight of progress toward specified goals also mitigates this risk.

CONCLUSION

Can science find the way to both find clean energy and improve the material well being of mankind? Or more realistically, what is the probability this can be done? How clear was the path to achieving useful atomic energy? What degree of confidence existed that in fact it could be done? The theory was there, but there were substantial issues of implementation. How does this compare to the degree of difficulty in achieving the goal posed here? Then the risk of failure was not germane because we were potentially in a race for survival. Some contend that we face similar, albeit longer range, issues of the same ilk. Can the best scientists reach some sort of consensus on whether such a goal can be achieved? If so, such a consensus, or lack thereof, should be put forward to those who must make decisions on these issues.

In an uncertain world, the improvement in welfare from lower energy costs is a certainty.

EPILOGUE

As should have been noted in the original presentation, U.S. Energy Secretary Chu has called for the creation of mini-Manhattan projects, and a program, "Advanced Research Projects Agency Energy" has been formed to pursue such an end. Of course this author applauds the effort, but remains concerned about the implementation of the project and the steadfastness of the financing. More funding from more nations would also be very important to further the effort.

http://virtualology.com/MANHATTENPROJECT.COM/costs.manhattanproject.net/

FAST REACTORS WHEN?

RICHARD L. GARWIN
Thomas J. Watson Research Center, IBM Research Division
Yorktown Heights, USA

The Planetary Emergencies Seminar has long considered nuclear power as an important resource for the future. These considerations have included various reactors, including the existing thermal reactors, high temperature gas-cooled reactors, and fast reactors of various types.

We have also taken up fuel cycle issues. Among these has been the availability of importance of improving our knowledge of the cost of acquisition of uranium from seawater, and also from low-grade terrestrial ores.

Furthermore, the disposition of spent nuclear fuel has been discussed extensively, highlighting the benefits of long-term interim storage of spent fuel and of rationalizing the structure for operating mined, geologic repositories. The proposal would be to allow such large commercial activities, in contrast with the national programs that are now the norm, although none is yet in operation except perhaps that of Finland.

The benefits of fast-neutron reactors are well known since the advent of nuclear power, and indeed fast reactors were assumed to be the normal progression, especially before substantial uranium resources were discovered. The benefit lies in the breeding of fissionable material from the 99.3% or natural uranium that is not fissile (fissionable with thermal neutrons)—namely the U-238 in contrast with the 0.7% of natural uranium that is U-235.

It has always been considered that a 50 to 100-fold sparing of uranium would be possible in this way, expanding uranium resources at any given cost by that same factor, and, furthermore, allowing a similar increase in the cost of natural uranium feed without impairing the economics of nuclear power. The paper by Tanju Sofu at this session yesterday, on "Passive Safety in Fast Reactors" showed the experience in one small sodium-cooled reactor that serves as a feasibility demonstration and goal for larger, commercial reactors.

In August 2009, this seminar was treated to an exposition of the Terrapower program for fast-neutron Traveling-Wave Reactors (TWR) with very attractive properties. At that time, however, there was no public documentation on which to base a critical evaluation of the proposal, although I did make an attempt to infer from the presentation what must have been assumed for fuel burn-up, initial U-235 load, and the like.

The 2009 TWR was considered an axially propagating burn wave, in which enriched uranium or Pu fuel at one end of a long cylinder of fuel in the reactor would be critical, breeding Pu-239 in the axially adjacent region, and then burning that Pu-239 as the breed-burn wave moved along the axis of the cylinder. I comment on this configuration (Figure 1) in my Beijing paper of March 15, 2010.[1]

[1] http://www.fas.org/rlg/3_15_2010%20Fast%20Breeder%20Reactors%201.pdf.

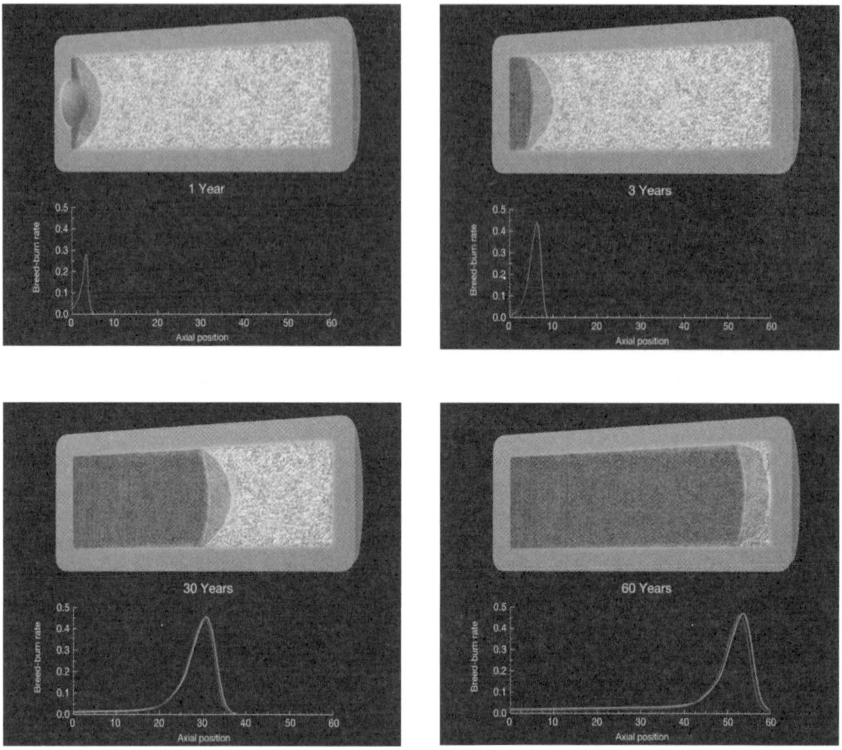

Fig. 1:

Further work by Terrapower has changed the baseline approach to a relatively static configuration of fission power, with fuel elements being shuffled into and out of that region[2], with the configuration exemplified in Figure 2.

---

[2] "Traveling-Wave Reactors: A Truly Sustainable and Full-Scale Resource for Global Energy Needs," by Tyler Ellis, et al, Proceedings of ICAPP '10, June 13-17, 2010, Paper 10189.

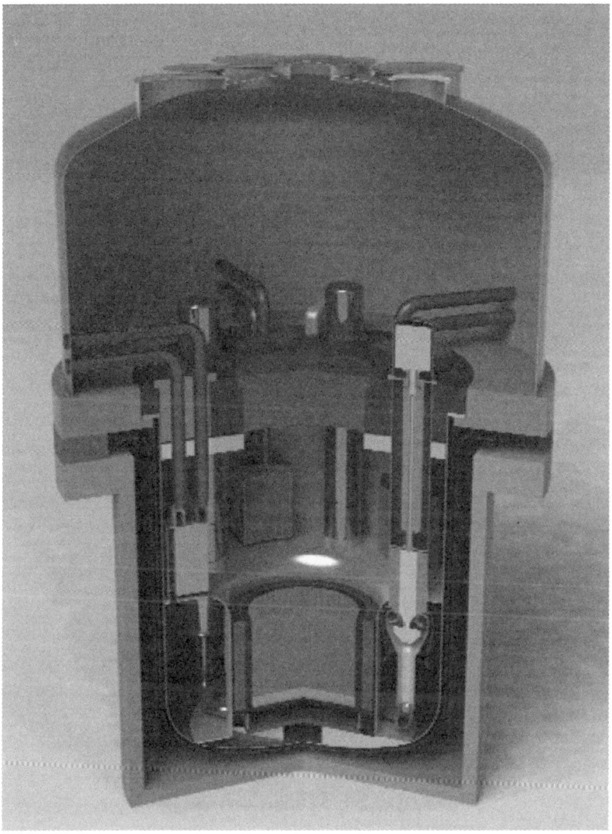

Fig. 2: "Possible practical engineering embodiment of a TWR" (from Ref. 2).

The 2010 TWR Report (Ref. 2) indicates a burn-up (BU) of the fuel of an average of 15%, noting that "1 at % is equivalent to 9.4 MWd/kgHM." A core that produces 1.25 GWe at 80% capacity factor (that is on for full-power production an equivalent of 80% of the days) fissions about 1 ton[3] of fuel per year, corresponding to a core mass of Yr/BU, where Yr is the residence time in years and BU the fractional burnup. For Yr = 6 years, the core mass would be about 6/0.15 = 40 tons.

Thus the cylindrical core contains approximately 40 ton of "heavy metal," which is at the beginning of operation mostly depleted uranium (DU) left over from enrichment of a vast amount of uranium already used for nuclear power in light-water reactors. The "starter fuel" could be Pu-239 from excess weapon plutonium, more complex or less neutronically reactive plutonium in the form of MOX from reprocessing the spent fuel of

[3] The unit "ton" in this paper is taken to mean "tonne"—1000 kg.

LWRs. Importantly, newly enriched uranium from newly mined ore, could be used, simply because the demand for electrical power and for an economical breeder reactor could be so large that could be filled only by enrichment.

One great virtue of a fast reactor is that it is not very sensitive to the enormous absorption cross-sections of certain fission products for slow neutrons. Furthermore, it is provided with a large amount of fertile material, so as to breed the next generation's plutonium starter fuel. According to Ref. 2, "the comparable TWR (1-GWe) requires an initial core load that in the early TWRs may contain on the order of two times as much fissile material as an LWR first core." The LWR has about 100 tons of 4.4% U-235, so an initial fissile inventory on the order of 4.4 tons. That for the TWR is thus stated to be on the order of 9 tons. In my August 24, 2009 commentary on Terrapower's linear TWR, I estimated that "For a large utility reactor with 60-yr core life, such a reactor might start with 60 t of fuel enriched to 15% U-235 (so 9 tons of U-235) plus 320 tons of DU. It would therefore generate 3 GWt for 60 years from 380 t of heavy metal, for a burn-up of 173 GWd/t. I noted there and here also:

> *"Although the impression might be given that the uranium content is 100% consumed, I figure that the 173 GWd/t compares with about 790 GWd/t for 100% conversion of uranium to fission products, so is a burn-up of about 22%."*

The estimated 9 t of fissile material for starting the linear TWR is consonant with the approximately 9 t inferred from Ref. 2. It is conventional in LWRs to do fuel shuffling at the time of refueling, in order that fuel elements should be downloaded at the typical 54 months, having had the same exposure to flux, and therefore in fuel burn-up. Unfortunately in LWRs, as is the case in the TWR, the ends of the fuel elements have much lower burn-up than does the center.

Furthermore, although the 15% burn-up is better than the 0.5% burn-up of raw uranium in a TWR, it is far from the 100% advertised by supporters of breeder reactors. The additional factor of 6 or so is to be obtained by "fuel repurposing," which term the authors prefer to "fuel recycling" with its implication of chemical separation and refabrication of fuel. Because the TWR is to use metal fuel, it is argued that physical processes can do an adequate job of separating out fission products that occupy volume in the fuel rod. Thus, fuel rods would be chopped and the ends which are not much depleted in fissile material could be used in more valuable positions than the centers, which are more highly burned. Ref. 2 indicates that they will have a "peak burn-up in the range of 28-32%," so that several such cycles (perhaps 6) would be planned.

For most of the past decade, I have been urging the creation of a world laboratory for breeder reactors, with the purpose of thoroughly exploring the design of three of four different types of breeder reactors, each with its own fuel form and fuel cycle. Evidently, the fuel cycle is a necessary part of the breeder, and the fuel form greatly influences the fuel cycle. My proposal builds on the great advances that have been made in computing in the Stockpile Stewardship Program for U.S. nuclear weapons. The advances include faster hardware, mostly by building very highly parallel machines of consumer electronics, and the pioneering of efficient means of having these "cores" operate effectively in parallel.

In fact, reactor design and safety analysis is a more difficult problem than is nuclear weapon design and maintenance, essentially because there are so many ways in which a system designed to do one thing can do something else by accident or intent. For stockpile stewardship, the important questions are whether the weapons maintain their intrinsic safety, and whether they will, with some reasonable probability, give the yield for which they were designed and assumed to function when they were first put into the stockpile.

Nuclear weapons also have certain specific safety requirements including that there should be less than one part in a million probability during an accident that the weapon will give a yield greater than that of 2 kg of high explosive equivalent, and that in non-accident scenarios, the probability should be less than one in a billion throughout the weapon lifetime.

The World Breeder Lab would have open technology and computing, and therefore would forego the industrial competition that is often a great spur to innovation. But the field of particle physics and great accelerators has long functioned in this way, with outstanding results.

In any case, Terrapower seems to be moving ahead with the substance of one of the sub-programs of a World Breeder Laboratory, in analyzing thoroughly not only the reactor itself, but the solid metal fuel, undersized for the ferritic steel sheath, with the gap at operating temperature filled with molten sodium. This is essential, especially with high-burn-up metal fuel, because the fuel density is so much greater than that of the fission products produced. The fuel would soon fail and the fuel rod cladding swell, if there were not extra space provided in this way. The sodium spacer also facilitates splitting the sheath in order that the metal fuel be available for melting, during which process most of the fission products will combine with a reagent (perhaps including the zirconium oxide of the crucible!) to allow purified metal fuel to be drawn off. In large part, the residual metal consists of the fissionable fuel material plus the residue of the depleted uranium that has not yet been transformed to Pu-239. Obviously, there must be makeup with fresh raw uranium or fresh DU in order to fabricate a new core for the TWR.

According to Ref. 2, the fuel could be withdrawn from the TWR at the end of life, with "a factor 3 multiplication" of the initial fissile content, so with the capability of fueling 3 new TWRs. Alternatively, Ref. 2 proposes removing fuel during the 15-20 year life to fuel another TWR. Assuming this doubling time of 20 years leads to a compound growth rate of 3.5% per year. The fueling of three reactors from a first-generation TWR after 40 years (thus, presumably, plus time for fuel processing and fabrication and loading into the reactor) would be an accomplishment, but it would correspond to a growth rate of the TWR population itself of about 2.7% per year. If one is planning an all-TWR future, then they must take over not only from the LWRs, but also from fossil fuels, and that will require a growth rate in excess of 10% per year. Thus, by far the most of the TWR population must be other than autogenous, and after the modest amount of separated HEU and weapon plutonium are exhausted, the TWRs must be started on enriched uranium from ore, creating a substantial and enduring requirement for enrichment capacity.

Looking to the long-term future, Ref. 2 places great emphasis, as do I, on the utilization of uranium from seawater, where there is 4,000 million tons. This resource would, of course, be valuable for fueling existing LWR-type reactors, and breeder reactors could afford essentially any cost for seawater uranium, given that it can be obtained in an environmentally acceptable fashion.

In summary, Terrapower may contribute to a bright future for safe, economical breeder reactors, as their baseline approach under intensive analysis has become more conventional than the linear traveling-wave reactor presented at Erice in 2009. It would be useful to have available the comparative analyses of the linear vs. the stationary breed-burn "wave" so that the world could learn from the great investments made in this work.

SECTION 5

SESSION 5

INFORMATION SECURITY

FOCUS: CYBER CONFLICTS AND CYBER STABILITY—FINDING A PATH TO CYBER PEACE

SCIENCE AS A PRIME MOVER IN THE CONSTRUCTION OF A SAFE AND STABLE ORDER OF CYBER SPACE

HENNING WEGENER
Ambassador of Germany (ret.), Information Security Permanent Monitoring Panel, World Federation of Scientists, Madrid, Spain

Refocusing on the role of science and applied science as the motor of progress, in an interdisciplinary setting, is nowhere more apposite than in the further expansion and development of the digital world. Information and Communication Technologies (ICT) increasingly become the new dominant paradigm of all aspects of human endeavour, providing the all-encompassing operative system of human societies.

The optimal functioning of this operative system and the realization of its development potential are thus ever more crucial.

In its work over the last few years, the PMP has concentrated on the significance of shielding information and vital systems from the impact of conflict and hostile intervention—through cybercrime, cyberterrorism, cyberwar. Their significance is undiminished indeed. Information security in this primary sense, the reliability, integrity and trustworthiness of information and networks and their effective defense, remain top global values, a public as much as a private good.

Lately our work has given emphasis to the incipient concept of *cyber peace*, a concept which the PMP perhaps more than others has helped to develop. In this work it has become increasingly clear that cyber peace not only presupposes the absence of attack and conflict. The concept is shaping up as broader, as going beyond damage wilfully inflicted.

Cyber peace requires an overall context that enables the functioning of digital systems and the generation of their benefits. As we have frequently stressed, important other ingredients include widely accepted codes of peaceful conduct, modes of cooperation, indeed, as our PMP has stipulated early on, a "universal order of cyber space". But risks and threats that arise from technological causes must also be assessed and harnessed. This gradual conceptual shift is clearly visible in our latest papers on Internet stability and cyber peace, as for instance in the Erice Declaration of last year. Stability and the concept of cyber peace are closely intertwined. We must be concerned not only with hostile intent and behavior, but also with the technical underpinnings of the digital world and threats emanating therein. Vulnerabilities stem from multiple causes.

Among the stability requirements are first the accessibility and technical prowess of digital systems, their performance characteristics and capabilities, interoperabilities and time-critical system resiliencies.

Cyber space becomes unstable not only as a consequence of deliberate attacks (often with unpredictable consequences and snow-ball effects) but also if important technical resources are unavailable, or fail through technical defects or operational errors, invalid data, or software aging. Performance, technical dependability, and integrity (as the absence of hostile intervention) are thus the triple hallmark of cyber stability, and beyond it, cyber peace.

Mastering these three major impact factors becomes more crucial and difficult as the complexity of digital systems and processes continues to grow exponentially, propelled by an accelerating rate of innovation, a revolution in connectivities, miniaturization, automation and advanced microprocessor technologies. We are witnessing a complexity explosion.

The scientific methods and technical tools for the mastery of this complexity explosion, but before all the top-of-the art conceptual scientific thinking they require, have not yet been fully developed. Essential steering functions, scientifically founded, need to be introduced and applied to key sectors of human endeavour, including the future world economy. The scientific management challenges include the transition to new control paradigms which will necessarily include autonomous and bio-analogous control systems. These may—or will—result in the emergence of system behaviors as yet unexplored, and in all probability leading to critical impacts on cyber stability. The protection of Critical National Infrastructures and the optimization of industrial control systems (SCADA) continue to present a permanent research challenge. The scientific agenda resulting from these developments is awe-inspiring.

One example of the need for sophisticated new approaches is the energy economy where the establishment of intelligent energy management nets—smart grids—is the prerequisite for grid stability and distribution efficiency in an energy economy which has to accommodate novel energy mixes, and to optimize the relationship between supply and demand in terms of energy efficiency; billions of euros could be saved in each national economy, and climate-sensitive emissions could be minimized. The advances of electromobility, indispensable in a future energy economy, depend on a new level of sophistication in embedded systems and on network connections between vehicles and their energy environment.

The emergence of huge new smart nets to cope with the economic and ecological challenges of the future generates the demand for an enhanced scientific focus on the requirements, functions and capabilities of such smart grids, but also on information security needs. The WFS's recent paper on Top Cyber Security Problems That Need Resolution in one of its chapters provides a list of the scientific challenges that need to be confronted: e.g., to enhance resilient control systems, provide network interoperabilities, more robust data structures for identification and authentification, and new methods of meeting the security challenges of mobile and wireless systems, as well as cloud and grid computing. The new ITU Focus Group on Smart Grids will have an important role to play in this regard.

Our plenary meeting today is of limited duration. Fortunately, Secretary General Touré has already made an important contribution to our topic in the Inaugural Session; he will speak to us again today. The other four presentations to follow—all by members of our Permanent Monitoring Panel—will no doubt also lead us further in illuminating our task to work towards cyber stability and cyber peace.

The author gratefully acknowledges the advice of Prof. Axel Lehmann on the technical developments described above.

FROM A DIGITAL DIVIDE TO A SECURITY DIVIDE

JODY R. WESTBY[1]
Global Cyber Risk LLC, Information Security Permanent Monitoring Panel, World Federation of Scientists, Washington, USA

BRIDGING THE DIGITAL DIVIDE AND CREATING THE SECURITY DIVIDE

On April 30, 1995, the U.S. National Science Foundation turned the Internet over to four commercial providers, igniting a digital revolution that has dramatically changed the way people around the world work, live, learn, and play.[2] By 1996, the Digital Divide between the connected "haves and have-nots" was one of the most pressing policy issues of the day. Donor organizations established telecenters and kiosks in developing countries and encouraged governments to develop e-government systems and liberalize telecommunications services as a means of driving connectivity.

In the fifteen years hence, great progress has been made in bridging the Digital Divide: today, there are 233 countries and territories connected to the Internet and 1.97 billion online users, with connectivity rapidly increasing in developing countries.[3] The World Summit on the Information Society's goal that more than half of the world's population will have access to information and communication technologies (ICTs) by 2015[4] was reached seven years ahead of schedule due to the dynamic growth in mobile telephony: in 2008, there were more than 4 billion mobile subscriptions. Although the *United Nations' Information Economy Report 2009* cautions against complacency and more progress needs to be made in bridging the Digital Divide—especially with respect to expanding access to broadband[5]—the momentum is unstoppable.

In the rush to bridge the Digital Divide, however, a Security Divide was created. Pleas by security experts that privacy and security should be incorporated into connectivity projects were largely ignored by donor organizations under the belief that connectivity was more important and security could come later. At that time, developing countries were struggling to get schools wired, teachers trained in the use of ICTs in education, and curricula updated in a push to become "knowledge economies." They had

---

1 Jody R. Westby is CEO of Global Cyber Risk LLC, based in Washington, DC and serves as Distinguished Fellow to Carnegie Mellon CyLab, She is co-chair of the World Federation of Scientists' Permanent Monitoring Panel on Information Security and chairs the American Bar Association's Privacy & Computer Crime Committee. She led the development of the *ITU Toolkit for Cybercrime Legislation* and served as a member of the ITU Secretary-General's High Level Experts Group on cyber security. She has co-authored and edited four books on privacy, security, cybercrime, and enterprise security programs.

2 Dave Kristula, "The History of the Internet," Aug. 2001, http://www.davesite.com/webstation/net-history4.shtml.

3 Internet Usage Statistics, "The Big Picture," http://www.internetworldstats.com/stats.htm.

4 "OECD and the World Summit on the Information Society (WSIS), "3. Background reading," The WSIS "Plan of Action, " http://www.oecd.org/document/21/0,3343,en_2649_37441_35282901_1_1_1_1,00.html.

5 *Information Economy Report 2009: Trends and Outlook in Turbulent Times,* United Nations Commission on Trade and Development (UNCTAD), UNCTAD/IER/2009, 2009 at xi-xii, www.**unctad**.org/en/docs/ier**2009**_en.pdf.

little information about privacy and security, much less access to information on best practices and standards for securing computer systems and networks. Thus, as connectivity to the Internet increased from 360 million people in 2000 (of which about two-thirds of those consisted of Americans and Europeans) to nearly two billion today, most of these new users now operate in an environment ill equipped to deal with cybercrime. Legal frameworks are inadequate, law enforcement personnel are not trained to assist with cyber investigations, and networks and systems operate without necessary security controls.

Increased connectivity also advanced globalization and pushed information technology outsourcing (ITO) and business process outsourcing (BPO) to developing countries with workers skilled in ICTs. Today, data critical to business operations is processed in countries that do not have cybercrime laws or legal frameworks that enable them to assist in cyber investigations, conduct search and seizure of electronic evidence, and cooperate in tracking and tracing cyber criminal activities.

Consider the growth rates for Internet usage from 2000-2009 in Brazil, Russia, India, and China, dubbed the BRIC countries by Goldman Sachs' chief economist James O'Neill, who predicts that their economies will overtake the six largest western economies by 2032[6]:

Country	**Growth Rate Internet Usage**[7]
Brazil	1,250.2%
Russia	1,359.7%
India	1,520.0%
China	1,500.0%

Although each country has some form of cybercrime law, none of the laws is considered to be harmonized with developed nations' cybercrime laws. Moreover, two of these countries—Russia and China—are considered havens for cyber criminals and have notoriously failed to assist in the investigation of significant cyber criminal activities originating from within their borders. India and China are also favored destinations for ITO/BPO, and corporate data in these countries is at greater risk due to inadequacies in their legal frameworks.

STATE OF CYBERCRIME

The state of cybercrime today is much more sophisticated than the early days of the Internet. The script kiddies who wrote hacking programs and attacked systems for ego gratification or to impress peers have given way to sophisticated cyber criminals who operate either as rogue actors or operatives who conduct cyber attacks at the behest of organized criminal rings, terrorist organizations, and nation states. Cybercrime today

---

6 Gillian Tett, "The story of the BRICs," *Financial Times,* Jan. 15, 2010, http://www.ft.com/cms/s/2/112ca932-00ab-11df-ae8d-00144feabdc0.html?SID=google.

7 "The Internet Big Picture: World Internet Users & Population Stats," Internet Usage Statistics, http://www.internetworldstats.com/stats.htm.

occurs 24/7 and consists of financial fraud, identity theft, spam and phishing scams, hacking, economic espionage, theft of intellectual property, web defacing and hijacking, and denial of service attacks. One thing is clear: the bad guys are winning.

Deloitte & Touche's report on the 2010 CSO CyberSecurity Watch Survey bluntly declared that:

> Threats posed to organizations by cyber crimes have increased faster than the potential victims—or cyber security professionals—can cope with them, placing targeted organizations at significant risk.[8]

The victims include everyone: individuals, businesses of all sizes, and governments. Individuals are targeted for identity theft and other financial fraud. The Javelin Strategy and Research Center, who has been tracking identity theft in the U.S. since 2004, reports a 27% increase between 2007 and 2008.[9] The Anti-Phishing Working Group (APWG) reports an unprecedented rise in phishing scams targeting primarily financial and payment services: the total number of infected computers rose more than 66% from the end of 2008 to mid-2009. The total number of computers infected was approximately 12 million, comprising 54% of the total sample of computers belonging to individuals and companies in over 100 countries. This was a 34% rise since the fourth quarter of 2008. Banking Trojan and password-stealing infections rose 186% during this period.[10]

The international aspect of cybercrime is one of the reasons it is such a successful crime: the criminals are often located in jurisdictions far away from the victims where other countries' cybercrime laws do not apply and where local law enforcement has other priorities. Increasingly, cybercrimes are directed toward businesses. For example, organized cyber gangs in Eastern Europe have begun exploiting small- and medium-sized businesses and committing funds transfer fraud. A U.S. school district had $700,000 taken from its bank account, and a Texas plumbing equipment company lost $1.2 million through malware planted on its computers.[11]

Web defacement remains the most common form of Web attack, but cyber criminals continue to take advantage of Web application vulnerabilities that enable them to plant malware on a site that will infect visitors to the Web page with viruses, Trojans, or root kits. This indicates that cyber criminals are beginning to prefer Web delivery of malware through business sites over email delivery to individuals. The sharpest rise in web attacks in 2009 involved Web 2.0 sites.[12] Cloud computing also has been under

---

8 *Cyber crime: a clear and present danger,* Deloitte & Touche LLP, Center for Security & Privacy Solutions, Jan. 22, 2010 at 3, http://www.deloitte.com/view/en_US/us/Insights/centers/Center-Security-and-Privacy-Solutions/article/bcdc005f1e056210VgnVCM100000ba42f00aRCRD.htm.

9 "2009 Identity Theft Statistics," Javelin Strategy and Research Center, 2009, http://www.spendonlife.com/guide/2009-identity-theft-statistics.

10 *Phishing Activity Trends Report: 1st Half 2009,* Anti-Phishing Working Group, January—June 2009 at 7-10, http://www.antiphishing.org/phishReportsArchive.html.

11 Brian Krebs, "European Cyber-Gangs Target Small U.S. Firms, Group Says," *The Washington Post,* Aug. 25, 2009 at A9, http://www.antiphishing.org/phishReportsArchive.html.

12 *The Web Hacking Incidents Database 2009,* Biannual Report, Breach Security, Inc., 2009 at 2, 4, http://www.breach.com/resources/whitepapers/index.html.

attack. Twitter had corporate information stolen from its Google cloud[13], and presenters at the 2009 Black Hat security conference revealed how Amazon's Elastic Compute Cloud can be manipulated for "resource theft attacks."[14] Trend Micro predicts that in 2010 "new attack vectors will arise for virtualized/cloud environments."[15]

Hacking and economic espionage continue to play a dominant role in cybercrime, with nation states figuring prominently as accusations fly across headlines. A few years ago, the *Washington Post* called the Russian Business Network (RBN) a "Conduit for Cybercrime".[16] RBN hosts companies' Web sites, many of which are engaged in cyber criminal activities. In its 2009 Threat report, Symantec boldly notes that the RBN "has been credited with creating approximately half of the phishing incidents that occurred last year" and that it is "also thought to be associated with a significant amount of the malicious activities on the Internet in 2007."[17]

RBN's activities provide a good case example of why cybercrime becomes more pernicious every year. The companies that buy hosting services from RBN target victims in other countries. Thus, when cybercrime investigators call, Russian authorities are being asked to help investigate foreign crimes, so their assistance is minimal, if at all. In fact, Russia's cooperation rate on international cyber investigations is close to zero. Even during the attacks on the government of Estonia and many of its critical infrastructures, Russia refused to cooperate in tracking and tracing the attacks. "They won't even pick up the phone," complained Rein Lang, Estonia's minister of justice regarding Russia's refusal to help end the attacks or investigate evidence that Russian state employees were behind them.[18] As one reporter noted:

> Interestingly, there has developed the widely-held belief that writing malicious viruses and code is no worse than working as a designer at a munitions company...And then you have the fact that it is almost inconceivable to get arrested in Russia for cyber-crimes, even if the Government becomes aware of the reality.[19]

---

13 Jon Brodkin, "Twitter/Google Apps hack raises questions about cloud security," *Network World,* July 15, 2009, http://www.networkworld.com/news/2009/071509-theft-twitter-docs.html.

14 "Black Hat demonstration demo aids: Amazon," *SensePost*, Aug. 8, 2009, http://www.sensepost.com/blog/3797.html.

15 *The Future of Threats and Threat Technologies: How the Landscape is Changing*, Trend Micro, Dec. 2009 at 5, http://us.trendmicro.com/us/search/?q=future+of+threats&search.x=0&search.y=0&search=search.

16 Brian Krebs, "Shadowy Russian Firm Seen as Conduit for Cybercrime," *The Washington Post,* Oct. 13, 2007, http://www.washingtonpost.com/wp-dyn/content/article/2007/10/12/AR2007101202461.html.

17 *Symantec Global Internet Security Threat Report: Trends for 2008,* Symantec, Apr. 2009 at 11, http://www.symantec.com/business/theme.jsp?themeid=threatreport.

18 Peter Finn, "Cyber Assaults on Estonia Typify a New Battle Tactic," *The Washington Post,* May 19, 2007 at 1, http://www.washingtonpost.com/wp-dyn/content/article/2007/05/18/AR2007051802122.html.

19 Kelly O'Connell, "Russian Company Outed as Mother of all Cybercrime," Internet Business Law Services, Inc., Oct. 24, 2007, http://www.ibls.com/internet_law_news_portal_view.aspx?id=1887&s=latestnews.

China has also been blamed for cyber criminal activities. In 2007, The Director-General of the U.K.'s counter-intelligence and security agency, MI5, posted a confidential letter to 300 CEOs and security officers on the Web site of the Centre for the Protection of National Infrastructure, warning them that their infrastructure was being targeted by "Chinese state organizations" and that the attacks were designed to defeat security best practices.[20] In 2009, the Information Warfare Monitor released a report detailing its investigation that uncovered a network of 1,295 infected computers in 103 countries. The computers were infected with a Trojan known as *gh0st RAT* that was "consistently controlled from commercial Internet access accounts located on the island of Hainan, People's Republic of China."[21] The compromised systems included the offices of the Dalai Lama and high-value diplomatic, political, economic, and military targets around the world.[22]

Cybercrime in 2010 began with Google's assertion that Chinese hackers had stolen some of its intellectual property and accessed or obtained information from the Gmail accounts of Chinese human rights activists. Google also revealed that it had discovered that at least 20 other companies from a wide range of industry sectors had been targeted in a similar manner.[23] Google later revised the number of additional companies that had been attacked to 34.[24] Reports have indicated that portions of the malware used to attack Google and other companies were written by Chinese security professionals, consultants, and temporary contractors, some of whom were based at Chinese and U.S. companies in China. The attacks were routed through the servers of two schools with strong technical capabilities and ties to the Chinese government.[25] Google has turned to the U.S. National Security Agency (NSA) for assistance on the investigation and to counter future attacks.[26] While this can be viewed as an unusual move for a private sector company, it also likely indicates that the attacks were more sophisticated than reported.

The Google-China confrontation was followed by the leakage of a restricted MI5 report that warned U.K. business executives of Chinese efforts to eavesdrop and steal confidential and proprietary data. The report warned business travelers about accepting trade show gifts of cameras and memory sticks because they contained spyware that

---

20 "MI5 warns over China spy threat," BBC News, Dec. 2, 2007, http://news.bbc.co.uk/2/hi/africa/7123970.stm.

21 *Tracking GhostNet: Investigating a Cyber Espionage Network,* Information Warfare Monitor, Mar. 29 2009 at 5, http://www.scribd.com/doc/13731776/Tracking-GhostNet-Investigating-a-Cyber-Espionage-Network.

22 *Id.* at 6.

23 "Statement from Google: A new approach to China," *The Washington Post,* Jan. 12, 2010, http://www.washingtonpost.com/wp-dyn/content/article/2010/01/12/AR2010011202903_pf.html.

24 Ariana Eunjung Cha and Ellen Nakashima, "Google attack part of vast campaign," *The Washington Post,* Jan. 14, 2010 at A1, A14, http://www.washingtonpost.com/wp-dyn/content/article/2010/01/13/AR2010011300359.html.

25 Ellen Nakashima, "Security pros, consultants wrote code to attack Google," *The Washington Post,* Feb. 20, 2010 at A9, http://www.washingtonpost.com/wp-dyn/content/article/2010/02/19/AR2010021902643.html?nav=hcmoduletmv.

26 Ellen Nakashima, "Google to enlist NSA to help it ward off cyberattacks," *The Washington Post,* Feb. 4, 2010, http://www.washingtonpost.com/wp-dyn/content/article/2010/02/03/AR2010020304057_pf.html.

would allow remote access to computer systems. The document also described how China has attacked U.K. defense, energy, communication, manufacturing companies, public relations companies, and international law firms.[27]

Headlines about cyber events no longer feature the name of viruses: today, they focus on the names of nation states. One recent *Financial Times* headline screamed, "Cyberwar breaks out as Iranians hack into Chinese search engine."[28] The Web page of Baidu, Google's Chinese competitor, had been defaced to include the Iranian flag and the words "Iranian Cyber Army," the name of the group that hacked into Twitter after its support for Iranian revolutionaries. Chinese hackers retaliated in similar fashion.[29]

With unnerving frequency, the Web sites and systems of governments are cyber targets. The 2007 attacks on government and private sector networks in Estonia were the watershed event that served as a government wake-up call. The attacks quickly escalated, seriously impacting Estonian government Web sites and systems and shutting down the country's newspaper and financial networks. The attacks demonstrated the rapid pace at which a cyber attack can become to a national security issue, involve other nation states, and raise the issue of collective defense.

A few months after the Estonia attacks, U.S. Pentagon computer networks were allegedly hacked by the Chinese military in what has been called "the most successful cyber attack on the U.S. defense department,"[30] shutting down parts of the Pentagon's systems for more than a week. Chinese hackers have also been blamed for attacks that compromised German government systems[31] and cyber espionage incidents against the United Kingdom's (U.K.) government systems.

The 2008 attacks on Georgian systems during the Russia-Georgia conflict over South Ossetia were a more obvious example of cyber warfare that demonstrated the degree to which governments are dependent upon computers and communications networks—especially during crisis management. A sequence of distributed denial of service (DDOS) attacks against Georgian government Web sites essentially shut down government communications.

The attacks on Google and other companies heightened the awareness of governments and business executives about potential attacks on critical infrastructure. The U.S. Federal Bureau of Investigation has warned that terrorists also have indicated an interest in the vulnerabilities of U.S. critical infrastructures to cyber attacks.[32] Although security experts have been predicting major infrastructure attacks for years, it has been hard to get private sector companies to take the threat seriously. A key finding of the

---

[27] David Leppard, "China bugs and burgles Britain," *TimesOnline*, Jan. 31, 2010, http://www.timesonline.co.uk/tol/news/uk/crime/article7009749.ece.

[28] Kathrin Hille, "Cyberwar breaks out as Iranians hack into Chinese search engine," *Financial Times*, Jan. 13, 2010 at 2, http://www.ft.com/cms/s/0/0ed1d424-ffe2-11de-ad8c-00144feabdc0.html.

[29] *Id.*

[30] Demetri Sevastopulo, "China 'hacked' into Pentagon defence system," *Financial Times*, Sept. 6, 2007 at 1.

[31] John Blau, "Chinese hackers infect German government with spyware," *it World Canada*, Aug. 27, 2007, http://www.itworldcanada.com/news/chinese-hackers-infect-german-government-with-spyware/00215.

[32] Siobhan Gorman, "FBI Suspects Terrorists are Exploring Cyber Attacks, *The Wall Street Journal*, Nov. 19, 2009, http://online.wsj.com/article/SB125850773065753011.html.

Project Grey Goose Report on Critical Infrastructure was that state and/or non-state actors from China, Russia, and Turkey "are almost certainly targeting the networks of energy providers and other critical infrastructures" in the U.S., Brazil, Russia, and the EU.[33] In April 2009, a former U.S. counterintelligence official warned that the Chinese had penetrated U.S. electrical grids and "our networks are being mapped."[34]

The reality is that the capabilities of cyber criminals have become so advanced that they threaten the national and economic security of every country. Dennis Blair, former U.S. Director of National Intelligence, noted in his statement for the record regarding the 2010 *Annual Threat Assessment of the U.S. Intelligence Community*:

> I am here today to stress that, acting independently, neither the U.S. government or the private sector can fully control or protect the country's information infrastructure....Malicious cyber activity is occurring on an unprecedented scale with extraordinary awareness....We cannot protect cyberspace without a coordinated and collaborative effort that incorporates both the U.S. private sector and our international partners.[35]

SOLVING THE PROBLEM: THE INTERSECTION OF SCIENCE, TECHNOLOGY, AND LAW

The reason cyber criminals have been able to advance so dramatically is that cybercrime has become the near-perfect crime: cyber criminals rarely get caught. They can operate independently, with organized criminals or terrorists, or at the behest of nation states, and it is very difficult to track and trace an attack to its origin. "Investigating cybercrimes across borders almost always presents problems for the investigator," notes Anthony Reyes, CEO of ARC Group of New York and former president of the International Executive Committee of the High Technology Crime Investigation Association. "You need to understand the native language and the cybercrime laws of the country involved and be familiar with any governmental agreements between the countries; it can be a nightmare," he adds.

There are technical issues with investigating cybercrimes but primarily, investigations and prosecutions are plagued with legal and policy problems. Successfully investigating, charging, and convicting a cyber criminal is a complicated and difficult process:

- The cyber activity has to be tracked and traced to its point of origin and requires communicating with contact points along the communication path on

---

33 *Project Grey Goose Report on Critical Infrastructure: Attacks, Actors, and Emerging Threats,* GreyLogic, Jan. 21, 2010, http://www.scribd.com/doc/25550091/Proj-Grey-Goose-report-on-Critical-Infrastructure-Attacks-Actors-and-Emerging-Threats.

34 Ellen Nakashima and john Pomfret, "China proves to be an aggressive foe in cyberspace," *The Washington Post,* Nov. 11, 2009 at A11, http://www.washingtonpost.com/wp-dyn/content/article/2009/11/10/AR2009111017588.html.

35 Dennis C. Blair, "Far-Reaching Impact of the Cyber Threat," *Annual Threat Assessment of the U.S. Intelligence Community for the Senate Select Committee on Intelligence,* Office of the Director of National Intelligence, Feb. 2, 2010 at 2-4, http://www.bespacific.com/mt/archives/023410.html.

a 24/7 basis. This usually involves tracing communications backward from point to point. This requires the cooperation of each provider along the route and may well involve multiple requests for assistance from governments and/or law enforcement. These people have to be available and skilled in assisting with cyber investigations.

- If the person conducting the activity is actually identified, it may be difficult to get assistance from local law enforcement if the cybercrime is not illegal in that country (or if the country was sponsoring the activity). Even if the activity is illegal, the search and seizure of electronic evidence may not occur if local law enforcement is not skilled in cyber investigations,
- Even if the digital evidence is seized, the capture may have been conducted in such as manner that the evidence is tainted or inadmissible in the courts of the country pursuing the investigation.
- Even if the evidence is gathered and is admissible and the suspect is located, it may not be possible to extradite the person due to treaty provisions or the laws of the country where the suspect lives.

Turning the tide on cybercrime will require multidisciplinary action in many areas. The following nine points identify the main issues that must be addressed if cyber criminal activity is to be curbed and some degree of cyber stability is to be attained for the 233 countries and 1.97 billion people connected to the Internet:

Problem Points

1. Many countries do not have cybercrime laws and, of those that do, there are varying definitions of what constitutes a cybercrime.
2. Procedural rules regarding investigations, collection and preservation of evidence, and mutual cooperation are non-existent in many countries and inconsistent in many others.
3. The laws of armed conflict for countries and the treaties of multilateral organizations, such as the United Nations Charter, Geneva Convention, and NATO Treaty, are not written to accommodate cyber warfare, making it difficult to hold nations accountable for aggressive cyber tactics (including cyber espionage) and subjecting citizens to potential harm from cyber attacks against critical infrastructure.
4. Very few law enforcement personnel are trained to manage cybercrime investigations, including the search and seizure of electronic evidence.
5. There has been very little coordination between law enforcement, the military, and the private sector in any country regarding responding to or defending against cyber attacks.
6. There is no global 24/7 points of contact registry for obtaining assistance in cyber investigations other than the G8 24/7 High-Tech Points of Contact database established in 1998, which has information for only about 50 countries.
7. Information sharing between public and private sectors within most countries is inadequate, and international sharing can be very difficult because trusted exchange protocols have not been established.

8. Judges, prosecutors, and attorneys in many countries are not experienced in handling cybercrime investigations and digital evidence, making cross-border requests difficult and time consuming when minutes matter.
9. There is inadequate funding for cyber security research and development (R&D) and cross-border collaboration is very limited due to requirements that the funds be spent within the jurisdiction of the funding source (e.g., EU money can only be used to pay EU researchers, U.S. government funding can only be used to pay U.S. researchers, etc.).

The Council of Europe Convention on Cybercrime (CoE Convention) was intended to jumpstart a harmonized legal framework by providing substantive and procedural provisions with respect to cyber criminal activities (points 1 & 2 above). However, after it was opened for signature on November 23, 2001, only 46 countries have signed the treaty and, of those, only 30 have ratified it, including four nonmember "partner states," Canada, Japan, South Africa, and the U.S. Since there are 233 countries and territories connected to the Internet, one can hardly conclude, now nine years later, that the CoE Convention is the answer to harmonizing the global legal framework. Although many countries have concerns with certain provisions of the Convention, it does provide a thorough framework for defining and responding to cybercrime—both substantively and procedurally—and it is a starting point from which to proceed.

The United Nations' International Telecommunications Union (ITU) recognized the need for an alternative approach to further advance global consistency in cybercrime laws and sponsored the development of the *ITU Toolkit for Cybercrime Legislation.* The project was led by the American Bar Association's Privacy & Computer Crime Committee, with global participation from technical experts, industry representatives, government officials, attorneys, and policy advocates. The *Toolkit* contains sample legislative language that may be used by any country in drafting or amending their cybercrime laws. In developing the *Toolkit*, the Committee undertook an examination of cybercrime laws in many developed nations, as well as India, China, Mexico, Singapore. It matrixed these laws against the CoE Convention and painstakingly arrived at sample language that is harmonized with developed countries' laws and the CoE Convention, but which could be adapted to local legal frameworks in a more flexible manner. The *Cybercrime Toolkit* was released in May 2009 and addresses points 1 and 2 above.

The ITU has made tackling the core issues of cybercrime a centerpiece of its Global Cybersecurity Agenda (GCA). In addition to the *Cybercrime Toolkit,* it sponsored the development of the *ITU National Cybersecurity / Critical Information Infrastructure Protection Toolkit* and also has programs addressing capacity building, organizational structures, and international cooperation, and training (points 1, 4, and 7). The GCA also has a banner program dedicated to the protection of children against online predators that has the unanimous support of every member country of the ITU.

Beyond the ITU, CoE, and G8, other multilateral organizations that are actively addressing cybersecurity and cybercrime issues include the Asia-Pacific Economic Cooperation Forum (APEC), the Organization of American States (OAS), the Organization for Economic Cooperation and Development (OECD), European Commission, and the Association of Southeast Asian Nations (ASEAN). Numerous organizations are providing cybercrime training for law enforcement and judges, including the U.S. Department of Justice (law enforcement and judges), the Internet Governance Forum, the Council of Europe, and the European Commission. These activities address points 3, 6, and 8 above.

A number of non-governmental organizations (NGOs) are contributing mightily to these issues. The World Federation of Scientists (WFS) was one of the first organizations to call for multilateral discussions to harmonize cybercrime laws and address gaps in the global legal framework pertaining to cyber warfare (points 3 and 5 above). The East-West Institute (EWI) has launched a three-year cybersecurity initiative designed to strengthen global information and communications infrastructure through the exchange of effective methods to detect, prevent, and respond to threats to information systems (points 6 and 7 above). The Center for Strategic and International Studies (CSIS) has been extraordinarily active in the cyber security space and its links to national security, particularly in the United States. George Mason University's Critical Infrastructure Protection Program has developed a close relationship with the NATO Cyber Defense Center in Tallinn, Estonia, and has established relationships with other NGOs and professionals working in the cyber security and critical infrastructure areas. This partial listing is by no means inclusive, but it is intended to illustrate the types of organizations around the globe that are linking and working with governments and private sector participants to address the issues identified above.

Information sharing and 24/7 points of contact (points 4 and 5 above) remain weak links in the security chain because without cooperation in the investigation of cyber criminal activities, the ability to prosecute the perpetrators of these crimes will remain difficult, if not impossible. One of the biggest gaps is in funding for developing countries to establish points of contact and train personnel. Plamen Vatchkov, until recently Bulgaria's National Cybersecurity Coordinator, noted that after dialogues with cyber personnel from Turkey, Russia, China, U.S., and many other countries, "It is amazing how ready every country is to enhance their cooperation with each other. What lacks is, as usual, enough funding; the U.S. and EU should lead an effort to help countries become cyber coordinators."

Minimal funding would be required to resolve the 24/7 points of contact issue. *Another organization needs to step in to build upon the G8 24/7 effort and develop a full database of contacts for all countries connected to the Internet.* It seems logical to require all countries who want to connect to the Internet to (1) provide points of contact, and (2) assist in the investigation of cyber criminal activities occurring in or transiting through their countries. A trusted model for global information sharing is also desperately needed; countries seem to get tangled in their own interests and multilateral organizations quickly become paralyzed in trying to meet the demands of all parties.

There is a critical need for increased funding of cyber security research and development (R&D), including easing the rules on funding cross-border collaboration (point 9 above). Many of today's security problems are caused by cyber criminals exploiting vulnerabilities in Internet protocols and technologies. Jeff Moss, one of the founders of the Black Hat conference, noted at the opening of the 2010 event, "We have not solved any fundamental problems" and called on the cyber security community to work together to catch up with criminals.[36] Serious vulnerabilities can be overcome by focused research. For example, in 2008 cyber security researcher Dan Kaminsky revealed a security weakness in the Domain Name System (DNS) that led to the current wave of web page hijackings and DNS exploitations associated with criminal activities.[37] The Internet Engineering Task Force (IETF) and research community responded with the development of DNSSEC, which ensures that DNS address queries are not compromised. Kaminsky endorsed DNSSEC and demonstrated the ease of deploying the solution at the recent Black Hat 2010 conference. Shortly thereafter, Rod Beckstrom, CEO of the Internet Corporation for Assigned Names and Numbers (ICANN), called DNSSEC "the biggest structural improvement to the Internet in 20 years, specifically, since the introduction of the world wide web."[38] It *is* possible to overcome serious security gaps with committed effort and resources.

What we have today on the global front is cyber chaos…and the criminals are taking advantage of this at every possible point. Ivailo Kalfin, member of the European Parliament, laments that, "Each country takes different measures to protect herself, but there is no coordinated effort. Therefore, the efficiency of the national efforts are very dubious; we need more multilateral coordination if we are to turn the tide on cybercrime."

An early 2010 article on the future of the Internet gloomily noted that, "Some experts fear the Internet will collapse under the weight of ever-increasing cyber attacks…"[39] One thing is certain…while countries continue to ponder and confer; cyber criminals are reducing the benefits of the Internet…or perhaps threatening the very existence of the Internet. There can be no greater priority than bridging the Security Divide through the intersection of law, science, and technology to address the identified problem points.

---

[36] Lauren Price, "DNSSEC Taking Center Stage at 2010 Black Hat," http://www.circleid.com/posts/20100803_dnssec_taking_center_stage_at_2010_black_hat/ (hereinafter "Black Hat").

[37] "Vulnerability Note VU#8000113: Multiple DNS implementations vulnerable to cache poisoning," U.S.-CERT, http://www.kb.cert.org/vuls/id/800113.

[38] Black Hat.

[39] Carolyn Duffy Marson, "2020 Vision: Why You Won't Recognize the 'Net in 10 Years," *CIO*, Jan. 4, 2010, http://www.networkworld.com/news/2010/010410-outlook-vision.html.

WEB SCIENCE FOR SECURITY AND TRUST ON THE WEB—SHORT INTRODUCTION

JACQUES BUS
DigiTrust.eu
Brussels, Belgium

Since the World Wide Web blossomed in the mid-1990s, it has exploded to more than 15 billion pages that touch almost all aspects of modern life. Today more and more people's jobs depend on the Web. Media, banking and health care are being revolutionized by it. And governments are even considering how to run their countries with it. Little appreciated, however, is the fact that the Web is more than the sum of its pages. Vast emergent properties have arisen that are transforming society. E-mail led to instant messaging, which has led to social networks such as Facebook. The transfer of documents led to file-sharing sites such as Napster, which have led to user-generated portals such as YouTube. And tagging content with labels is creating online communities that share everything from concert news to parenting tips.

Few investigators are studying how the Web's emergent properties have actually blossomed, how we might harness them, what new phenomena may be coming or what any of this might mean for humankind. A new branch of science—Web Science—aims to address such issues.

Web Science looks to explain the evolution of the web as a complex organism and an independent ecology, through multidisciplinary analysis of the range of knowledge areas that converge in the web. These are mainly technological (how does the web evolve technologically), organizational (what standards and specifications support this evolution) and social (what use do users make of this technology).

There are also other areas involved in these changes, given that many economic, political and legal interests do not want to be left out of the evolution of the information society. Web science looks to provide a reference framework to ensure the proper analysis of all of these events from a range of perspectives and different levels of resolution, taking into account all the elements involved and the relationships established between them, with a clear multidisciplinary spirit to promote the participation of the widest possible selection of members of the scientific community.

The Web Science Research Initiative, WSRI, was created in November of 2006 with the main aim of proposing a new discipline, Web Science, to observe the web and all that that surrounds it. It defends the need to analyze what is going on inside and outside the web and, thus, be able to propose improvements and corrections. This idea requires combining disciplines that have, to date, been very disperse, such as IT, psychology, law or economics. This has led to a new professional profile, the Web Scientist, and, likewise, new academic needs. See: A Framework for Web Science (Berners-Lee et al. 2006)—the basic reference to Web Science. (http://eprints.ecs.soton.ac.uk/13347/1/1800000001%5B1%5D.pdf)

WSRI (http://webscience.org/home.html) was an initiative of Tim Berners Lee and colleagues at the Massachusetts Institute of Technology together with Wendy Hall, Nigel Shadbolt and others at the University of Southampton in England.

Thorough understanding of scale-free networks, gleaned by analyzing the Web, has also prompted experts to analyze other network systems. They have since found power-law degree distributions in areas as far-flung as scientific citations and business alliances. The work has helped the U.S. Centers for Disease Control and Prevention improve its models of sexual disease transmission and has helped biologists better understand protein interactions.

One of the last (and first) texts published is the provocative book *The Web's Awake*[i] (Tetlow, 2008) (http://www.thewebsawake.com/), which presents the web as a new way of life that we can no longer control. It provides a description of new web characteristics and types of behavior that we have not created but have emerged of their own accord.

TRUST ON THE WEB

Trust and trustworthiness are concepts which are at the basis of human existence. We use them intuitively and their assessments are invariably context dependent. But when we transpose these concepts to a digital environment, we can easily run into trouble.

O'Hara[ii], Luhmann[iii], Hardin[iv], Fukuyama[v], Putnam[vi], Nissenbaum[vii], O'Hara and Hall[viii] and a number of others have written about the concept of trust in general, and about its problems in the digital environment. Nevertheless, it is still insufficiently understood how trust on the Web can be made operational.

When analyzing trust in relation to technology we must make a difference between:

- Trust between persons who make extensive or exclusive use of digital technology for communication and transactions.
- Trust or confidence of people in the infrastructure of digital networks and systems they use for services, communication, data storage, services etc.

The problems with trust between persons in the digital society in comparison with the "old society" relate to:

- The new ways and uses (incl. profiling and new business cases) of data collection, storage and processing.
- Identification, reputation, authentication and accountability have a different meaning on the Internet, also due to different jurisdictions involved in transactions.

---

i Tetlow, 2008—http://www.thewebsawake.com/

ii Kieran O'Hara Trust: from Socrates to Spin, Icon Books, Cambridge (2006).

iii Niklas Luhmann—Trust: A mechanism for the reduction of Social Complexity, in TRUST and Power: Two works by Niklas Luhmann (1979)

iv Russell Hardin (2002) —Trust and trustworthiness; Russell Sage Foundation, NY

v Francis Fukuyama, Trust: The social virtues and the creation of prosperity (1995)

vi Robert D. Putnam, Making democracy work: Civic traditions in modern Italy (1993)

vii Helen Nissenbaum—Securing Trust Online: Wisdom or oxymoron? (2001)

viii Kieran O'Hara and Wendy Hall—Trust on the Web; Some Web Science Research Challenges (http://www.uoc.edu/uocpapers/7/dt/eng/ohara_hall.html)

- Increase of complexity through incomprehensible technology with insufficient assurance through certification and standardisation, and lack of transparency.

But such trust between people can only be obtained in our technological world if one also can have trust in the technology used to communicate and interact. Networks, systems and services must be *trustworthy* to a certain level, i.e., a person can have a certain degree of justifiable trust that the system or service will deliver in accordance to its description and promises, and that it will not perform actions that are not described, independent of the circumstances. Justifiable trust can be given through accountability (product liability), transparency on data processing and storage, technical system certification, and ex-post auditability, and authentication and identification, as well as tools to support this.

SECURITY OF THE WEB

The ICT based societal revolution will lead to essential changes in the balance of power, at the national level where citizens are having abundant information on the political processes which will be used in the democratic process, but also at the international level. Access to the Internet is empowering citizens to be better included in economic and political life, and to understand situations and ways of life in other cultures. We have seen the way Obama used the social networks in his campaign and we may expect that similar activities will be developed in the future to support governmental policy making.

ICT also allows international companies to organize themselves in ways that make optimal use of opportunities all over the world. This can all give a strong boost to economic development and growth globally, and particularly in low-cost countries. We see already large developing countries taking advantage and become important economic and political players.

However, as with every revolution in history, together with the opportunities and benefits, there always comes a downside.

The information and communication infrastructures and services have become a critical part of our economies. They are extremely vulnerable, as the many attacks reported almost daily demonstrate. Most of our other critical infrastructures, e.g., energy, water, transport, financial systems are heavily dependent on ICT for communication and control. There is therefore a high risk of accidents or deliberate attacks on these critical infrastructures that may potentially lead to chaos and enormous economic losses. This includes intrusion and attack on systems and databases of National Security Agencies.

This vulnerability of our societal ICT systems and the systems dependent on it make these systems an easy target for attacks in times of war or by terrorists. Cyber crime is becoming a very worrying issue. The number of malicious and criminal code threats is increasing exponentially, with 1.6 million threats detected by Symantec in 2008. There is general agreement that there is reason for alarm about the security of the Internet. Current trends risk increasing fear for and rejection of the new digital world by citizens. It may have huge economic consequences if politics and technology are not able to deal with these negative societal developments.

World leaders are facing enormous and unparalleled challenges. Climate change, rapid changes in global economic power and energy security, to name a few other issues, need to be managed at the same time as the transfer to a digitally connected global world. There are fascinating opportunities (e.g., in the use of ICT for climate change or energy economy and security) and fearful risks.

It will need a solid multi-disciplinary scientific basis and strong and visionary global political leadership in the interest of all to build a future of freedom, security, creativity and prosperity.

ACKNOWLEDGEMENT

I am grateful to Prof. Nigel Shadbolt (Univ. Southampton) for allowing me to use some of his material.

THE NEW WAVE OF GLOBAL COMMUNICATION—THE FUTURE INTERNET (A New Challenge to Interdisciplinary Research)

ZENONAS ROKUS RUDZIKAS
Vilnius University, Institute of Theoretical Physics and Astronomy
Vilnius, Lithuania

ALGIMANTAS JUOZAPAVIČIUS
Vilnius University, Faculty of Mathematics and Informatics
Vilnius, Lithuania

ABSTRACT

Science is the must for human culture of the third millennium. The exceptional role in this process is played by information and communication technologies (ICT), and the high demand for ICT leads to many forms of it: Internet, Mobile Internet, Internet of Things (IoT), Grid Computing, Cloud Computing. The article discusses influences and reflections to our life, caused by recent and future forms of ICT: a centralized and decentralized IoT regime, human privacy protection and protection of personal data, risk of cyber terrorism, communication reflection for the small country (like Lithuania), "silence of chips", some others. IoT is based on the idea of a worldwide, wireless, integrated network of smart facilities and devices ("things") as well as a whole range of different sensors and actuators, in which, using standard protocols, the "things" communicate with each other, and with people. Technically, the road to IoT covers device integration, certification and maintenance, service provisioning, as devices participating in IoT need to talk to each other, not just to people. Such phenomena are presented in the article as a challenge to interdisciplinary research. Future generations will have a smart live in a global environment, if we care more about sustainable development and planetary emergencies.
Keywords: information and communication technologies, Internet of Things, grid computing, cloud computing, interdisciplinary research.

INTRODUCTION

The Internet has become the backbone for communications of our global world. It is a powerful tool that assists us in our efforts in social, economic, exploration areas, in academics, it plays global role to promote peace and security, as well as development and human rights. It is an essential infrastructure component of Information and Communication Technologies (ICT), which form a vital part of the economy and society.[1] One key characteristic of Internet and ICT is their rapid, indeed meteoric growth, of which the Internet is a stage in the development. In virtually a single generation, Internet and ICT have moved out of isolated scientific laboratories into the public domain. Even more, now ICT can be seen as the queen (or some would say "servant") of sciences and technology, repaying it's "debt of gratitude" by placing research methods, equipment and other tools at their disposal and also by improving the day-to-day lives of the public at large.[2,3] Given the tremendous potential of the Internet to change our lives, it is no wonder that people take an interest in how it is being run and managed. In essence, Internet development and governance are based on collaboration between all stakeholders.

The different types of Internet are defined by functional objectives coming from human needs as driving forces for Internet:

- Internet of Things
- Cloud computing
- Critical Information Infrastructures

INTERNET OF THINGS

As it was already mentioned, IoT is based on the idea of a worldwide, wireless, integrated network of smart facilities and devices ("things") and a whole range of different sensors and actuators, in which, using standard protocols, the "things" communicate with each other and with people. Such a network will connect billions of people (IoT for people).

Given the specific features involved in the growth of information and communication technologies and their particular importance on various fronts, for a country's development and that of the lives of its people, the Internet of Things seeks to create a new and broad paradigm: the transition from an Internet that connects people, to one that connects people with things or things with each other.[4]

By its very nature, the IoT will inevitably encircle us with "smart or intelligent objects", that are set:

- To comprehend their own properties and potentialities, at least at a certain point (and those in the locality around them).
- To make autonomous decisions and take proactive steps to meet the targets they have been set or to fulfil the remit assigned to them.
- To carry out all manners of activities and tasks and, at a certain point, react to their environment, i.e., adapt to the world around them, alter their configuration, repair any faults themselves—and even decide who can access them or switch owners.

There are a number of the other definitions of the IoT:

1. An ever-changing mix of relations between the physical body of a person, his or her objects and a "smart" environment.
2. Internet was born connecting people through machines, IoT—information generated by people vs information generated by things.
3. "Anywhere, anytime, by anyone…and with anything".
4. IoT—a global, mental, political, bartering, educational, technological and spiritual revolution.
5. IoT—when every object in the world will be imbedded with a tiny microchip that will feed the information into cyberspace (e.g., IoT/RFID technology).
6. IoT—the next step in "always on" computing, promising a world of networked and interconnected devices.

7. IoT—Internet of Everything.

It is expected the IoT will greatly contribute to addressing societal challenges in areas such as health monitoring, ecology and environmental protection, consumer transport and other areas of human activity.[4] Networked communications using IoT applications will have profound effects on our society and gradually result in a genuine paradigm shift in this field.

Another characteristic of IoT is that it is predominantly application-based, and will benefit greatly from the rapid development of ICT-supported devices and facilities.[5] One need think only of the mobile communications boom, the pace of change in computer properties, the development of algorithmic languages or the expansion of the Internet. As for basic issues, IoT will benefit from directed research and development topics: a) intelligent object identification and addressing, b) object networking architecture, c) level of "intelligence" of objects, including their virtualization, interoperability, interfaces and protocols, d) middleware for IoT, e) data organization, transmission and security, f) application systems for IoT, and g) business models for IoT.

To meet such challenges, efforts of many stakeholders are needed and have to be jointly developed.[5] Each country will add suitable functionality to the stake. For example, Lithuania has a potential to elaborate processes in engineering of computer network, information systems and software, computer systems design and data security, "smart living environment" procedures, artificial intelligence technologies.[6]

IoT applications will improve people's quality of life, yet they may also have adverse impacts. These include, among others, the threat to privacy, the risk of cyber-terrorism and the use of the Internet to disseminate pornography and spread homophobia, racism, etc.[5] Moreover, young people are in particular risk, becoming addicted to the Internet, with reality to a large extent replaced by life in a "virtual" world, and this may raise after introducing IoT functionality.

It is worth to note that the key progress has already been achieved on a number of fronts. Devices are becoming noticeably smaller and will soon be invisible to the human eye. Appliances increasingly have wireless connections and are mobile. Systems are becoming ever more heterogeneous and complex. The latest technologies are becoming used ever more widely, including radio frequency identification devices (RFID), Near Field Communication (NFC), Internet Protocol version 6 (IPv6) and ultra-broadband connections.[4] Quantum computing and quantum cryptography are not so far away.

The other danger comes from the "right to silence of chips" (i.e., that individuals should be able to disconnect from the networked environment). The situation now does not provide sufficient safeguards for privacy protection or object safety. Thus, for instance, switching off a mobile phone does not prevent certain interested parties from obtaining information on the owner. It is not enough simply to "launch a debate", further practical steps are also needed.

From the point of view of standards, IoT is now in a cybernetic "Tower of Babel" case. The importance of standards may be illustrated by instance, as a refrigerator in China has to be able to "communicate" successfully with a shelf of Dannon yogurt pots in a supermarket in France. However, any moves towards standardisation must take in account the diversity and specific characteristics of languages, cultures and traditions of any individual country involved. Therefore, there is the urgent need for the cybernetic "Kyoto protocol".

The IoT obviously will lead to much higher level of "electronic pollution", as the pulses emitted by IoT systems are weak, the number of radiation sources is set to increase exponentially. There is a need for a more detailed examination of the impact of electromagnetic waves on humans. Modern science has yet to determine conclusively whether there is a threshold above which harmless levels of radiation become dangerous, and what the cumulative impact of such exposure is. Sometimes, in fact, a single electromagnetic pulse at quantum level is enough to trigger unchecked cancerous growth in a cell. If the genie will let out of the bottle, humans will be stuck with the consequences.

To function properly, an IoT requires elaborated, structured information and complex algorithms. Clearly, it is made up of modules of centralised and solitary "smart" objects. The processes involved here may be similar to those used by the European Organisation for Nuclear Research (CERN) where data are collected, analysed, stored and used via a grid technology infrastructure based on the EGEE project and other initiatives. In Lithuania, for instance, projects such as BalticGrid and the national schemes of grid, LitGrid and GridTechno were implemented.[7] In the IoT, however, the data processing involved is much more complicated. Hence, the EGEE project can only be seen as the first stage in the development, planning and launching of IoT components.

CLOUD COMPUTING

Let us mention that grid technologies are connecting geographically distributed clusters into heterogeneous distributed/parallel computing system via communication network and provide the efficient infrastructure for heterogeneous computing/modeling procedures as well as for the comprehensive data maintenance functionality. The Cloud is the enabler of the IoT. The cloud and IoT are best considered as a continuum of internet connectivity. No two clouds alike, no "one-size-fits-all".

There are also various definitions of cloud computing (CC):

1. CC—Internet + Computing (computation over internet)
2. Access to networks, storage elements, software series etc., on demand
3. Everything You can use over the internet
4. Virtual servers available over the internet
5. CC describes information stored and processed on the computers remotely and brought back to your desktop.

Main features of the CC:

- Simplicity, scalability and flexibility; renting the computer power and storage from a cloud service provider;
- Enhancing the cloud security, confidentiality and privacy mechanisms;
- Supporting the data and software federation in clouds;
- Computing on demand, endless amounts of storage space and processing power;

- Access to the information anytime and anywhere; however, risk of information theft, control over data location, etc.

Cloud computing is an emerging issue, most likely to have a huge impact to Internet and to many components of ICT.[8,9] It is understood as the "technology on tap" where people can access computing services on demand from any location without worrying about how these services are delivered and where they are hosted. Cloud computing is a revolutionary Internet-based technology which enables businesses, especially small and medium ones to enter the market without upfront costs and operate entirely without a large IT department. From this point of view cloud computing will become increasingly important access and distribution model for the Internet.[8] While in many ways cloud computing represents an extension of existing technologies, the expanded scope and spread of cloud computing use, plus the new roles it enables for individuals as creators and distributors of content and developers of applications may pose challenges for existing policy and governance structures. These enabling technologies can empower individuals and contribute to growth and societal benefit when correctly fostered and implemented.

Nevertheless cloud capabilities have not yet been developed to meet all potential opportunities of usage (compare[10]):

- There are technical and non-technical gaps.
- The vision of a Cloud usable "anytime anywhere" depends on advances in the design of (distributed) computer systems.
- Current popular Clouds (Amazon Elastic Computing, Google Aps, Microsoft Azure) usually deliver poor performance for compute-intensive applications.
- Transferring data from customers into the Cloud is expensive and time-consuming.
- Legal boundaries have still to be established for Cloud-based operation of critical services.
- The social and economic implications have not yet been fully understood.

As the name of clouds is a metaphor for the Internet and its combination with computing access to networks, storage elements, software services, cloud compting raises demanding issues to Internet: a) cloud systems will explore Grids, Service Oriented Architectures, server virtualization, b) clouds will require enhanced cloud security and privacy mechanisms, c) adapted economic management and legal laws will be developed for clouds, d) the special issiues of data and software federations in clouds will be elaborated, e) special attention will be paid to clouds and social contextualisation.[11,12,13]

For each of stakeholders of Internet, id est countries cloud computing is a promising and beneficiary initiative. As for Lithuania cloud computing is a core component for the development:

- Science and technology parks (involving most active universities, research institutes, national and international business companies).
- National research and development programs.
- Special governmental measures to stimulate research-business collaboration in

ICT.

- Special carrier program for highly educated and active researchers (especially in IT).

It is worth to note, that Lithuania has a suitable foundation for the successful development of clouds:

1. Penetration of Internet in Lithuania is very high (covering more than 60% of households).
2. The special project RAIN provides an equally developed advanced e-infrastructure as well as ample resources of electronic content, and for business.
3. Broadband Internet and fiber optics network are dense, even in rural regions.
4. Infrastructure of WiMAx, 4G/LTE, "Last Mile" solutions are under intensive development.
5. The Computer Emergency Response Team (CERT) is established for security issues.
6. The network of interconnected open access IT centres is under development, to synchronize compting and e-services efforts.
7. The special cloud initiative will be established under the Digital Science and Computing Center (DiSCC) at Vilnius University, providing cloud computing for education, research and for business, for joint projects with public sector (e-Health projects in medical signal processing and imaging, cyber security projects, etc.), for cooperation with universities and businesses of the Republic of Belarus, for cooperation with European research centers: CERN in Switzerland, Finland's CSC, Netherland's SARA, ITER and for joint projects with prominent scientists abroad of Lithuanian descent.

CRITICAL INFORMATION INFRASTRUCTURES

As ICT is increasingly intertwined in our daily activities, some of these ICT systems, services, networks and infrastructures are regarded as critical information infrastructures (CIIs). Their disruption or destruction may lead to serious impact on vital societal functions, and the World Economic Forum estimated in 2008 that there is a 10 to 20% probability of a major CII breakdown in the next 10 years, with a potential global economic cost of approximately 250 billion U.S.$.

The risks due to man-made attacks, natural disasters or technical failures are often not fully understood and/or sufficiently analyzed. Consequently, the level of awareness across stakeholders is insufficient to devise effective safeguards and countermeasures. Cyber-attacks have risen to an unprecedented level of sophistication. Simple experiments are now turning into sophisticated activities performed for profit or political reasons. The recent large scale cyber-attacks on Estonia, Lithuania and Georgia are the most widely covered examples of a general trend. The huge number of viruses, worms and other forms of malicious software, the expansion of botnets and the continuous rise of spam confirm the severity of the problem.[14,15]

The high dependence on CII, its cross-border interconnectedness and interdependencies with other infrastructures, as well as the vulnerabilities and threats they face raise the need to address their security and resilience in a systemic perspective as the frontline of defense against failures and attacks.

It is worth to note that more and more ICT services are based on or transferred to Internet. Substantial investments are being made by operators and equipment manufacturers in what are often referred to as *IP-Enabled Next Generation Networks* (NGN). IP-enabled NGN can be seen as a logical progression from separate Public Switched Telephone Network (PSTN), mobile and IP-network infrastructures to unified networks for electronic communications based on IP. The fundamental difference between NGN, today's data networks and PSTN networks is the shift towards a converged, packet-switched multi-service platform. IP-enabled NGN are already being deployed by numerous service providers around the globe. Good example of it is WiMAX/4G technologies, already taken for deployment in Lithuania. Then the prevention, preparedness and awareness appear to be even more essential reasons of immediate actions to strengthen the security and resilience of CII.

ICT are also pervasive for the work of governments and public administrations: the uptake of eGovernment services at all levels, as well as new applications such as innovative solutions related to health, energy and political participation, make the public sector heavily dependent on ICTs. Last, not least, citizens increasingly rely on and use ICTs in their daily activities: strengthening CII security would increase citizens' trust in ICTs, not least thanks to a better protection of personal data and privacy. eGovernment considered as a "brave" experiment a few years ago, now is one of important tools in public sector, assuring more effective participation of citizens in the governance of state and regional administration. Lithuania in this aspect is one of the leading countries, ranked together with USA, Australia, Japan, Denmark, Germany, and Netherlands.

To meet challenges for CII, the EU highlights the urgent need to coordinate efforts to build trust and confidence of stakeholders in electronic communications and services. It strengthens the role, on tactical and operational levels, of the European Network and Information Security Agency (ENISA), and in parallel the European Program for Critical Infrastructure Protection (EPCIP), including another important element of EPCIP, the Critical Infrastructure Warning Information Network (CIWIN). Simultaneously, this initiative takes into account NATO activities on common policy on cyber defense, i.e., the Cyber Defence Management Authority and the Cooperative Cyber Defence Centre of Excellence.

In order to ensure that ICT infrastructures are used to their maximum extent, all stakeholders must have a high level of confidence and trust in them. This depends on various elements, the most important of which is ensuring their high level of security and resilience. Diversity, openness, interoperability, usability, transparency, accountability, auditability of the different components and competition are key drivers for security development and stimulate the deployment of security-enhancing products, processes and services. This is a shared responsibility: no single stakeholder has the means to ensure the security and resilience of all ICT infrastructures and to carry all the related responsibilities.

Five pillars are proposed to tackle these challenges:

1. Preparedness and prevention: to ensure preparedness at all levels.

2. Detection and response: to provide adequate early warning mechanisms.
3. Mitigation and recovery: to reinforce defense mechanisms for CII.
4. International cooperation: to promote EU priorities internationally.
5. Criteria for the ICT sector: to support the implementation of the Directive on the Identification and Designation of European Critical Infrastructures.

The Lithuania's strategy for cyber security is being developed, by suitable organizational and technical measures designed and related for cooperation within EU and NATO organizations. The additional important point is that the broad participation of public organizations, NGOs, as well as academic ones is foreseen and encouraged.

It is important to notice that usually in many countries about 80-85% of criminal activities against CII are not fixed at all, nevertheless these activities are making influence on information wars, deformation of information and discrediting one. It has to be emphasized that usually the most part of publicly sensitive information is concentrated in the government sector, so the strategies for cyber, information and data security must encounter suitable measures and activities to defend the CII.

In conclusion, the modern information and communication technologies will essentially contribute to the sustainable future of the mankind. Indeed, there is no doubt that the future generations will have a smart life in a smart globalised and computerized world, if we all, World Federation of Scientists included, care more about the Sustainable Development and Planetary Emergencies.

Rapid ICT development and dissemination requires properly trained experts. University lecturers must continuously update their teaching content so that the students have access to the latest information and are able to help shape and use the IoT, CC and the other future ICT. Schools pupils too need to be given appropriate preparation and further training for adults is also required. Lifelong learning and distance learning technologies are thus particularly appropriate in this field. Action is at all costs needed to bridge the geographical digital divide. Society and the various bodies of which it is made up have a key role to play in the practical success of these endeavours.

REFERENCES AND SOURCES

1. ICT Shaping the World (A Scientific View), Wiley, 2009.
2. EC communications on transforming the digital divide into social benefits and economic growth (2009).
3. EC communications on a public-private partnership on the Future Internet (2009).
4. Z.R. Rudzikas (Rapporteur). Internet of Things—An action plan for Europe (European Economic and Social Committee, 2009)
5. The 2nd Annual Internet of Things Europe 2010 (A Roadmap for Europe). Conference Program and Delegate Handout, 1,2 June 2010.
6. A. Juozapavičius, D. Mažeika, E. Kutka, and R. Naujikas. "Grid Infrastructures For Lithuanian Research And Public Sectors", Proceedings of TERENA Conference TNC2010, Vilnius, 5 pp.
7. A. Juozapavičius. "Collaborative Environment for Applications in BalticGrid", EGEE User Forum /Open Grid Forum, Manchester, 8-12 May 2007.
8. L. Schubert. The Future of Computing. Opportunities for European Cloud

Computing Beyond 2010, http://cordis.europa.eu/fp7/ict/ssai/docs/cloud-report-final.pdf.

9. Wikipedia, 'Cloud Computing'—available at http://en.wikipedia.org/wiki/Cloud_computing

10. Sutter, H. (2005) "The Free Lunch Is Over: A Fundamental Turn Toward Concurrency in Software", in *Dr. Dobb's Journal*, 30(3).

11. Members of EGEE-II (2008), "An EGEE comparative study: Grids and clouds—evolution or revolution." Technical report, Enabling Grids for E-sciencE Project'—available at https://edms. cern.ch/document/ 925013/.

12. Vaquero, L.M., Rodero-Merino, L., Caceres, J., and Lindner, M. (2009) "A break in the clouds: towards a cloud definition", *SIGCOMM Comput. Commun. Rev.* 39(1), 50-55.

13. Harris, D. (2008) "Grid vs. Cloud vs. What Really Matters"—available at http://www.ondemandenterprise.com/blogs/Grid-vs-Cloud-vs-What-Really-Matters.html.

14. Catteddu, D. and Hogben, G eds. (2009) "Cloud Computing—Benefits, risks and recommendations for information security", European Network and Information Security Agency (ENISA)—available at http://www.enisa.europa.eu/act/rm/files/deliverables/cloud-computing-risk-assessment/at_download/fullReport.

15. Kincaid, J. (2009) "T-Mobile Sidekick Disaster: Danger's Servers Crashed, And They Don't Have A Backup", available at http://www.techcrunch.com/2009/10/10/t-mobile-sidekick-disaster-microsofts-servers-crashed-and-they-dont-have-a-backup/

ENSURING CYBERPEACE IN A NEW WORLD ORDER

DR. HAMADOUN I. TOURÉ
Secretary-General, International Telecommunication Union
Geneva, Switzerland

A NEW DOMAIN: CYBERSPACE, SECURITY & WARFARE

- Digital communications, networks and systems provide vital resources and indispensable infrastructure
- Cyberspace—the physical and conceptual realm in which ICTs operate
- Security—critically important as global reliance on ICTs grows
- Recent attacks against computer systems in Georgia, Estonia, South Korea and the United States have been linked with cyber warfare

CYBERWAR

- Cyberwar—targeting military capabilities and systems, as well as all the vital infrastructure of a society that allows it to function and defend itself.
- Tactics involve data collection or infiltration of computerized systems to cause damage to critical systems.

- Other/new targets and threats include:
  - SCADA
  - Smart Grids
  - Nuclear weapons control facilities
  - Power generators (and other costly replacement parts)
  - Financial systems

Cyberwar—Features and Impact

- A network-based attack would cause failure beyond a single system and often beyond national boundaries.
- Cyber offenders need not be present at the origin nor the destination of the attack.
- Use of software tools to automate attacks.
- One computer can attack thousands of computers in a single day.
- Deterrence and retaliation are difficult due to attribution and collateral damage problems.

National Policies and Approaches to Cyberwar

- Incorporating cyber capabilities into conventional warfare strategy—building up cyber offensive weapons and defensive capabilities.
- Increasing funding, research and tactical and diplomatic resources to improve cyber security.

- Cultivating cyber tactics as a national resource or strategic asset.
- Building cyber military outfits.
- Using cyber tactics to level the playing field.
- Educating citizens and raising awareness of cybersecurity problems
- Less connected and developing countries also have a stake in the future of cybersecurity

International Responses to Cyberwar

- Arms control treaty and agreements
- North Atlantic Treaty Organization (NATO)
- UN Economic and Social Council (ECOSOC)
- United Congress on Crime Prevention and Criminal Justice (Salvador Declarations, April 2010)
- European Convention on Cybercrime
- Memoranda of understanding between countries
- ITU-SG 17 Standard Setting [Focus group on Smart Grid]

NECESSITY OF AN INTERNATIONAL FRAMEWORK

- Non-viability of deterrence
  "The fundamental characteristics of cyberspace undermine the efficacy of deterrence as an approach to cyberpeace."

- To harmonize national legal frameworks to combat cybercrime
  "An international framework committing countries to the protection of cyberspace and providing for information sharing, warning systems, coordinated responses and post-incident restoration will best accomplish these goals."

PROPOSALS FOR INTERNATIONAL CYBERPEACE

5 Principles:

1. Every government should commit itself to giving its people access to communications.
2. Every government will commit itself to protecting its people in cyberspace.
3. Every country will commit itself not to harbor terrorists/criminals in its own territories.
4. Every country should commit itself not to be the first to launch a cyber attack on other countries.
5. Every country must commit itself to collaborate with each other within an international framework of cooperation to ensure that there is peace in cyberspace.

ITU AND CYBERSECURITY

- 2003 World Leaders at the World Summit on the Information Society (WSIS) entrusted ITU as sole facilitator for WSIS Action Line C5—"Building Confidence and Security in the use of ICTs".
- Since 2006 ITU Plenipotentiary Conference, ITU has set cybersecurity as one of its top three priorities.
- On 17 May 2007, in response to the decisions of ITU Membership and ITU role as sole facilitator for WSIS Action Line C5, ITU Secretary-General launched the **Global Cybersecurity Agenda (GCA)**—a framework for international cooperation in cybersecurity.
- ITU is also promoting cybersecurity through a range of its activities related to standardization and technical assistance to developing countries tailored to their specific needs.

GLOBAL CYBERSECURITY AGENDA (GCA)

- GCA is designed for cooperation and efficiency, encouraging collaboration with and between all relevant partners, and building on existing initiatives to avoid duplicating efforts.

- GCA builds upon five pillars:
  - Legal Measures
  - Technical and Procedural Measures
  - Organizational Structures
  - Capacity Building
  - International Cooperation

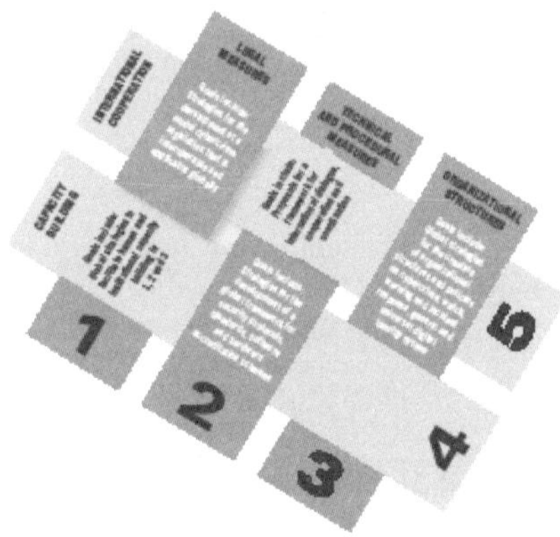

- **H.E. Blaise Compaoré**, President of Burkina Faso and **H.E. Dr. Óscar Arias Sánchez**, Former President of the Republic of Costa Rica and Nobel Peace Laureate, are both Patrons of the GCA.

GCA: From Strategy to Action

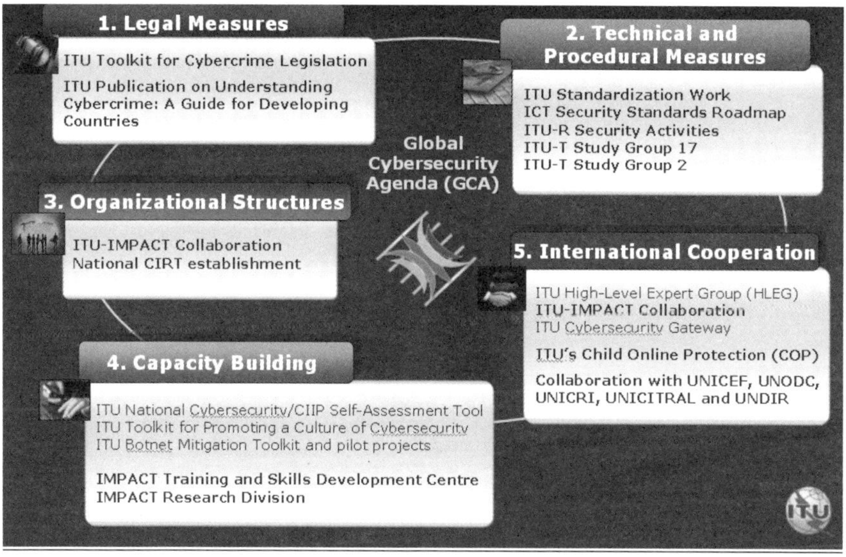

CONCLUSION

Curbing Cyberthreats Globally

- The threat of cyberwar now looms larger than ever.
- We must protect our cyberspace for the continued flourishing of infrastructure and new technologies, and ensure lasting peace.
- ITU takes very seriously its responsibility for WSIS Action Line C5 "Building confidence and security in the use of ICTs."
- ITU's GCA provides its 191 Member States and others with the expertise, facilities and resources to effectively address the world's most serious cyber threats.

Thank you for your attention.

SECTION 6

SESSION 6

POLLUTION

FOCUS: AIR, WATER AND WASTE
SCIENTIFIC CRITERIA

DETERMINING WHICH TYPES OF FINE PARTICLES IN AMBIENT AIR HARM HUMAN HEALTH

THOMAS J. GRAHAME
Office of Fossil Energy, U.S. Department of Energy
Washington DC, USA

This article summarizes and expands upon a presentation made at the 2010 World Federation of Scientists seminars in Erice, Italy.

ABSTRACT

For many decades, it has been accepted that particles in ambient air can harm human health. However, only within the last decade have advances in air pollution statistics and science made it possible to evaluate which specific types of tiny airborne particles might be most harmful to human health, or perhaps have limited toxicity.

Population-based epidemiology is the standard method by which an air pollutant's harm to populations is estimated. It is important to include a large number of potentially harmful $PM_{2.5}$ species in a given epidemiology model, so researchers can be assured that any associations found are not due to the absence of potentially important $PM_{2.5}$ species. However, due in part to a lack of standardized available monitoring information for many different $PM_{2.5}$ species, epidemiology studies rarely included many $PM_{2.5}$ species in models until relatively recently. Epidemiology studies from the 1990s rarely included more than one or two $PM_{2.5}$ species in models of health effect associations (although they often included gases). The only $PM_{2.5}$ species consistently included in older studies was sulfate, which has been monitored for several decades. Perhaps unsurprisingly, sulfate was frequently associated with the health effects examined in the models.

Because several epidemiology studies with 6 to 20 $PM_{2.5}$ species have been published since 2008, there is now a "critical mass" of such at least 9 such studies. Cardiovascular disease daily emergency hospital admissions, daily mortality, or survival since enrollment in prospective cohort studies are the health endpoints examined.

In 8 of the 9 studies, black carbon or elemental carbon (BC/EC) is significantly associated with the health endpoint of the study. In 8 of the 9 studies, sulfate is not significantly associated with the health endpoint considered. Despite the fact that the metals vanadium (V) and nickel (Ni) are each a tiny fraction of $PM_{2.5}$ species, every study which includes Ni, all but one which include V, find health associations with each.

These studies, by themselves, may not be adequate to make policy determinations as to which types of emissions are most harmful, and which emissions may pose little health threats. The consistency of findings, however, is quite surprising. It is doubtful that many air pollution health experts would have predicted such consistent findings. These results may suggest that $PM_{2.5}$ species capable of causing particular biological effects, such as oxidative stress and inflammation, might be the ones which cause significant portions of chronic and acute mortality and morbidity via these mechanisms. Such a possibility should spur additional toxicology research.

To provide added confirmation of these epidemiological findings, it is important to utilize toxicology and human panel studies, where the exposure to specific emissions is known accurately. Doing so will allow us to see if biological mechanisms of harm found in toxicology, from exposure to particular $PM_{2.5}$ species, would also be found in studies of humans using the same $PM_{2.5}$ species. Adverse effects associated with particular $PM_{2.5}$ species in population-based epidemiology studies, and effects caused by the same species in human panel studies, could then confidently linked. We will briefly review results of some human panel studies which allow such comparison of effects among different types of $PM_{2.5}$ species. Because effects of BC/EC, and of traffic emissions, have been studied comprehensively, we will also briefly examine findings of two review articles of such studies.

INTRODUCTION

In London in 1952, a wintertime inversion lasting several days trapped local pollution near emission sources. The combination of exhausts from diesel buses of the time, residential burning of coal without pollution controls, and industrial emissions, created a particulate smog in London with concentrations two to three orders of magnitude higher than would occur today (peak of 4 to 460 $\mu g/m^3$ $PM_{2.5}$, in contrast to a range of daily concentrations in the U.S. of roughly 3 to 50 $\mu g/m^3$ $PM_{2.5}$). Tissue autopsies decades later showed elevated levels of solid carbonaceous material in the lungs of victims. Over 4,000 deaths in excess of the expected numbers occurred during the five days of the episode, with a total of 12,000 excess deaths occurring over the thirteen months (Hunt et al. 2003) during and after the episode. This "killer smog" made it clear that very high concentrations of tiny airborne particulates could be deadly, especially for susceptible populations such as the elderly.

Research decades later found that even the lower levels of particulate matter found in the U.S. and Europe in the 1980s and later were associated with mortality (Dockery et al. 1993; Pope et al. 1995).

Over time, researchers have come to understand that smaller particles are the most dangerous, because they can penetrate further into the lungs. Regulations in the U.S. changed over time to recognize these findings. Particulate matter (PM) 10 micrometers or less in size (PM_{10}) were explicitly regulated in 1987, and even smaller particles ($PM_{2.5}$) were regulated in 1996. However, regulation of PM on the basis of chemistry or biological toxicity has yet to occur in the U.S.

U.S. scientific advisory bodies such as the Clear Air Scientific Advisory Committee (CASAC, 1996) and the National Research Council (NRC, 1998 and 2004) suggested that different types of particles have different toxicities, and that researchers should establish biological mechanisms by which different $PM_{2.5}$ species cause harm in humans, as a means of determining which particles are most or least harmful, and of confirming findings from epidemiological studies.

Determining which particles are most harmful requires data on concentrations of many $PM_{2.5}$ species in many different localities. Without such data, epidemiological studies associating different types of particles with different health endpoints would not be possible. These $PM_{2.5}$ species include both primary and secondary emissions. Primary emissions are particles when initially emitted. Secondary particulates are first emitted as a gas, and then only secondarily become particles after transformation by atmospheric chemical reactions. Here are some $PM_{2.5}$ species found in ambient air, most of which are included in newer monitoring protocols using a central monitor in an urban location:

Primary particles:

- Black carbon (BC)/elemental carbon (EC), primarily from on and off road diesels, also from fire (forest fires, agricultural burning, fireplaces). BC has a core consisting mostly of elemental carbon, with many different organic species (such as PAHs) adsorbed onto the core during and rapidly after emission.
- Dusts, either natural in origin [marked by silicon (Si), aluminum (Al), and calcium (Ca)], or roadside dusts, which also include iron (Fe) from engine wear, zinc (Zn) from tire wear and lubricating oils, copper (Cu) from break wear, and trace amounts of other metals mixed with natural dusts.
- Trace amounts of nickel (Ni) and vanadium (V) from burning of residual oil, primarily "bunker" oil in shipping in the U.S., but also from electricity generation and commercial heating. Nickel also is emitted from smelting and some industrial practices.
- Trace amounts of selenium (Se) are emitted from burning of coal and residual oil.
- Sea salt (in localities near oceans), marked by sodium.
- Other trace metals, mostly from industrial sources.

Secondary particles:

- Sulfates (which originate as sulfur dioxide emissions, mostly from coal fired power plants, transformed by atmospheric chemistry to sulfuric acid and then to neutralized secondary sulfate particles). A very small portion of sulfates are primary emissions.
- Nitrates, which originate as gaseous oxides of nitrogen, mostly from vehicles in major cities but also from power plants.
- Secondary organic aerosols (SOA), which originate as either gases from vegetation, such as isoprenes, or from vehicles, such as toluene and xylene, and are transformed by atmospheric chemistry to organic particles.

Human panel studies also require the ability to distinguish among particulate species. While some such studies utilize data from a central monitoring site, this practice will usually lead to subject exposure misestimation—for example, in a study of people living in a retirement center—because some pollutants have strong spatial variability, even within a few blocks. This means that a monitor a mile from a residence may record a very different concentration of a locally variable emission that study subjects would actually be exposed to.

BC/EC and other vehicular emissions exhibit this variability. Thus, if the central monitor is not near a major highway, but the retirement center does have such proximity, then the actual exposure to BC/EC will be understated. Further, if subjects live in several different locations, each location could have different BC/EC concentrations, but the central monitor would assign each subject the same exposure. Thus human panel studies which monitor for different $PM_{2.5}$ species close to residences will use more accurate exposure figures for spatially variable species, and therefore will have more reliable findings.

RECENT POPULATION-BASED EPIDEMIOLOGICAL STUDIES

Because of a number of new studies published in 2008 or 2009, there is now a "critical mass" of nine mostly new epidemiological studies which include from 6 up to 20 different $PM_{2.5}$ species as pollution variables, and which have either mortality or cardiovascular disease (CVD) morbidity as health endpoints. Four of these are multi-county studies, two are multi-city studies, and three are of Atlanta, a city which does not have heavy industry or residual oil sources, and thus enables a clearer comparison of the effects of remaining emissions. For comparison, we also show results from two studies from the 1990s which include only one of the main $PM_{2.5}$ species. We exclude one new study (Ostro et al. 2010) which appears to be overly sensitive in that each of eight $PM_{2.5}$ species in the model is associated with every health effect in the model. We also exclude near duplicates of studies by identical or nearly identical groups of authors. Table 1 shows the results of the multi-city and multi-county studies, and Table 2 shows the results of the Atlanta area studies.

The first two studies in Table 1 are the two older multi-city studies (Dockery et al. 1993; Pope et al. 1995). Note that sulfate is the only $PM_{2.5}$ species utilized in one study, and one of the only two species (along with acidity) used in the other. Sulfate is significantly associated with lower survival since enrollment in each study.

The next two studies in Table 1 are the two newer multi-city studies, and include larger number of pollutants and sources. Franklin et al. (2008) includes 18 $PM_{2.5}$ species, including 13 from the Speciation Trends Network (STN). The STN network monitors for over 50 different $PM_{2.5}$ species in many U.S. localities. In this study of 25 U.S. cities, the authors find daily mortality associations with Si and Al (representing dusts), with the metals As and Ni, and with sulfate. No associations were found with BC or with V. In each of these results (except for Ni), the findings of this study are at variance with the results of the other eight studies which include many $PM_{2.5}$ species.

Janssen et al. (2002) examines the associations with daily cardiovascular disease (CVD) hospitalization of several $PM_{2.5}$ species (in this case, species used to represent sources) in residents of 14 U.S. cities. The study is also one of the few which examines pollution associations while taking into account effect modification due to air conditioning. In southerly areas of the U.S., where the great majority of homes have central air conditioning, windows are shut and the building "tight" on high temperature days when the central air conditioning is in use and people are indoors. This means that people are likely to breathe less pollution in warm months. In contrast, in northerly areas, people often open their windows to cool a residence in the summer, and hence are more exposed to emissions than those in the south.

Janssen et al. (2002) finds significant, positive associations for CVD hospital admissions with emissions from highway diesels, highway vehicles, and traffic density (all of which would be correlated with BC); with residual oil combustion (marked by Ni and V); and with metal processing. No associations were found with wood burning or with coal combustion (usually marked by sulfate). Significant and *negative* associations were found with fugitive dust (marked by Si and Al).

Table 1. Multi City Studies and Multi County Studies: $PM_{2.5}$ Species Associated With All Cause Long Term and Daily Mortality, or Emergency Department Visits for Cardiovascular Disease

Study	**Geographic Basis**	**Temporal Basis**	**Particles Components Studied**	**Associations Found**
Multi City Studies (older studies)				
1. Six Cities Study (Dockery et al. 1993)	6 Eastern and Midwestern localities	Prospective cohort study, survival since enrollment	$PM_{2.5}$, sulfate, acidity	**Sulfate** significantly associated with reduced survival since enrollment
2. ACS Study (Pope et al. 1995)	151 U.S. localities	Prospective cohort study, survival since enrollment	$PM_{2.5}$, sulfate	Significant mortality risks found for all-cause, cardiopulmonary mortality with **sulfate**
Multi City Studies (newer studies)				
3. Franklin et al. 2008	25 U.S. cities	Daily mortality	BC, OC, nitrate, sulfate, ammonium ion, and 13 elements (mostly metals)	Associations with **Al, Si, As, Ni, sulfate** (no BC or V associations)
4. Janssen et al. 2002	14 U.S. cities	Hospital admissions for CVD	PM_{10} from highway vehicles; diesels; coal; oil; wood combustion; fugitive dust and metal processing; also traffic and population density	Associations with PM_{10} from **highway vehicles, highway diesels; oil combustion; metal processing; traffic density**. No associations with coal combustion, wood burning. Fugitive dust associations significant and negative. In multi-variate analysis, PM from oil combustion, highway vehicles/diesels retain significance
Multi County Studies				
1. Bell et al. 2009	106 counties	Daily CVD hospital admissions	20 $PM_{2.5}$ components (those in Peng et al. (2009) below, plus 13 elements which are minor components of $PM_{2.5}$, mostly metals	Significant associations in cardiovascular hospital admissions found for interquartile range increase in **vanadium, nickel**, and **elemental carbon** components of $PM_{2.5}$, but not for other $PM_{2.5}$ components
2. Peng et al. 2009	119 counties	Daily CVD emergency hospital admissions	7 largest $PM_{2.5}$ components (sulfate, nitrate, silicon, elemental carbon, organic carbon, sodium and ammonium ions	Significant associations in cardiovascular hospital admissions found only for **elemental carbon**, not for other $PM_{2.5}$ components

Table 1 (continued). Multi City Studies and Multi County Studies: $PM_{2.5}$ Species Associated With All Cause Long Term and Daily Mortality, or Emergency Department Visits for Cardiovascular Disease

Study	Geographic Basis	Temporal Basis	Particles Components Studied	Associations Found
Multi County Studies (continued)				
3. Lipfert et al. 2009	206 rural and urban counties	Prospective cohort study, survival since enrollment	Twelve HAPs (incl. Ni but not V), sulfate, NOx, BC, traffic density (surrogate for traffic emissions)	Significant associations in two pollutant models with traffic density are for: **benzene, formaldehyde, diesel particulate, NOx, BC, Ni; Traffic density** significantly associated in 14 of 16 two pollutant models. Sulfate associations are significant and negative.
4. Lipfert et al. 2006	206 rural and urban counties	Prospective cohort study, survival since enrollment	15 elements (mostly metals), EC, OC, nitrate, sulfate, $PM_{2.5}$, traffic density (surrogate for traffic emissions)	Significant mortality risk associations with **traffic density** in 9 of 12 single and multi-pollutant models[1]; **nitrate, EC, V, and Ni** associated with elevated mortality risk in single pollutant, occasional multi-pollutant models; sulfate not associated with elevated mortality risks

1. In the 3 models of 12 in which traffic density was not significantly associated with elevated mortality risks, EC was also a variable. Like traffic density, EC is a proxy for vehicular (mainly diesel) emissions.

Table 2. Atlanta Studies: $PM_{2.5}$ Species Associated With Emergency Department Visits for Cardiovascular Disease.

Atlanta Area Studies				
1. Metzger et al. 2004	31 hospitals in Atlanta area	Time series study (daily CVD emergency department admissions)	PM_{10}, $PM_{2.5}$, 10 to 100 nanometer particle count, water soluble metals, sulfate, acidity, OC, EC	Emergency department admissions for cardiovascular disease associated with **EC, OC, and oxygenated hydrocarbons, in single pollutant models**, but not with other $PM_{2.5}$ components, including sulfate and water soluble metals
2. Tolbert et al. 2007	41 hospitals in Atlanta area	Time series study (daily CVD emergency department admissions)	$PM_{2.5}$, sulfate, OC, EC, total carbon, water soluble metals (1998-2004 period of study)	Emergency department admissions for cardiovascular disease associated with **EC, OC, total carbon**, but not with other $PM_{2.5}$ components, including sulfate and water soluble metals
3. Sarnat et al. 2008	27 hospitals in Atlanta area	Time series study (daily CVD emergency department visits)	EC, OC, Se, nitrate, sulfate, K and Zn	PM components associated with CVD hospital admissions were **EC, OC, and K** (a tracer for wood smoke). Sulfate, nitrate, Se, and Zn not associated.

Turning to the multi-county studies, let us first examine the paired studies of Peng et al. (2009) and Bell et al. (2009). Both used the seven $PM_{2.5}$ species most abundant in U.S. atmospheres (sulfate, nitrate, silicon, elemental carbon, organic carbon [OC], sodium and ammonium). In addition, Bell et al. (2009) used 13 $PM_{2.5}$ species, almost all metals, from the STN. Both studies examined associations between daily CVD hospital admissions and the different $PM_{2.5}$ species. Peng et al. utilized data from 119 counties, while Bell et al. looked at 106 counties.

Peng et al. (2009) found associations with CVD emergency hospital admissions with only one of the seven most abundant $PM_{2.5}$ species, EC. Bell et al. (2009) also found associations for CVD hospital admissions with EC, and in addition found associations with two metals, Ni and V.

Lipfert et al. (2006, 2009) are studies of a cohort of veterans which look at relationships between $PM_{2.5}$ species and survival since enrollment, the same endpoint as the two older multi-city studies. The 2006 study includes STN data for $PM_{2.5}$ species (as do the Franklin et al. (2008) and Bell et al. (2009) studies). The 2009 study, instead of using STN data, uses data from EPA's Hazardous Air Pollutants (HAPs) data base. Both utilize data from 206 counties and include the "normal" $PM_{2.5}$ species such as sulfate, EC, and nitrate as well as the more specialized ones just noted.

Lipfert et al. (2006) found associations between reduced survival since enrollment in the cohort and four $PM_{2.5}$ species: Ni, V, nitrate, and EC. Traffic density was also significantly associated with lower survival since enrollment, as with Janssen et al. (2002). In Lipfert et al. (2009), vehicular HAPs such as benzene, formaldehyde, and diesel particulate emissions were associated with lower survival, as were Ni, EC, nitrate and traffic density. V was not included in the 2009 study.

The Atlanta area studies didn't include V or Ni among pollutants monitored, presumably because no residual oil sources such as shipping or residual oil-fired power plants are nearby. Metal working industry such as steel and coking operations are not nearby, either. Since these sources also emit sulfate, there is not the possibility in Atlanta of finding a sulfate association because sulfate might be correlated with harmful emissions from such sources, a possibility in earlier decades and places when emissions from these sources were higher.

The three Atlanta area studies, all of which use emergency room CVD daily hospital visits as the health effect considered, make generally similar findings as do the other studies. Metzger et al. (2004) finds EC, OC (organic carbon), and oxygenated hydrocarbon associations. Tolbert et al. (2007) find EC, OC and total carbon associations. Sarnat et al (2008) find EC, OC, and K (potassium, a marker for wood burning) associations. As with all the studies reviewed above except that of Franklin et al. (2008), sulfate is not associated with the health endpoint in question.

Of the nine new studies considering many $PM_{2.5}$ species, only Metzger et al (2004) and Tolbert et al (2007) include water soluble metals among $PM_{2.5}$ species, and neither finds a health effect association. This is an important finding because sulfates *per se* have been found not to be harmful in a large number of tests with concentrations well above ambient levels (Schlesinger and Cassee, 2003). Some speculate that ambient levels of sulfate could cause metals such as iron to become more bioavailable and thus more harmful. If this were true, it might explain the sulfate associations found in the early epidemiological studies reviewed above—it wouldn't be the sulfate itself that was harmful, but rather that the sulfate caused some other $PM_{2.5}$ species to become harmful. Thus these two studies suggest that for the metals normally found in an urban airshed—such as the most abundant such metal, Fe—the theory that sulfates cause metals to become more harmful may be in practice overstated or incorrect.

In sum, these studies provide evidence that if a number of important $PM_{2.5}$ species are included in epidemiological studies of mortality and CVD morbidity, then (1) emissions from diesel engines (BC/EC) and (2) emissions from burning of residual oil and other sources of Ni and V appear to be the emissions in ambient air which are most dangerous to susceptible people. Sulfate is associated with only one of these health endpoints, and only in one of the nine studies reviewed. However, if these specific $PM_{2.5}$ species are not in the epidemiological model, then it would appear that other emissions such as sulfate are then associated with some of these endpoints, as with the two older studies which didn't include any of the $PM_{2.5}$ species (other than sulfate) included in the nine newer studies.

Is it appropriate to compare results of studies which have different health endpoints, such as long term survival, daily mortality, and daily CVD hospital admissions? Is this an "apples and oranges" comparison? An analogy might be a search for the license plates of vehicles seen driving at high speeds from different crime scenes: to solve the crime, it is important to know if the vehicles have many different license plates, or whether the escape cars in most instances have the same license plates.

More directly, if it appears that only certain $PM_{2.5}$ species are statistically associated with several different endpoints—some of which are mortality *per se*, some of which would be expected to lead to mortality—then a hypothesis for research is suggested. That hypothesis is that perhaps these particular $PM_{2.5}$ species might cause particular effects at the cellular level—e.g., oxidative stress, inflammation, lipid peroxidation, atherosclerotic buildup, issues with vascular function—which are not caused by other $PM_{2.5}$ species. This possibility is briefly examined in the next section.

TOXICOLOGY AND HUMAN PANEL STUDIES

BC/EC has been the subject of considerable analysis in toxicology studies, and has also been part of several human panel studies which have good exposure information, and also include several different $PM_{2.5}$ species, allowing comparison of effects among such $PM_{2.5}$ species.

In addition to these studies, there is also a subset of population-based epidemiological studies which compares mortality and cardiovascular or cardiopulmonary health outcomes for those people who live in close proximity to a freeway or a major urban road away (typically, within 100 meters of a freeway or within 50 meters of a major urban roadway), vs. those who live further away. After taking into account socioeconomic characteristics, these studies almost uniformly find that those living in close proximity to major roads have significantly higher risks of mortality, or of more advanced cardiovascular disease (e.g., more advanced atherosclerosis, higher blood pressure). Some of these studies include BC, and find BC associations as well.

A recent review study examined:

- The body of these "highway proximity" studies;

- *In vitro* and *in vivo* toxicology studies of BC/EC and/or diesel exhaust; and
- Human panel studies which have good subject exposure information and which allow comparison of effects among several different $PM_{2.5}$ species, including BC/EC.

By examining toxicology and human panel studies, biological mechanisms can be found which could explain cardiovascular morbidity and mortality. The morbidity and mortality can be examined at the population level in the epidemiology reviewed in Tables 1 and 2, as well as in the "highway proximity" studies.

This review study (Grahame and Schlesinger, 2009) found that BC/EC and/or diesel exhaust is linked at the *in vitro* and/or *in vivo* stages, at the level of human panel studies, and finally in population based epidemiological studies, with the following CDV health measures:

- Oxidative stress
- Inflammation
- Atherosclerosis and lipid peroxidation (a cause of atherosclerosis)
- ST-segment depression (an indicator of ischemia, or lack of ability to get oxygen to muscles)
- Heart rate variability changes (mostly reductions), an indicator of oxidative stress and also thought to be a predictor of future heart attacks for those with prior heart problems)
- Arrhythmias
- Vascular function (higher blood pressure, increased vasoconstriction)

A few brief illustrative examples will be shown, rather than review the body of evidence for any of these effects comprehensively in this paper.

After taking into account socioeconomic factors which might influence mortality risks, the "highway proximity" studies of Finkelstein et al. (2004, 2005) found that for those living in close proximity to major roads in a Canadian city, all-cause mortality and circulatory disease mortality risks were significantly elevated (relative risks of 1.18 and 1.40, respectively). The 2004 study found that those living in close proximity to highways would on average live 2.5 years less than those living further from highways, everything else equal. Hoffmann et al. (2006, 2007) found significantly increased rates of coronary heart disease (2006) and increased coronary artery calcification (2007) with decreased distance between residences and major roads.

In the human panel studies of Delfino et al. (2008, 2009), monitors were placed inside and outside subjects' residences. The studies find associations in elderly subjects with fresh vehicular emissions—marked by BC/EC and primary organic carbon, but not by secondary organic carbon—for biomarkers of inflammatory, oxidative stress, and atherosclerotic mechanisms. These and other studies thus provide the linkages between findings in studies of animals and at the cellular level, on the one hand, and epidemiological findings associating BC/EC with elevated mortality and cardiovascular morbidity risks on the other hand.

A study of changes in heart rate variability (HRV) allows us to compare how BC/EC affects HRV changes, vs. how regional air masses containing sulfates and secondary organic carbon may do so. Reductions in various measures of HRV indicate increased oxidative stress, and appear to predict increased risk for a heart attack among those with heart disease.

Schwartz et al. (2005) took HRV measurements from patients who lived in close proximity to a major road. The monitor was also located in close proximity to the major road. Increasing BC concentrations were associated with monotonically decreasing HRV measurements (in this case, SDNN, the standard deviation of normal RR intervals). Figure 2 from Schwartz et al. (2005) shows this relationship:

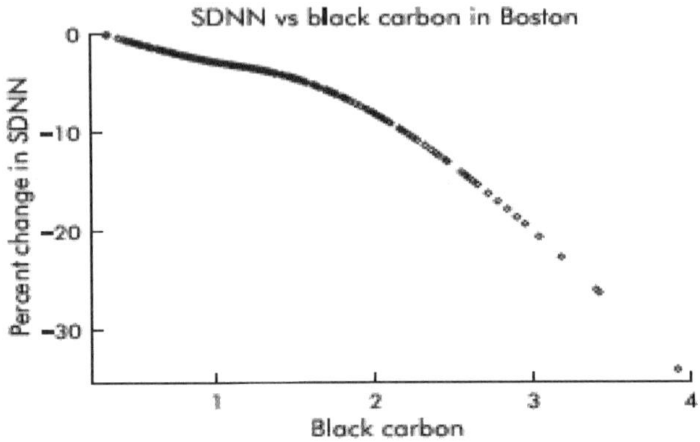

Figure 2 Smoothed plot of the percentage deviation from predicted SDNN (based on the model with all other covariates) versus black carbon concentrations in Boston. The association is nearly linear, despite being fit with about 3 degrees of freedom.

BC and $PM_{2.5}$ concentrations were measured hourly. BC, an urban emission, was subtracted from $PM_{2.5}$ for each hour, and the remaining $PM_{2.5}$ was termed secondary $PM_{2.5}$, since regional air masses without urban emissions contain mostly secondary sulfate, secondary organic carbon, and to a lesser extent, secondary nitrate. When reduction in HRV (SDNN) was then graphically associated with $PM_{2.5}$, reduction in HRV with increasing $PM_{2.5}$ occurred only at low $PM_{2.5}$ concentrations (up to about 18 μg/m^3), where BC and $PM_{2.5}$ were highly correlated. At higher levels of $PM_{2.5}$ (between 20 and 50 μg/m^3) where $PM_{2.5}$ was not correlated with BC but was instead correlated with secondary $PM_{2.5}$, HRV was unchanging, suggesting no effects on HRV and thus on oxidative stress, when $PM_{2.5}$ increased from 20 to 50 μg/m^3. Figure 3 below from Schwartz et al. (2005) illustrates these relationships.

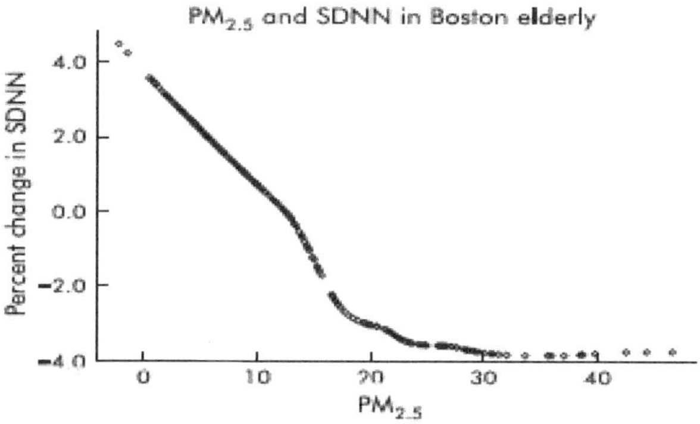

Figure 3 Smoothed plot of the percentage deviation from predicted SDNN (based on the model with all other covariates) versus $PM_{2.5}$ concentrations in Boston. The association flattens out at high concentrations where the correlation between $PM_{2.5}$ and black carbon disappears.

A broader point is that in studies which do not have good exposure information for BC/EC, there are fewer significant associations with measures of HRV. In the extreme case of a study using a monitor which is several hundred feet in the air and thus unable to measure ground level concentrations of BC, there are no associations with BC, but significant associations with sulfate (Grahame, 2009), thus possibly indicating "transference" of the association from the $PM_{2.5}$ species with large exposure measurement error to the species with little measurement error (Goldberg and Burnett, 2003).

Although there have been other comprehensive reviews of the health effects of traffic emissions, linking epidemiology with toxicology and human panel studies (Health Effects Institute, 2010), there do not appear to be any such studies for sulfate, Ni, or V. As reviewed above, the epidemiology and human panel studies to date appear to favor the interpretation that sulfate is relatively unharmful, perhaps even benign, and that Ni and V appear to be harm human health. However, given the longstanding belief that atmospheric sulfate is harmful to human health, a belief formed when sulfate was the only major $PM_{2.5}$ species extensively monitored and studies, more research is required. Specifically, human panel studies using ambient air and susceptible subjects, with the ability to compare effects of different $PM_{2.5}$ species, need to be done to further verify these interpretations. Additionally, it would be useful to perform the same type of comprehensive reviews done for BC (Grahame and Schlesinger, 2009) and for vehicular emissions (Health Effects Institute, 2010) for sulfate, Ni, and V.

CLIMATE CHANGE IMPLICATIONS

It is well known that black carbon emissions increase warming, both by warming the atmosphere, and by decreasing the reflectivity (albedo) of snow and ice, increasing the melting rate of each, exposing dark water beneath ice which absorbs heat rather than reflecting it back towards space.

Shindell et al. (2009) discusses the cooling effects of sulfate. Figure 1 in Shindell et al. (2009) illustrates this effect of sulfates, based upon abundance in the atmosphere, and also upon emissions. Ramanathan and Carmichael (2008) discuss the warming properties of black carbon.

CONCLUSIONS

Given recent improvements in understanding the health effects of different chemical species of particles, it appears that a proverbial and elusive "win-win" situation may exist. To control emissions that clearly harm the public health, and clearly contribute to warming, black carbon emissions should be controlled wherever emitted. However, in those parts of the world where acid rain problems are largely diminished, further control of sulfate emissions may do little to protect public health, but will contribute to warming in their absence. Therefore control strategies to protect the public health should shift toward controls of emissions such as black carbon.

REFERENCES

1. Adar, S.D., Gold, D.R., Coull, B.A. et al. (2007) "Focused exposures to airborne traffic particles and heart rate variability in the elderly." *Epidemiology* 18:95–103.
2. Bell, M.L., Ebisu, K., Peng, R.D. et al. (2009) "Hospital Admissions and Chemical Composition of Fine Particle Air Pollution." *Am J Respir. Crit. Care Med.* 179:1115-1120.
3. CASAC (Clean Air Scientific Advisory Committee). (1996)"Closure" letter from CASAC Chairman Dr. George T. Wolff to U.S. EPA Administrator Carol M. Browner, June 13, 1996.
4. Dockery, D.W., Pope, C.A., Xu, X. et al. (1993) "An association between air pollution and mortality in six U.S. cities." *N Engl. J Med.* 329:1753– 1759.
5. Delfino, R.J., Staimer, N., Tjoa, T., Polidori, A., Arhami, M., Gillen, D.L., Kleinman, M.T., Vaziri, N.D., Longhurst, J., Zaldivar, F., and Sioutas, C. (2008) "Circulating biomarkers of inflammation, antioxidant activity, and platelet activation are associated with primary combustion aerosols in subjects with coronary artery disease." *Environ. Health Perspect.* 116:898–906.
6. Delfino, R.J., Staimer, N., Tjoa, T., Gillen, D.L., Polidori, A., Arhami, M., Kleinman, M.T., Vaziri, N.D., Longhurst, J., and Sioutas, C. (2009) "Air Pollution Exposures and Circulating Biomarkers of Effect in a Susceptible Population: Clues to Potential Causal Component Mixtures and Mechanisms." *Environ. Health Perspect.* 117:1232-1238.
7. Finkelstein, M., Jerrett, M., and Sears, M.R. (2004) "Traffic air pollution and mortality rate advancement periods." *Am J Epidemiol.*160:173-177.
8. Finkelstein, M., Jerrett, M., and Sears, M.R. (2005) "Environmental inequality and circulatory disease mortality gradients." *J Epidemiol. Community Health* 59:481–487.

9. Franklin, M., Koutrakis, P., and Schwartz, J. (2008) "The Role of Particle Composition on the Association Between $PM_{2.5}$ and Mortality." *Epidemiology* 19:680-689.
10. Gent, J.F., Koutrakis, P., Belanger, K., et al. (2009) "Symptoms and Medication Use in Children with Asthma and Traffic-Related Sources of Fine Particle Pollution."*Environ Health Perspect* 117:1168-1174.
11. Goldberg, M., and Burnett, R. (2003) *Revised analysis of the Montreal Time-Series Study. In Revised analyses of time-series studies of air pollution and health.* Health Effects Institute Special Report, Boston, MA, pp 113–131.
12. Grahame, T.J. (2009) "Does improved exposure information for $PM_{2.5}$ constituents explain differing results among epidemiological studies?" *Inhal. Toxicol.* 21: 381-393.
13. Grahame, T. and Hidy, G.M. (2007) "Pinnacles and Pitfalls for Source Apportionment of Potential Health Effects from Airborne Particle Exposure." *Inhal .Toxicol.* 19: 727-744.
14. Grahame, T.J. and Schlesinger, R.B. (2009) "Cardiovascular health and particulate vehicular emissions: a critical evaluation of the evidence." *Air Qual. Atmos. Health.* 2010 March 3(1):3-27. Epub 2009 Jun 30.
15. Health Effects Institute. 2010. Traffic-Related Air Pollution: A Critical Review of the Literature on Emissions, Exposure, and Health Effects. HEI Special Report 17. Health Effects Institute, Charlestown, MA
16. Hoffmann, B. et al. (2006) "Residence close to high traffic and prevalence of coronary heart disease." *Eur Heart J* 27:2696– 2702.
17. Hoffmann, B. et al. (2007) "Residential exposure to traffic is associated with coronary atherosclerosis." *Circulation* 116:489–496.
18. Hunt, A., Abraham, J.L., Judson, B., Berry, C.L. (2003) "Toxicologic and Epidemiologic Clues from the Characterization of the 1952 London Smog Fine Particulate Matter in Archival Autopsy Lung Tissues." *Environ. Health Perspect.* 111:1209–1214.
19. Janssen, N.A.H., Schwartz, J., Zanobetti, A. et al. (2002) "Air Conditioning and Source-Specific Particles as Modifiers of the Effect of PM_{10} on Hospital Admissions for Heart and Lung Disease." *Environ. Health Perspect.* 110:43–49.
20. Lipfert, F.W., Baty, .J.D., Wyzga, R.E. et al. (2006) "$PM_{2.5}$ constituents and related air quality variables as predictors of survival in a cohort of U.S. military veterans." *Inhal. Toxicol.* 18:645–657.
21. Lipfert, F.W., Wyzga, R.E., Baty, J.D., et al. (2009) "Air Pollution and Survival within the Washington University—EPRI Veterans Cohort: Risks Based on Modeled Estimates of Ambient Levels of Hazardous and Criteria Air Pollutants." *J Air Waste Manage. Assoc.* 59:473-489.
22. Metzger, K.B., Tolbert, P.E., Klein, M., et al. (2004) "Ambient air pollution and cardiovascular emergency department visits." *Epidemiology* 15:46–56.
23. NRC (National Research Council). (1998) *Research priorities for airborne particulate matter: I: Immediate priorities and a long-range research portfolio.* Washington, DC: National Academy Press.

24. NRC (National Research Council). (2004) *Research priorities for airborne particulate matter: IV: Continuing research progress.* Washington, DC: Pre-publication edition. National Academy Press.
25. Ostro, B., Lipsett, M., Reynolds,. P, Goldberg, D., Hertz, A., Garcia, C., Henderson, K.D., and Bernstein, L. (2010) "Long-Term Exposure to Constituents of Fine Particulate Air Pollution and Mortality: Results from the California Teachers Study." *Environ. Health Perspect.* 118:3633-369.
26. Peng, R.D., Bell, M.L., Geyh, A.S., et al. (2009) "Emergency Admissions for Cardiovascular and Respiratory Diseases and the Chemical Composition of Fine Particle Air Pollution." *Environ. Health Perspect.* 117:957-963.
27. Pope, C.A., Thun, M.J., Namboodiri, M.M., Dockery, D.W., Evans, J.S., Speizer F.E., and Heath, C.W. (1995) "Particulate air pollution as a predictor of mortality in a prospective study of U.S. adults." *Am. J. Crit. Care Med.* 151:669–674.
28. Ramanathan, V. and Carmichael, G. (2008) "Global and regional climate changes due to black carbon." *Nature Geoscience* 1:221-227.
29. Sarnat, J.A., Marmur, A., Klein, M., et al. (2008) "Fine particle sources and cardiorespiratory morbidity: an application of chemical mass balance and factor analytical source-apportionment methods." *Environ. Health Perspect.* 116:459–46.
30. Schwartz, J., Litonjua, A., Suh, H., et al. (2005) "Traffic related pollution and heart rate variability in a panel of elderly subjects." *Thorax* 60:455–461.
31. Schlesinger, R.B. and Cassee, F. (2003) "Atmospheric secondary inorganic particulate matter: The toxicological perspective as a basis for health effects risk assessment." *Inhal. Toxicol.* 15:197–235.
32. Shindell, D.T., Faluvegi, G., Koch, D.M., Schmidt, G.A., Unger, N., and Bauer, S.E.. (2009) "Improved Attribution of Climate Forcing to Emissions." *Science* 326:716-719.
33. Tolbert, P.E., Klein, M., Peel, J.L., et al. (2007) "Multipollutant modeling issues in a study of ambient air quality and emergency department visits in Atlanta." *J Expos Sci. Environ. Epidemiol.* 17, S29–S35.

ACTING TO AVOID A WORLD WATER CATASTROPHE

WILLIAM J. COSGROVE
Ecoconsult, Inc.
Québec, Canada

This presentation addressed three questions:

- What is the state of water resources in the world?
- What are the pressing global water related issues people face today?
- Who should respond and how?

The World Water Vision (2000) concluded that there was not a global crisis of water shortage, but a crisis of management of these resources. It analysed three scenarios of what the future might hold. It concluded that under a business as usual scenario limited investments in water infrastructure would slow irrigation expansion and limit water scarcity but result in food scarcity. There would be a global grain deficit of 200 million tons with major deficits in North Africa and Middle East while India would be self-sufficient.

Under a scenario that put the emphasis on technology and investments primary water supply would increase by 24%. China and India would be water short to meet irrigation expansion. Many countries would face economic water scarcity. There would be global surpluses of 70 million tons with a surplus in OECD and higher-income developing countries but growing deficit in low-income countries.

Finally, under a scenario focusing on sustainable development in low-income countries, both water scarcity and food scarcity would be reduced. Through closing the yield gap, there would be rising production in low-income countries reducing the food deficit accompanied by lower population growth and more concern about the environment.

The World Water Vision pictured that in 2025 we will be living in a world with a population of 7.5 billion people where everyone will have access to safe water supplies. Agriculture would produce enough food so that no one need go hungry. Reduced global consumption by industry will accompany substantially higher economic activity in the emerging and developing countries. Similar concern for freshwater and the environment will have reduced the volume of waste from human activity and led to the treatment of most solid and liquid wastes before their controlled release into the environment.

Citizens will participate in decisions affecting their lives and livelihoods.

The 3rd World Water Development Report (2009) *Water in a Changing World* described the current situation. There has been some progress since 2000.

- The importance of IWRM was recognised at Johannesburg Summit 2002.
- A significant number of countries have WRM plans, strategies and legislation.
- Globally the world is on track to meet drinking water objectives of MDGs (with increased investment in sub-sector following Camdessus report).
- There is increased awareness and knowledge of groundwater stocks and quality among water managers.

- Industry is increasingly aware of dependence on water, implementing water conservation and pollution reduction plans.
- Approaches gathering data using satellite technology and modelling can facilitate monitoring trends.
- Community action often produces better results than government.
- Public contribution to decision-making uses new information technology.
- Some less wealthy countries are making better than average progress by establishing financially and politically autonomous, effective and efficient institutions.

But WWDR3 shows that this is not enough:

- The world is not on track to meet household sanitation targets.
- Where there is household sanitation, frequently wastes discharged to environment without treatment along with industrial waste and non-point pollutants. Water is the universal solvent. Even solid wastes degrade resulting in water pollution. What does not settle in beds of rivers and lakes flows to oceans. Oceans are the sink for land-based wastes.
- Global data mask lack of progress in many of poorest countries
- MDGs do not implicitly reflect important role of water in meeting all of them, including poverty reduction through economic development.
- WRM strategies and legislation often not translated to action through institutional, financial and cultural change.
- Investment in water infrastructure continues to be inadequate and single-purpose in most cases.
- Abuse and uninformed use of groundwater continues.
- Leading actors from business and government are often not involved in public consultation.
- When we need to know more about water resources and their uses we are collecting less data.
- Those knowledgeable about the impact on water of decisions in other sectors are often not at the table when decisions are taken.

In addition a rapidly changing world poses threats, offers opportunities.

- Climate change is now a fact—and will continue.
- There is an accelerated in development of new technology.
- Socio-economic and cultural changes are also accelerating.
- The human population now majority urban.
- Virtual connectedness is ordinary.
- It is accepted that pre-eminence of the West could end, while vigour and energy are transforming the Third World.
- Governments worldwide outsource public duties, while private firms turn to public coffers to socialize their risks.
- There is growing awareness that present human consumption levels amount to a massive redistribution of wealth from future generations to ours.
- Globalisation of both communications and the economy continue.

Climate change and other factors external to water management (such as demography, technology, politics, societal values, governance and law) are demonstrating accelerating trends or disruptions. Combined with an inadequate database on water quantity, quality and use, these create new risks and uncertainties for water managers and for those who determine the direction of water actions.

The World Water Assessment Program of UNESCO (WWAP) is responding. The 4th edition of the World Water Development Report will deal with managing water under risk and uncertainty. WWAP has also launched a new World Water Scenarios project in four phases:

1. Review and analysis of principal drivers including identification of linkages, considering applicability of drivers depending on major distinguishing characteristics of groups of countries.
2. Review of drivers by Scenario Focus Group (SFG) and representatives of countries to outline set of about four scenarios (possible futures) to be developed through qualitative and quantitative analysis (modelling) and used as background material for the preparation of scenarios by local actors.
3. Development of scenarios for selected transboundary and country basins and for some countries and states; review by the SFG of the global scenarios to take account of learning at local level.
4. Dissemination/outreach/training to strengthen the capacity of water managers and professionals as well as people in other sectors at the local, national, transboundary and regional levels. Will also inform political decision-making and address risks and uncertainties linked to global changes.

The first phase has examined the following drivers:

- Climate change and variability
- Water resources, including groundwater and ecosystems
- Governance and Institutions (including the right to water)
- Technology
- Economy and Security
- Agriculture
- Infrastructure
- Demography
- Ethics, society and culture (includes questions of equity)
- Politics

These drivers obviously will have different importance in different regions. And will interact. These interactions at global and sub-global scales will be examined in the later phases of the project. Logic and science will be important to the analysis. At least one of the scenarios will recognise that mankind is smart enough to control the negative effects of progress (sustainable development).

Thus although the World Water Vision will not be achieved by 2025, it can still be achieved through:

- Recognition of the crisis and the need for action.
- Stakeholder representation in integrated water resource management.
- Full-cost pricing of water services for all human uses.
- More public funding for research and innovation.
- Increased cooperation in international water basins.

The figure below illustrated the roles of the many actors involved.

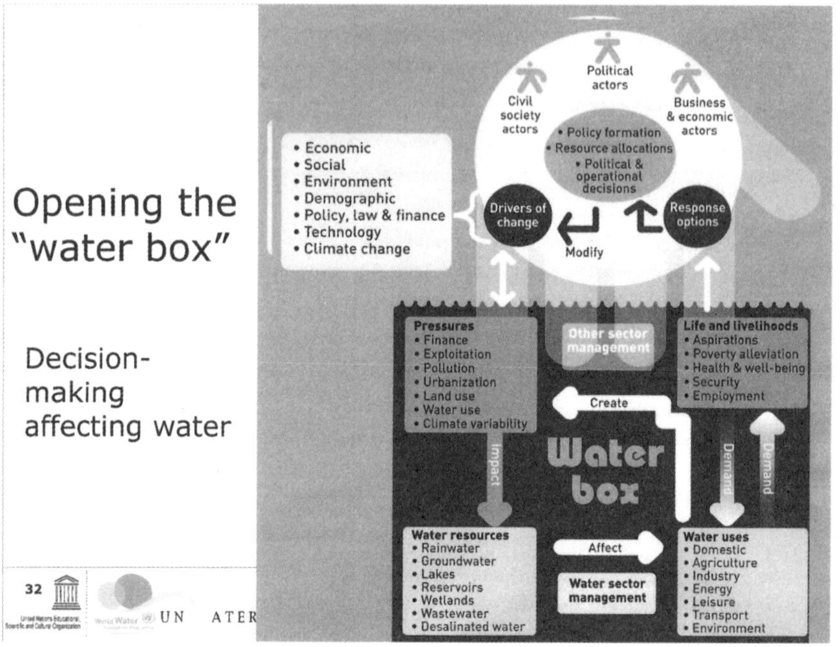

The scientific community will play an important role. Among possible scientific contributions:

- Continual refinement of GIS with ability for real-time monitoring of agricultural crops and water quality and quantity;
- Information technology permitting a global collective intelligence system, hopefully public, to facilitate knowledge management and decision-making;
- Nanotechnology to replace current water sensors, water purification and desalination;
- Biotechnology to grow food plants, biofuels and trees using saline or brackish water and to increase the yield, disease and drought resistance of crops;
- Seawater-based food and biomass, including algal production; and
- Plant-based meat substitutes and (in vitro) cultured meat.

The American Geosciences Institute has identified the following areas for the contribution of science to global challenges in the 21st Century:

- Energy and Climate Change
- Water
- Waste Treatment and Disposal
- The Food-Energy-Water Nexus

The World Federation of Scientists identified is already focusing on the role of science in the third millennium in two of these areas:

- World Energy Crisis
- Global Monitoring of the Planet

My hope is that this presentation has encouraged you to look also at the challenges presented by the management of the world's limited water resources.

OUR NUCLEAR FUTURE: THE MARRIAGE OF SCIENCE AND SOCIETY

TOM ISAACS
Center for International Security and Cooperation (CISAC),
Stanford University, Stanford, USA

INTRODUCTION

Beginning with the end of World War II, nuclear power has had a remarkable and growing impact on many of the most important aspects of modern life. Energy adequacy and sustainability, national security and international stability, public health and safety, environmental quality, climate change, and economic health are all deeply dependent on the application of nuclear policies and expertise. The continuing development of nuclear science and technology and the policies that frame them will help shape our future.

Nuclear power is in 30 countries today and supplies approximately 20% of the world's electricity. In France it supplies 75%. After many years of relative stagnation, nuclear power is undergoing a new wave of growth in existing nuclear nations. Literally dozens of additional countries have expressed recent interest in beginning to add nuclear power to their energy supply mix. While it is unlikely that all of them will build plants in the near future, the spread of nuclear power to additional countries is all but certain.

Nuclear techniques are also essential in the treatment of many medical conditions and provide a key diagnostic tool through the application of CAT scans, magnetic resonance imaging (MRI), and PET scans. Nuclear power has important space applications including the use of Pu-238 as a source for power and heat on extended space missions. Food and home safety rely increasingly on nuclear procedures, for example to verify food suitability and to warn of home fires. Accelerator mass spectrometry identifies the isotopic composition of substances, key to archaeologists in dating items of interest, in determining atmospheric chemistry, in forensic dating, and in aiding paleoclimatology.

As we all know too well, nuclear power is a controversial topic and events over these past decades have demonstrated that a number of important challenges remain. Nuclear power is a dual use technology; the same facilities that are used to prepare fuel for a nuclear power plant to provide electricity can be used to fashion the materials for nuclear weapons. Concerns about the proliferation of nuclear weapon capability to a growing array of nations is, both an immediate concern and a longer term worry. Prospects of nuclear terrorism have grown bringing legitimate concern over coming prospects. Incidents at Three Mile Island in the U.S. and Chernobyl in the former Soviet Union have colored many people's perceptions of the safety risks. Cold war weapons programs in both countries resulted in crash programs that caused substantial unnecessary environmental ground and water contamination and atmospheric testing instrumental resulting in atmospheric contamination. The slowing of energy demand coupled with unfavorable economics, high financing hurdles, and regulatory uncertainties contributed to the marked slow down of nuclear power growth after the initial surge in the 1960-70s.

Two other interrelated challenges remain and are central to the future success of nuclear power. One is the need to demonstrate the scientific and societal capability to mange and ultimately dispose of the wastes that arise from nuclear power. Today there are no operating geological repositories for the permanent disposal of high-level radioactive waste anywhere in the world. The other is the ability to garner the public and political acceptance necessary for the full promise of nuclear power to be realized. The focus of this report is on the notion that responsible and successful conduct in waste management can provide a template for not just "what" we need to do but, as importantly, "how" we need to do it. Together the equal and combined consideration of science and society may provide a lynch pin for addressing most of the key challenges outlined above and provide a template for solving what some see as an intractable problem.

NUCLEAR WASTE

High-level radioactive waste (HLW) arises from the operation of nuclear power plants, most of which are light water reactors (LWRs). Fuel elements are loaded with uranium that has been enriched (in most cases) in the isotopic concentration of U-235 from its naturally occurring 0.7% in mined ore to the 3-5% required for efficient operation in the reactor. The remaining uranium is the isotope U-238. U-235 is fissile meaning that when struck by a neutron, it absorbs it and fissions, releasing a large quantity of heat, fission products, and several more neutrons. If for every fissioning event initiated by a neutron another such event is initiated by one of the neutrons released by that event you have a chain reaction and the continual production of heat that can be transformed into electrical energy.

The much larger quantity of U-238 in the fuel is not fissile, it's fertile. This means that if a U-238 atom is struck by a neutron, it too can absorb it but the result is not a fission of the atom. Rather the atom is absorbed and through a rapid radioactive decay process, the U-238 atom is transformed into a plutonium-239 atom. Pu-239 is fissile, so it is now capable of fission if it is struck by a neutron, releasing heat, fission products, and more neutrons as well.

Eventually, the efficiency of the fuel degrades as the process builds in more fission products which inhibit the reaction. At these times, usually about once a year, the reactor is shut down and about one-third of the reactor core is removed and replaced by fresh fuel. The used or "spent" fuel is highly radioactive and thermally hot and is transferred under water to spent fuel pools of water which cool the elements and provide a radiation barrier. It is these spent fuel elements that make up the great majority of the HLW. Because there are no operating repositories for final disposal, the spent fuel pools adjacent to nuclear power plants have been filling up and an increasing use of dry casks has emerged to handle the overflow. Most of the remaining HLW arises from operations supporting national security missions involving nuclear weapons, submarines, and surface ships though there are other sources of HLW as well.

The spent fuel elements provide an intriguing option. They contain large quantities of unused uranium, plutonium that has been created during operation and fission products which are largely seen as not of great value. Techniques exist to reprocess the waste; that is to chemically dissolve the waste and then separate it into streams of uranium, plutonium, and fission products. This has been conducted in a few countries to date, notably France, the U.K. and Japan. This reprocessing allows the plutonium to be recycled into fresh fuel elements stretching the use of the original mined ore. The remaining uranium can also be reused and the fission products can be vitrified into glass logs for ultimate disposal. With the eventual introduction of new generation of "fast" neutron reactors, the spent fuel could theoretically be recycled many times, allowing the existing reserves of uranium to fuel power plants for not just decades, but at least many centuries.

However few countries will be investing in reprocessing any time soon as the plants are very expensive (the Japanese Rokkasho reprocessing plant has cost more than $20 billion USD so far and is still not in operation), uranium is relatively plentiful and affordable, and the recent growth of nuclear power has been quite modest. New reserves of uranium are being discovered in several countries. Research is being conducted on extracting uranium from sea water, and if that can be developed at a reasonable cost, we will have an essentially unlimited supply of uranium and the potential desire for reprocessing could be eliminated. The question of whether to dispose of the spent fuel elements directly (called the once through fuel cycle) or to reprocess the spent fuel to extract and reuse the uranium and plutonium (called the closed fuel cycle) remains a contentious issue, but in either case, a repository for the eventual disposal of the wastes will be required.

That provides a major challenge, largely because the waste is highly radioactive and will remain so for an exceedingly long time and because the siting and conduct of repository programs have underestimated the societal and political challenges and have operated in ways that have caused most programs to either be substantially delayed or stopped all together. Without evidence of the commitment and ability to solve the waste problem, the future expansion of nuclear power remains in doubt in many people's minds.

WHY IS NUCLEAR WASTE SPECIAL?

Nuclear waste provides both a technical and institutional challenge. The principal technical challenge is to find a suitable geologic site in which to design and construct a deep underground (>500 meters) repository that can be shown to isolate the waste from the accessible environment for the many millennia during which the waste remains a hazard. The time period of concern can reach hundreds of thousands of years. Thus proof in the absolute sense is not possible and our confidence must come from extensive scientific and technical investigations coupled with careful, conservative design and the modeling and computer simulations of the complex environments underground. Convincing ourselves, the independent nuclear regulator, and an often skeptical public remains a difficult task.

The non-technical or institutional challenge was under-appreciated for many years and is now considered to be even more difficult to achieve by many. The extraordinary time frame, nuclear fears, the concerns over a nuclear stigma, political consequences in local, regional, and national politics, differing values, and ethical dilemmas all contribute to a track record of great difficulty and setbacks in trying to site a repository facility. It has been difficult, though not always impossible, to convince communities and their elected politicians to be the nuclear burying ground for essentially forever.

Fortunately there are also some unique advantages in disposing of HLW in a repository that can be exploited. A repository is passive when completed; no active management or technical features are required. In fact, that is why a repository has been the preferred choice for disposal—given the extraordinary time frames involved, the consensus is that a final solution must be one which does not require continuing active administrative oversight and control (though it may allow it for an extended period of time). Any occurrences are likely to be slow and in a properly designed repository, there should be no inherent energy that could release the materials back into the environment.

The facility can be designed to maintain extended retrievability, preserving the capability of future generations to remove the waste for any reason. This retrievability feature has been argued as contradicting the very nature of the facility's purpose by some, but public sentiment to preserve the flexibility and to be able to monitor the repository performance have led many countries to adopt retrievability over the past couple decades.

And perhaps most importantly, the repository only becomes a repository upon closure, when future generations are comfortable with the decision. Until then the process is reversible and the waste is retrievable, providing succeeding generations with a solution—and a choice.

THE STATUS OF NUCLEAR WASTE REPOSITORIES

In all programs devoted to ultimate permanent disposal, the decision has been to pursue a deep mined geological repository in a stable rock formation. Interestingly a variety of different rock types are being investigated, demonstrating the ability to site a suitable facility in many different geologic media.

As previously mentioned there are no operating repositories for the permanent disposal of HLW or spent nuclear fuel. However, many countries are pursuing programs and some are making good progress. It is important to note, however, that almost all countries have experience difficulties and delays and many programs have had substantial setbacks or been stopped altogether.

Finland, Sweden, and France all have identified candidate sites and are proceeding to verify site adequacy and develop an appropriate design and operating scheme to then request a construction and operating license from an independent regulatory authority. Other countries including Canada, the UK, and Japan are actively searching for sites and volunteer communities, while other nations including Belgium, Switzerland, South Korea, China, and many others are in various stages of progress. Many other nations have small nuclear programs (15 of the 30 current nuclear power nations have 5 or fewer plants) and some are sensibly investigating the possibilities of shared regional repositories, though these are in early stages of development.

The United States has identified a candidate site at Yucca Mountain, Nevada, but state elected officials have been adamantly against the site for the past 20 years and the current Administration has determined that the site is "not workable" and has requested that the license application be withdrawn from the regulator. A blue-ribbon commission has been established to make recommendations on other options for the U.S. Similar reevaluations have been conducted by many nations, including France, Canada, and the UK after programs were stopped. It is noteworthy that the U.S. has had an operating repository for transuranic wastes (non-heat bearing waste with low levels of economically non-recoverable actinides such as Pu in contaminated gloves, wipes, etc.) called the Waste Isolation Pilot Plant (WIPP) in Carlsbad, New Mexico that has recently celebrated its 10th anniversary. After a very contentious beginning, the facility has been accepted by the state and WIPP is highly favored by the local population who would like to see their mission expanded to other wastes, including possibly HLW.

The important lesson is that while scientific and technical challenges are significant, waste management programs around the world have suffered and continue to suffer in many cases more from social, political, and institutional difficulties. Attempts to improve prospects by better public information programs and dialogue have proven important but insufficient.

Nuclear programs in general and waste management programs in particular have most often been led by scientific and technical leaders who view the non-technical dimensions as perhaps less central to success and more as something that can be addressed by non-technical specialists later. What experience is beginning to show us is that a true integration of the technical and societal dimensions may offer best prospects for success. Programs that explicitly derive legitimacy from extensive, early, and continuing engagement about societal issues such as ethics, fairness, transparency, local control, and just compensation may enhance prospects for success in nuclear waste management and perhaps provide a template for other emerging scientific and technical initiatives that have a galvanizing effect on society. A reemerging repository program in Canada may be one current example of putting this new perspective into action.

LESSONS FROM CANADA

In 1998, the "Seaborn Panel" (named after its chairman Blair Seaborn, issued a report after a number of years spent studying the Canadian nuclear waste management program. The key conclusions were:

"From a technical perspective, safety of the AECL concept has been on balance adequately demonstrated for a conceptual stage of development. But from a social perspective, it has not.

As it stands, the AECL concept for deep geologic disposal has not been demonstrated to have broad public support. The concept in its current form does not have the required level of acceptability to be adopted as Canada's approach for managing nuclear fuel wastes."

After consideration by the Canadian government, the AECL was relieved of its waste management responsibilities and in 2002 the Parliament passed the Used Nuclear Fuel Waste Act. The act called for the establishment of a new organization, the Nuclear Waste Management Organization (NWMO) by the nuclear waste generators - principally the nuclear utilities - and gave them three years to bring a recommendation to the government on how Canada should manage its used nuclear fuel. The NWMO program is funded by the waste generators in line with "the polluter pays" principle.

The NWMO undertook a markedly different process for reaching its recommendations. It raised societal engagement, transparency, ethical considerations, and fairness to a level of importance at least as great as that of the technical options. NWMO operated on the assumption that the "how" was as important as the "what". The development of the ultimate management approach was informed by a suite of techniques for broad engagement across Canadian society. These included among others:

- An ongoing roundtable on ethics
- Regional and national citizen dialogues
- Public attitude research
- National citizen's dialogue on values
- Nuclear community dialogues
- E-dialogues
- An international panel
- Expert papers and workshops
- Scenarios exercises
- AN NWMO website
- Public information and discussion sessions
- An assessment team of diverse individuals to help develop the options based on the insights and information gathered

The recommendation report was preceded by a report titled, "Are We Asking the Right Questions?" which sought to elicit key issues from Canadian citizens. And the final recommendation report was also preceded by a draft report to give interested parties an opportunity to have their comments reflected in the final report. That report was titled, "Choosing a Way Forward," and delivered to the Canadian government in 2005. It was approved as the Canadian approach in 2007.

The foundation of the approach rested on the following findings:

> "...this generation of citizens which has enjoyed the benefits of nuclear energy has an obligation to begin provision for managing that waste."
>
> and
>
> "...our obligation is to give them (succeeding generations) a real choice and the opportunity to shape their own decisions while at the same time not imposing a burden which future generations may not be able to manage."

The central element of the recommendation was called Adaptive Phased Management (APM) which consisted of both a technical method and a management approach. APM management approach calls for sequential decision making and flexibility in the pace and manner of implementation. It is designed to be highly adaptive in the near term, the period in which there are reasonable expectations that one will be able to rely on strong active management and institutional oversight. It includes commitment to a program of continuous learning and research and development, with an expectation that learning and innovation will be welcome and incorporated into the program to improve performance, reduce uncertainties, and adapt to changing values and circumstances. It provides meaningful options and contingency plans if early stages do not proceed as envisioned, particularly given the uncertainties regarding the future development of nuclear power in Canada. And it provides opportunities for collaborative decision-making with citizens, including future generations, to influence the way in which the fuel is managed.

The technical method requires ultimate centralized isolation in a deep geologic repository with an option for an interim step of shallow underground storage at the central site and a transportation system designed to safely and efficiently transport the waste from the power plants to the final repository site. It calls for monitoring with a potential for retrievability. It concluded that fairness called for initial siting focus on those provinces that had ongoing nuclear activities. Importantly, it seeks to find an informed and willing host community and preserves the right for the community to withdraw for any reason during the early stages of program development.

The APM approach reflected both what NWMO had heard from its extensive engagement program with Canadian citizens and groups along with insights gleaned from social research and best practices in other comparable programs. The NWMO has just finished a two-year phase to work with Canadians to develop a fair and ethical siting approach and the siting phase has begun.

The Canadian approach recognizes the dual technical and societal nature of the endeavor. Finding a scientifically promising candidate site, evaluating its suitability, designing a conceptual repository, preparing a license application to the regulatory authority, constructing a facility after receiving the license, operating the facility for the several decades it will take to fill it, and eventually closing it will rely on a daunting variety of scientific and technical disciplines of which much is demanded. These include, but are not limited to:

- Geology
- Hydrology
- Seismology
- Climatology
- Vulcanology
- Geochemistry
- Material Science
- Radiation Science
- Modeling and Simulation

Together the gathering of extensive data in these disciplines must be merged into a "safety case". This case is an integrated understanding of the suitability and likely performance of the site in combination with the designed repository, waste packages, and any buffer materials in isolating the waste for geologic time periods and a conclusion that there is sufficient confidence in that performance that a license is appropriate. The final judgment must be based on sufficient confidence that the waste, once emplaced, will not find its way back into the "accessible environment." This requires a thorough evaluation of all the conceivable pathways, the most likely of which would be water contacting the waste packages over very long periods and ultimately dissolving hazardous components of the waste which would eventually be carried back into ground water. Other pathways must also be evaluated including the effects of potential earthquakes, volcanic events, erosion, climate change, and human intrusion. In this regard, the program is much like the previous program and the programs of many other nations.

What the Seaborn panel report inspired was the recognition that the non-technical aspect was as essential and had to be fully integrated with the scientific and technical program. The many dimensions of social science include:

- Ethics
- Sociology
- Psychology
- Public engagement
- Risk Assessment and Risk Communication
- Socio-economics
- Cultural understanding and appreciation
- Sustainability
- Political Science

These factors are not simply add-ons or afterthoughts to the scientific and technical program. They can intimately affect the content, pacing, milestones, and even the technical scope and design of the program.

THOUGHTS ON ETHICS AND SOCIETY

NWMO commissioned a number of background papers by experts to inform their decision-making. Among these was a paper by Peter Timmerman, titled, "Ethics of High-Level Nuclear Fuel Waste Disposal in Canada," (2003). In it, Timmerman argues that in the basic trajectory of modern society, nuclear waste is an "ethical magnet," while our tendency has been to compose the waste management program with a technical mindset. Concerns over the nuclear age and the potentially inequitable redistribution of benefits and risks as wastes are moved from their current locations at operating nuclear power plants to a centralized location are magnified by the additional generational redistribution as we pass along risk and responsibility for our wastes to untold future generations. What is fair in making such determinations? What constitutes an ethical approach? What is our obligation to future generations? How do we balance the continuing need for more and more electricity and the prospect of a new generation of nuclear plants and resulting waste with impending climate change and the chance to use nuclear power to limit the future release of seemingly harmful greenhouse gases into the environment?

Timmerman argues that there is no common ethical framework that provides clear and consensual answers. Rather there are "ethical intuitions" that should inform the process by which we reach such decisions.

It is as much the process by which we engage society that can build the trust and confidence necessary to earn the social license to proceed. We cannot know the answers or even the path at the outset. What we can and must do is commit to the management approach that empowers the social engagement, that values and integrates social norms and insights, and that delivers on promises again and again.

The Seaborn panel recognized this as well when it stated that for a concept to be acceptable it must:

- Have broad public support;

- Be safe from both a technical and a social perspective;
- Have been developed within a sound ethical and social assessment framework;
- Have the support of Aboriginal people;
- Be selected after comparison with the risks, costs and benefits of other options; and
- Be advanced by a stable and trustworthy proponent and overseen by a trustworthy regulator.

These insights and principles are not unique to the Canadian program. There is a great similarity in research, lessons learned, best practices and conclusions in many studies in a variety of nuclear nations. The Nuclear Energy Agency, an international organization with many nuclear nations as members created the Forum on Stakeholder Confidence (FSC) which has provided an opportunity to share such ideas for the past decade. There is no substitute for stable and inspired leadership in conducting such programs, but the continued research, experience, and application of social and political science will enhance prospects for success.

AN EMERGING OPPORTUNITY AT THE BACK-END OF THE FUEL CYCLE

One relatively recent development is the consideration of new collaborative ventures among nations concerned with the back-end of the nuclear fuel cycle, particularly waste management and disposal. The international dimension may present some new and novel possibilities. In the winter 2010 issue of *Daedalus*[1], Dr. Charles McCombie and this author wrote:

> "Today, developed and emerging countries are striving to maintain or improve their standards of living by assuring a sufficient supply of energy; at the same time, they are striving to deal responsibly with global warming. Accordingly, prospects for a substantial growth and spread of nuclear power and associated facilities are increasing. For this growth to be successful, however, there are a number of concerns that need to be addressed, some technical and some economic. The potential for a systems approach to technical and economic optimization should certainly be examined, explicitly taking into consideration the holistic nature of the fuel cycle. The technical and economic challenges associated with the expansion of nuclear power are, however, outweighed by the institutional concerns that need to be addressed.
>
> Because the nuclear fuel cycle is global and because the consequences of misuse of nuclear materials are also global, all nations can be affected by an expansion of nuclear power. Multinational cooperation is essential for ensuring safety, security, and protection of the environment during expansion. This cooperation must extend to the back-end of the nuclear fuel cycle.
>
> Recent policy initiatives have focused on incentives to nations in the form of fresh fuel assurances in return for promises by recipient nations not to pursue

---

[1] McCombie, Charles and Isaacs, Tom, "The Key Role of the back-end in the nuclear fuel cycle," *Daedalus, Journal of the American Academy of Arts and Sciences*, winter, 2010.

indigenous enrichment or reprocessing (the sensitive elements of the fuel cycle). These offers have met with less than popular acceptance. To many in the emerging world, fresh fuel assurances by the developed nuclear nations look like the start of a nuclear fuel cartel, something they oppose. The assurances appear to perpetuate a division between nuclear haves and have nots, and ask emerging nuclear states to put themselves in a political situation that they believe might threaten their access to fuel in the coming decades. Many would prefer a continuation of what they feel they already have: access to a healthy nuclear fuel marketplace.

Nonetheless, revisiting the nuclear bargain established by the NPT (Non-Proliferation Treaty) and related agreements is being pushed - for different reasons - by both the nuclear weapons states and the emerging nuclear nations. These efforts present both a concern to many that the NPT may be fraying at the edges, but also a possible opportunity to build a new set of understandings and behavior that will better meet the energy, proliferation, and environmental needs of all concerned.

...Effectively integrating a successful approach to spent fuel and high-level radioactive waste management is a crucial component of pursuing such an agenda. The lack of a credible, sustained program to provide an ultimate solution to the disposal of these materials is a serious hindrance to a healthy nuclear power program. The growth and spread of nuclear power may well lead to more countries accumulating more spent fuel. The subsequent buildup of this material in an increasing number of nations will provide a reservoir of plutonium that could later be accessed through reasonable quick and simple, and possible covert, reprocessing techniques. Along with the spread of expertise and necessary technical knowledge, this buildup can bring countries closer to weapons creation and potentially set off regional instabilities as neighbors begin to hedge their nuclear bets as well.

Creating an international initiative to explore the prospects for multinational spent fuel storage, with eventual multinational disposal of spent fuel of the high-level waste resulting from reprocessing, can begin a win/win process for solving the waste issue in a manner that addresses proliferation, energy, and waste management issues simultaneously. Companion efforts could pursue multinational enrichment facilities and, as needed, reprocessing facilities with opportunities for financial participation by emerging nuclear nations.

Established nuclear nations, particularly the nuclear-weapons states, should lead by example. As leaders, they can transform waste management and disposal from issues of "nuclear garbage" to integral elements of an internationally accepted system. This system would not only provide for the resurgence of nuclear power, but in doing so would simultaneously reduce proliferation, regional instability, and waste management concerns."

THE ROLE OF SCIENCE IN THE THIRD MILLENNIUM

"About 1100 AD, the crossbow was developed in Europe and outlawed, in 1139, by the second Ecumenical Lateran Council, 'as humankind's first formal attempt at arms control' (O'Connell 2002:64)".

It is most timely to investigate the role of science in the coming millennium. It was about 1000 years ago when Ibn al-Haythem became in some sense the "first scientist." Al-Haythem developed what today we call the scientific method based on experiments and observation. So we may say without too much of a stretch of imagination that we have just brought a close to the first millennium of science.

Of course, any attempts to depict the role of science—and technology—in the next millennium must be approached with the greatest sense of humility. Thermonuclear weapons, the personal computer, the internet, men walking on the moon, the cloning of animals, and cell phones are but some examples of the transformative application of science and technology in the recent past. It is probably fair to say that none of them was fully anticipated well ahead of time and their impacts are still unfolding. The science of the next century, let alone the next millennium is unimaginable.

That is not to say that this is the only period of rapid advances: Morton (anesthesia), Kelvin, Darwin, Mendel, Mendeleev, and Maxwell all did their defining work within a 30 year period in the mid-19th century.

Earlier, Leonardo da Vinci famously made careful observations and magical drawings of anatomy and nature. Leonardo set up controlled experiments in water flow, medical dissection, and the study of movement and aerodynamics. Fritjof Capra in "The Science of Leonardo," classified him as the "father of modern science" concluding that Leonardo's science was more "in tune with holistic non-mechanistic and non-reductive science which are becoming popular today." The focus on empirical evidence and mathematics influenced Copernicus and Galileo. As seems often the case, the more we progress, the more we understand how much we have to learn from Leonardo.

But the remarkable growth in the speed and content of communications and transportation, have severely reduced the time and increased the impact of new transformative advances, often with unanticipated and sometimes harmful societal effects. Globalization, modernization, and the movement of ideas and people are improving the lives of billions of people while causing fault lines in global harmony.

We may well be in the infancy of new advances in hand-held smart devices, synthetic biology, genetic engineering, electric cars, space travel and many ideas that are today only a gleam in someone's eye. We may also be at the beginning of major planetary emergencies such as climate change, population growth, and access to clean water. In each case we must call on the best science and technology to understand and effectively address issues of global importance. What may be new, except to Leonardo, is the growing recognition that we must simultaneously address the social and political elements as well—the "holistic" science. The future of the civilian nuclear enterprise and the disposition of nuclear wastes may provide a unique opportunity to learn and then apply this lesson.

SECTION 7

SESSION 7

CLIMATE

FOCUS: PALEOCLIMATE NATURAL VARIATIONS AND IMPLICATIONS — NON-GHG INFLUENCES ON CLIMATE

GLOBAL CLIMATE IN THE INDUSTRIAL ERA

RICHARD ARNOWITT
Department of Physics, Texas A&M University
College Station, USA

ABSTRACT

A brief review is given of some of the current ideas concerning global climate since the beginning of the industrial era. While it is generally agreed that there has been a global warming, the cause of this is not universally agreed upon, particularly that it has been caused by greenhouse gases (primarily carbon dioxide) of anthropogenic origin. The arguments implying anthropogenic origin of the warming are discussed, as well as some of the questions raised. What additional work is needed to help clarify the issues is discussed.

INTRODUCTION

A great deal of data now exists concerning the change of global temperature over the last century and a half. It is generally agreed that the surface temperature of the earth has warmed significantly with the onset of the industrial era. An example of this data is shown in Figure 1.

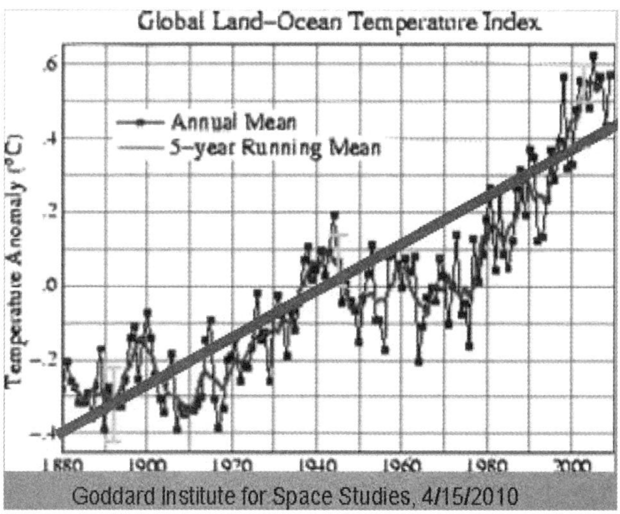

Fig. 1: Global land-ocean temperature relative to 1951-1980 base. Black dots are annual mean and the smooth (red) line is a five year running average. (The straight (green) line exhibits the trend from 1880 to the present.) [NASA Goddard Institute for Space Studies (GISS), 4/15/2010]

While the data shows a great deal of "noise" (i.e., short-term variability) implying the complexity of forces governing the global climate, in recent years the warming has become quite apparent. The existence of global warming implies that its effects should also be visible in a number of other ways, e.g., the sea surface temperature should be rising, the sea level should be rising, Arctic sea ice coverage should be decreasing, etc., and indeed all these effects and many others have been observed, as seen in Figure 2. Figure 3 shows the volume of Arctic Sea ice from 1979-2010 (relative to 1979-2009 mean). The volume has been following a decreasing line that appears to have steepened over the past four years.

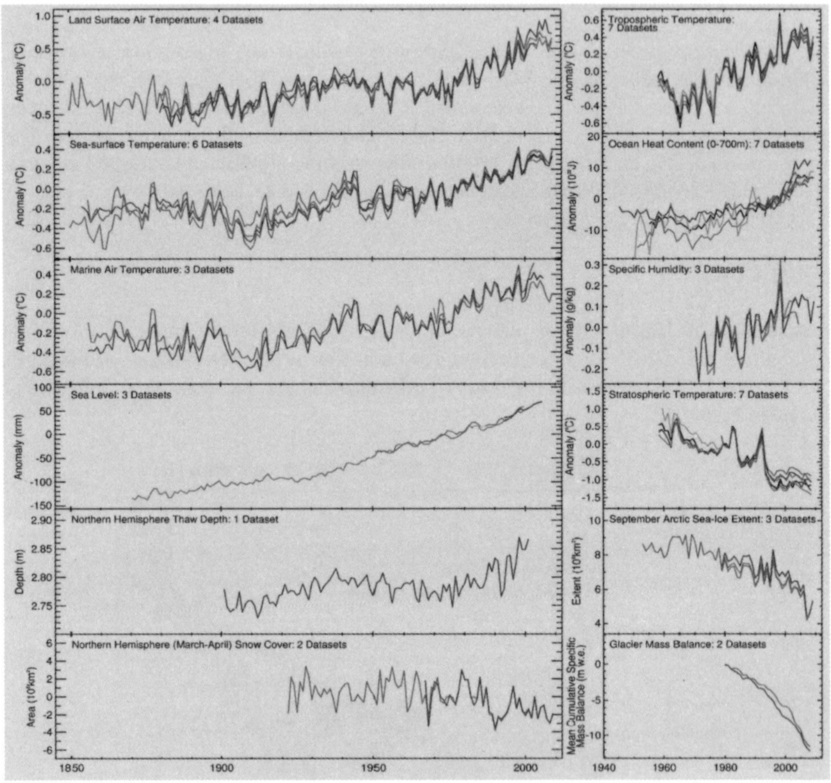

Fig. 2: Twelve climate quantities (involving 48 data sets and the work of more than 100 scientists at over 30 institutions) each exhibiting the effects of global warming. [Courtesy of G. North]

The short vertical (green) lines in Figure 1 are the estimated errors in the temperature measurements. Thus the 1σ error in recent times is about 0.05°C rising to twice that at the earlier times. The data shows a significantly greater variability than these errors indicating the existence of short time duration forces.

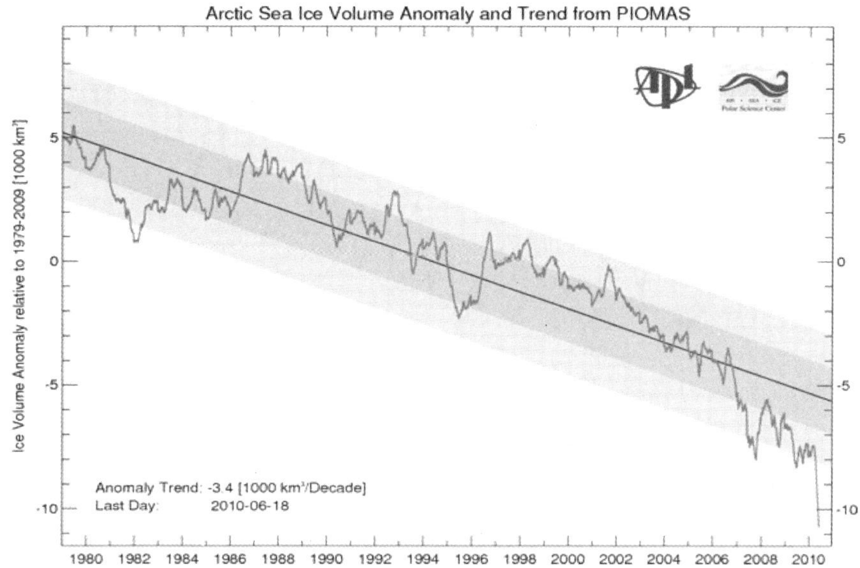

Fig. 3: Arctic ice volume relative to 1979-2009 mean. The blue line is -3.4(1000 km^3/decade). [Polar Science Center, Pan-Arctic Ice-Ocean and Assimilation System (PIOAS 06/18/2010)]

Over the past decade some of the shorter time duration forces that give rise to the jittery behavior of the yearly mean data have been examined. A 12 month running mean (to eliminate the bias of the time of year) is shown in Figure 4. Here one can see the effects of major volcanoes (which put sulfates into the atmosphere causing cooling for about two years). The effects of the el Niño–la Niña oscillations (ENSO) (causing increased warming and cooling respectively) correlate well with the short-term peaks and valleys of the temperature curve.

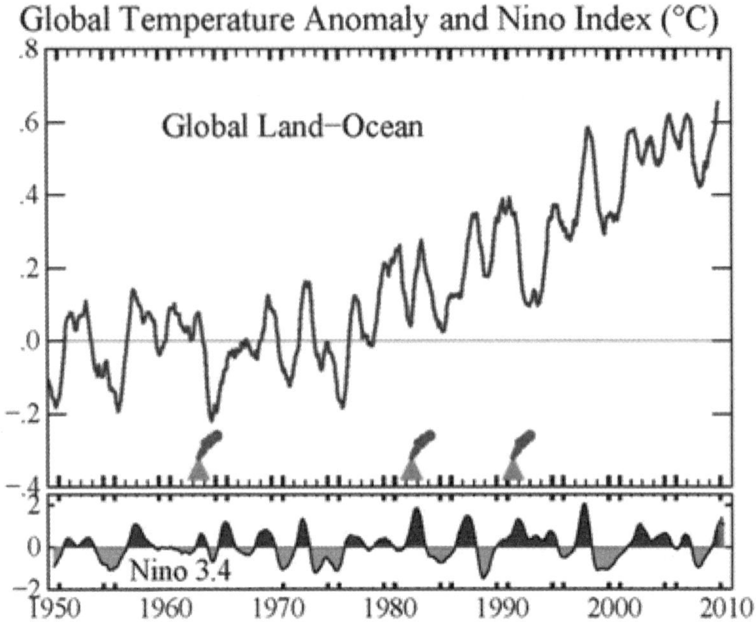

Fig. 4: Twelve month running mean global temperature (blue) curve 1950-2010 relative to base period 1951-1980. The three major volcanoes during this period give rise to minimums in the temperature curve for a two year period (due to the sulfate aerosols produced). The temperature curve also shows correlations with the el nino and la nina maximum and minimums. [J. Hansen, et al. Rev. Geophys., *in press, doi:10.1029/2010 RG000345]*

If one takes a five-year running mean, one averages over the Niño oscillations, and if one further takes an 11-year running mean one also averages over the solar output oscillation. This is shown in Figure 5. One sees a significant reduction in the noise compared to the yearly averages of Figure 1. Figure 5 also shows that since about 1970, the temperature has been rising at a rate of (0.15-0.20)°C/decade, and the suggestion that the warming has plateaued over the past decade (as one might have read into the annual data of Figure 1) disappears.

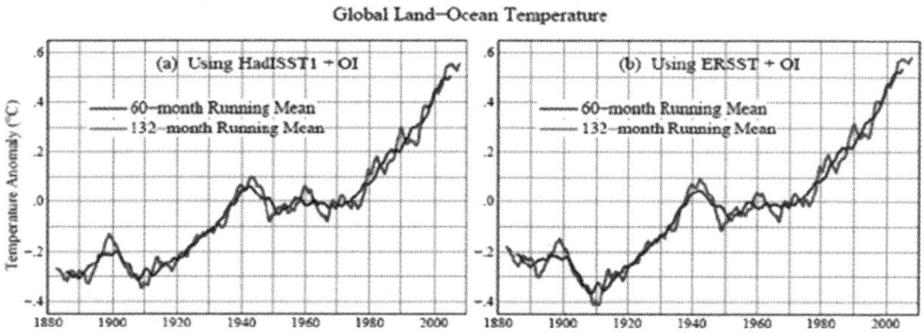

Fig. 5: Sixty month (red) and 132 month (blue) running mean global temperature for two ocean data sets. The 60-month running mean averages over the nino oscillations, and the 132 month running mean further averages over the 11 year solar irradiance oscillation. The latter curve is smoother, indicating the origin of some of the short time variations seen in Figure 4. Note also that the temperature rise continues through the last decade. [J. Hansen, in press, doi:10.1029/2010 RG000345]

THE GREENHOUSE GAS HYPOTHESIS

While there is general agreement that the global temperature is rising, there is not full accord as to the cause. The greenhouse gas hypothesis argues that the temperature rise is due to the ever increasing amount of greenhouse gases (mainly CO_2) put into the atmosphere by human industry, and thus the temperature rise is mainly of anthropogenic origin. The correlation between the amount of atmospheric CO_2 and the global temperature is shown in Figure 6 relative to a base period taken to be 1961-1990. As the amount of CO_2 (black line) began to rise significantly starting in the mid 20th century, the global temperature (blue and red bars) began to increase significantly. Of course a correlation does not by itself imply a causal connection. For example, one alternate explanation of Figure 6 could be that the temperature rise is just natural climate variation. That is, the 19th century represents the end of a global cold period (the "little ice age"), and so as one moves into the 20th century one would expect a warming period to continue. That this happened to occur at the same time that the industrial era began is accidental, and if so there could be no causal connection between warming and CO_2 increases.

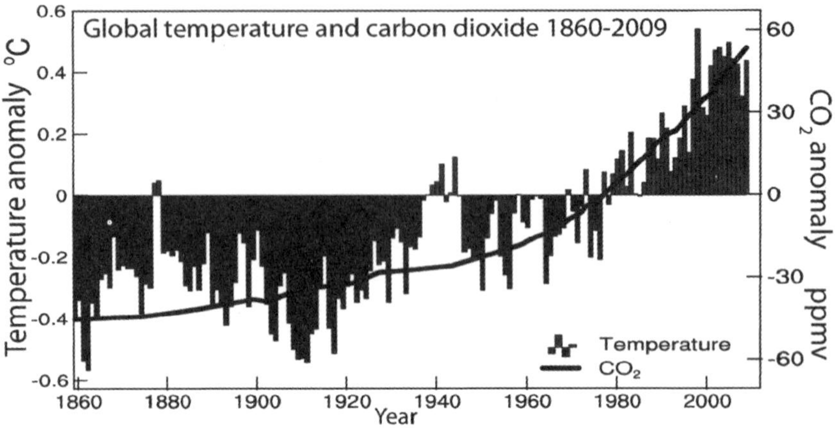

Fig. 6: Global temperature and atmospheric carbon dioxide 1860 to 2009 relative to base period 1961-1990. Temperature is the (blue and red) bars, and the solid line is the carbon dioxide amount. [National Center for Atmospheric Research (NCAR)] Note that 50% of the increase in atmospheric carbon dioxide occurs after 1980, where the sharpest temperature increase occurs.

However, one does expect CO_2 to produce a greenhouse warming, and this can be checked from explicit data and has been known for over 40 years. Figure 7 shows the radiation at high altitude as a function of wavelength escaping upward on a clear day (no clouds). The solid curves with temperature labels are Planck distributions which would be the distribution if the earth were a simple black body. One sees reductions from this due to the greenhouse gases (GHG): the measured large dip in radiation escaping is due to CO_2, a smaller one from ozone, as well as other reductions from water vapor (which also is a greenhouse gas). There is a well established theory of radiative transfer of radiation from the surface of the earth to high altitudes, and also the properties of the CO_2 and H_20 molecular spectra are well known. This allows the calculations of what might be expected to be seen at high altitudes. The theory result is given by the dotted line on Figure 7, and is in excellent agreement with the data.

If one were to increase the amount of CO_2, there would be less radiation escaping into space, and so the global temperature would increase. An accurate calculation shows that if one doubles the amount of CO_2 in the atmosphere (keeping everything else unchanged) the amount of radiation escaping would be decreased by about 3.9 Wm^{-2}, which would correspond to a temperature rise of about 1.2ºC. (This result is generally agreed upon by all workers in the field.)

However, the global climate is a very complex non-linear system, and if one changes one aspect, other changes occur. For example, if the temperature is increased, more water vapor is evaporated from the oceans, which would lead to more water vapor

GHG in the atmosphere, and hence more warming. Alternately, the extra water vapor might lead to cloud formation that could reflect some of the incoming solar radiation leading to cooling, etc. Similarly, the amount of surface ice might change, affecting the albedo of the earth (since ice reflects incoming solar radiation). Thus there are feed backs that can modify the final result. If one assumes that the various feedbacks depend only on temperature, one can phenomenologically characterize the feedbacks as follows:

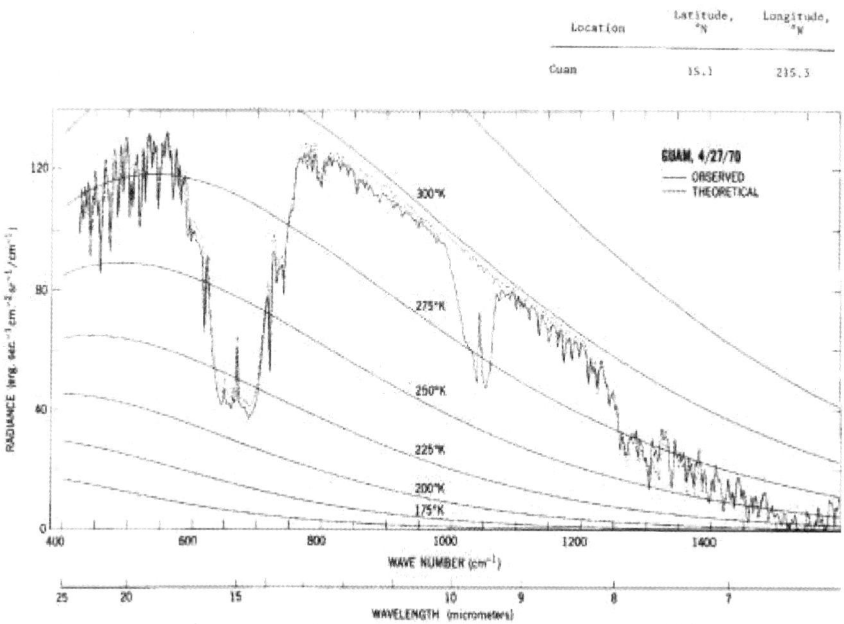

Fig. 7: Upward radiation as a function of wavelength at an altitude of 20 km near Guam on a clear day. The thin smooth solid lines (labeled with numbers) are Planck black body distributions at different temperatures. The superimposed solid line is the experimentally observed radiation. The large dip on the left hand side is due to carbon dioxide, and the smaller dip in the middle is due to ozone. The other parts are due to water vapor. The dashed curve is the theoretical expectation, in good agreement with what was observed. (Ozone was not included in this calculation.) [Courtesy of G. North)

$$(\Delta T)_{2xCO_2} = \frac{(\Delta T)_{f=o}}{1-\frac{f}{3.2}} \tag{1}$$

where (as discussed above) $(\Delta T)_{f=0} = 1.2°C$, and f (measured in W m^{-2}/°C) represents the feedbacks from different sources e.g., water vapor (WV), albedo (A), clouds (C), etc.:

$$f = f_{WV} + f_A + f_{cloud} + \ldots. \tag{2}$$

A positive feedback will decrease the denominator of Eq.(1) and hence increase the temperature rise when the amount of CO_2 in the atmosphere is doubled, while a negative one will reduce the effect CO_2 has on global warming. A similar equation can be written for the other greenhouse gases (e. g. methane, etc.), and since the feedback factors f are universal, one can add the equations from all the GHG to get a similar equation where the amount of CO_2 is replaced by "CO_2eq" which includes the effects of all the GHG in terms of an equivalent amount of CO_2. (The other GHG contribute about (25-30) % to CO_2 eq and so increase the effective amount of CO_2 by that amount.)

One can determine the explicit values of the feedback factors using a climate model. Climate models have been the subject of much research for the last 35 years. The current global climate models have become quite sophisticated, taking into account a large number of parameters, and their validity have been checked against many climate phenomena. However, a number of key quantities have not yet been fully explored, either because the basic physics describing them is not fully known or experimental data is not available. An example of this involves the behavior of clouds, whose upper surface will reflect the short wave incoming radiations from the sun (and thus produce global cooling), but are opaque to the incident long wave radiation coming from the earth (which will act to produce global warming). What clouds do if GHG raise the global temperature depends on the details of the climate model being used. (Further, since clouds contain water vapor, it is not clear that one can treat clouds and water vapor feedbacks as independent phenomena). An example of results from a number of models is given in Figure 8. In these cases, the water vapor (WV) gives a large positive feedback, the albedo from ice coverage (A) gives a smaller positive feedback, and the clouds (C) evaluate (mostly) to an additional positive contribution. The total contribution to the feedback is given on the right hand side (labeled ALL), and using Eq. (1) one gets a global change of temperature from doubling the atmospheric concentration of CO_2 of between 2.3 and 6.4°C. This is a wide range, but in all these models, the resultant increase is not small, particularly since most estimates predict nearly a tripling of atmospheric CO_2eq by 2100 if no mitigation to its increase occurs. (Thus the Intergovernmental Panel on Climate Change (IPCC) in their 2007 AR4 report estimates a mean temperature rise by 2100 of about 4°C while a more recent MIT report [MIT Joint Program on Science and Policy of Global Change (Report 180), 2009] estimates a mean temperature rise of 5.1°C.)

However, if the cloud contribution to the feedback should be negative, (as argued by some) it could cancel the positive contributions from WV and A, greatly reducing the expected temperature rise. This illustrates the importance of understanding the nature and effects of clouds.

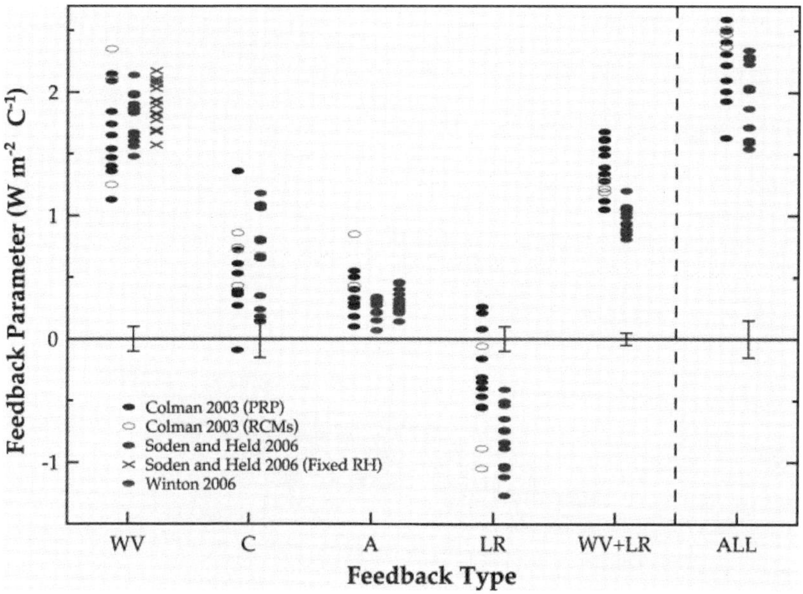

Fig. 8: Feedbacks for water vapor (WV), clouds (C), albedo (A), lapse rate (LR), and the total feedback (ALL) for five climate models. In these models the clouds give mostly positive, though moderate contributions, so that the total feedback is large and positive, producing a large amplification of the change in temperature when one doubles the amount of carbon dioxide [see Eqs (1) and (2)]. (One would need a significant negative cloud contribution to reduce this amplification.) [Intergovernmental Panel on Climate Change (IPCC) report AR4 WG1, 2007, Figure 8.14]

The current climate change models can give a fairly good reproduction of the known global temperature changes from the pre-industrial time to the present. This is exhibited in Figure 9. The black line is the actual yearly temperature changes relative to the mean of the 1901 to 1950 global temperature. Note the cooling dips that coincide with the major volcano eruptions due to the sulfates they put into the atmosphere for two to three years (which reflect incident solar radiation). The thin (yellow) lines in the upper graph are the expectations from 14 different models which include both natural and anthropogenic climate forcing, and the thicker (red) line is the global temperature averaged over all models. This line agrees reasonably well with the actual data (and it seems to be a general case in a number of phenomena that an average over many global models reproduces the actual data much better than any individual model). In the lower graph of Figure 9, the thin (blue) lines represent the predictions from five models using only natural causes of climate change, and the thicker (blue) line is the mean of these. One sees a disagreement in this case between the model predictions and the data from about 1960 on, thus suggesting that anthropogenic causes are indeed needed to explain the observed temperature rise over the past half century. There is, however, one caveat to this conclusion. In order to obtain the good agreement between the models and the data in the upper graph, the models have to assume a certain amount aerosol sulfates exist (or else too much warming would occur). The amount of this assumed aerosol forcing ranges from -1.7 Wm^{-2} to -0.1 Wm^{-2}. Efforts at direct determination of this quantity give a range of -2.2 Wm^{-2} to -0.5 Wm^{-2} which is consistent with what is assumed, and so the assumptions are not unreasonable. However, more information concerning atmospheric aerosols is needed to clarify these assumptions. Since aerosols make a different geographical pattern then CO_2, it may be possible to separate the two. (See e.g., Global Aerosol Climatology Project, NASA (GISS).)

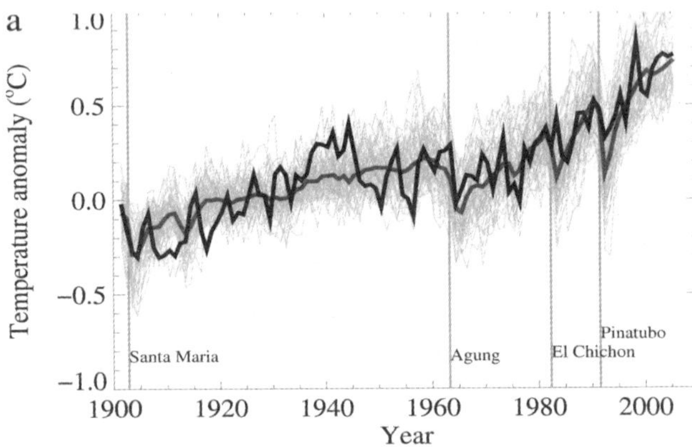

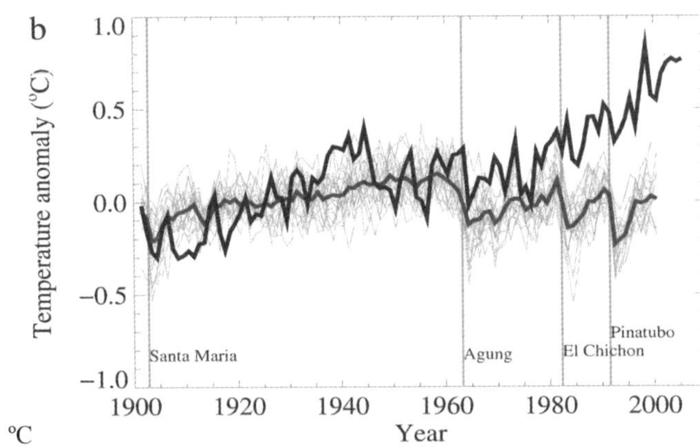

Fig. 9: Temperature anomaly from 1900 to 2005 relative to a base of 1901-1950. The solid lines are the observed global temperatures. In the upper graph (a), the thin (yellow) lines are results of 14 global climate models where natural and anthropogenic forcings are included. The thicker (red) line is the average over the model predictions. The ensemble average of models interpolates the experimental values reasonably well. In the lower graph (b), the thin (blue) lines are the results for five global climate models where only natural forcings are included, and the thicker (blue) line is the ensemble average. Here the models significantly deviate from the observed temperatures after about 1960. [IPCC report AR4 WG1, 2007, Figure 9.5]

CONCLUDING REMARKS

What actions might the above discussion suggest? For the scientists it is clear. They will continue to study the important subject of global climate until a full understanding is obtained. As discussed above for the question of predicted temperature rise, the effect of clouds and of aerosols (among other things) is of particular significance. For the policy makers, the situation is more complicated. It may be decades before a global climate model is textbook perfect. However, if the current models are reasonably accurate, a temperature rise of much more than 2°C could have serious consequences of disrupting agriculture leading to famines, sea level rises flooding low lying land and more violent storms, and most current models predict temperature rises much more than 2°C. Further, once these global warming effects are underway, it will become difficult to stop them, as the carbon dioxide causing them can stay in the atmosphere for a millennium or more [S. Solomon et al. PNAS 106, 1704(2009)]. On the other hand, if the climate critics are correct, and the effect of GHG lead to a small or negligible increase in temperature, all the money spent to mitigate climate change will have been wasted. In a democracy, policy makers have an additional problem to face. The constituencies that elect them are the ones that will have to agree to pay for any mitigation policies, while they will be long out of office before any benefits that might have accrued from this will have occurred. Thus taking action will require political courage, while doing nothing will have small political consequences, at least while they are in office.

ACKNOWLEDGMENTS

It is a pleasure to thank Gerald North for his great help and for clarifying for me many aspects of global climate and climate models. I should also like to thank Prof. Antonino Zichichi for bringing to my attention the questions about what to do about global warming (even though I may not always agree with his conclusions!).

THE ROLE OF WATER IN THE FATE OF CARBON: IMPLICATIONS FOR THE CLIMATE SYSTEM

JÁN VEIZER
University of Ottawa, Department of Earth Sciences
Ottawa, Ontario, Canada

It is a general public perception that the science of climate change is settled. It isn't. And the issue is not whether there has been an overall warming during the past century. There has, although it was not uniform and none was observed during the past decade. The geologic record provides us with abundant evidence for such perpetual natural climate variability, from icecaps reaching almost to the equator to none at all, even at the poles.

In a recent article Lacis et al. (2010) acknowledge that the water vapour greenhouse effect accounts for at least ¾ of the temperature response in the Intergovernmental Panel on Climate Change (IPCC, 2007) models. Yet, despite this overwhelming impact on the outcome of the models, these authors argue that the water cycle (including clouds) "plays only a passive role, responding to a driver which is the non-condensing greenhouse gases", principally CO_2. According to this scenario it is the stabilizing role of these non-condensable greenhouse gases in the atmosphere that prevents the collapse of the entire planetary greenhouse, including its "positive water vapour feedback". In short, it is the tiny carbon cycle that is driving the huge water cycle. Just how realistic is this scenario? In an everyday analogy, this would be akin to a claim that the economy of the European Union is sustained by one of its tiny members, because it is its export/import balance that drives and stabilizes the huge, but passive, German economy. In the subsequent text I will argue that it is the other way around, with the tiny carbon cycle "piggybacking" on the huge water cycle, and the models are therefore reversing the cause and effect relationship.

The climate debate is in reality about a 1.6 Watts/m^2, or a 0.5 % (!) "discrepancy" in the poorly known planetary energy balance (Foster et al. 2007). This 1.6 Watts/m^2, apparently an outcome of the industrial revolution, is a rather uncertain number by itself, with some entries, such as the aerosols, having error bars almost as large as the entire "discrepancy" and others, such as the clouds, assigned only a -0.7 Watts/m^2 impact. Yet clouds, on balance, produce a cooling effect of -13 Watts/m^2 (Ramanathan et al. 1989; Latham et al. 2008). Moreover, some physicists (Gerlich and Tscheuschner, 2009) question even the physical reality of the planetary greenhouse concept. It is beyond my expertise to dwell on this controversy and I will therefore tentatively accept the 1.6 Watts/m^2 as given for the subsequent discussion.

The above energy "discrepancy" is claimed to have resulted from the compound greenhouse impact of anthropogenic CO_2 over the last century (IPCC, 2007). Note nevertheless that it is not the CO_2 itself, but its supposed energy equivalent, called "prescribed CO_2", that mostly serves as the model input, due to the fact that our understanding of the coupling of the biologically controlled carbon cycle to climate models is rather rudimentary (Denman et al. 2007). Considering the overwhelming importance of the water cycle for the planetary greenhouse (Figures 1 and 2), it is clear that the input of energy from any source will result in a warmer and wetter climate. From the point of view of the water cycle the imputed energy is just that, energy, regardless of its source. Because of this overwhelming importance of water vapour for the greenhouse effect, and the inherent ambiguity as to the source of energy, the existing climate models are unlikely to yield a definitive answer as to the role of CO_2 vs. the sun, for example.

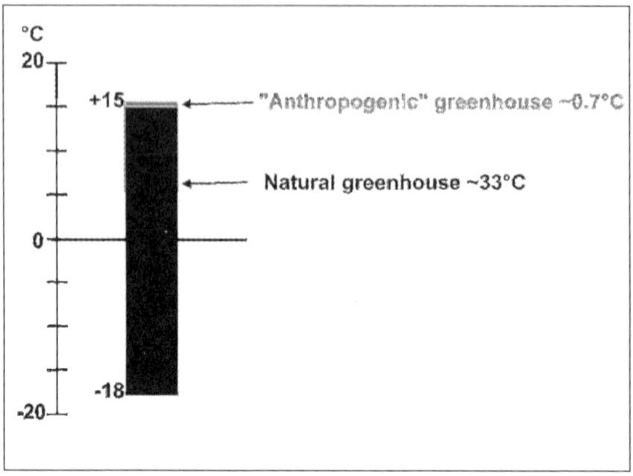

Fig. 1: The relative amplitudes of natural and "anthropogenic" greenhouse effect (Ruddiman, 2001). Note that the earth would be a frozen ball without its overall atmospheric natural greenhouse of ~ 33°C.

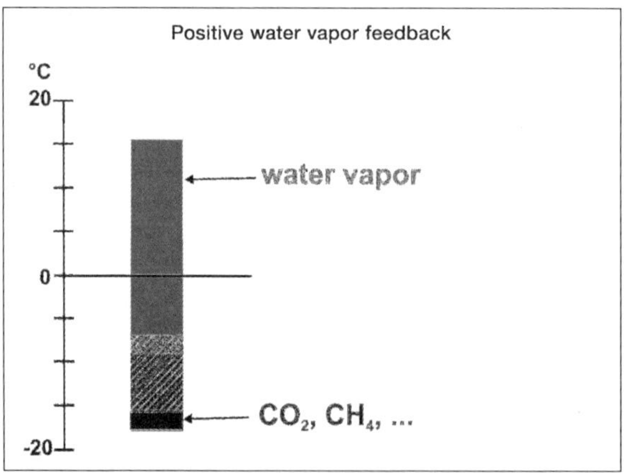

Fig. 2: The relative role of water vapour to "non-condensing greenhouse gases" in generating the temperature in climate models. Note that the "positive water vapour feedback" is responsible for some 60% (IPCC, 2007) to perhaps 95% of the entire temperature effect.

But this is an assumption, an attribution by default, not an actual empirical or experimental proof that carbon dioxide is the driver, an assumption that remains open to challenge not only in its spirit but also in substance: 1) the correction of Froehlich and Lean that transformed a slightly positive TSI trend into a slightly negative one has been strongly disputed by both principal investigators of the satellite experiments (see emails of R.C. Wilson and D. Hoyt in Scafetta, 2010); 2) the positive water vapour feedback amplifier is there for any driver, not just the CO_2; and 3) potential amplifiers to solar activity, mostly via direct or indirect cloud modulation (e.g., Svensmark, 2007; Kirkby, 2007), have been proposed. Yet the above attribution is then taken as a fact in the subsequent complex model calibrations of climate sensitivity to CO_2.

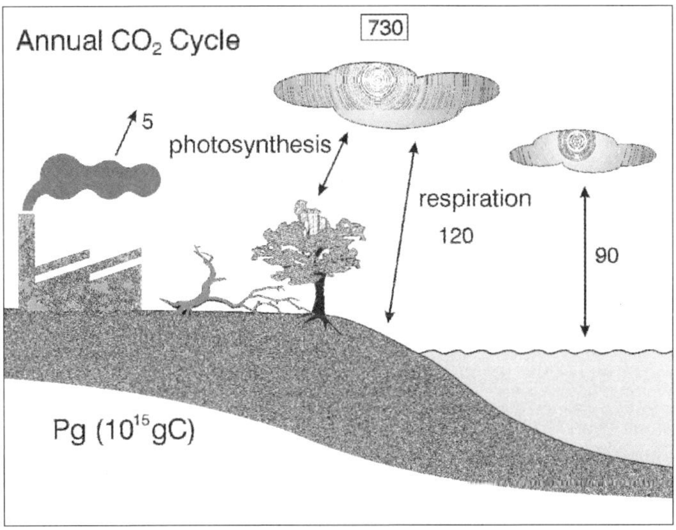

Fig. 3. Simplified presentation of the modern carbon cycle (based on IPCC, 2001).

If an amplifier to solar output does exist, and empirical observations detailed below argue for its existence, the need to attribute the energy input to man-made greenhouse gases would diminish accordingly. So how realistic is the basic model assumption that the tiny—biologically controlled—carbon cycle drives the climate via the passively responding huge water cycle? Nature tells us that it is the other way around. Surely, the blossoming of plants in the spring is the outcome, not the cause, of the warming sun and abundant rain?

Our atmosphere contains 730 billion tons of carbon as CO_2. Each year about 120 billion tons of carbon is cycled via plants on land and 90 billion tons via oceans (Figure 3). On time scales relevant to the discussion of the anthropogenic greenhouse, years to centuries, it is the interaction of atmosphere and land plants (photosynthesis/respiration) that is the dominant flux controlling the atmospheric CO_2 budget, as attested to by the observations that atmospheric CO_2 concentrations are higher, and seasonal oscillations larger, on the northern hemisphere, because this is where the bulk of the land mass is situated. Note that an equivalent of the entire atmospheric carbon budget is cycled via the biosphere on land in about 6 years. On time scales of millennia and beyond, however, it is likely the ocean that controls the atmospheric CO_2 levels.

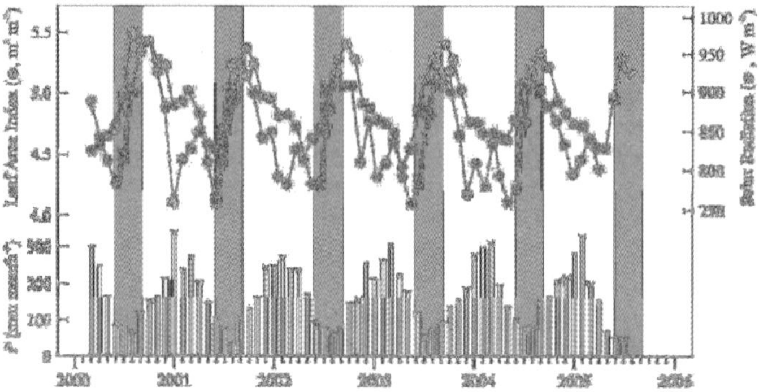

Fig. 4: The intensity of solar radiation, photosynthesis (Leaf Area Index) and precipitation (P) in the Amazon biome. Based on data in Myneni et al. (2007) and reproduced from Ferguson and Veizer (2007).

Human emissions account for about 7 to 10 billion tons, or less than 5 % of the annual CO_2 flux, but they are likely the main source of the increase in atmospheric CO_2 concentrations from the "preindustrial" levels of ~280 ppm to ~380 ppm today. Note, however, that only about one half of this anthropogenic CO_2 is present in the atmosphere, with perhaps one quarter sequestered into the oceans and one quarter still not quite accounted for. This large "missing" sink, lately renamed the "residual land" sink (Denman et al. 2007) because it is postulated that this CO_2 disappeared within the land biomass, attests to the level of our ignorance in regards to the intricacies of the global carbon cycle.

Taking into consideration that the modern carbon cycle is almost entirely biologically mediated, what is it that drives it? Clearly, it is the energy of the sun via photosynthesis, as seen in the tropical biomes, such as the Amazon (Figure 4), where photosynthesis is most active during the dry seasons because the incident radiation is enhanced by diminished cloud cover. In the tropical biomes, the most productive biological factories, the system operates at its maximum efficiency (~6–8 %) with respect to incident solar radiation, defining a Net Primary Productivity (NPP) plateau at ~1200 gCm^2yr^{-1} (Figure 5). The broad nature of this plateau arises from the combined effect of the variable quality of experimental observations and the superimposed phenomena, such as nutrient or water deficiency, cloudiness, and other secondary factors. At higher latitudes, however, the NPP is water limited. CO_2 deficiency or fertilisation is unlikely to play any significant role in this overall pattern because CO_2 concentrations are more or less the same all over the globe.

While the driver of the biological cycle of carbon is clearly the solar energy, photons, water is the medium that modulates and limits the intensity of its impact. In the process of photosynthesis, for every unit of CO_2 sequestered from the atmosphere the plant has to transpire almost 1000 units of water from the roots back into the atmosphere (Figure 6). This ratio, called the Water Utilisation Efficiency (WUE) depends on the type of the biota, C3 or C4, but in any case it is in the range of many hundreds to one. Note that it is only one molecule of water that is required to react with 1 CO_2 to generate the organic matter of the plant. All the remaining hundreds of molecules of water are therefore required not for actual photosynthesis but as a conveyor belt for delivery of food (nutrients) to the plant. The maintenance of this water conveyor belt, which appropriates almost the entire incidental solar energy, is a precondition for the operation of the carbon cycle, not its consequence. Thus the carbon cycle is not driving the water cycle, but is "piggybacking" on it.

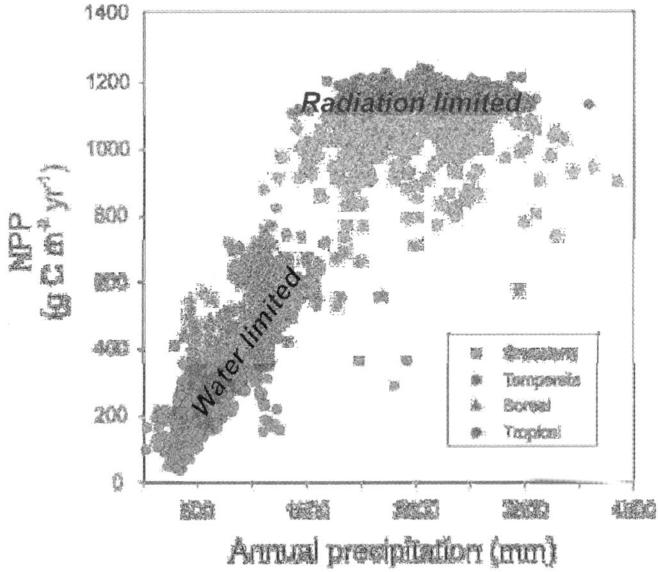

Gridded data from Global Primary Productivity Data Initiative (GPPDI)

Fig. 5: Plot of annual precipitation vs. Net Primary Productivity (NPP) for tropical to boreal biomes. Based on the GPP data in Zheng et al. (2001) and modified from Ferguson and Veizer (2007).

Water, in its different states, is therefore the medium for energy redistribution on this planet, and all other cycles are only "piggybacking" on it. This is as true of the animate as it is of the inanimate world. No water cycle, no food, no life, no CO_2 cycle! Similarly, without erosion, weathering, transport, etc. there would be no inorganic cycle of carbon. While the water cycle on this planet would operate even with no carbon present, albeit with some reactions modified, the opposite is not true.

Can this postulate be tested empirically? The working hypothesis mandates that, for a given spatial unit, the amount of water involved in transpiration would have to be about 1000 times greater than the amount of carbon sequestered. Watersheds of larger rivers can serve as examples of such spatial units because their water balance can be quantified and compared to independent estimates of the NPP (Ferguson and Veizer, 2007). Studies based on 15 watersheds between Canada and Australia (Figure 7), examples that cover most of the typical biomes, from boreal to tropical, wet to dry, indeed yield the same transpiration vs. precipitation pattern for water as was observed for NPP, only 1000 times larger (Figure 8). This coherency of the NPP and water balance estimates, despite the fact that they were assembled by entirely independent research groups and experimental protocols, is remarkable and entirely in accord with the concept of the carbon cycle "piggybacking" on the water cycle.

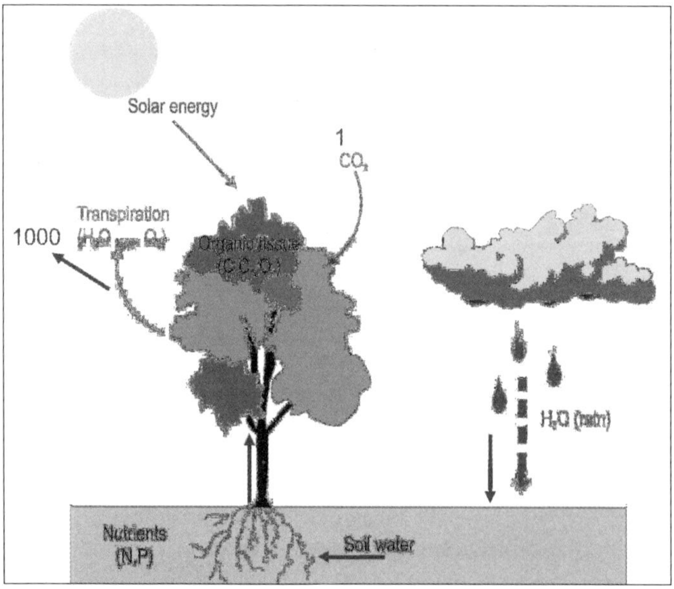

Fig. 6: Schematic representation of the water and carbon cycles in the process of photosynthesis.

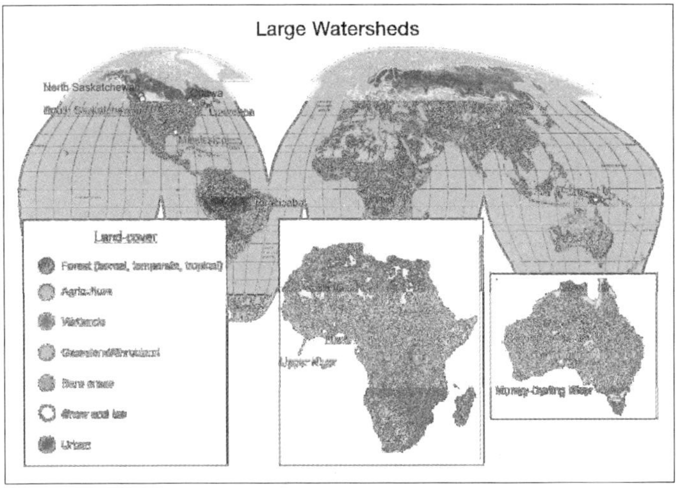

Fig. 7: The watersheds studied for their water and carbon budgetary balances. Reproduced from Ferguson and Veizer (2007).

What would be the consequences of this concept for the global water and carbon cycles? Accepting that the entire unified system is driven by the sun, enhanced solar input would result in more evaporation plus resulting advection of more water to higher latitudes, thus generating more humid and warmer climate across the globe. At the same time the NPP tropical plateau (Figures 5 and) will shift upwards and the water limitation at higher latitudes will become less stringent, both increasing the global NPP. Pictorially the entire pattern in Figure 8 will move diagonally up and to the right. With the diminishing solar activity the pattern will shift downward and to the left. Hence the unified system oscillates in response to variations in solar energy input. The centennial record of precipitation, and related NPP, for the contiguous United States (Figure 9) appears to be indeed following this pattern, with all but one minima coincident with the minima in solar activity as reflected in sunspot numbers.

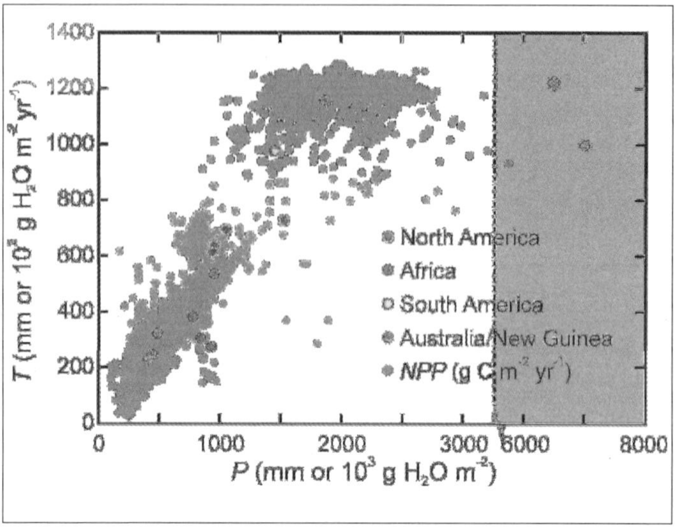

Fig. 8: Plot of annual precipitation vs. the amount of water involved in transpiration (photosynthesis) for the watersheds in Figure 7. Note the underlying pattern for NPP from Figure 5. Reproduced from Ferguson and Veizer (2007).

There may also be additional consequences to this scenario. In contrast to present perception, an increase in global bioproductivity may result not in sequestration of atmospheric CO_2, but in its rise. This is because higher NPP means also more litter (dead leaves) that within a few years will respire into CO_2. The atmosphere is only a small reservoir for carbon, the latter commuting between two large reservoirs, the living biosphere and the litter plus soil. In analogy to two banks that have money floating in transition (mortgages, etc.), the bigger the banks, the more money floating. Moreover, the enhanced solar activity invigorates not solely the carbon cycle on land, but also affects the oceanic carbon cycle, with warmer oceans degassing their CO_2. The overall outcome is an increase in CO_2 levels at warm times.

Atmospheric CO_2 is thus the product and not the cause of the climate, as demonstrated by past records where temperature changes precede changes in atmospheric CO_2 concentrations and fluxes: ice cores, the 1991 Mt. Pinatubo volcanic eruption in the Philippines, or seasonal oscillations are all instructive examples, spanning time scales from seasons to millennia.

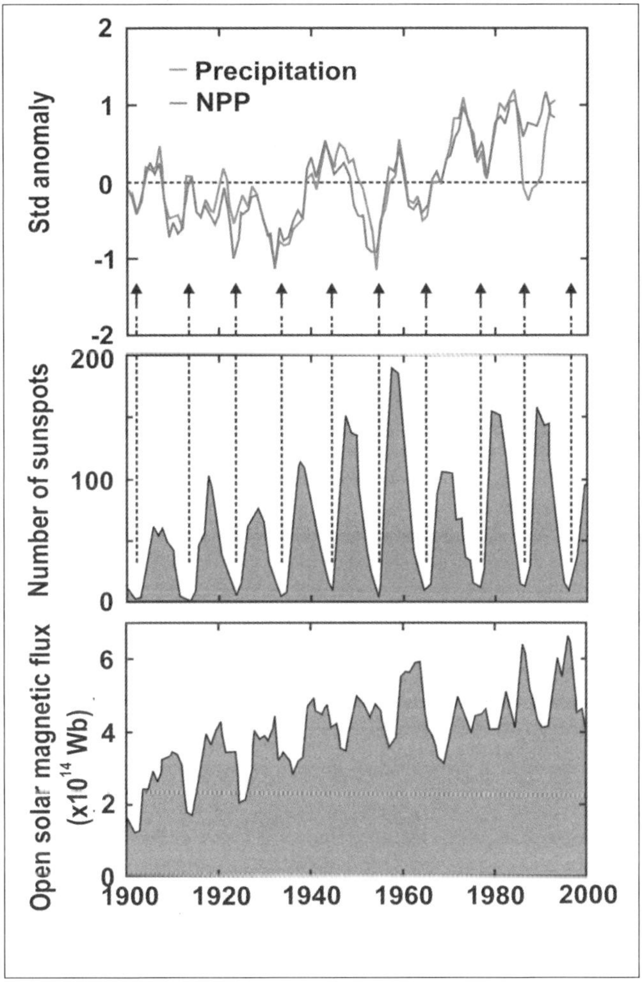

Fig. 9: Variation in mean annual precipitation and NPP for conterminous United Stated over the last century (based on Nemani et al. 2002) and variations in solar activity. Reproduced from Ferguson and Veizer (2007).

Considering that the water vapour greenhouse accounts for almost the entire temperature effect in the climate models, it is possible to envision the input of energy into the water cycle from either the sun or the CO_2 greenhouse, or both. These alternatives can be designated as "top down" and "bottom up", respectively (Figure 10). Because the models themselves are not specific in pinpointing the source of incoming energy, the discrimination has to be based on past records. For the "bottom up" alternative, the past climate record would have to correlate with coeval CO_2 trends, such as those embedded in the ice cores (Siegenthaler et al. 2005). For the "top down" alternative, the past climate would have to correlate with the proxy for solar activity, such as the cosmogenic nuclides (^{14}C, ^{10}Be, ^{36}Cl) that are generated by interaction of incoming cosmic rays with the planetary atmosphere. These nuclide then rain down to the Earth's surface and can be measured in ice, trees, rocks and minerals.

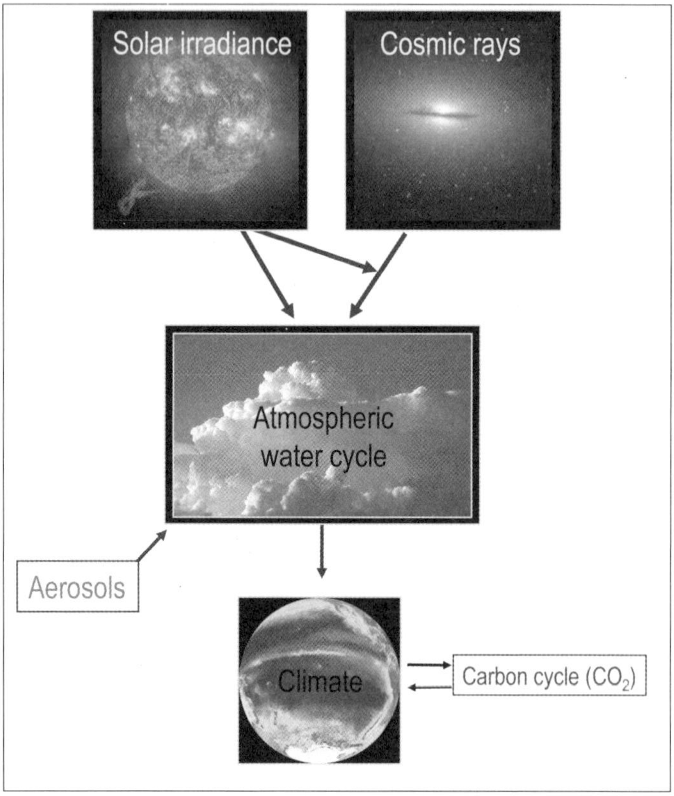

Fig. 10: Conceptual model of solar and CO_2 interactions with the climate system.

The amplifying connection to the sun comes via its electromagnetic envelope, called the heliosphere, and a similar envelope around the Earth, the magnetosphere. These act as shields that screen the lethal cosmic rays from reaching our planet. A less active sun is not only colder but its heliospheric envelope shrinks, allowing more cosmic rays to reach our atmosphere, and vice versa. Indeed, satellite data from the NASA IBEX mission for the past decade shows a 25 per cent shrinking of the heliosphere that is coincident with the halt, or even decline, in planetary temperature since 1998: a trend at odds with the ever rising levels of atmospheric carbon dioxide.

The records of atmospheric CO_2 levels and temperatures in the ice cores cover almost one million years of planetary history (Siegenthaler et al. 2005) and indeed show a strong correlation of these two variables. However, in all cases the CO_2 lags temperature by several centuries, in accord with the above postulate that the atmospheric CO_2 levels are a product of the climate and not its cause.

a.

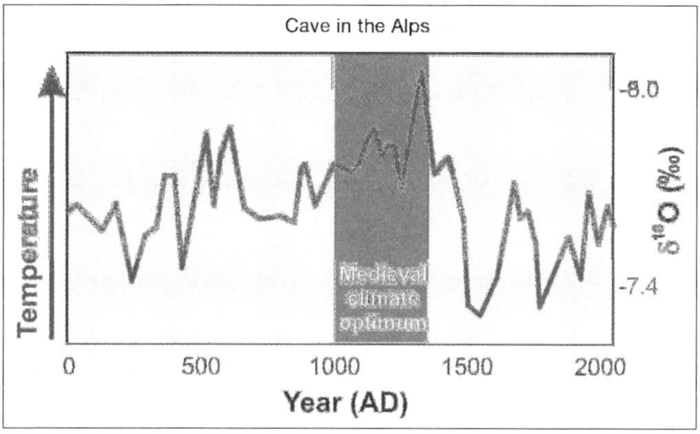

b.

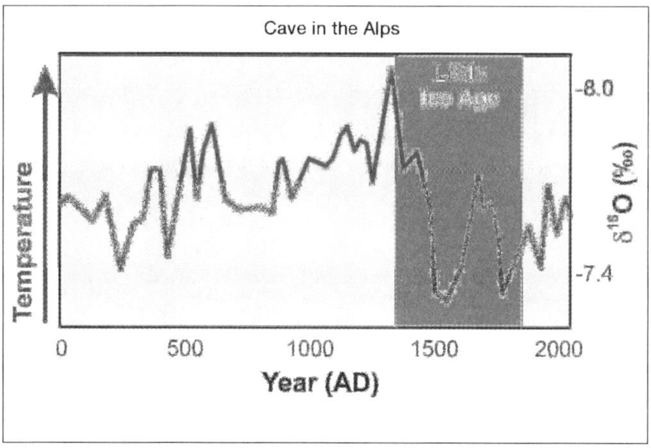

Fig. 11: The variation in $\delta^{18}O$, climate proxy, in the stalagmite from the Spannagel cave over the last 2000 years; during a) the warm Medieval Climate Optimum and b) the Little Ice Age. Adapted from Mangini et al. (2005).

In contrast to CO_2, the reliable records for cosmogenic nuclides are available for less than 100,000 years. For the time spans where both records, CO_2 and cosmogenic nuclides, are available, which is the case mostly for the period following the retreat of the ice sheets about 10,000 years ago, the observational data show clear correlations with solar proxies. This is the case for marine (Bond et al. 2001) as well as terrestrial (Mangini et al. 2005) records. The CO_2 levels for almost the entire 10,000 year interval were marooned around the "preindustrial" levels of ~280 ppm and show no correlation at all with the climate. An illustrative example is the $\delta^{18}O$ record in stalagmites of the high alpine Spannagel cave. In the last 2000 years, it clearly reflects the warm medieval climate optimum at ~950–1350 AD followed by the Little Ice Age that lasted until about 1850 AD (Figure 11). This climate trend is clearly coincident with ^{14}C (Figure 12) and ^{10}Be variability over the same time span, as opposed to essentially flat coeval CO_2 levels. Additional examples of such correlations are listed in the review publications of Veizer (2005), Scherer et al. (2006) and Kirkby (2007). Unfortunately, the atomic tests of the last century pre-empted utilisation of this tracer as solar proxy for the times of rapid growth of atmospheric CO_2 levels due to anthropogenic sources. Nevertheless, the relationships in Figure 9 and the publications of the Paris group (e.g., Le Mouël et al. 2010) argue for a similar solar driven scenario even for the last century. Data for the deep geologic record of the Phanerozoic, while at this stage rather preliminary, also appear to show better relationship of climate with the estimates for the cosmic ray flux (Shaviv and Veizer, 2003) than with the proxy-derived CO_2 levels (Veizer et al. 2000).

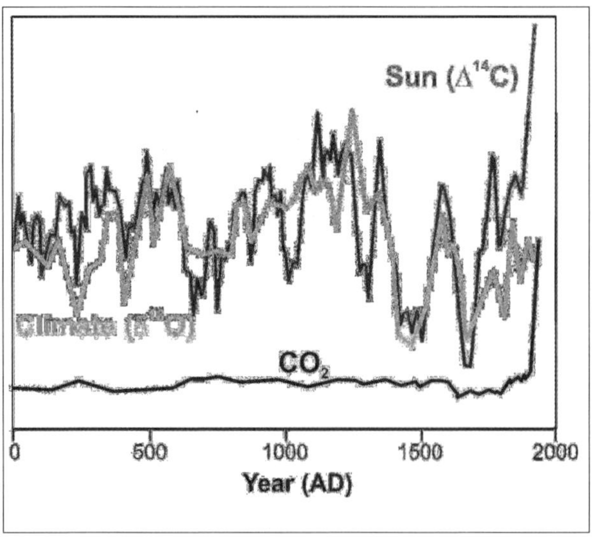

Fig. 12: The variations in $\delta^{18}O$ (climate) in the Spannagel stalagmite superimposed against the record of coeval ^{14}C and CO_2 levels. Data from Mangini et al. (2005) and IPCC (2001). Modified from Scherer et al. (2006).

Clouds are an integral part of the sun-driven water cycle; however, formation of water droplets requires seeding and this is where solar amplification likely comes into play. Empirical and experimental results suggest that cosmic rays hitting the atmosphere may generate such initial seeds, particularly over the oceans. While the actual mechanisms are still debated (Kirkby, 2007), the correlations between cloudiness and cosmic ray flux have already been published (e.g., Svensmark, 2007). Clouds are a mirror that reflects solar radiation back into space. The amount of solar energy reflected by the Earth is about 77 Watts/m^2 and the difference between cloudless and cloudy skies may alter the energy balance by ~13 Watts/m^2 (Ramanathan et al. 1989). Therefore a change of just a few per cent in cloudiness easily can account for the disputed energy discrepancy.

The above cited centennial to millennial correlations, coupled with direct observations of coincidence of cloudiness with cosmic rays and temperature argue that the sun and its amplifying mechanism must play a leading role in climate control even if the cosmic ray signal proves no more than an indirect measure of solar variability. The observation that solar radiation reflected from clouds and the atmosphere to space (~77 Watts/m^2) equals the amount of energy necessary to drive the water cycle, with evapotranspiration and precipitation each accounting for ~78 Watts/m^2 (Figure 13), may not be accidental. It may simply be the proportion of solar energy that leaves our planet after accomplishing its task of driving the water cycle, and climate, with all "lesser" cycles "piggybacking" on the water conveyor belt.

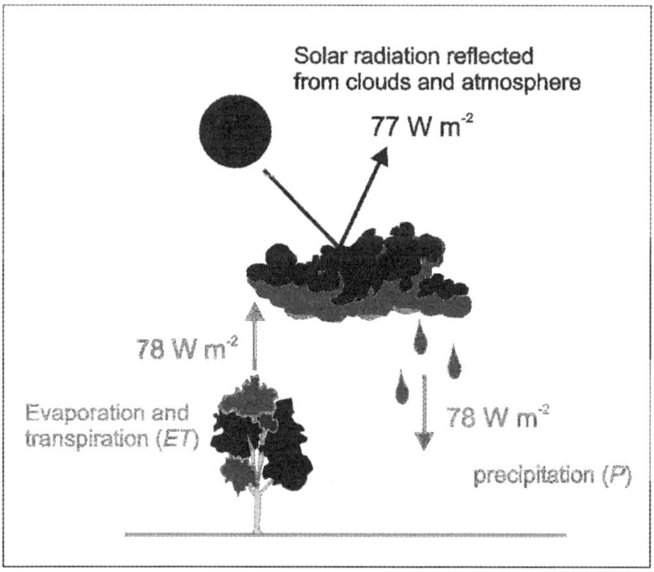

Fig. 13: Simplified version of global energy fluxes related to the water cycle. Based on IPCC (2001).

I wish to emphasize that the "top down" and the "bottom up" are complementary, not mutually exclusive, concepts. Both are involved and it is only a matter of relative importance that is at issue. While the IPCC maintains that CO_2 accounts for about 2/3 and the sun up to perhaps 1/3 of the driving force, I maintain that empirical records favour the opposite interpretation. The science of climate change continues to evolve and regardless of the outcome of the climate debate, observational data suggests that we may be served well by basing out climate agenda, scientifically and economically, on a broader perspective than that in the IPCC outlined scenarios. Our pollution abatement and energy diversification goals could then be formulated, and likely implemented, with less pain.

REFERENCES

1. Bond, G., Kromer, B., Beer, J. et al. (2001). "Persistent solar influence on North Atlantic climate during the Holocene." *Science* 294:2130-2136.
2. Denman, K.L., Brasseur, G., Chidthaisong, A. et al. (2007). "Couplings between changes in the climate system and biogeochemistry." In: Climate Change 2007: The Physical Science Basis. Contribution of the working group 1 to the Fourth Assessment Report of the IPCC. Edited by S. Solomon et al. pp. 501-587, Cambridge University Press, Cambridge.
3. Ferguson, P.R. and Veizer, J. (2007). "The coupling of water and carbon fluxes via the terrestrial biosphere and its significance to the Earth's climate system." *J. Geophys. Res.—Atmosph.* 112:D24S06, doi:10.1029/2007JD008431.
4. Foster, P., Ramasvarny, V., Artaxo, P., et al. (2007). "Changes in atmospheric constituents and in radiative forcing." In: Climate change 2007: The Physical Science Basis. Contribution of the working group 1 to the Fourth Assessment Report of the IPCC. Edited by S. Solomon et al. pp. 130-234, Cambridge University Press, Cambridge.
5. Froehlich, C. and Lean, J. (2004). "Solar radiative output and its variability: Evidence and mechanisms." *Astron./Astrophys. Rev.* 12:273-320.
6. Gerlich, G. and Tscheuschner, R.D. (2009). "Falsification of the atmospheric CO_2 greenhouse effect within the frame of physics." *Int. J. Modern Phys.* 23:275-364.
7. IPCC (2001). IPCC Third Assessment Report: Climate Change 2001. Cambridge University Press, Cambridge.
8. IPCC (2007). IPCC Fourth Assessment Report: Climate Change 2007. Cambridge University Press, Cambridge.
9. Kirkby, J. (2007). "Cosmic rays and climate." *Survey Geophys.* 28:333-375.
10. Lacis, A.A., Schmidt, G.A., Rind, D. and Ruedy, R.A. (2010). "Atmospheric CO_2: Principal control knob governing earth's temperature." *Science* 330:56-359.
11. Latham, J., Rasch, P., Chih-Shieh, J., et al. (2008). "Global temperature stabilization via controlled albedo enhancement of low-level maritime clouds." *Trans. R. Soc. A* 366:3969-3987.
12. Le Mouël, J.-L., Blantner, E., Shirman, M. and Courtillot, V. (2010). "Solar forcing of the semi-annual variation of length-of-day." *Geophys. Res. Lett.* 37:doi:10.1029/2010GL0431185.

13. Mangini, A., Spotl, C. and Verdes, P. (2005). "Reconstruction of temperature in the Central Alps during the past 2000 yr from a $\delta^{18}O$ stalagmite record." *Earth Planet. Sci. Lett.* 235:741-751.
14. Myneni, R.B., Yang, W., Nemani, R.R. (2007). "Large seasonal swings in leaf area of the Amazon rainforest." *Proc. Nat. Acad. Sci.* 104:4820-4923.
15. Nemani, R.R., White, M.A., Thornton, P.E., et al. (2002). "Recent trends in hydrologic balance have enhanced the terrestrial carbon sink in the United States." *Geophys. Res. Lett.* 29, doi:10.1029/2002GL014867.
16. Ramanathan, V., Cess, R., Harrison, E. et al. (1989). "Cloud radiative forcing and climate: results from the earth radiation budget experiment." *Science* 243:57-63.
17. Ruddiman, W.F. (2001). Earth's Climate: Past and Future. Freeman and Comp., New York, 465 p.
18. Scafetta, N. (2010). Climate change and its causes: A discussion about some key issues. SPPI original paper, 1-55.
19. Scherer, K., Fichtner, H., Borrmann, T. et al. (2006). "Interstellar-terrestrial relations: the variable cosmic environments, the dynamic heliosphere, and their imprints on terrestrial archives." *Space Science Reviews* 127/1-4:327-465.
20. Shaviv, N.J. and Veizer, J. (2003). "Celestial driver of Phanerozoic climate?" *GSA Today* 13/7:4-10.
21. Siegenthaler, U., Stocker, T.F., Monnin, E., et al. (2005). "Stable carbon cycle-climate relationship during the Late Pleistocene." *Science* 310:313-1317.
22. Svensmark, H. (2007). "Cosmoclimatology: A new theory emerges." *Astronomy and Geophysics* 48:1.18-1.24.
23. Veizer, J. (2005). "Celestial climate driver: a perspective from four billion years of the carbon cycle." *Geoscience Canada* 32:13-28.
24. Veizer, J., Godderis, Y. and François, L.M. (2000). "Evidence for decoupling of atmospheric CO_2 and global climate during the Phanerozoic eon." *Nature* 408:698-701.
25. Zheng, D.L., Prince, S.D. and Wright, R. (2001). NPP multi-biome: Gridded estimates from selected regions worldwide, 1989-2001. Oak Ridge Nat'l. Lab. (http://www.daac.ornl.gov).

OCEAN CIRCULATIONS AND CLIMATE[1]

WILLIAM KININMONTH
Australasian Climate Research
Kew Victoria, Australia

ABSTRACT

The Earth's climate system is dominated by the need for poleward transport of energy. The transport, necessary to bring about near global radiation balance, is achieved largely by the atmospheric circulation (about 80 percent, including the hydrological cycle), although the ocean circulations are also significant contributors. The magnitude of the atmospheric transport is controlled by the rate of exchange of heat and latent energy across the interface between the tropical surface (particularly the ocean) and the atmosphere. The rate of energy exchange from the ocean to the atmosphere is regulated by ocean surface temperature and varies with the rate of mixing of cold interior water into the tropical ocean surface mixed layer.

The impact of varying tropical ocean-atmosphere exchange of energy on global climate is demonstrated through El Niño events that recur on interannular timescales. However ocean circulations that modulate mixing of cold ocean interior water into the surface mixed layer have natural periods ranging from multi-decadal, through centennial, to millennial timescale. Climate variations on these timescales are to be expected as the natural ocean periods are excited through varying surface wind stress and through variations of gravity forces associated with the orbits of the Moon and major planets.

INTRODUCTION

Life on Earth has evolved through hundreds of millions of years of changing climate. As climate changed so the pattern of biosphere diversity also altered. Some species migrated as their favoured climatic characteristics shifted to new regional locations. Some species failed to adapt or migrate with the changing climate and local biodiversity and disappeared. Other species did adapt to their new regional climate and biodiversity by evolving special characteristics or finding their existing characteristics advantageous in the new regional climate and biodiversity.

Humankind has come to a recognition that a changing climate poses particular challenges for the future. While humans are now settled on all but the harshest climate regimes there is a recognition that existence, especially for the production of food, is fragile and subject to the vagaries of climate patterns. Storm, tempest and flood are destructive to habitats and community infrastructure; drought and temperature extremes often produce famine despite a pan-continental ability to transport food and provide relief.

Our understanding of the extent of climate hazards has come about from research that links prosperity with particular climate regimes and society setbacks to regimes that

---

[1] Presented at the International Seminars on Planetary Emergencies, 43rd Session at Erice, Sicily 19-24 August 2010 (Sponsored by the World Federation of Scientists).

stress food production and social infrastructure. Some of the research is reliant on archaeological evidence because very limited documented history has survived from earliest times. Megalithic structures, such as the pyramids of Egypt and the temples of the Americas, point to organised civilisations that were once prosperous but have since declined. More recent written records from China and the Mediterranean regions clearly identify waxing and waning of civilisation with varying climate. As an example, the prosperity of the Mediterranean and Western Europe maximised during the warmth of the Greco-Roman Period about 2,000 years ago, the Medieval Period of about 1,000 years ago and the current period that commenced about 250 years ago. These periods of prosperity were separated by the cold of the Dark Ages and the Little Ice Age, where harsh climate brought scarcity of food and increased conflict.

The actual reasons for changing climate are slowly emerging but many aspects are uncertain or remain unknown. This ignorance as to cause has, in the past, given opportunity for mysticism and delusion. It was not uncommon during the Little Ice Age for calamitous events to be ascribed to the work of the devil operating through malevolent individuals and groups; tens of thousands of people were executed over the centuries and witchcraft was not outlawed in many European legislatures until the early 1800s (with 'the warmth of the enlightenment'!).

What we do understand about the climate system is that there are some 'constants' that put broad constraints on the climate system and regional climates. These include the Sun as an energy source, the orbit of the Earth that generates seasonal characteristics, and the role for the ocean and atmospheric circulations in transporting energy from the tropics to polar regions. What is less understood is the role of the variability of the ocean circulations and how this has affected and determined the characteristics of past climates.

The purpose of this discussion is to set out some important changes that have taken place in the ocean circulations over the past 100 million years and how these have affected the evolution of Earth's climate. Over this epoch, a small fraction of Earth's history, the climate has changed from overall warmth to the current one of extensive polar land glaciation that regularly extends to middle latitudes in parts. The discussion will be in three parts: the energy controls of climate patterns; how continental drift has changed the ocean circulations and energy transports; and why the prevailing thermal structure of the oceans makes climate sensitive to external variations.

EARTH'S ENERGY REGIME

The Earth's mantle has cooled and, because conduction of heat is very slow, there is little energy from the hot interior. The energy source of significance to warm the Earth's surface is the Sun. Solar energy is absorbed by the atmosphere, the land surfaces (including plants) and the ocean surface layer and this input is in steady state with the infrared radiation emitted to space from the land and ocean surfaces and the atmosphere. The steady state does vary because the Earth's temperature, and hence the emission to space, is known to have varied in the past.

The seasonal cycle of climate is driven by the characteristics of the Earth's orbit around the Sun, including the obliquity of the axis of rotation. The orbit and obliquity result in an enormous annual cycle in the amplitude of the intensity and accumulation of solar radiation over polar regions; from permanent darkness and no absorption at midwinter to a daily total (or insolation) in excess of the tropics at midsummer. The energy loss to space is much less variable in time such that over polar regions there is excess absorbed solar radiation during the summer months but excess emission over the remainder of the year. Over tropical regions there is always excess solar radiation over the emission of radiation to space.

The latitudinal and seasonal radiation imbalances mean that there is need for transport of energy from the tropics to polar regions, especially in winter. The transport is by the ocean and atmospheric circulations and varies not only with the seasonal cycles but also with fluctuations of the respective circulations. The strength of the atmospheric circulation is regulated by prevailing temperature gradients and responds rapidly to summer warming and winter cooling over middle and high latitudes. In contrast, the ocean circulations have much more inertia and have much less seasonal variation.

The regional radiation imbalances and need for energy transports mean that local temperatures and polar temperatures in particular, vary with the intensity of transport. More intense poleward transport will sustain warmer wintertime temperatures over high latitudes while a reduced seasonal transport will lead to lower temperatures as energy lost by radiation to space is not replenished. High latitude continental temperatures are particularly susceptible to variations in atmospheric circulation while ocean surface temperatures and coastal margins are more affected by variations in ocean circulations.

The fact that the annual radiation deficit over high latitudes is offset by energy transport from the tropics means that the magnitude of the overall transport varies with the partitioning of the transport between each of the ocean and atmospheric circulations. There is no physical reason why any reduction in ocean transport should be taken up by an increase in atmospheric transport. There is direct heat, latent energy and momentum exchanges between the ocean surface and atmosphere but the magnitudes of these exchanges are regulated by local processes.

Temperatures over high latitude and polar regions vary with the transport of energy from the tropics and also with the partitioning of the transport between the ocean and atmospheric circulations. Sea surface temperature and the propensity for winter sea ice formation is largely a function of the ability of ocean currents to transport warm waters into the high latitudes. In contrast, winter temperature over high latitude continental regions is regulated by variations of the atmospheric circulation. In looking for the reasons behind long term trends and cycles in high latitude it is necessary to identify how and why the ocean and atmospheric circulations vary.

CONTINENTAL DRIFT AND OCEAN OVERTURNING

Over the last 60 million years Earth's climate has cooled. Not only have extensive ice sheets formed over high latitude land areas but the ocean's interior has also cooled. There is no evidence to suggest that the Sun's radiation has changed over this period and it is logical to look to changes in the way Earth has processed absorbed solar radiation in order to explain the shift in steady state characteristics. One plausible explanation is based on the breakup of the former super continent of Gondwanaland that once straddled the South Pole.

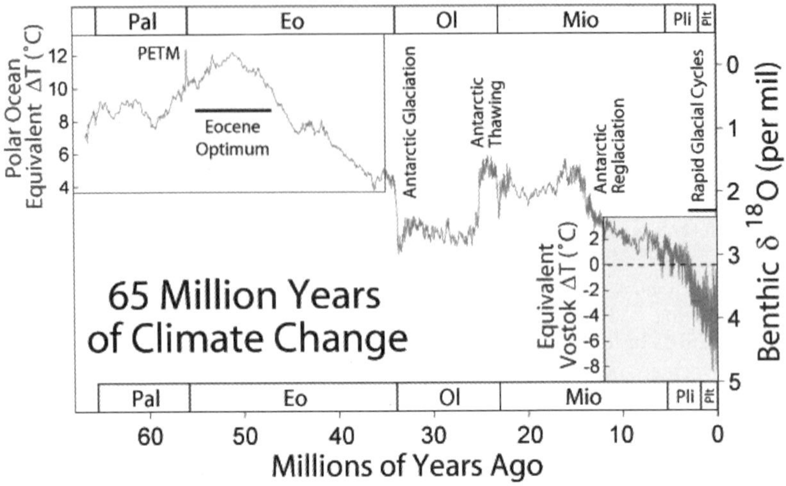

Fig. 1: A representation of temperature change across the Cenozoic Era as determined from sediment benthic foraminifera (from Zachos et al. 2001). There is not a regular cooling trend and the glaciation and thawing of Antarctica between 20 and 10 million years ago represents a major unexplained period of hiatus before the continuation of the cooling trend.

The land mass of Gondwanaland extended from the South Pole into the tropics. Warm tropical surface ocean currents washed the shores, even those of high latitudes, and the coastal margins were ice-free in winter. Heat and moisture from the coastal margins were transported inland by the winds and the biosphere was characterised by wet forests. The breakup of Gondwanaland significantly altered the ocean circulations with lasting consequence.

By 90 million years ago Africa, South America and India had separated from what is now Antarctica but Australia was still attached. The protrusion of Australia into the middle latitudes meant that relatively warm surface currents continued to maintain ice-free coastal conditions. Wet rainforests continued to flourish over most land areas.

The tectonic event with most impact on global climate was the separation of Australia from Antarctica about 50 million years ago. This separation provided for a continuous ocean girdling Antarctica at the latitudes of the predominant westerly winds. The wind stress acting on the ocean surface generated the Antarctic Circumpolar Current to effectively isolate coastal Antarctica from the warmer tropical and subtropical waters. Without the heat source of adjacent coastal waters the continental interior cooled, especially over winter.

The critical importance of the Antarctic Circumpolar Current was the ocean overturning generated by wind stress. In the southern hemisphere Ekman turning causes the surface current generated by wind stress to drift to the left of the prevailing wind. As a consequence, there is a northward drift in the surface layer of the Antarctic Circumpolar

Current. The source of the water mass for the surface drift is upwelling of cooler subsurface water on the poleward margins of the circumpolar current.

The Antarctic Circumpolar Current and the associated upwelling of cooler subsurface water provided thermal insulation for Antarctica and was also a trigger for the slow global cooling of the Cenozoic Era that continues to the present, as depicted in Figure 1. Winter cooling allowed the formation of Antarctic coastal sea ice and as the sea ice formed it expelled salt into the water below to increase its density. Having negative buoyancy the cold saline water sank to the sea floor to form Bottom Water. With time cold saline water has accumulated on the sea floor and gradually filled the ocean interior.

The ocean overturning of the Southern Ocean that commenced with the formation of the Antarctic Circumpolar Current continues today and is illustrated at Figure 2. Today there are three principal sources of Bottom Water formation to replenish the cold interior: the Weddell and Ross Seas off Antarctica and the North Atlantic-Arctic Ocean.

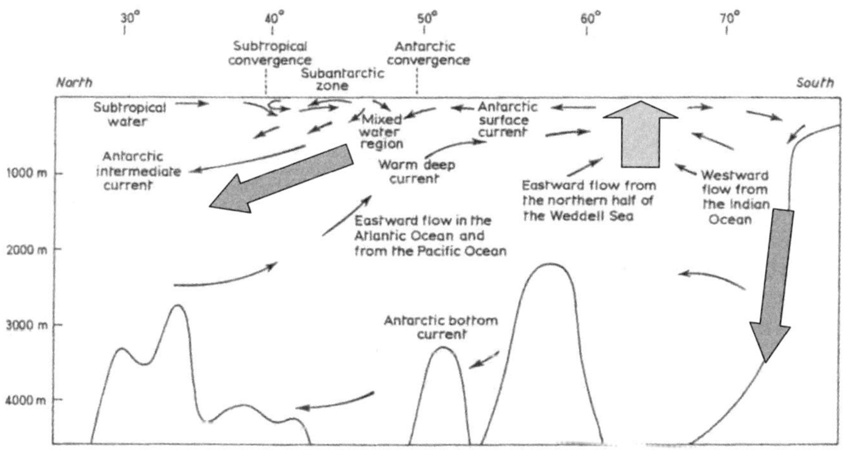

Fig. 2: A schematic meridional depth section across the Southern Ocean illustrates the main overturning circulations associated with the wind-driven Antarctic Circumpolar Current located between 50°S and 60°S (Adapted from a graphic of the Scientific Committee for Antarctic Research). The block arrows highlight upwelling to the south of the current that draws cooler subsurface water into the surface layer; the surface water drifts northward before plunging to form Intermediate Water under the warmer subtropical surface waters. The density of the Antarctic coastal surface waters increases as salt is expelled under winter sea ice and the dense water sinks along the continental shelf to form Bottom Water. Over time the very cold Bottom Water has accumulated to fill the ocean interior.

The extent of heat loss from the climate system during the Cenozoic Era can be gauged from the cooling of the ocean interior. At the Eocene Optimum about 50 million years ago the temperature near the ocean floor (mostly below 4,000 metres) was about 10°C. As Bottom Water near 0°C formed and accumulated the 10°C isotherm rose and can now be found at a depth of about 400 metres with colder waters filling the ocean interior. Excess radiation to space over polar regions generated cold saline water under the polar sea ice and allowed accumulation of Bottom Water to replace the enormous mass of ocean interior water. The energy from cooling the ocean interior is lost from the climate system.

Analysis of ocean floor sediments from the most recent 5 million years reveals a climate that has taken on a characteristic of great fluctuations (Figure 3). The fluctuations first commenced about 3.5 million years ago as the Isthmus of Panama formed and separated the ocean connection between the tropical Atlantic and Pacific oceans. Initially the glaciations were relatively weak and recurring regularly with a 41,000 year periodicity but, as the glaciations increased in magnitude, about 1 million years ago it was the 100,000 year periodicity that emerged as dominant.

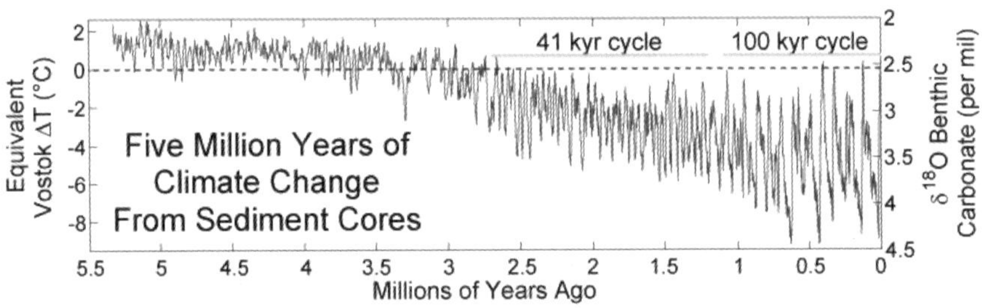

Fig. 3: Details of the ocean cooling trend of the most recent 5 million years as determined from benthic foraminifera of 57 widely distributed deep sea sediment cores (Lisiecki and Raymond, 2005). As the oceans got colder the amplitude of variation increased and the periodicity switched from a 40,000 year to a 100,000 year cycle. Note that peak warming during the fluctuations has remained relatively constant and the current temperature is near the warmest phase of the glacial cycle.

A characteristic of the last million years of the contemporary glacial epoch has been the permanence of ice sheets over Antarctica and Greenland and the regularity of land ice formation and contraction over North America and northern Europe. Water is transferred from the oceans during the long ice building phase then the land ice rapidly melts as the ice sheets disappear. The cycling of water between ocean and land during glacial events has been accompanied by a regular rising and falling of sea level exceeding 130 metres.

Ice core analysis from Greenland and pollen spore analysis from ocean sediment cores taken south of Greenland suggest that the maximum ice recession of the current Holocene Optimum was between 8 and 4 thousand years ago. Despite the millennial scale fluctuations evident in historical accounts and the apparent current warmth there are indications that Earth has again entered a period of slow ice sheet expansion.

CONTEMPORARY CLIMATE

The systematic analysis of climate using instrument observations with global coverage is only available for the last 150 years and suggests a warming trend. In the broad canvas of climate this is a brief and misleading glimpse of the totality. Nevertheless the record does allow a measure of some factors affecting variability, including the important processes affecting the biosphere and the security of human civilisation.

The most significant contributor to variability is the El Niño phenomenon that is on the interannular timescale and has its roots in changes to the magnitude of upwelling of cold subsurface water across the central and eastern equatorial Pacific Ocean. The upwelling is regulated by the strength of the offshore easterly Trade Winds from the Americas and the associated variability of the ocean surface stress. Equatorial ocean surface temperature is normally about 5°C colder over the eastern Pacific Ocean than over the west due to the upwelling. However, when there is a prolonged easing of the Trade Winds there is a reduction in upwelling and warm waters from the west tend to flow eastwards.

The importance of an El Niño event to climate is that warming and cooling of the ocean surface temperature regulates the pattern and magnitude of heat and latent energy exchange between the ocean and atmosphere. An immediate effect is to rearrange the patterns of tropical convection and precipitation; there is a general shift of focus to ocean regions and a significant reduction in precipitation over many tropical land areas. Tropical drought, famine and a range of associated health impacts are associated with El Niño events.

The changing pattern of tropical convection causes a shift in the preferred longitudes of middle latitude waves in the westerly wind flow. These Rossby Waves strongly influence the regions of preferred middle latitude cyclonic development and movement. Regions that are usually under preferred tracks of cyclones may no longer receive expected rainfall and suffer drought while other areas have more frequent or more intense cyclonic events and suffer flooding.

A third impact of El Niño events is through regulation of poleward transport of energy by the atmospheric circulation. The warmer tropical ocean surface temperature during El Niño events enhances the exchange of heat and latent energy to the atmosphere. Some of this excess energy is transported to middle and high latitudes and is available to melt sea and land ice and to warm the land surface.

The reason that the El Niño phenomenon is important for contemporary climate is that the tropical ocean mixed layer represents a thin lens of relatively warm water overlying the much colder ocean abyss. The thermocline separating these layers is generally at the base of the mixed layer but is drawn upward as surface wind stress increases; as the thermocline rises more cold water is mixed into and cools the surface layer. As the wind stress subsides so too the thermocline lowers and less cold subsurface water is mixed into the surface layer. Figure 4, a depth section across the equatorial Pacific Ocean, clearly shows how the thermocline and surface layer temperatures vary from the El Niño state (July 1997) to a more active Trade Wind situation (July 1998).

The recent instrumental climate record also has recurring cycles on decadal and longer time scales. The origins of these cycles are not established but are likely connected to other characteristics of the oceans and their interaction (through changing upwelling and ocean-atmosphere energy exchange) with regional wind regimes. However what the El Niño phenomenon clearly reveals is the sensitivity of climate to the patterns and magnitudes of ocean surface temperatures.

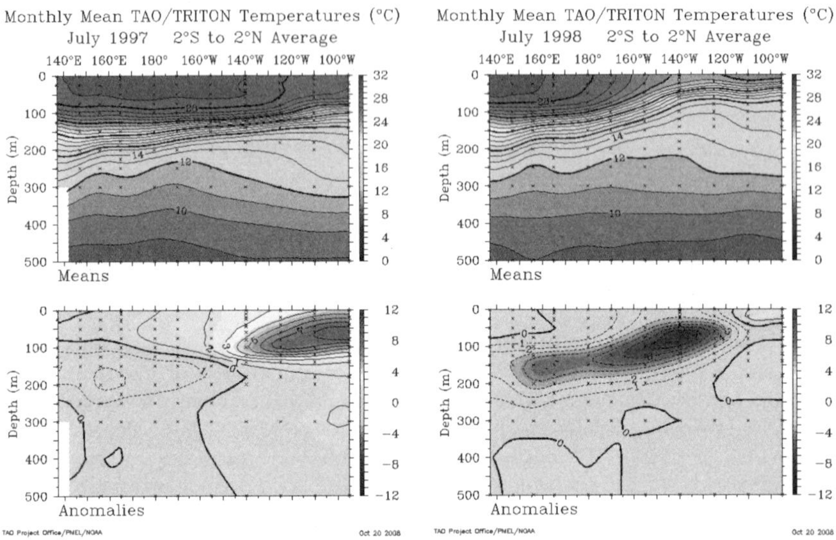

Fig. 4: The panels show equatorial surface layer depth-sections of temperature and temperature anomaly across the Pacific Ocean (Source: U.S. NOAA Pacific marine Environmental Laboratories http://www.pmel.noaa.gov/tao/jsdisplay/). Over millions of years the ocean interior has filled with cold water originating under the sea ice of the Arctic Ocean and surrounding Antarctica. There is a marked thermocline of strong temperature gradient at about 300 metres separating the warm surface water from the cold interior. The panels contrast conditions during an El Niño event (1997–left, with suppression of upwelling and warmer surface layer temperature over the central and eastern Pacific Ocean) and strong upwelling (1998–right, raising of the thermocline and mixing of cold subsurface water into the surface layer). Variations in regional upwelling regulate temperature of the surface layer, the exchange of energy with the atmosphere, and global climate.

OCEANS CIRCULATIONS AND GLACIAL CYCLES

Many of the time-varying characteristics of the subsurface ocean circulation are only now being revealed from the near decade long ARGOS buoy programme and its systematic data. The data are already pointing to complexity in the circulations of the ocean interior not previously expected. In particular, the data establish the importance of eddy motions for transport of mass rather than the steady currents previously envisaged. The existence of eddies on a range of scales establishes the reality of temporal and spatial variability in the transport of heat by the ocean circulations.

The correlation established by Milankovitch between the periodicity of the Earth's orbital cycles and glacial cycles has naturally drawn attention to changing solar insolation as a cause for the latter. However the processes linking changing solar insolation to glaciation have not been well established. Firstly, the precession cycle of 20,000 years is dominant in varying the solar insolation over polar regions but this period is very weak in the glacial cycle. Secondly, in the dominant eccentricity cycle of 100,000 years there is very little difference in the annual solar energy intercepted by the Earth's disk as Earth moves between a near circular orbit to one of large eccentricity. Thirdly, there is no plausible explanation for the change in glacial frequency from the dominant 40 thousand years to the current 100 thousand year cycle about a million years ago.

The reason for the changing orbital characteristics of Earth is the influence of the gravitational force exerted by the large planets of the solar system: Jupiter and Saturn but also Uranus and Neptune. The changing gravitational attraction on Earth's mass as each planet orbits the Sun at its own periodicity, occasionally being aligned together but then drifting out of alignment, is sufficient to alter the Earth's orbital characteristics on a regular basis.

The relative orbital speeds also lead to an exchange of angular momentum between the planets and Sun. One impact is small variations to their respective rotation speeds. For Earth the change of angular momentum is partitioned between the solid mass that determines length of Day, the liquid ocean and the gaseous atmosphere. This is a potentially rich vein for research in respect of climate variability on long time scales. Changes in ocean surface upwelling on the scale of El Niño have significant impact on climate. These same processes will be reflected in variation of the magnitude of atmospheric angular momentum.

The main influences on Earth's tides are the gravitational attractions of the Sun and the Moon. The main tidal currents are in the ocean surface layer where changes in vertical mixing have most influence on surface temperature and ocean-atmosphere exchange of heat and latent energy. A fundamental question that needs to be resolved is whether or not the small changes to ocean circulation generated by inter-planetary exchanges of angular momentum are sufficient to change the poleward transport of energy and shift the steady state characteristics of Earth's emission to space? The growth and melting of ice sheets indicates that the amplitude of global emission variation is only of the order of 0.1 watts/metre^2 over a complete glacial cycle and relatively small but systematic changes in ocean circulations and overturning may be sufficient to generate such changes.

SUMMARY

The oceans are the inertial and thermal flywheels of the climate system. Changes in the spatial and temporal distribution of ocean surface temperature do vary the exchange of heat and latent energy with the atmosphere and regulate climate. Ocean surface currents and local overturning are largely driven by wind stress at the surface. Although the ocean and atmospheric circulations primarily respond to the need for redistribution of energy from regions of solar excess in the tropics the relative motions and exchange of heat, moisture and momentum generate additional variability on a range of timescales.

The current climate state of relative glaciation can plausibly be linked to ocean circulation changes associated with the slow rearrangement of tectonic plates. The breakup of Gondwanaland and the location of Antarctica over the South Pole have allowed the relative thermal isolation of Antarctica in a region of excess radiation to space. Also, a continuous surrounding ocean domain within the westerly wind belt has generated massive overturning ocean circulations. With time the once warm oceans have been cooled to leave a relatively thin lens of surface water that is warmed by the Sun and separated from the colder interior by a sharp temperature gradient. The temperature of tropical surface waters is sensitive to local upwelling and the mixing of cold subsurface water into the layer.

The El Niño phenomenon, with events recurring every few years, is an extreme example of how climate responds to variations in regional ocean upwelling. Variations in heat and moisture exchange from the tropical oceans have profound impact on tropical and middle latitude weather patterns; there is also significant impact on the rate of poleward energy transport by the atmospheric circulation. This stark example underscores the potential for natural climate variability across a range of timescales, either triggered by ocean-atmosphere interactions or slowly varying external factors.

The cooling of Earth through the 50 million years of the Cenozoic Era and the concurrent evolution of the prevailing large-scale ocean overturning circulation underscores that Earth is now locked into a glacial state. The processes causing regular expansion and melting of northern hemisphere ice sheets are not known but are likely linked to varying ocean circulations that respond to forcing from relative motions of the large planets.

At this point in time the prevailing Holocene interglacial is deceptive, leading some to conclude that human emissions of carbon dioxide might push Earth's climate across a 'tipping point' and a return to much warmer conditions as experienced earlier in the Cenozoic. This is delusional and takes no account of the tectonic rearrangement and ocean energy loss that have led to the current glaciation. Without major tectonic disruption the most probable direction of climate is for a resumption of expansion of the northern hemisphere ice sheets as climate progresses on the next glacial cycle.

REFERENCES

1. Lisiecki, L.E., and M.E. Raymo (2005). "A Pliocene-Pleistocene stack of 57 globally distributed benthic $\delta^{18}O$ records," *Paleoceanography*, 20:1003.

2. Zachos, James, Mark Pagani, Lisa Sloan, Ellen Thomas, and Katharina Billups (2001). "Trends, Rhythms, and Aberrations in Global Climate 65 Ma to Present." *Science*, 292(5517):686–693.

NATURE—NOT HUMAN ACTIVITY—RULES THE CLIMATE

S. FRED SINGER
Science & Environmental Policy Project
Arlington, USA

SUMMARY

Santer and sixteen(!) coauthors (2008; hereafter S08) claim that observed temperature trends (in the tropical troposphere) are "consistent" with trends derived from General Circulation Models (GCM). This result, *if correct,* would seem to support the validity of greenhouse (GH) models—and thus the claim of substantial anthropogenic global warming (AGW).

However, their claim disagrees not only with an earlier analysis (Douglass et al. 2007), but also with the conclusions of an independent study (NIPCC, 2008), and with a recent report of the U.S. Climate Change Science Program (CCSP-SAP 1.1) (Karl et al. 2006), which was written by some of the same authors as S08! The crucial CCSP Chapter 5 had identified a "potentially serious inconsistency" between modeled and observed trends in tropical lapse rates (Santer et al. 2006).

But S08 now claim that there are problems with the temperature data used in the CCSP report and call attention to "new observational estimates of surface and tropospheric temperature trends." They state that "there is no longer a serious discrepancy between modeled and observed trends..." Their key graph (shown here as Figure 4) conveys a (misleading) visual impression of an overlap between observations and GH model results.

I show here that no credible analysis supports the S08 claim of "consistency": The "new observational estimates" are spurious and do not agree with satellite data (as claimed by S08). I show that the wide 'spread' (range) of modeled trend values, calculated by S08, is a pure artifact that reflects the particular models' chaotic and structural uncertainties (which had been completely ignored). My conclusion, therefore, is that the claimed "consistency of modeled and observed temperature trends" does not exist.

INTRODUCTION

The principal issue here is the *cause* of warming observed in the past 30 years, the era of satellite temperature data: ***Is it mainly naturally caused or is it due to human-produced greenhouse (GH) gases?*** The question is of obvious importance since natural causes cannot be influenced by policies that limit emissions of GH gases, such as CO_2. But resolving the question is a difficult scientific task. Natural causes are plausible; the climate has been warming and cooling for billions of years on many different time scales (see, e.g., Singer and Avery, 2007; Loehle and Singer, 2010); there is no reason to assume that such natural climate fluctuations would suddenly cease. On the other hand, anthropogenic global warming (AGW) is also plausible, since the concentration of GH gases has been increasing due to human activities (see, e.g., IPCC, 2007).

All parties seem to agree that both natural and anthropogenic factors have contributed to climate warming in the past 30 years; at issue is whether the human contribution has been preponderant and will grow in future—or whether it has been minor and will remain unimportant. All parties agree also on the proper *methodology* to distinguish between natural and human causes: It is the "fingerprint" method, which compares the *pattern* of temperature trends calculated from GH models with the pattern observed in the tropical atmosphere. By "pattern" we understand the distribution of temperature trends with latitude and altitude in the troposphere—with the tropical region being the crucial one.

This matter has been a contested issue since 1990 when early satellite results first showed no significant warming trend in the troposphere (Spencer and Christy 1990)—contrary to all expectations from GH models. The debate has involved the UN-sponsored IPCC (Intergovernmental Panel on Climate Change)—and more recently, the independent NIPCC (Nongovernmental Panel on Climate Change).

In 2004, Douglass, Pearson, and Singer (DPS, 2004) carried out such a comparison using the best then-available data and models; they concluded that there was a real disparity and that the observations did not confirm GH models. Later, a major U.S. government-funded effort, the Climate Change Science Program, arrived at essentially the same result. CCSP-SAP-1.1 (Karl et al. 2006) found that while models predicted increasing trends with altitude in the tropical zone, observations actually showed a slight decrease [See Figures 1 and 2, which duplicate Figures 1.3F and 5.7E of the CCSP report and Figures 7 and 8 of the 2008 NIPCC report (NIPCC 2008).]

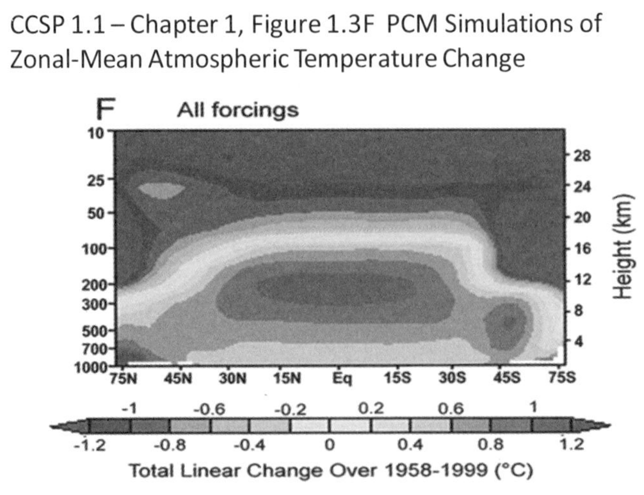

Fig. 1: Model-predicted temperature trends versus latitude and altitude (this is Figure 1.3F from CCSP 2006, p.25 and Figure 7 from NIPCC 2008). Note the increasing trends in the tropical mid-troposphere. All GH models show an amplification of temperature trend with altitude, up to about a

factor 2, over the equator at 10 km. Also note: There is little variation with latitude in surface trends; NH is about the same as SH; but warming is expected in both polar regions—contrary to observations.

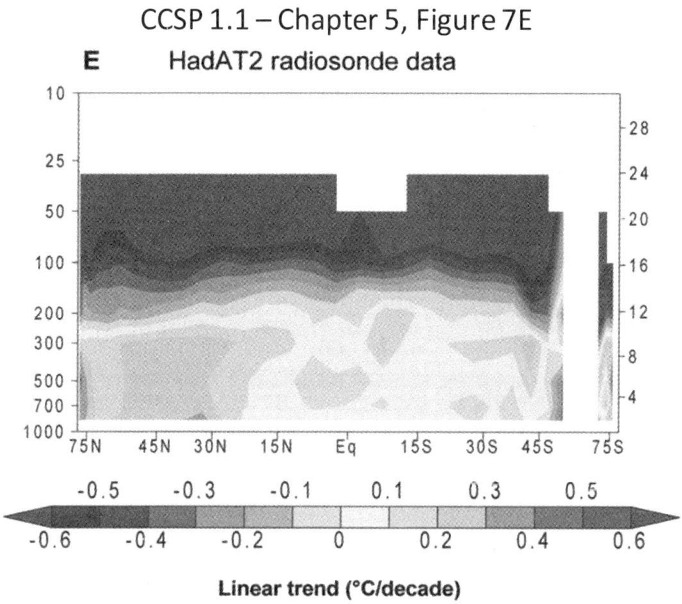

Fig. 2: Observed temperature trends versus latitude and altitude (this is Figure 5.7E from CCSP 2006, p.116 and Figure 8 of NIPCC 2008). Note the striking absence of any increasing trends in the tropical troposphere and the missing "hot spot" of Figure 1.

In 2007, Douglass, Christy, Pearson and Singer (hereinafter DCPS07) extended the DPS study by comparing in greater detail the published model-calculated trends with observed trends from balloons and satellites, in the tropical zone. They again concluded that the observations could not confirm the GH model results (Figure 3). In 2008, a group of independent scientists—the Nongovernmental International Panel on Climate Change (NIPCC, 2008)—published its summary report and simply compared the modeled (Figure 1) and observed (Figure 2) patterns of trends shown in the CCSP report.

NIPCC then stated the obvious logical conclusion—if one accepts the published CCSP results: Since GH models cannot explain the observed temperature trends in the tropical troposphere, then the warming of the past 30 years must be due predominantly to causes other than GH gases. In other words, ***the human contribution to the warming trend since 1979 is minor***—a conclusion which is contrary to that of the IPCC (2007).

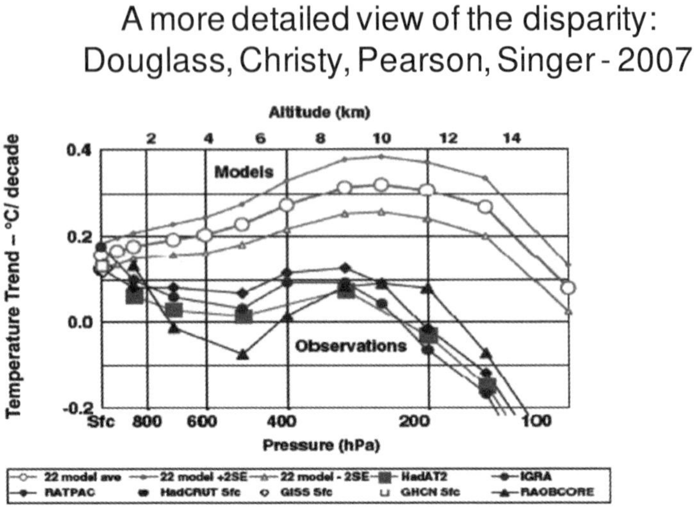

Fig. 3: Observed temperature trends versus latitude and altitude (this is Figure 1 from DCPS 2007 and Figure 10 of NIPCC 2008). Note the disparity between (increasing) model trends and the decreasing observed trends in the tropical mid-troposphere.

Deciding they could no longer ignore this issue, Santer and 16 coauthors (2008—hereinafter S08) have critiqued the conclusions of both DCPS and NIPCC. They assert (see Figure 4):

1. The observational data have changed; "new observational estimates" now show greater tropospheric warming trends.
2. The uncertainties of both models and data are greater than previously estimated.
3. Therefore, the discrepancy between models and atmospheric data in the tropics no longer exists.

In the present paper I address these issues and conclude that S08 have failed to demonstrate the claimed consistency—which is their main objection to DCPS and NIPCC.

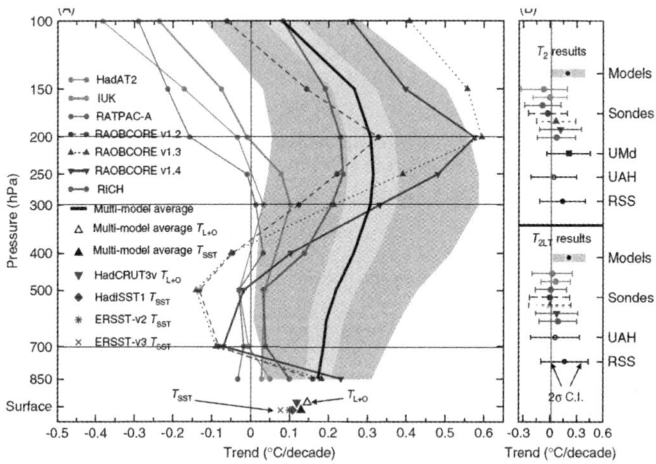

Fig. 4: This is Figure 6 from S08. Panel A shows tropospheric observed and modeled temperature trends vs. pressure-altitude (1979-1999; tropics 20°S-20°N). The RATPAC (NOAA) analyses agree well with Hadley and IUK, but are quite distinct from RICH and the RAOBCORE versions. See discussion in text that shows the latter sets to be spurious.

The grey area shows model uncertainty according to S08; the yellow area represents the Standard Error of DCPS (with which I no longer agree). See discussion in text that shows the grey area to be an artifact, dependent on the use of models having an insufficient number of runs and therefore a large chaotic uncertainty.

Panel B suggests good agreement of all observed and modeled tropospheric trend values (suitably weighted) with satellite data for both middle (T_2 or MT) and lower troposphere (T_{2LT} or LT). See, however, the discussion in the text, which demonstrates that S08 did not use a suitable Discriminant.

CONFLICTING SETS OF TROPOSPHERIC TEMPERATURE TRENDS

Figure 4 (which is Figure 6 in S08) shows basically two sets of temperature trends. One set is derived from balloon-borne radiosondes, independently analyzed by the NOAA group and Hadley group; they agree fairly well with each other. This is the set used by DCPS07 and NIPCC.

The other set, showing drastically different trend values, especially in the upper troposphere, stems from the papers of Haimberger (2007 and "in press"). S08 give no indication as to which of these two conflicting sets is credible, nor is there discussion of the cause of disagreement. The NOAA/Hadley results are based on direct observations, while the Haimberger results are derived from a reanalysis of temperature data, involving also a certain amount of modeling. Sakamoto and Christy (2009) have criticized his procedures.

S08 claims that both the NOAA/Hadley data set and the Haimberger set are compatible with satellite data (see Figure 4B, their Figure 6B, and discussion). This indicates to me that their comparison is not a very sensitive one and that we need a better 'Discriminant.' In order to refine the comparison, I have therefore used the difference (of temperature trends) between the satellite results for the middle and lower troposphere. I have constructed a new 'weight factor,' which is the difference between the published 'weight factors' for MT and LT (Christy, private communication 2010). I then apply this factor to each of the observed trend values. As can be seen from Figure 5, temperature trends for both LT and MT are near zero, or even negative, up to 1997—with the MT trend consistently less than LT's. This result is clearly inconsistent with the "new datasets" introduced by S08 but supports the RATPAC-NOAA and HadAT2 datasets.

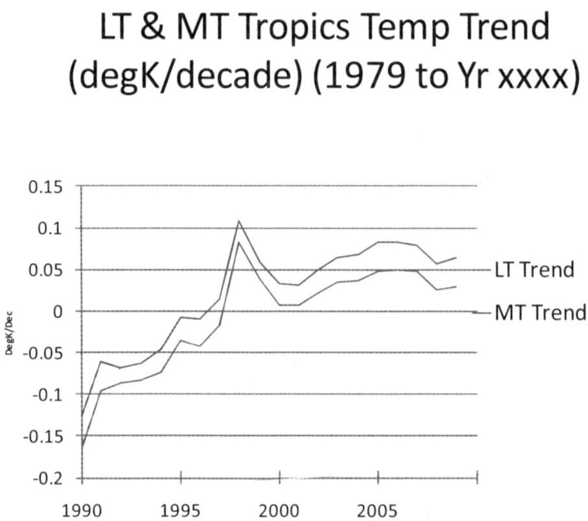

Fig 5: Temperature trend (K per decade) for the time period 1979 to end of year shown, for MSU-UAH-LT and MT data in the tropical region. Note absence of a positive trend until 1997 and the steep rise as a result of the 1998 super-El-Nino "warm spike." The RSS analysis of satellite data shows a similar relationship between LT and MT trends (but somewhat higher trend values).

CLIMATE MODELS

In addition to the well-known problems of accurately defining various radiative forcings and simulating the relevant atmospheric, oceanic, and terrestrial processes, model problems include the lack of homogeneity, the basic issue of the chaotic nature of (non-linear) models, and the fundamental problems of 'averaging' a group of model results. This critique applies not only to S08, but also to the "multi-model ensemble averages" of the IPCC (2007), and to my earlier publication (DCPS, 2007). I try to show here that the estimates of uncertainty, indicated by the 'grey area' in Figure 4A (Figure 6A of S08), are an artifact. S08 considers them to be of a statistical nature; in reality, they reflect the chaotic nature of models.

Chaotic Variability

The chaotic nature of trends from climate models presents a fundamental problem. Successive runs of a model often lead to vastly different results. I show this in Figure 6, which is based on the five 'runs' of a particular GCM (Japan MRI model). According to the results shown there, a single 'run' of a particular model can give trends spanning nearly a factor of ten. In actuality, had there been more than five runs, the range of trend values would have been even greater.

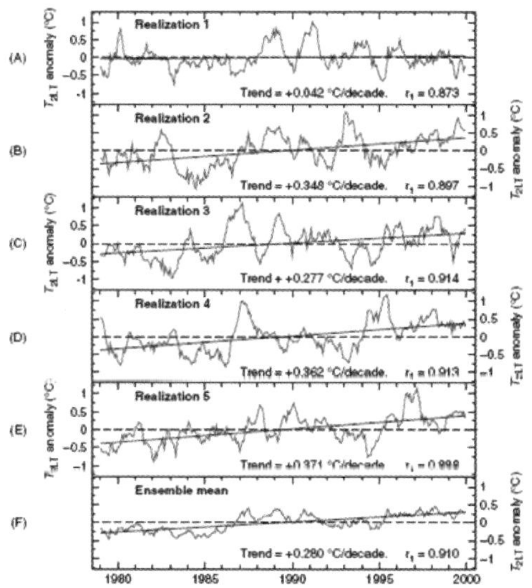

Fig. 6: Illustrating the chaotic nature of model trends, using the results of 5 runs (sometimes referred to as "realizations" or "simulations") from 1979 to 1999 of a particular GCM (Japan MRI), as presented in Figure 1 of S08. The OLS trends of the runs range from +0.042 to +0.371 (K per decade). (The range of trends is likely to be even larger if more runs had been displayed.) Selecting any 2 runs allows 10 possible means; there are 5 possibilities for the 4-run mean ensemble trend, and just one for a 5-run mean. It suggests convergence to a "true" trend, but as discussed in the text, one needs to show that the cumulative mean approaches an asymptotic value as the number of runs increases.

However, as the number of runs increases, their *cumulative* average trend should converge to a single value. We don't know where that is, but I have developed a synthetic approach that can increase the number of runs up to 25, and much greater; I then note when the cumulative trend approaches an asymptotic value (Singer 2010). Of course, it becomes extremely expensive to increase the number of runs of an actual GCM; so in a practical sense one never carries out such an experiment. But it is intuitively obvious that an 'ensemble-mean' trend based on averaging a large number of runs should be less subject to chaotic variability and more reliable than a trend based on fewer or even just a single run.

This point becomes particularly relevant when we consider that the 22 models used by the IPCC have runs ranging from one up to nine (see Table II in DCPS07). How then should one average these models? Shouldn't we give greater weight to models that have more runs? The paper by S08 does not delve into this important matter; the dispersion shown in their Figure 6 (grey area in Figure 4) makes no allowance for the chaotic nature of models.

In Figure 7, I plot trend values vs. altitude for all models. More or less as expected, the single-run models seem to show the widest dispersion. When I superimpose the limits of the 'grey area of Figure 4, I find that they show a similar dispersion. One may conclude, therefore, that the model uncertainty, shown by the grey area, largely reflects chaotic variability—a factor completely ignored in the S08 analysis. If the IPCC compilation excluded single-run models, the grey area would shrink. I have demonstrated this in a synthetic analysis (Singer, 2010).

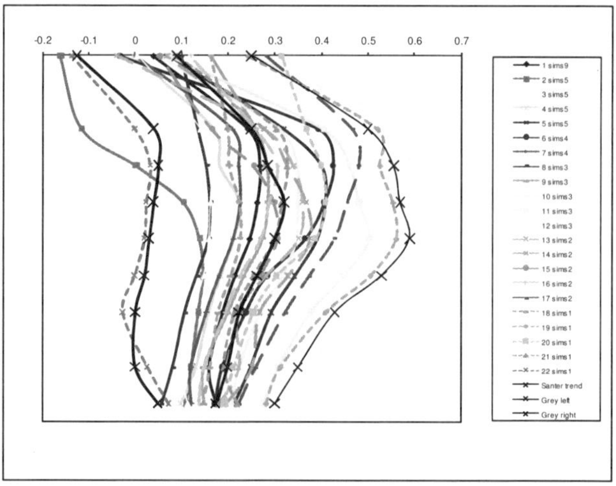

Fig. 7: Temperature trends (x-axis) and pressure-altitude (y-axis) for the twenty-two 20CEN models of IPCC shown in Table II of DCPS (2007). Models #18 to 22 are single-run models and are shown here by short dashes. Models #13 to 17 are 2-run models and shown by long dashes. Models #1

to 12 have 3 or more runs and are shown by solid lines. The mean and the limits of the grey area of S08 (Fig, 5) are indicated by lines carrying X-crosses.

It is difficult to draw any firm conclusions from this graph, since the individual models also incorporate different structural uncertainties. But the single-run models show the widest dispersion, as expected, and also appear to cover the same area as the 'grey area' of S08.

The problem of averaging: over model runs or over model ensemble-means?

From a statistical point of view, one could consider each run as if it were an independent model—albeit using the same forcings and parameterizations. Therefore, one choice would be to average all of the runs (of all models), thereby giving greater weight to models with more runs and less weight to models with few runs. The alternative is to average first all of the runs of a particular model to obtain a "model ensemble-mean", and then average these model means. But this latter method, chosen by S08, gives equal weight to each model irrespective of the number of runs— and therefore undue weight to "outliers." More important even, it exaggerates the effects of chaotic variability.

This problem has not been successfully tackled. Yet the S08 claim of 'consistency' hinges on the degree of agreement between observed and modeled trends. By ignoring the chaotic and structural uncertainties, the analysis by S08 has served to increase the uncertainties in the model trends, and thus any 'overlap' with observed trend values: (Compared to this, the issue of auto-correlation discussed in S08 is relatively unimportant.)

CONCLUSIONS

I conclude that the S08 claim that observed and modeled trends are "consistent" is invalid. Specifically, the "new observational evidence" presented in S08 does not hold up to scrutiny. The claimed uncertainties of the modeled temperature trends, although seemingly supported by elaborate statistical analysis, fail to consider the more important chaotic and structural uncertainties. I hope that my critique will serve to avoid the kind of disastrous mistake that led to the 1997 Kyoto Protocol.

REFERENCES

1. Douglass, D.H., Pearson, B.D., and Singer, S.F. (2004) "Altitude dependence of atmospheric temperature trends: Climate models versus observations." *Geophysical Research Letters* **31**: L13208, Doi:10.1029/2004/GL020103.
2. Douglass, D.H., Christy, J.R., Pearson, B.D., and Singer, S.F. DCPS (2007) "A comparison of tropical temperature trends with model predictions." *International Journal of Climatology* **27**: Doi:10.1002/joc.1651. In printed form: *IJC* 28:1693-1701 (2008).
3. Free, M., Seidel, D.J., Angell, J.K., Lanzante, J., Durre, I., and Peterson, T.C. (2005) "Radiosonde Atmospheric Temperature Products for Assessing Climate

(RATPAC): A new data set of large-area anomaly time series." *J. Geophys. Res. 110*, doi:10.1029/2005JD006169.

4. Haimberger L. (2007) "Homogenization of radiosonde temperature time series using innovation statistics." *Journal of Climate* 20:1377–1403.
5. IPCC AR4. *Climate Change 2007: The Physical Science Basis. Contribution of Working Group I to the Fourth Assessment Report of the Intergovernmental Panel on Climate Change*; Solomon, S., Qin, D., Manning, M., Chen, Z., Marquis, M., Avery, K.B., Tignor, M., and Miller, H.L., Eds.: Cambridge University Press: Cambridge, UK and New York, NY.
6. Karl, T.R., Hassol, S.J., Miller, C.D., and Murray, W.L. (2006) *Temperature Trends in the Lower Atmosphere: Steps for Understanding and Reconciling Differences*; U.S. Climate Change Science Program Synthesis and Assessment Product 1.1(CCSP-SAP-1.1); Washington, DC.
7. Loehle, C. and S.F. Singer (2010) "Holocene temperature records show millennial-scale periodicity." *Can J Earth Sci* 47:1327-1336.
8. Mitchell, J.F.B., et al. (2001) "Detection of climate change and attribution of causes." In *Climate Change 2001: The Scientific Basis*, Contribution of Working Group I to the Third Assessment Report of the Intergovernmental Panel on Climate Change, Mitchell, J.F.B., Karoly, D.J., Hegerl, G.C., Zwiers, F.W., Allen, M.R., Marengo, J. (eds). Cambridge University Press: Cambridge, New York; 881.
9. NAS-National Research Council (2000) *Reconciling Observations of Global Temperature Change*. National Academy Press, Washington DC.
10. Sakamoto, M., Christy, J.R. (2009) "The influences of TOVS radiance assimilation on temperature and moisture tendencies in JRA-25 and ERA-40." *J. Atmos. Ocean. Tech.*, 26:1435.
11. Santer, B.D., Penner, J.E., and Thorne, P.W. (2006) "How well can the observed vertical temperature changes be reconciled with our understanding of the causes of these changes?" In *Temperature Trends in the Lower Atmosphere: Steps for Understanding and Reconciling Differences*, Karl, T.R., Hassol, S.J., Miller, C.D., Murray, W.L. (eds). A Report by the U.S. Climate Change Science Program and the Subcommittee on Global Change Research, Washington DC.
12. (S08) Santer, B.D., Thorne, P.W., Haimberger, L., Taylor, K.E., Wigley, T.M.L., Lanzante, J.R., Solomon, S., Free, M., Gleckler, P.J., Jones, P.D., Karl, T.R., Klein, S.A., Mears, C., Nychka, D., Schmidt, G.A., Sherwood, S.C., and Wentz, F.J. (2008) "Consistency of modelled and observed temperature trends in the tropical troposphere." *Int. J. Climatol.* doi:1002/joc.1756.
13. Singer, S.F. (2000) Climate Policy— From Rio to Kyoto A Political Issue for 2000—and Beyond. Essays in Public Policy No. 102. Hoover Institution, Stanford, California.
14. Singer, S.F. and D.T. Avery (2007) *Unstoppable Global Warming: Every 1,500 Years*. Rowman & Littlefield Publishers, Inc. Lanham, MD.
15. Singer, S.F. (2008) *Nature, Not Human Activity, Rules the Climate: Summary for Policymakers of the Report of the Nongovernmental International Panel on Climate Change (NIPCC)*, Heartland Institute: Chicago, IL.

16. Singer, S.F. (2010) Overcoming chaotic variability of climate models (submitted for publication).
17. Spencer, R.W. and J.R. Christy (1990) "Precise monitoring of global temperature trends from satellites." *Science* 247:1558–1562.
18. Thorne, P.W., Parker, D.E., Santer, B.D., McCarthy, M.P., Sexton, D.M.H., Webb, M.J., Murphy, J.M., Collins, M., Titchner, H.A.and Jones, G.S. (HadAT2 2007) "Tropical vertical temperature trends." *Geophys. Res. Lett. 34*, doi:10;1029/2007GL029875.

THE CLIMATIC ROLE OF THE SUN—HOW, HOW MUCH AND WHAT DOES IT MEAN?

NIR J. SHAVIV
Hebrew University of Jerusalem, Mount Scopus Campus
Jerusalem, Israel

ABSTRACT

The extent of the solar variability's climatic role is a topic of intense debate. The main reason is that if the sun is important, it can explain part of the 20th century warming, and it would imply that the climate sensitivity is on the low side, such that 21st century temperature increase due to human activity will be moderate. Here I summarize the evidence for how the sun affects the climate (primarily through modulation of the cosmic ray flux), how large the solar forcing is, and its implications to the understanding of current climate issues.

1. INTRODUCTION

Different empirical pieces of evidence suggest that the sun affects the climate. But the size of the effect and its implications are still part of a highly contentious debate. On one side stand those who claim that the sun does not affect the climate by mechanisms other than the relatively small changes in the solar irradiance. In such a scenario, the sun could not have played an important role in the 20th century warming. On the other side, there are those who argue that the sun is affecting the climate more than one could expect from changes in the solar irradiance. This in turn implies that the increased solar activity over the 20th century should have played a role in the 20th century global warming. I will argue here that the evidence suggests the latter.

We begin in §2 by summarizing the evidence that the sun has a large effect on climate. We will then continue in §3 by showing that the large effect appears to arise from the climate's sensitivity to the amount of atmospheric ionization, which is governed by the flux of cosmic rays. In §4 we will show that the effect can be quantified by several means, in particular, by looking at the 11-year solar cycle in the ocean heat content. In §5 we will show that the amplified solar effect helps us understand 20th century warming and also predict the possible range of global warming over the 21st century.

2. DOES THE SUN AFFECT THE CLIMATE?

It is already more than 200 years since Sir William Herschel[1] claimed that variations in solar activity affect climate on Earth. Since he did not have any reliable temperature measurements, Herschel looked for indirect proxies. He compared the price of wheat in the London wheat exchange to the solar activity as mirrored in the sunspot number, and found a correlation between them.

In the 1970's, Eddy[2] pushed the idea that solar activity may be affecting the terrestrial climate. He found a correlation between long-term variations in solar activity and different climate indicators. For example, he found that the nadir of the period called "the little ice age" in Europe during the latter half of the 17th century took place in a period during which solar activity was very low. This low activity culminated in a several decade period during which there were almost no apparent sunspots, which was called the "Maunder minimum". On the other hand, there were other periods, such as the end of the middle ages, during which solar activity was as high as the latter half of the 20th century, and the temperatures were roughly as warm as today. During the "medieval optimum", Vikings could settle in Greenland (and call it a "green land") and catholic monks adopted sandals suitable for warm climates.

Presently, there are a large number of different empirical indicators showing that changes in solar activity have a non-negligible climatic effect. Changes in solar activity manifest themselves as changes in the strength of the solar magnetic field, changes in the sunspot number, in the strength of the solar wind (which is responsible for the impressive tails of comet) and in other phenomena. These changes can be divided into three time scales.

The basic variation is an activity cycle of about 11 years, which arises from quasi-periodic reversals of the solar magnetic dipole filed. On longer time scales (of decades to millennia) there are irregular variations, which modulate the 11-year cycle. For example, during the middle-ages and during the latter half of the 20th century, the peaks in the 11-year cycles were notably strong, while these peaks were almost absent during the Maunder minimum. On the other hand, solar eruptions may appear on the time scale of days. Today there is evidence linking solar activity to the terrestrial climate on all these scales.[3-11]

Since the work of Eddy, many empirical results show a correlation between different climatic reconstructions and different solar activity proxies.[5-7] In particular, one of the most beautiful results is of a correlation between an Indian Ocean temperature proxy and solar activity.[5]

It is much harder to see climate variations over the 11-year solar cycle. There are two reasons for that. First, if we study the climate on short time scales, we find that there are large annual variations (for example, due to the el-Ñino oscillation), which introduce cluttering "noise", hindering the observation of solar related signals. Second, because of the large oceanic heat capacity, it takes decades until it is possible to see the full effects of given changes in the radiative budget, including those associated with solar variability. It is for this reason, that climates of continental regions are typically much more extreme than their marine counterparts.

If, for example, a given change in solar forcing is expected to give rise to a temperature change of 1°C after several centuries, as is observed[10], then the same radiative forcing varying over the 11-year solar cycle is expected to give rise to temperature variations of only 0.1°C or so.[11] This is because on short time scales, most of the energy goes into heating the oceans, but because of their very large heat capacity, large changes in the ocean heat content do not translate into large temperature variations.

Nevertheless, if the global temperature is carefully analyzed (for example, by folding the global temperature of the past 120 years over the 11-year solar cycle), it is possible to see variations of about 0.1°C in the land temperature, and slightly less in the ocean surface temperature.[3,4,11]

Clearly then, there is ample evidence to suggest that solar variability has a large effect on climate. The question is therefore how does this link arise and whether it can be quantified.

3. HOW DOES THE SUN AFFECT THE CLIMATE?

Grossly speaking, there are two types of mechanisms that can amplify solar activity. The first type is hypersensitivity to one of the non-thermal solar components. One such mechanism was proposed by Joanna Haigh from the UK, and it is hypersensitivity to variations in the UV.[12] This kind of sensitivity can arise because UV is almost entirely absorbed in the stratosphere and although it only includes about 1% of the solar output, the stratospheric structure (and thus the tropospheric-stratospheric interface) is determined by this 1%. Numerical simulations have shown that by including the variations in the UV and their effects on the stratosphere, one can amplify the surface climate variations by as much as a factor of two[13], namely, it can be a large effect. But as we shall see below, it still cannot explain for example the large amounts of heat seen entering the oceans every solar cycle.

The second type of mechanisms is indirect, through the solar modulation of the cosmic ray flux and the effect that the latter may have on the climate. Cosmic rays are high energy particles (primarily protons) which appear to originate from supernova remnants (the leftovers from the explosive death of massive stars). A possible climatic link through cosmic rays was first suggested by Ney already in 1959.[14] It was well known that the solar wind decreases the flux of these high energy particles and that these particles are the primary source of ionization in the troposphere (which is the lower part of the atmosphere). Ney proposed that the changing levels of ionization can play some climatic role.

In the 1970's, Dickinson proposed the possibility that the atmospheric ion density could play a role in the formation of cloud condensation nuclei.[15] When air reaches saturation, that is, 100% humidity, the preferred equilibrium state is that of liquid water. However, if the water vapor has nothing to condense upon, it will not do so. In fact, under very clean environments, it is possible to reach 400% humidity before the vapor condenses spontaneously. In order to get clouds at 100%, as we see in nature, we need cloud condensation nuclei (CCNs). Over land, there are many natural sources for CCNs, however, this is not the case over the oceans, were the CCNs must be grown out of something. Dickenson suggested that this growth process of CCNs could be affected by the amount of atmospheric charge.

In the 1990's, Svensmark and his colleagues found empirically that clouds, and in particular low altitude clouds, appear to vary in sync with the solar activity.[16-19] Since Svensmark's work, more evidence was found to support this link, the full picture of which is the following. When the sun is more active, it has a stronger solar wind. The stronger wind slows the cosmic rays as they propagate into the inner solar system. As a consequence, the amount of atmospheric ionization is reduced. Less ions reduce the efficiency with which new cloud condensation nuclei can grow, especially over the oceans, such that the clouds that later form have fewer but larger droplets. These clouds are less white, they reflect sunlight less efficiently and therefore cause more warming.

The evidence to this particular link comes from experimental results and from correlations between independent cosmic ray flux variations and climate changes on different time scale. Just by itself, a cosmic ray climate correlation over the 11-year solar cycle does not necessarily imply a causal link. One could imagine that the solar activity affects both the cosmic ray flux and the climate, making it appear that there is a causal relation between the latter two. Nevertheless, there are indications that it is not just an apparent link. For example, the dependence of the relative cloud cover variations with the magnetic latitude is the same as the latitudinal dependence of the relative change in the atmospheric ionization, over the solar cycle. Another important fact is that the full solar cycle is not that of 11 years, but 22 years instead. It takes 11 years for the magnetic field to flip, but 22 years for it to return to the original state. However, all the solar activity proxies are "blind" to the polarity of the magnetic field, all except the cosmic ray flux which exhibits a clear asymmetry between odd and even solar cycles. This asymmetry is seen in the change of the low altitude cloud cover[19], implying that the cloud cover variations originate from cosmic ray flux variations.

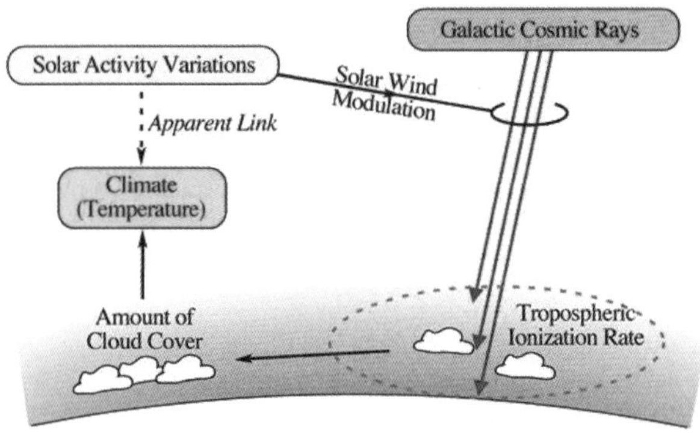

Fig. 1: The cosmic ray link between solar activity and the terrestrial climate. The changing solar activity is responsible for a varying solar wind strength. A stronger wind will reduce the flux of cosmic rays reaching Earth, since a larger amount of energy is lost as they propagate up the solar wind. The cosmic rays themselves come from outside the solar system (cosmic rays with energies below the "knee" at 10^{15}eV, are most likely accelerated by supernova remnants). Since cosmic rays dominate the tropospheric ionization, an increased solar activity will translate into a reduced ionization, and empirically (as shown below), also to a reduced low altitude cloud cover. Since low altitude clouds have a net cooling effect (their "whiteness" is more important than their "blanket" effect), increased solar activity implies a warmer climate. Intrinsic cosmic ray flux variations will have a similar effect, one however, which is unrelated to solar activity variations.

On short time scales, the sun can undergo flaring activity, which is caused from the reconnection of magnetic loops. These flares are accompanied by a strong solar wind "gust" which later causes a decrease in the cosmic ray flux for several days. If the cosmic ray flux has an effect on clouds, then cloud properties should change following these events, known also as Forbush decreases. Several results indicate that clouds are affected during Forbush decreases. In particular, a recent study of Forbush decreases has shown the cosmic ray mechanism at work. Not only was a cloud signal observed, the intermediate step of affecting the aerosol size distributed was detected as well.[9]

Over longer time scales, of decades to millennia, there are the aforementioned solar climate links, however, even though they demonstrate a clear causal link between solar activity and climate change, it is hard to prove with them that this link is specifically due to solar modulation of the cosmic ray flux. If we however go to longer time scale still, there is evidence from cosmic ray flux variations, which are not associated with solar activity.

On the time scale of tens of thousands of years, Earth's magnetic field varies and with it the flux of cosmic rays which can penetrate the atmosphere. However, because the magnetic field can only prevent the penetration of cosmic rays which are anyway severely attenuated by the atmosphere, changing the magnetic field, and even altogether switching it off is not expected to give rise to significant climate effects. A rough estimate gives that switching the magnetic field will only cool the Earth by typically 1°C. However, over the time scale that the magnetic field changes, Earth witnesses variations that are 5 times larger from other natural causes. In other words, it is not easy to detect the terrestrial field effects, but it was claimed to be detected nonetheless.[20]

Over geological time scale, the cosmic ray flux changes because of our motion around the Milky Way and the changing solar neighborhood. Because the cosmic ray flux originates from the death of the short living massive stars, in events called supernovae, passage through regions with a higher star formation rate is associated with an elevated cosmic ray flux level. The largest variations actually originate from the solar system's passage through the Milky Way spiral arms.[21,22]

As it turns out, it is possible to reconstruct the changes in the cosmic ray flux originating from these passages, using iron meteorites. This reconstructed flux shows variations by as much as a factor of 3 between the flux between the spiral arms and the flux in them. And indeed, when the global climate is studied over this time scale, it is possible to see all the past seven passages of the solar system through the arms of the galaxy over the past billion years. Every spiral arm passage, the increased cosmic ray flux manifested itself as a cold epoch during which Earth's poles were glaciated. When the solar system was in between the arms, it was much warmer than the present climate.[21-23]

In addition to the empirical evidence, an experiment was carried out by Svensmark's group. This experiment was carried out to simulate marine air conditions and study how the changed atmospheric ionization affects the growth of condensation nuclei under controlled laboratory conditions. The experiment demonstrated that elevated ionization rates give rise to a more efficient formation of condensation nuclei.[24] Today, this experiment is carried out in a mine in the UK to see how the total removal of ions affect the formation of condensation nuclei, and also at CERN, to corroborate the effects of ionization.

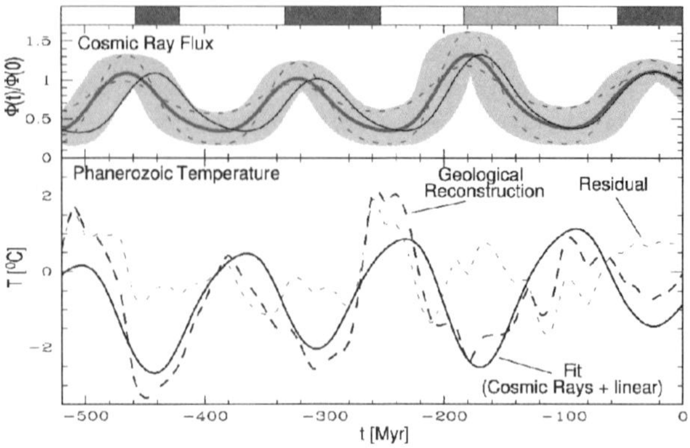

Fig. 2: The cosmic ray flux (Φ) and tropical temperature anomaly (ΔT) variations over the Phanerozoic. The upper curves describe the reconstructed CRF using iron meteorite exposure age data. The heavy gray line depicts the nominal CRF, while the shading delineates the allowed error range. The two dashed curves are additional CRF reconstructions that fit within the measurement errors. The solid black curve describes the nominal CRF reconstruction after its period was fine-tuned, within the measurement error, to best fit the low-latitude temperature anomaly. The bottom dashed curve depicts the temperature reconstruction, measured with a 10 Myr bin and smoothed with a 5-bin top hat averaged. The solid line is the predicted ΔT based on the nominal CRF model above while also taking into account a secular long-term linear contribution. The light dashed line is the residual. More details can be found in Shaviv and Veizer, 2003.[23]

4. HOW LARGE IS THE SOLAR/CLIMATE EFFECT?

After demonstrating that cosmic ray modulation amplifies solar activity, the next step is to quantify the size of the effect. Here we will concentrate on three methods (out of more which exist). In the first two, we will consider at the changes in the cloud cover and the changes in the heat going into the oceans, both over the solar cycle. In the third method, we will consider the 20th century warming. Since the latter is relevant to the understanding of 20th century warming, will address it in the next section.

Cloud cover variations

If one works under the assumption that the cloud cover variations which are observed to be in sync with the 11-year solar cycle are the mechanism through which solar variability is amplified, then satellite observations can be used to estimate the forcing associated with solar variability.

It was shown using the ERBE satellite data that the observed 1.6±0.4% low altitude cloud cover variations correspond to a radiative forcing of 1.0±0.35 W/m^2.[17,4] Note that this implicitly assumes several assumptions. For example, it assumes that the radiative forcing changes associated with the total 27% low altitude cloud cover is the same, per 1% change, as the 1.6% change associated with the solar cycle. This is not necessarily the case because we should expect cloud cover of the oceans to be more susceptible to change in the atmospheric ionization than cloud cover land (where cloud condensation nuclei are naturally abundant).

Ocean Heat Content

Another method for determining the radiative forcing variations associated with the solar cycle is by studying the total ocean heat content. This is because the total ocean heat capacity is by far larger than that of any other component in the climate system. As a consequence, changes in the total amount of heat can be used to derive the changes in the radiative budget of Earth.

Such an analysis reveals that the amount of heat going into the oceans is 1.05±0.25 W/m^2, of which only 0.17 W/m^2 can be explained by changes in the solar irradiance.[11] This large flux can be seen in three independent data sets – direct measurements of the ocean heat content, the sea surface temperature, and the thermal expansion derived from tide gauges. Although this analysis cannot point to the actual mechanism amplifying the solar climate, it is interesting to note that it is consistent with the changes in the energy budget associated with the cloud cover variations. The tide gauge based data is depicted in Figure 3.

Effect on the global temperature

If we wish to translate the changes in the radiative forcing into changes in the temperature, we require the Earth's climate sensitivity, namely, how large is the temperature change expected from a given change in the energy budget.

The canonical IPCC range, obtained primarily from global circulation models is a temperature increase of 1.5 to 4.5°C per CO_2 doubling, i.e., 0.4 to 1.2°C/(W m^{-2}). The reason for the large uncertainty is the very limited understanding of different climate feedbacks, with cloud cover being the primary one. Several different analyses have shown empirically that the climate sensitivity is low (e.g., Lindzen and Giannitis, 1998[25]) In particular, Shaviv 2005[4] has shown that once one includes the CRF/climate link, different empirical estimates become more consistent with each other, and give a sensitivity of 0.35±0.09°C/(W m^{-2}).

If we use the latter value and assume that the long term solar activity variations are similar in amplitude to the variability over the 11-year cycle, then about 0.4±0.2°C of the 20th century warming should be attributed to the increased solar activity.

On shorter time scales, the expected variations are smaller. This is because of the large heat capacity of the oceans, which forces the terrestrial climate system to behave as a low pass filter for variations in the radiative budget. In particular, the 1 W m^{-2} variations over the 11-yr solar cycle translate into ~0.1°C variations, as is indeed observed.

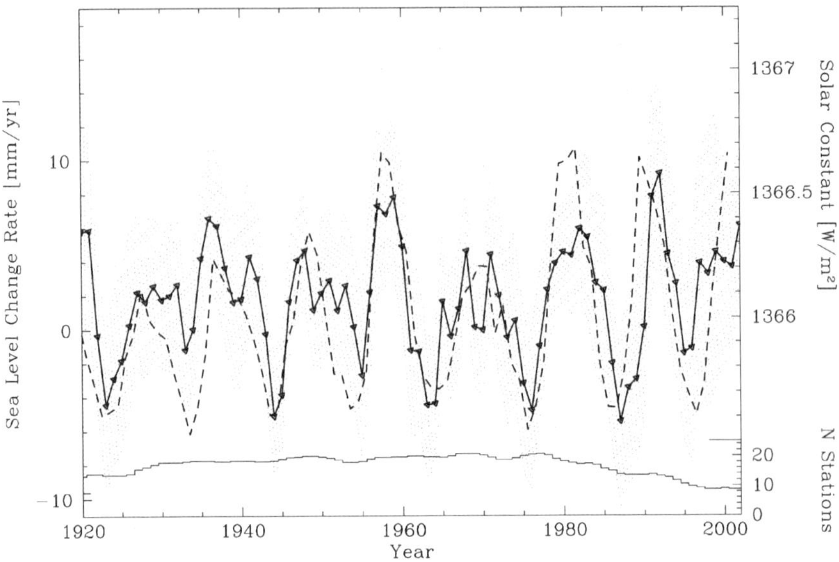

Fig. 3: Sea Level vs. Solar Activity. Sea level change rate is based on 24 tide gauges for the stringent criteria they satisfy (solid line, with 1-σ statistical error range denoted with the shaded region). The rates are compared with the total solar irradiance variations (dashed line, with the secular trends removed). More details in Shaviv 2008.[11]

5. WHAT DOES IT MEAN?

<u>Understanding the 20th century warming</u>

The fact that the sun has a large effect on the climate implies that the increased solar activity over the 20th century should have played a role in the 20th century warming, one which cannot be neglected. As shown above, the solar forcing associated with the 11-year solar cycle can be used to estimate this contribution to be 0.4±0.2°C. Since the total temperature increase is about 0.8°C, the contribution from other forcings, in particular that of anthropogenic greenhouse gases, is comparable.

This conclusion can be obtained more directly by studying the actual 20th century temperature increase. Ziskin[26] modeled the 20th century climate using a multi-box diffusion model with various free parameters. The most important were the climate sensitivity, the size of the indirect solar forcing and the size of the indirect aerosol effect. By systematically searching for the region in parameter space for which the model gave temperature and ocean heat content variations which best fit the observed data over the past century, it was possible to obtain probability distribution functions for the different climate model parameters.

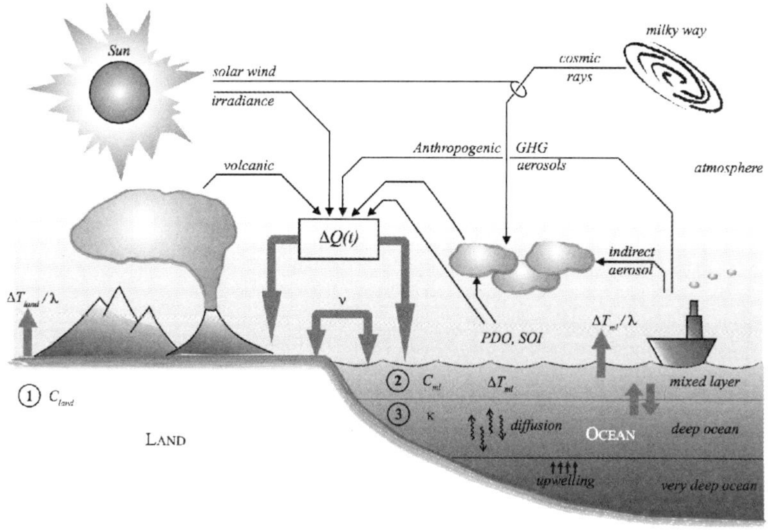

Fig. 4: A heuristic description of the energy model employed. The black lines with arrows denote the different forcings, which are either known or described with the different model parameters. The heavy gray arrows denote interaction between the different components in the systems: Land, Ocean Mixed Layer and Ocean Deep Layer. The set of equations in described in detail, in Ziskin and Shaviv.[26]

It was found that by allowing the sun to have a climatic effect that is larger than just the changes in the solar irradiance one obtains that the r.m.s of the residual between the predicted temperature and the observed temperature is typically 0.1°C. In addition, one finds that the increased solar forcing explains 0.3±0.1°C of the warming, while the anthropogenic caused increase is 0.4±0.1°C. Last, the most important result is that the best fit requires a climate sensitivity of 0.25±0.09°C/W m^{-2}. The fit and the residual are depicted in Figure 5.

These results should be compared with fits obtained using the same climate model or with full GCMs that do not include additional solar forcing, where the r.m.s. residuals are typically 0.2°C. The reason for the degraded fit is that in order to explain the observed warming without the extra solar forcing, the climate sensitivity has to be higher. But the higher sensitivity forces the climate to overreact to different radiative forcings, more than the real climate does; for example, following volcanic eruptions (e.g., see Lindzen and Giannitis, 1998[25]).

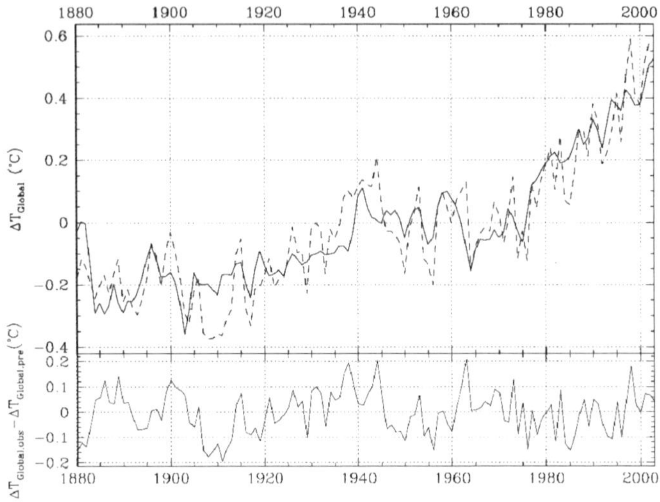

Fig. 5: Top: The predicted (solid) and observed (dashed) global temperature.[26] *Bottom: The residual temperature.*

Predictions for the 21st century temperature increase

As demonstrated above, the climate sensitivity obtained by different empirical methods is typically comparable to that expected for a feedback-less black body Earth, when allowing the sun to have a large effect on climate. This sensitivity can be used to estimate the temperature increase over the 21st century, based on the chosen emission scenario.

Business as usual scenarios typically give that the amount of tropospheric CO_2 will roughly double over the 21st century. This corresponds to a radiative forcing of about $4W/m^2$, or, an equilibrium warming of about 1°C. This is significantly less than the much larger predictions of the GCM based IPCC predictions.

REFERENCES

1. W. Herschel. (1756) "Some remarks on the stability of the light of our sun." *Royal Society of London Philosophical Transactions Series I*, 86:166–226.
2. J.A. Eddy. (1976) "The Maunder Minimum." *Science*, 192:1189–1202.
3. W.B. White, J. Lean, D.R. Cayan, and M.D. Dettinger. (1997) "Response of global upper ocean temperature to changing solar irradiance." *J. Geophys. Res.*, 102:3255–3266.
4. N.J. Shaviv. (2005) "On climate response to changes in the cosmic ray flux and radiative budget." *J. Geophys. Res.*, 110:8105.

5. U. Neff, S. J. Burns, A. Mangini, M. Mudelsee, D. Fleitmann, and A. Matter. (2001) "Strong coherence between solar variability and the monsoon in Oman between 9 and 6kyr ago." *Nature*, 411:290–293.
6. G. Bond, B. Kromer, J. Beer, R. Muscheler, M.N. Evans, W. Showers, S. Hoffmann, R. Lotti-Bond, I. Hajdas, and G. Bonani. (2001) "Persistent Solar Influence on North Atlantic Climate During the Holocene." *Science*, 294:2130–2136.
7. D.A. Hodell, M. Brenner, J.H. Curtis, and T. Guilderson. (2001) "Solar forcing of drought frequency in the Maya lowlands." *Science*, 292:1367–1370.
8. B.A. Tinsley. (1996) "Correlations of atmospheric dynamics with solar wind-induced changes of air-earth current density into cloud tops." *J. Geophys. Res.*, 101:29701–29714.
9. H. Svensmark, T. Bondo, and J. Svensmark. (2009) "Cosmic ray decreases affect atmospheric aerosols and clouds." *Geophys. Res. Lett.*, 36:15101.
10. S.P. Huang, H.N. Pollack, and P.-Y. Shen. (2008) "A late Quaternary climate reconstruction based on borehole heat flux data, borehole temperature data, and the instrumental record." *Geophys. Res. Lett.*, 35:13703.
11. N.J. Shaviv. (2008) "Using the oceans as a calorimeter to quantify the solar radiative forcing." *J. Geophys. Res.*, 113(A11).
12. J.D. Haigh. (1994) "The Role of Stratospheric Ozone in Modulating the Solar Radiative Forcing of Climate." *Nature*, 370:544.
13. A. Larkin, J.D. Haigh, and S. Djavidnia. (2000) "The Effect of Solar UV Irradiance Variations on the Earth's Atmosphere." *Sp. Sci. Res.*, 94:199–214.
14. E.P. Ney. (1959) "Cosmic radiation and weather." *Nature*, 183:451.
15. R.E. Dickinson. (1975) "Solar Variability and the Lower Atmosphere." *Bull. Am. Met. Soc.*, 56:1240–1248.
16. H. Svensmark. (1998) "Influence of Cosmic Rays on Earth's Climate." *Phys. Rev. Lett.*, 81:5027–5030
17. N. Marsh and H. Svensmark. (2000) "Cosmic Rays, Clouds, and Climate." *Sp. Sci. Rev.*, 94:215–.
18. E. Palle Bago and J. Butler (2000) "The influence of cosmic terrestrial clouds and global warming." *Astron. Geophys.*, 41:18–22.
19. H. Svensmark. (2007) "Cosmoclimatology: a new theory emerges." *Astronomy & Geophysics*, 48(1):1–18.
20. M. Christl, A. Mangini, S. Holzkämper, and C. Spötl. (2004) "Evidence for a link between the flux of galactic cosmic rays and Earth's climate during the past 200,000 years." *Journal of Atmospheric and Terrestrial Physics*, 66:313–322.
21. N.J. Shaviv. (2003) "The spiral structure of the Milky Way, cosmic rays, and ice age epochs on Earth." *New Astron.*, 8:39–77.
22. N.J. Shaviv. (2002) "Cosmic Ray Diffusion from the Galactic Spiral Arms, Iron Meteorites, and a Possible Climatic Connection." *Phys. Rev. Lett.*, 89(5):051102.
23. N.J. Shaviv and J. Veizer. "A celestial driver of phanerozoic climate?" *GSA Today*, pages 4–11, July 2003.
24. H. Svensmark, J.O.P. Pedersen, N.D. Marsh, M.B. Enghoff, and U.I. Uggerhøj. (2007) "Experimental evidence for the role of ions in particle nucleation under

atmospheric conditions." *Royal Society of London Proceedings Series A*, 463:385–396.

25. R.S. Lindzen and C. Giannitsis. (1998) "On the climatic implications of volcanic cooling." *Journal of Geophysical Research. D. Atmospheres*, 103:5929–5941.
26. S. Ziskin and N.J. Shaviv. "Quantifying the role of solar radiative forcing over the 20th century." submitted to *J. Atmos. Sol. Terr. Phys.*

TWENTIETH CENTURY SOURCES OF METHANE IN THE ATMOSPHERE

TOM QUIRK
Institute of Public Affairs
Melbourne Victoria, Australia

ABSTRACT

Present global and national schemes for carbon regulation often include methane alongside carbon dioxide. It is therefore important to understand the sources, sinks and control of methane in the atmosphere and then consider if methane should be part of any carbon regulation scheme. Atmospheric measurements over the last 50 years show substantial changes in methane concentration. Natural gas leakage from pipelines has been the major contributor up to 1990. For the last 15 years there has been little increase in concentration and natural climate variability has been the dominant control in changing methane concentrations.

The role of methane in the atmosphere has been emphasized by the IPCC to the point that many governments regard methane as almost as important as carbon dioxide amongst the greenhouse gases. The result is that emissions from natural gas pipelines, coal seams and agricultural livestock have been included in schemes to limit the growth of greenhouse gas concentrations. Analysis of changes to atmospheric methane within the last one hundred years suggests that the annual increases from 1930 to 1990 were due to losses from the production, transmission and distribution of natural gas that have now been reduced. Measurements over the last fifteen years show only natural variability. The data provide no justification for any attempts to reduce methane from industrial or agricultural activity.[1]

The time development of methane concentrations is shown in Figure 1 from measurements at the South Pole. These direct in air measurements are part of a network that extends up to the Arctic Ocean. The data covers the period from the 1980s to the present. Figure 1 shows seasonal variations in methane concentration with minima in summer and maxima in winter. Methane is removed from the atmosphere by its interaction with OH radicals derived from ultra-violet sunlight, ozone and water vapour. These are at a maximum concentration in summer with higher levels of sunlight and water vapour in the atmosphere. The OH radicals combine with methane to produce a methyl radical through the reaction:

$$CH_4 + \text{-}OH \rightarrow \text{-}CH_3 + H_2O.$$

Thus methane is removed from the greenhouse gas inventory with a lifetime of 11 years and the methyl radical through further chemical reactions is finally reduced to CO_2.

In addition to this it is possible to get methane measurements from sampling air trapped in bubbles in ice cores. Extensive measurements have been made by a CSIRO group from ice cores in Antarctica. Figure 2 shows an example of atmospheric methane concentrations measured in ice cores at the Law Dome in Antarctica. Where there is overlap, the measurements agree with the direct atmospheric results from Cape Grim, a CSIRO station in Tasmania, Australia.

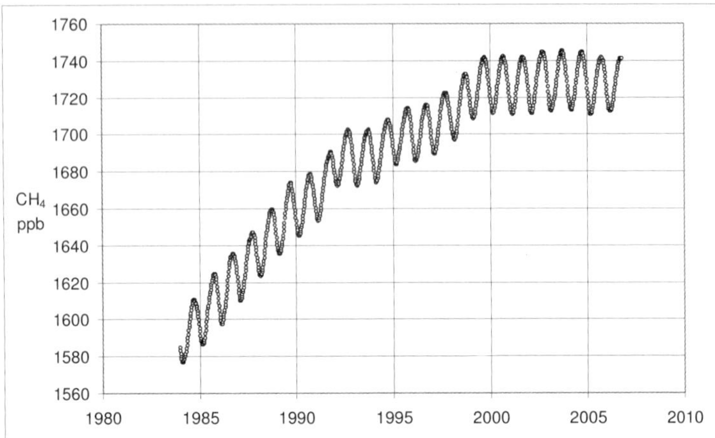

Fig. 1: Methane measurements at the South Pole from NOAA-ESRL Dataset.

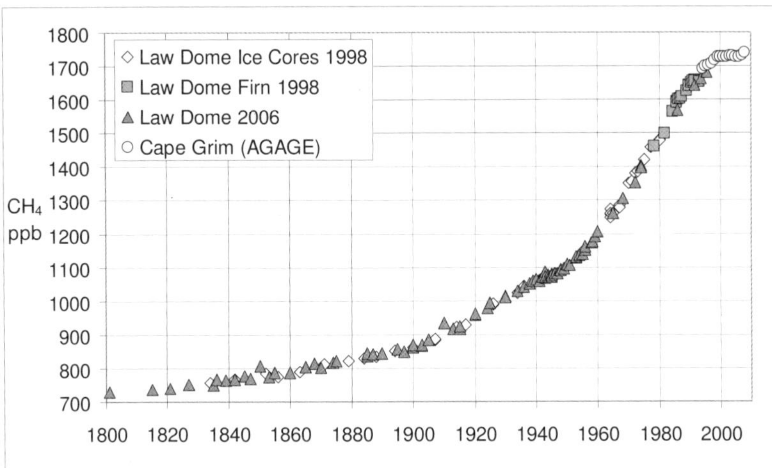

Fig. 2: Atmospheric methane concentrations from ice cores and Firn samples at the Law Dome. The ice core sampling uncertainty is estimated at 4 ppb standard deviation from multiple samples and re-sampling in 2006 while the Firn measurements have a 1 ppb standard deviation. Cape Grim are direct atmospheric measurements. Note that ice core measurements in this period sample the atmosphere with about a 12 year window.

These measurements may be used to derive the annual increases in atmospheric methane and this is shown in Figure 3. The ice core derived data has been averaged in ten year intervals and the differences have then given the annual methane changes in the atmosphere. The errors are derived from the stated measuring errors and the scatter of values within each 10 year interval.

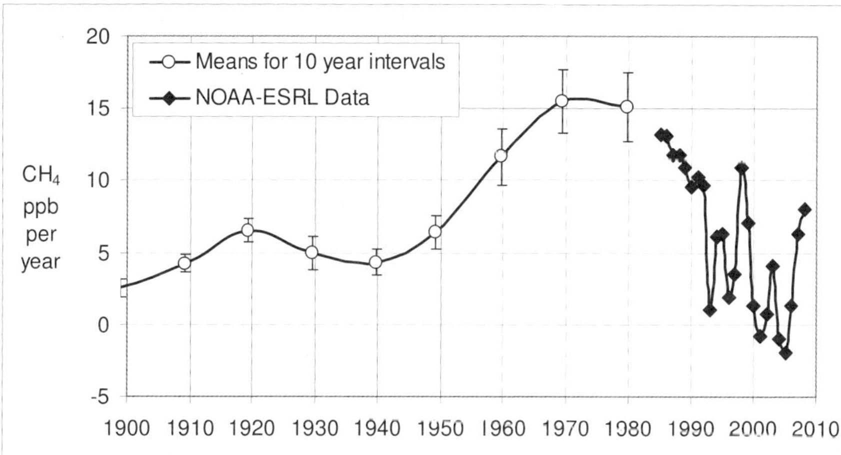

Fig. 3: Annual changes in atmospheric methane derived from the Law Dome measurements and direct annual atmospheric measurements from NOAA-ESRL data.

The fall in methane emissions during 1984-1990 implied by the plateau of Figure 1 and clearly shown in Figure 3 may be explained by reduced gas pipeline losses and a major part if not all due to reduced losses in the old Soviet Union and Eastern bloc countries[2] as shown in Figure 4. The magnitude of the uncertainties in these estimates is large. In terms of the database definitions[3] it is "medium" with a 50% uncertainty! It is more likely that the uncertainty is a systematic error rather than a random error as the uncertainties arise from assumptions made in deriving emissions from well documented production or consumption statistics.

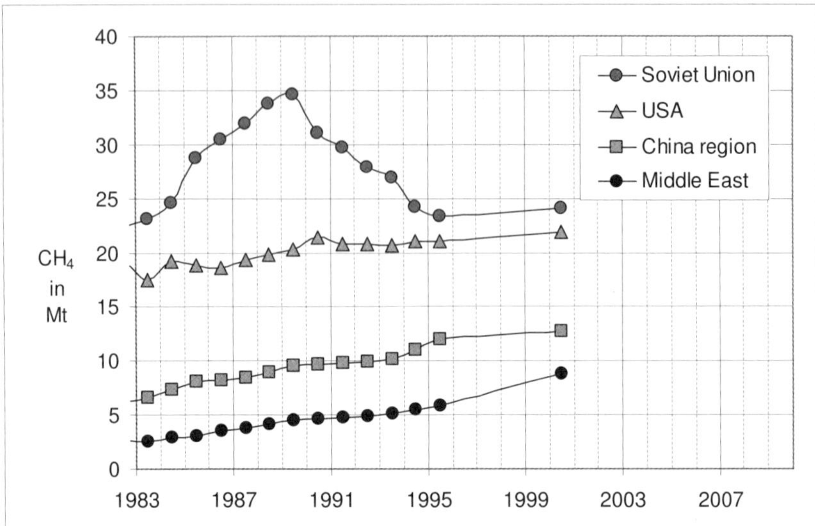

Fig. 4: Annual regional emission of methane from fossil fuel production and delivery. Source: Databases EDGAR-HYDE 1.4-CH4 and EDGAR 32FT2000

The data have been replotted in Figure 5 to show the relationship with the consumption of natural gas. The natural gas estimates come from BP Statistical Review of World Energy June 2008[4] for the period 1970 to 2006 while the estimates for 1930 to 1970 come from C J Campbell.[5] The Campbell data overlaps the BP data and agree to within 5%. An estimated background of an annual increase of 3 ppb is shown. This accounts for natural variability of El Niño induced changes, continuing modest natural gas losses into the atmosphere and volcanic effects. The volcanic effects are uncertain. In the twentieth century, there were 4 volcanic eruptions estimated to have had substantial impact on the atmosphere. Analysis shows that Pinatubo created a pulse of some 26 Mt of methane in 1991. The amount and timing of methane pulses of Santa Maria in 1902, Agung in 1964 and El Chichon in 1983 would add to the uncertainty in the non-natural gas contribution to the atmosphere from natural causes.

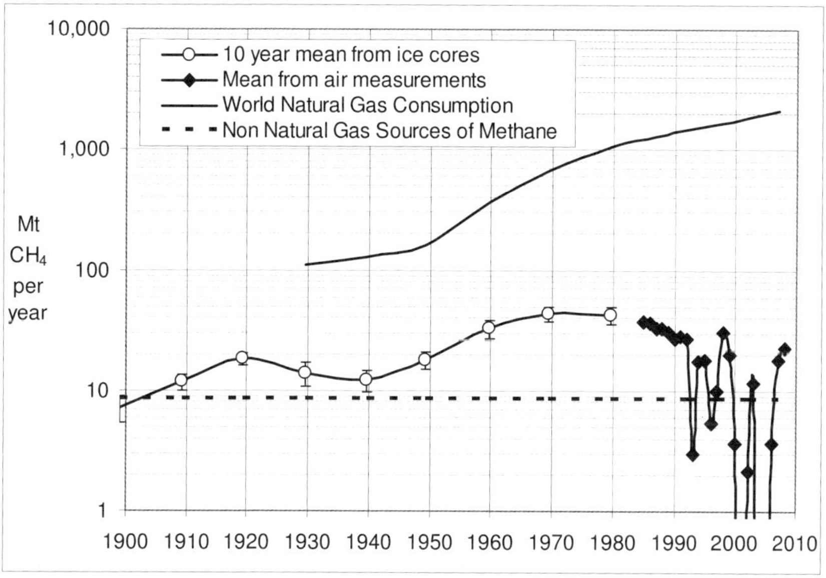

Fig. 5: Annual changes in atmospheric methane averaged in 10 year intervals from ice core derived data and year intervals for direct air sampling. The world natural gas consumption is from the BP Statistical Review and estimate from C.J. Campbell. The non-natural gas sources are estimated at 8.4 Mt per year (3 ppb per year) from 1900-1910 and 1993-2008.

Figure 6 shows the estimated natural gas losses to the atmosphere measured by the adjusted annual increase in atmospheric methane as a percentage of annual natural gas consumption.

Estimation of pipeline losses is notoriously difficult.[6] Flow meter reading errors, unmetered off-takes, bypassed meters and the loss and use of gas in compressors all contribute to the uncertainty of as much as 1.5%. From the late 1950's on, there was a changeover from the use of coal or town gas to natural gas worldwide. There would have been substantial leakage from old town networks of cast iron pipes. These losses were reduced by their slow replacement with more secure piping. It is also clear from discussions with those familiar with the gas supply industry and particularly Soviet practice[7] that large pipeline losses of up to 10% are possible in old gas transmission and reticulation networks. In fact troubles on the Trans-Siberian pipeline led to a massive explosion in June 1982 estimated at 3,000 tonnes equivalent TNT.[8] It was the largest non-nuclear explosion and fire ever seen from space.

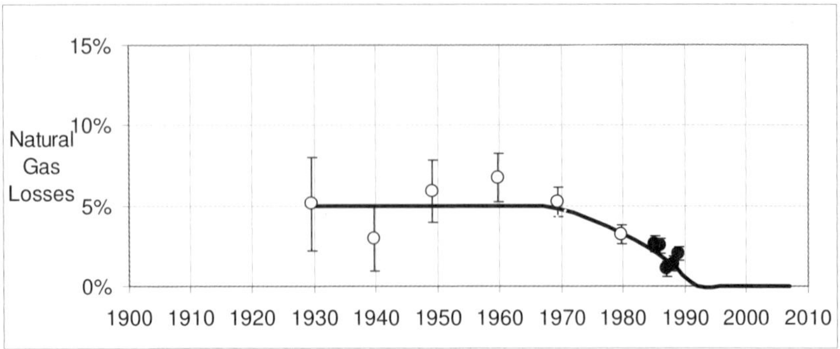

Fig. 6: Estimated natural gas losses to the atmosphere measured by the adjusted annual increase in atmospheric methane as a percentage of annual natural gas consumption. Estimates derived from 10 year intervals of ice core data and annual air measurements. The solid line is the average value from 1930 to 1970 and then follows the averages from 1970 to 1990.

If the bulk of increased atmospheric methane is from natural gas leakage, can the mechanisms for the residual variations of concentration be understood?

Variations in methane concentrations are subject to both source and sink variations as illustrated in the upper part of Figure 7. The annual changes from 1985 to 1992 include a source, natural gas leakage, the 1987-88 El Niño that provides both a source and a sink and a reduced sink from the Mt Pinatubo eruption of 1991. Volcanoes produce SO_2 so the pulse of methane was caused by the reduction of the number of OH radicals through the competing reaction with the massive amounts of SO_2 produced in the eruptions with 17 M tonnes coming from Mt Pinatubo.

One of the controlling variables for OH radicals is humidity. Humidity in turn is influenced by El Niños. Specific humidity at 1000 mb in the equatorial zone is about 16 g/kg and variations in humidity are shown in the lower part of Figure 7.

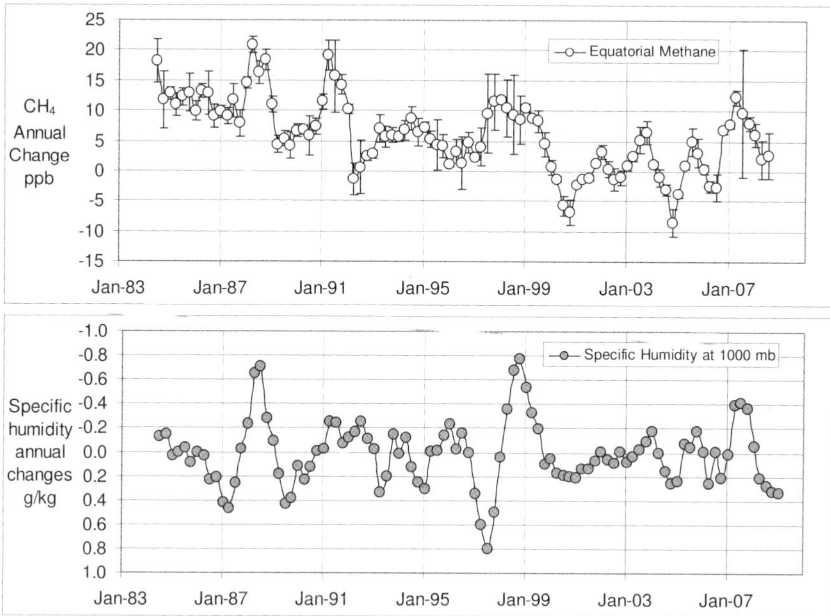

Fig. 7: Upper Year on year changes for equatorial *quarterly average methane concentrations from global data (NOAA-ESRL and AGAGE),* Lower *Annual changes in specific humidity in the equatorial latitude band 22.5 S to 22.5 N at 1000 mb. Note that the scale is inverted as decreasing humidity and OH radical concentration implies more methane remains in the atmosphere.*

Much of the pattern of methane variations follows humidity variations. This can be seen in the upper part of Figure 8 where the humidity variations have been normalized to the methane variations from 2000 to 2007. This is a variation of the sink for methane.

The lower part of Figure 8 shows the variations of methane and an inferred source contribution following the variations in CO_2. El Niños have a major influence on CO_2 with droughts and fires. In addition there are studies that suggest methane emissions occur where El Niños trigger bush, brush and swamp fires.

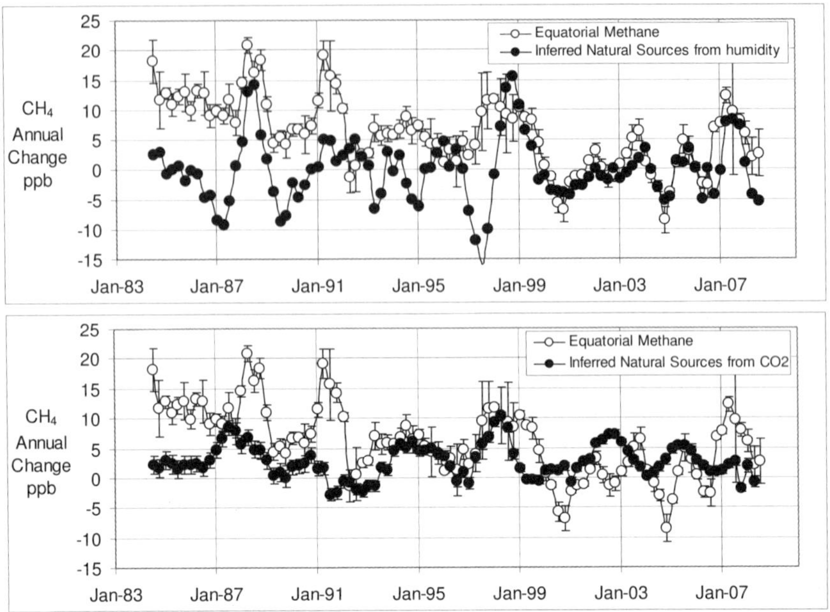

Fig. 8: Upper: Equatorial and inferred natural variations from variations in humidity and Lower: Equatorial and inferred natural variations from variations in CO_2.

Thus the El Niños are acting as a trigger for variations of both sources and sinks for methane.

CONCLUSIONS

- Natural gas leakage from gas pipelines was largely responsible for increased atmospheric methane concentrations in the twentieth century.
- Through better pipeline management and maintenance, the leakage was greatly reduced in the late 1980s and early 1990s.
- In 1991 volcanic emissions of SO_2 from Mount Pinatubo reduced methane removal giving an apparent spike in methane emissions.
- The main determinant of variations in atmospheric methane from 1992 on is variations of the sink not the sources of methane.
- Variations of the sink are controlled by humidity and subject to El Niño and La Nina events, and
- There are also source methane emissions from bush, brush and swamp fires triggered by El Niños.

There is no strong reason to regulate methane emissions that appear subject to natural variability alone.

REFERENCES

1. See also *Energy & Environment* Vol. 21 p 251-266, 2010.
2. Database EDGAR-HYDE 1.4; Van Aardenne et al. (2001) adjusted to Olivier and Berdowki (2001), and EDGAR 32FT2000 - EDGAR Fast Track 2000 dataset
3. Database definitions of uncertainty see: http://www.mnp.nl/edgar/documentation/disclaimer/
4. BP Statistical Review of World Energy June 2008, http://www.bp.com/statisticalreview
5. C.J. Campbell, and J.H. Laherrère. (1995) *The world's oil supply -1930-2050* Petroconsultants report, Oct . 650p, CD-ROM data.
6. David A. Kirchgessner, Robert A. Lott, R. Michael Cowgill, Matthew R. Harrison, and Theresa M. Shires, (1997) *Estimates of Methane Emissions from the US natural gas industry*. U.S. Environmental Protection Agency Air Pollution Prevention and Control Division.
7. Francis W. Carter and David Turnock (2002) *Environmental Problems of East Central Europe*, Routledge p 94 "...substantial pipeline leaks meant the loss of 10-15 per cent of all oil and gas produced." and Private Communication Bob Foster.
8. Thomas C. Reed (2004) *At the Abyss: An Insider's History of the Cold War*. New York: Ballantine Books.

SKILL BEFORE SENSITIVITY: TOWARDS GALILEAN VALIDATION FOR CLIMATE SIMULATIONS

KYLE SWANSON
Department of Mathematical Sciences, University of Wisconsin-Milwaukee
Milwaukee WI, USA

Successful prediction of climate change requires correctly capturing how cloud and radiative fields respond to changes in the Earth's surface temperature. In general, these fields respond to changes in surface conditions on time scales similar to that of the radiative relaxation time scale of the troposphere, i.e., on the order of one month, but are strongly coupled to the underlying dynamical fields through both radiative heating and heating due to phase changes of water. The complexity of these interactions lies at the root of the uncertainty in the magnitude of expected climate change over the 21st century.

While cloud/radiative properties have a definitive stochastic element over short time periods and at small spatial scales, such processes will ultimately come into quasi-equilibrium with the underlying surface (Emanuel et al. 1996). Averaging over time-scales of a month or longer should allow these cloud and radiative processes to be captured in a now-cast setting. In particular, this should be the case for advanced atmospheric reanalysis products, where the dynamical fields are allowed to evolve in a fashion consistent with the underlying atmospheric dynamical system.

To this point in time, there has been little formal examination of the skill of reanalysis cloud/radiative fields. The reasons for this are many: these fields are much more spatially and temporally variable than winds, temperature, and pressure, so measurements made at individual points are less representative of the grid-column mean produced by the forecast and used as performance metrics within the meterological centers. Further, observations of clouds and broadband radiation are only now beginning to be used by operational assimilation systems, so that cloud- and radiation-related quantities are only loosely constrained in the resulting analyses. In addition, though the distribution of clouds, radiation, and precipitation may be of importance for specific applications, it has a minor impact on the short- to medium-term evolution of the flow on which forecast skill scores are based.

However, the latest generation of reanalysis radiative fields appears to possess skill at levels that are useful for more in-depth climate motivated investigation. Moreover, that skill appears to be increasing as models become more sophisticated. As an example, Figure 1 shows a comparison of the mean skill score

$$SS = 1 - \frac{\sum_N (F_{\mathrm{mod}} - F_{obs})^2}{\sum_N {F_{obs}}^2}$$

for modeled (observed) radiative anomalies F_{mod} (F_{obs}) on seasonal and longer time-scales.

Comparison of a first generation reanalysis product, namely the NCEP/NCAR Reanalysis against a current generation product (ECMWF Intermediate Reanalysis) shows a stepwise improvement in skill in capturing radiative fluxes on monthly time scales, where verification is against the CERES TERRA ERBE-like fields, over the period March 2000–February 2005.

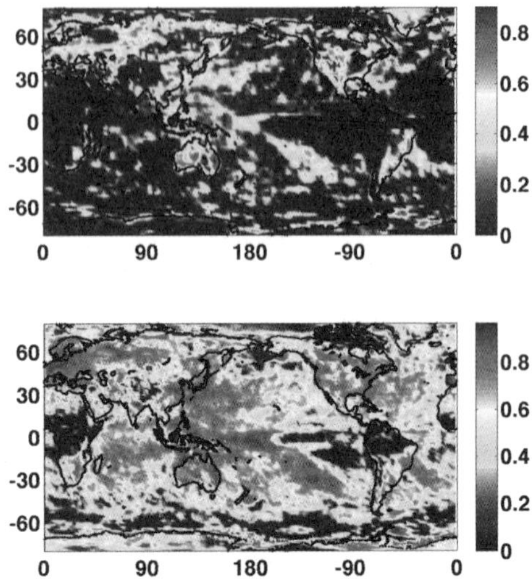

Fig. 1: Skill score for (top) NCEP reanalysis and (bottom) ECMWF Intermediate Reanalysis outgoing shortwave anomalies verified against the CERES/TERRA measurements.

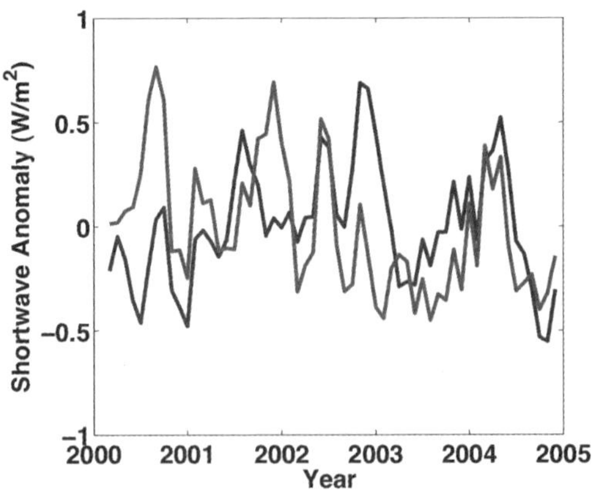

Fig. 2: Time series of global average shortwave fluxes for the ECWMF Intermediate Reanalysis (blue) and the CERES/TERRA observations (green).

The presence of skill leads to an inquiry about climate sensitivity, which involves spatial averaging of radiative fields. Figure 2 shows that the surprising outcome is that global averaged quantities appear to be more difficult to capture than grid-column or zonal average quantities. The ECMWF Intermediate Reanalysis shortwave anomalies shown in Figure 1 have significant local skill, but that skill disappears upon global averaging. This raises questions about the role of non-locality in determining global averaged radiative fluxes.

Both positive and negative messages are apparent from this analysis. Models are getting better at mimicking nature as they become more sophisticated. However, the question of how good models have to be before their skill becomes useful for the larger purpose of correctly capturing climate sensitivity remains very much open. Galilean validation of climate simulations along the lines proposed here must play an increasing role in evaluating models and in the continued development of climate theories.

REFERENCE

1. Emanuel, K.A., J.D. Neelin, and C.S. Bretherton (1996) "On large-scale circulations in convecting atmospheres." *Q. J. Royal Met. Soc.,* **120**:1111-1143.

SECTION 8

SESSION 8

COSMIC OBJECTS

COUNTERMEASURES TO AVOID ASTEROID AND COMET NUCLEUS IMPACTS ON EARTH: A SUMMARY OF PROGRESS

WALTER F. HUEBNER AND THE SWRI-LANL TEAM*
Southwest Research Institute
San Antonio, USA

ABSTRACT

We report on political realizations, technical and scientific progress, and international participation in countering Potentially Hazardous Objects that have the potential of colliding with Earth and inflicting enormous damage. We cover the areas of the most frequent, small (i.e., Tunguska-sized objects) as well as the most dangerous objects, the long-period comets that are usually ignored.

POLITICAL REALIZATIONS

Several countries, among them the USA, the European Union, Russia, and Japan, and the United Nations have been concerned about the potentials of asteroid and comet nucleus impacts on Earth. Such objects have impacted Earth since the formation of the planetary system and these impacts will continue far into the future. Although rare, impacts by Near-Earth Objects (NEOs, i.e., asteroids and comet nuclei that orbit the Sun and approach or cross Earth's orbit) can have catastrophic effects. For the first time in history of humanity one can consider countermeasures against such impacts. These countermeasures to protect Earth's population against such impacts are a long-time project that will evolve as technology advances. Asteroid and comet nucleus impacts are an international problem and countermeasures need international participation and international agreements.

Great strides have been and are still being made in detecting, tracking, cataloguing, and characterizing the physical properties of NEOs with new and powerful telescopes. However, development of countermeasures lags far behind. While it is reassuring from improved observations and analyses that known objects have little chance of colliding with Earth,[1] it will be of little comfort to identify an object far in advance before it collides with Earth if countermeasures are not available. Just as in medicine, early detection is only useful if a cure is at hand, and these cures must be developed and tested now. Therefore we concern ourselves here primarily with countermeasures and not as much with advances in detecting, tracking, cataloging, and characterizing NEOs.

Contributions from the USA

The NASA Authorization Act of 2005 directs NASA to transmit an initial report to the USA Congress that provides: (1) an analysis of possible alternatives that NASA

---

* D.C. Boice, S. Chocron, A. Ghosh, R. Goldstein, J. Mukherjee, W. Patrick, M. Tapley, and J.D. Walker from the Southwest Research Institute, San Antonio, TX and P. Bradley, J.A. Guzik, C. Plesko, R.P. Weaver, and K. Wohletz from the Los Alamos National Laboratory, Los Alamos, NM.

may employ to carry out a survey program of Near-Earth Objects (NEOs), (2) a recommended option and proposed budget to carry out the survey program; and (3) an analysis of possible alternatives that NASA could employ to divert an object on a likely collision course with Earth. The objectives of the NEO Survey Program are to detect, track, catalogue, and physically characterize NEOs equal to or larger than 140 meters in diameter with a perihelion distance of less than 1.3 AU, achieving 90 percent completion of the survey within 15 years after enactment of the NASA Authorization Act. The 140 m diameter is a new, much smaller size limit then was the case in previous surveys. This is referred to as the second phase of Spaceguard.

We participated in a conference that was held to discuss various options. NASA's response to the request is contained in the report to Congress dated March 2008 "Near-Earth Object Survey and Deflection Analysis of Alternatives." Among key findings of the survey program are:

- The goal of the Survey Program should be modified to detect, track, catalogue, and characterize, by the end of 2020, 90 percent of all Potentially Hazardous Objects (PHOs) greater than 140 meters whose orbits pass within 0.05 AU of the Earth's orbit (as opposed to surveying for all NEOs).
- The Agency could achieve the specified goal of surveying for 90 percent of the potentially hazardous NEOs by the end of 2020 by partnering with other government agencies on potential future optical ground-based observatories and building a dedicated NEO survey asset assuming the partners' potential ground assets come online by 2010 and 2014, and a dedicated asset by 2015.
- Together, the two observatories potentially to be developed by other government agencies could complete 83 percent of the survey by 2020 if observing time at these observatories is shared with NASA's NEO Survey Program.
- New space-based infrared systems, combined with shared ground-based assets, could reduce the overall time to reach the 90 percent goal by at least three years. Space systems have additional benefits as well as costs and risks compared to ground-based alternatives.
- Radar systems cannot contribute to the search for PHOs, but may be used to rapidly refine tracking and to determine object sizes for a few NEOs of potentially high interest. Other missions currently oversubscribe existing radar systems.
- Determining an NEO's mass and orbit is required to determine whether it represents a potential threat and to provide required information for most alternatives to mitigate such a threat. Beyond these parameters, characterization requirements and capabilities are tied directly to the mitigation strategy selected.

The study team assessed a series of approaches that could be used to divert a NEO potentially on a collision course with Earth. Nuclear explosives, as well as non-nuclear options, were assessed. Key findings for diverting a PHO are:

- Nuclear standoff explosions are assessed to be 10-100 times more effective than the non-nuclear alternatives analyzed in the study. Other techniques involving the surface or subsurface use of nuclear explosives may be more efficient, but they run an increased risk of fracturing the target NEO. They also carry higher development and operations risks.
- Non-nuclear kinetic impactors are the most mature approach and could be used in some deflection and mitigation scenarios, especially for NEOs that consist of a single small, solid body.
- "Slow push or pull" mitigation techniques (such as the gravity tractor) are the most expensive, have the lowest level of technical readiness, and their ability to both travel to and divert a threatening NEO would be limited unless mission durations of many years to decades are possible.
- 30-80 percent of potentially hazardous NEOs are in orbits that are beyond the capability of current or planned launch systems. Therefore, planetary gravity assist swing by trajectories or on-orbit assembly of modular propulsion systems may be needed to augment launch vehicle performance, if these objects need to be deflected.

Congress followed up with a second request asking the National Research Council (NRC) to conduct a study of NEO surveys and hazard mitigation strategies in June 2008. The committee consisted of a steering committee, a panel on survey and detection, and a panel on mitigation. The statement of task requires the committee to include an assessment of the costs of various alternatives, using independent cost estimating. Options that blend the use of different facilities (ground and space-based) or involve international cooperation were considered. We participated in the mitigation panel. Congressional staffers who were present stressed the importance of international participation. Two key tasks, the NEO survey and NEO hazard mitigation were to be addressed. The NEO survey task was concerned primarily about the roles of ground- and space-based facilities in addressing survey goals, specifically the potential contributions of the Large Synoptic Survey Telescope (LSST) and the Panoramic Survey Telescope and Rapid Response System (Pan STARRS). The NEO hazard mitigation task was concerned about the optimal approach to developing an orbit deflection capability, including options with a significant international component. The final version of the NRC report "Defending Planet Earth: Near-Earth Object Survey and Mitigation Strategies," is about to be published.

<u>European contributions.</u>

ESA has considered the Don Quixote mission. This concept is a test of a mitigation pre-cursor mission to further our understanding of the NEO impact hazard. It is a two-spacecraft mission consisting of Sancho, a science spacecraft, and Hidalgo, an impactor spacecraft. Sancho is intended to arrive at the target NEO a few months ahead of Hidalgo to perform reconnaissance and measure size, shape, mass distribution, bulk density, large-scale mineralogical composition, the internal structure using seismology, the ratio of the moments of inertia, thickness of the regolith, and other regolith properties. Hidalgo will impact the NEO. The impact will be observed by Sancho, which will remain with the NEO for several months to provide information on the regolith properties, interior structure, and mechanical properties, in addition to making a direct measurement of the dynamical response (change in velocity, Δv, of the asteroid) to the kinetic impactor. The results will provide crucial information for all further development of mitigation strategies, including numerical modeling. The time frame for launch is 2011 to 2018.

Russian contributions.

The Russian Academy of Sciences sponsored international conferences "100 Years since Tunguska Phenomenon: Past, Present, and Future," in June 2008 in Moscow and "Asteroid–Comet Hazard—2009," in September 2009 in St. Petersburg. Issues about discoveries of PHOs and about countermeasures were discussed openly and extensively. Papers presented at the Moscow conference were published in Solar System Research. The proceedings of the St. Petersburg conference are in press and will be published by the Russian academy of Sciences, St. Petersburg.

An important topic of the St. Petersburg conference was what can be done and what should be done to enlarge contributions of Russian research to help solve the problem of the asteroid-comet nucleus impact hazard. In connection with transition to the second stage of the Spaceguard survey they would like to develop a national program for Russian participation in NEO monitoring. An important element of their national program would be upgrading the radio telescope RT-80 in Ussurijsk in the Far East of Russia. This radio telescope could be used starting 2011 and could also become part of the very long baseline interferometer (VLBI). As compared with classic radar, the VLBI enables position determination of an object on the celestial sphere with very high precision; on the order of 0.01" for a single measurement and on the order of 0.001" for a series of observations. The radio telescope RT-80 together with three other radio telescopes is now included in the Russian VLBI network Quasar with the control center and the center of data processing in St. Petersburg. After completing the RT-80 upgrade, observations of small bodies will be available in the RADAR and VLBI regimes using telescopes of the Quasar network and possibly also other radio telescopes.[2]

Another important task would be organizing in Russia a regional center for NEO monitoring. Such a center would take over the task of coordination of Russian observatory efforts on NEO observations and would also be involved in orbit computations of NEOs.[2]

With the breakup of the Soviet Union, a few of the observatories that remained within Russia were in difficult circumstances. Their cooperation with observatories beyond present Russian boundaries drastically dwindled. The fate of the Crimean group of minor planet and comet observers, previously being a part of the Institute of Theoretical Astronomy in Leningrad, is typical in this respect. A few decades ago this group was among the world's three most productive observers of minor planet and comet positions. Currently the situation with respect to observations of minor planets and comets in Russia is getting gradually better. Some southern observatories, now in independent states, develop their activities in cooperation with astronomical institutions of Russia. Among them are the Simeiz Observatory, Terskol Observatory, and Maidanak Observatory. Turkey and Russia at Tübitak National Observatory, in Antalya, have established a joint observational program. Some new observatories and observational stations aimed at observations of small bodies are created on the territory of the Russian

Federation. Among them are the North Caucasus Observational station of Pulkovo Observatory, Kourovka Observatory of the Ural University, and the South station of Moscow University in the North Caucasus. The project of the 2-m telescope intended for discovery of NEOs, and that of the twin telescope with a lesser field of view intended for follow-up observations have been developed.

For a large country such as Russia, having a very large territory and a long coast line and, consequently, being exposed to a relatively high probability to suffer an asteroid or comet nucleus impact on its territory or in a nearby ocean, the availability of special services for keeping track of possible collisions with space bodies is a prime necessity. The Ministry of the Russian Federation displays an interest in such services for Affairs of Civil Defense, Emergencies, and Disaster Relief. We note that Russia has experienced three impacts from cosmic bodies in less than one hundred years [Tunguska on June 30, 1908; Sikhote-Alin on February 12, 1948; and Vitim River basin near Bodaybo (Mamsko-chuisky district, Irkutsk Oblast) on Septemnber 25, 2002], more frequently than any other country in modern times.

In the second stage of the Spaceguard program, after putting PanSTARRS (the Panoramic Survey Telescope & Rapid Response System), the LSST (Large Synoptic Survey Telescope), and the DCT (Discovery Channel Telescope) in service, the number of PHOs that need monitoring and the predictions of possible collisions with Earth and other planets will grow rapidly. Creation of a regional center for hazardous space body monitoring will be useful not only for Russian national purposes, but also for counteraction of asteroid-comet hazards world-wide. The Russian center could take over some part of the work necessary for all objects or the whole work for some objects. It is not unreasonable to perform the work in various centers in parallel with completely original and independent software and theory of motion of major planets and the Moon (EPM) that has been developed and is maintained at the Institute of Applied Astronomy of the Russian Academy of Sciences. A comparison of independently obtained results will be useful and necessary in critical cases.

In the area of countermeasures against asteroids and comet nucleus impacts on Earth, the `Citadel' program is being planned and developed at Planetary Defense Center at Khimki. Plans for this program have now been expanded to include two levels of response: A level for rapid response, Citadel-1, and a level for deliberate and longer-term reactions, Citadel (Kovtunenko and Zaitsev[3], Zaitsev[4], Zaitsev et al.[5]). They envision development of a reconnaissance spacecraft and an intercepting spacecraft. The reconnaissance spacecraft is intended for investigation of NEOs and the intercepting spacecraft for their deflection or destruction. For Citadel-1 they have investigated the feasibility to identify a PHO up to about hundred meters in size two to three days before its impact on Earth and deflecting or destroying it using nuclear explosives or in some cases kinetic impactors.

Plan of construction and operation of Citadel-1 call for incorporating international ground- and space-based surveillance services, two regional segments of the reconnaissance and interception services—the European-Asian, "East", and the American, "West", and two regional Centers of Planetary Defense (Kovtunenko and Zaitsev,[3] Zaitsev et al.[5]). Zaitsev will discuss details of this plan at this conference.

TECHNICAL AND SCIENTIFIC PROGRESS

As pointed out by Harris[1], we can expect that as a rule large asteroids, larger than hundred meters, can be discovered many years or decades before colliding with Earth. The discovery of asteroid Apophis is one such example. Smaller asteroids of the size from tens (Tunguska type) to a few hundred meters will be discovered primarily during their close approaches to Earth. The protection of our planet from such small objects can be ensured only through the creation of a rapid response system, e.g., Citadel-1. It must be ready for immediate action against a space threat just a few days after a sudden discovery.

Huebner et al.[7] presented a comprehensive plan for countermeasures against all potential collisions of asteroids and comet nuclei with Earth. Figure 1 summarizes the plan visually in general terms of countermeasure responses versus the size and frequency of the objects. The *ultimate goal* is to react to the shortest possible warning times.

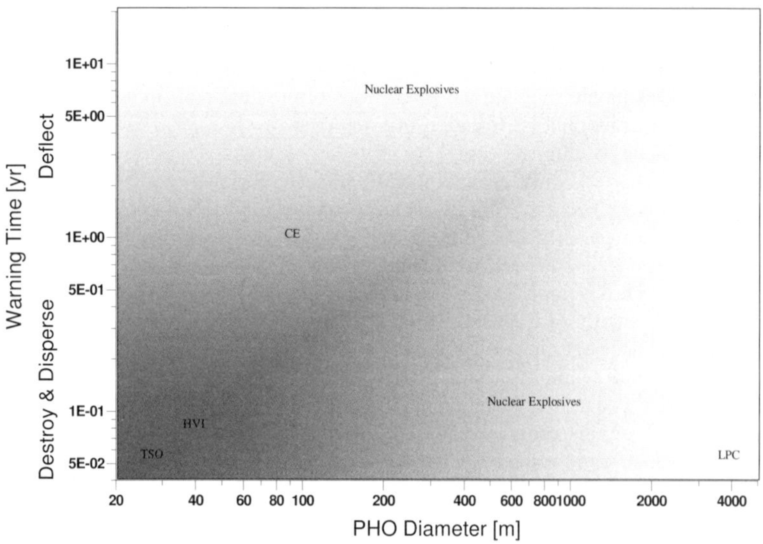

Fig. 1: The darker shading indicates high abundance and short warning times for small PHOs. The label TSO in the bottom left corner means Tunguska sized objects and LPC in the right bottom corner means long-period comet nuclei. The dividing line between destruction and deflection on the ordinate needs to be determined through further investigations of the engagement space. HVI means hyper-velocity impactor and CE chemical explosives.

Input from observations of a PHO, such as the time available for reaction, the size of the object, its spin state, whether it is a binary or a single object, its shape, the elements of its trajectory, its chemical composition, and its physical structure will determine the type of the countermeasures. The types of countermeasures we consider include hypervelocity impacts, conventional explosives for surface and subsurface applications, and nuclear explosives for standoff, surface, and subsurface applications. Walker and Plesko discuss these methods in more detail in this conference.

Decisions to select the most effective type of countermeasures for any situation depend on engagement space variables. Pre-investigated scenarios collected in a `playbook' will be useful as a guide, particularly if the warning time is short, as might be the case for some objects. The playbook that we plan to develop will define specific countermeasure scenarios for a series of variables. In Figure 2, we outline the beginning of a Decision-Making Chart for the playbook. The most important variable is the warning time, which will be considered first. Once a warning time has been determined, a trajectory will be calculated how best to reach the PHO. If the time needed to rendezvous with the object is longer than the warning time, then the mission will have to be an intercept (flyby type) mission. If time permits a rendezvous mission, then this option will be preferred. If the warning time is long enough to allow for a rendezvous mission, then the next test is to determine the spin state of the PHO. If the object is spinning fast or the spin state is complicated, for example the object spins about two axes, then placement of the countermeasures on or below the surface of the PHO may be difficult. In this case the decision should revert to an intercept mission. In either case of an intercept or a rendezvous mission, the size of the object must be considered. If the PHO is larger than a certain size, size (a) in case of an intercept mission or size (b) in case of a rendezvous mission, then nuclear explosives would be preferred. If the PHO is smaller than size (a) then a kinetic impactor may be appropriate for an intercept mission. If it is smaller than size (b), then chemical explosives strategically placed on or below the surface may be the preferable choice. Both size (a) and size (b) must be determined from more detailed analyses of the engagement space variables. More detailed discussions were presented by Huebner et al.[8]

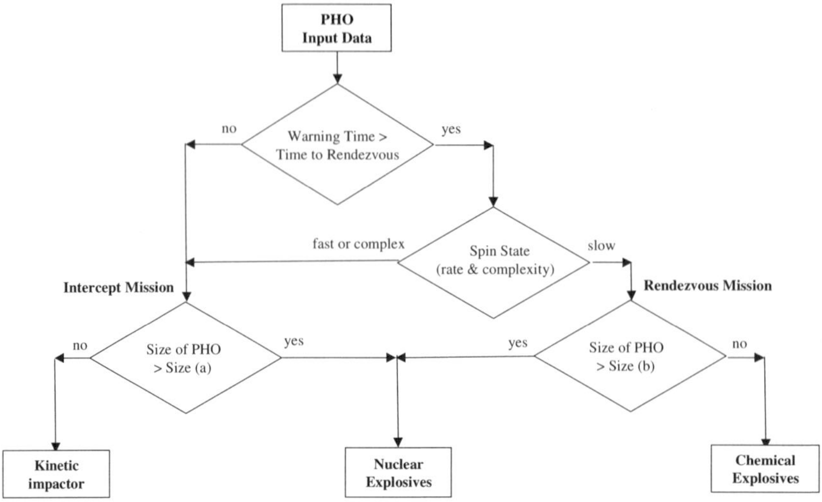

Fig. 2: Decision-making chart. More specific values for size (a) and size (b) will be determined from engagement space variables.

Not shown in Figure 2 are decisions appropriate for a PHO that is a binary system. A number of known impact craters on Earth and other planets and moons have a companion crater nearby. It appears that they were formed by the nearly simultaneous impacts of similar-sized objects, suggesting impacts of a binary set of asteroids. One possible approach to counter binary asteroids is to disrupt and disperse one component of the binary system at the right moment in their mutual orbit. With one component eliminated, the other component will be deflected in its orbit. However, a detailed analysis of such a scenario is needed.

Several reports about countermeasures have been issued or will be issued soon. While these reports and plans are well intended, we find it our duty to point out some weakness in them. Asteroids and comet nuclei have impacted Earth since the formation of the planetary system and these impacts will continue far into the future. Although rare, such impacts can have catastrophic effects. Thus, countermeasures to protect Earth's population against such impacts are a long-time project that will evolve as technology advances. While the NRC report mentioned earlier lists several research objectives for the immediate future, it does not project the needs and development of capabilities to achieve the ultimate goals for countermeasures needed in the more distant future: Very short warning times and some very energetic PHOs (e.g., long-period comet nuclei). While small asteroids with very short warning times may be quite frequent, large (typically tens of km in diameter), long-period comets are not. However, this does not mean that we should ignore them. Within several hundred years three long-period comets came very close to Earth: Comet C/1983 H1 (IRAS-Araki-Alcock), for example, has an orbital period of 963.22 years, was discovered 28 April 1983, and had a close encounter with the Earth only two weeks later (11 May 1983) at a distance of 0.0312 AU. Two other long-period comets came even closer to Earth: D/1880 L1 Lexell: 0.0151 AU and 55P/1366 U1 Tempel-Tuttle: 0.0229 AU. Aggressive methods and continuous alertness will be needed to defend against energetic objects with short warning times. Long-period comets could be real civilization killers. They should not be "swept under the rug."

As a general remark, `mitigation' means making less severe. But should the goal not be to completely avoid impacts by using effective countermeasures rather than making the effects less severe? Our responses should center on *countermeasures to avoid impacts* rather than on mitigating impacts.

Specifically, the NRC report mentions, "For events of sufficiently low energy, the methods of civil defense in the broadest sense are the most cost effective approach for saving human lives and minimizing property damage." While evacuation from seashores when an impact-generated tsunami can be expected is a very effective and important mitigation strategy and evacuation may also be possible if the expected impact is small, can be predicted with sufficient accuracy at a specific location and in sufficient time to evacuate. However, we all know about the inability to evacuate New Orleans a few years ago during hurricane Katrina. Evacuation may be possible if evacuation routes are open in many directions and transportation needs such as open roads, fuel, supplies, etc. are available. There are, however, locations where this is not the case. Southern Florida, Sicily, or southern Italy for example are such cases. The need to supply transportation (e.g., gasoline going south) while the population is fleeing north on a very limited number of highways from high-density population centers will be very problematic. While it may save lives, it will not protect the infrastructure, which will be exposed and abandoned for destruction. If the impacting object is large, evacuation is unthinkable and if it is small an aggressive countermeasure to deflect the object may be much more effective.

The NRC report states "Unless a very long warning time before impact is available, the practical application of slow push-pull techniques is limited to NEOs that are predicted to pass through a keyhole and to small NEOs near the limit for which civil defense alone might be adequate." However, small NEOs are difficult to detect, making long warning times very unlikely. *We need effective short warning time responses.*

The NRC report states further "Kinetic impactors are adequate to prevent impacts on Earth by moderately sized NEOs (many hundreds of meters to 1 kilometer) with decades of advance warning. The concept has been demonstrated in space, but the result is sensitive to the properties of the NEO and requires further study." We note that the algorithm used for targeting the kinetic impactor for NASA's Deep Impact Mission to comet Tempel 1, although not aimed at maximizing the transfer of momentum, had a circular error, at the 3-σ level of uncertainty, with a radius of about 200 m based on simulations. The actual delivery was within this limit according to M.A Hearn, PI of the mission. In order to deflect a small NEO (say only a few hundred meters in diameter or less), the accuracy of the targeting will need to be improved significantly. We note that the object that caused the Tunguska explosion is now thought to have been as small as 25 to 30 m in diameter.[9] Improving the targeting may not be an insurmountable obstacle but considerable effort will need to be applied. Targeting techniques and their performance have been described by Mastrodemos et al.[10] and by Kubitschek et al.[11] However, improving targeting techniques is not mentioned in the NRC report.

We agree with many comments made in the NRC report about nuclear explosives although the report refers to "hydrodynamic simulations presented here" that are neither there nor have they been published. Such calculations should be published because such information would be most useful for further work. The NRC report discusses primarily deflection. While destruction is considered, it is not recommended. Destruction with maximum dispersal of fragments has not been considered. This needs further investigation.

The NRC report states that predicted short-term (few years or less) impacts by a civilization- or species-destroying NEO are apocalyptic possibilities, they are extraordinarily unlikely to happen in our lifetime, thus traditional approaches to preparing for such a disaster become irrelevant. However, such an event may indeed shorten everyone's lifetime! Thus, 'throwing in the towel' instead of suggesting using every means to destroy the object is a fatalistic approach.

The Citadel-1 plan to deflect or destroy small, suddenly appearing objects is excellent. However, when applied to objects as large as 100 m in diameter two to three days before impact on Earth will need further investigation. Destroying a 100-m object close to the Earth may require extraordinary efforts for dispersal of fragments. This has not been discussed. However, it may be an excellent long-term goal.

No plans exist in any report to counter long-period comets. Such comets are usually large (tens of kilometers in diameter) and can come from any direction from the outer solar system. Because their orbital periods are very long (hundreds of thousands to millions of years) they cannot be predicted and since they come from the outer solar system, they have high speeds if they approach the Earth. Because their orbits are random, they can even come from behind the Sun. It should be our long-term goal to develop countermeasures against such long-period comets.

While deflection techniques will always be preferred, the primary option should always be for destruction of a PHO in emergency situations, such as extremely short warning times. In all cases one should investigate the threshold where any aggressive technique might lead to unacceptably large fragments remaining on a collision course with Earth. To avoid domestic and international political objections for the use of nuclear demolition munitions, using non-nuclear methods such as kinetic impactors or conventional chemical explosives should be preferred over nuclear explosives.

The first task is to define the engagement space for destruction vs. deflection based on size, composition, internal structure, shape (aspect ratio of length to width and thickness of object), and warning time of the PHO. For the purpose of discussion here, we have to amend the concept of warning time. Reaction time is usually defined as the time from discovery that a PHO may impact Earth to the impending impact. Here, warning time will mean true warning time: reaction time minus the time to implement a countermeasure, which includes the time to get ready to launch, the launch itself, getting there, and execution of the countermeasures. Conventional explosives that are politically more acceptable may be useful for countermeasures of small PHOs while nuclear demolition munitions may have to be used for larger objects. Long-period comets are typically objects in the 10 km size range. If such an object is discovered to have a short warning time before impacting the Earth, then nuclear demolition is almost certainly the only choice.

A small PHO with short warning time may have to be destroyed. Such a PHO must be fragmented in such a way that the fragments are small enough so that they will not have any significant detrimental effects on Earth. Since the PHO is already small to begin with, fragmentation may not be quite so big a problem. Part of such investigations will explore destruction using conventional explosives. Deflection of some PHOs may also be possible by using conventional explosives if the warning-time is sufficiently long. Deflection means imparting momentum through the center of mass of the PHO to cause a small change in its orbital velocity to delay it or speed it up so that it will not collide with Earth. Explosives form craters and release explosives gases and ejecta. The released materials may provide the PHO with a "rocket effect" sufficient to push it off a collision course.

Experiments need to be performed to augment our understanding of the momentum enhancement by both explosive loads and by hyper-velocity impact loads on representative PHO materials and structures. The experimental effort will begin by defining different PHO-simulating materials to represent a wide variety of asteroids and comet nuclei. These materials, aggregates, and porous structures will then be prepared as small surface samples to be tested with explosives and hyper-velocity impacts.

The first tests will be with small amounts of explosives in an evacuated chamber. The samples will be attached to a pendulum device to measure the loads that are transferred. The measurements will include both the load transferred by the expanding explosive product gasses and the excavated and ejected debris. These experimental results will be compared to computational results.

Next, experiments will be performed with hyper-velocity impacts. Southwest Research Institute (SwRI) has both a two-stage light gas gun and an explosive based hyper-velocity impact facility where the International Space Station shielding was characterized (impacts at over 11 km/s). These tests will examine the momentum transfer to the surfaces. Again, the experimental values include the momentum enhancement caused by the ejected debris. These experimental results will again be compared to the computational simulations to determine the accuracy of the simulations.

Asteroid geophysics and geology (seismologic characterization of asteroids) needs to be carried out to define their internal structure. PHOs have widely differing compositions, ranging approximately from (1) iron—nickel, (2) rocks, (3) carbonaceous materials, to (4) dust and ice. Knowledge of the composition and internal structure of a PHO and, therefore, its likely strength, is crucial in determining the best way to destroy or divert it without shattering it. To aggressively counter a collision of a PHO with Earth, one needs to know the structural integrity and its material strength. Ground-based and spacecraft observations and measurements have shed some light on internal structure and properties of asteroids. The physical properties of asteroid surfaces appear to be layers of regolith, consisting of small irregular particles with porosity higher than that of the whole object. However, judging the interior properties and structure of a PHO from surface appearances and visible and infrared broadband spectra may be misleading. Space weathering changes surface properties significantly. Bulk densities of over twenty asteroids and a few comet nuclei have been derived from spacecraft rendezvous and flyby missions, from mutually orbiting binary asteroids, and non-gravitational forces (e.g., outgassing) acting on comet nuclei. In all cases the densities so obtained were lower than equivalent materials on Earth would suggest. Typically they varied from 200 kg/m^3 to

3000 kg/m^3, indicating a huge diversity in the population. It is not known whether the low densities are caused by micro-porosity, as a result of the formation in low gravity environments, or whether they are caused by macro-porosity, implying fractures, loose assemblies (e.g., rubble piles) caused by collisional evolution, compositional inhomogeneities, or several of these effects combined. We can expect PHOs to vary from small monolithic objects to bigger fractured objects or gravitational aggregates as well as porous but nevertheless robust objects formed of cohesive materials. Fast spinning objects are probably not loose aggregates because the centripetal force would overcome their gravity and pull them apart. Spin periods obtained from light curves suggest that small objects are more likely monolithic because they spin fast while larger objects tend to be more likely fractured or rubble piles because they spin slowly, below the limit where centripetal acceleration overcomes gravitational attract of a mass on the surface.

A precise knowledge of the interior structure and of the regolith properties—of major importance to assess the consequences of an impact and develop efficient countermeasure strategies—can only be provided by *in situ* rendezvous missions, aimed at significantly different objects, representative of different categories of physical properties.

Artificially activated seismic measurements, using an array of miniaturized (MEMS) seismometers on the surface of a PHO, are one of the best methods to deduce this information. Seismic exploration is difficult because asteroids have negligible gravity, mostly unexplored surfaces, no atmosphere, and can experience large diurnal and seasonal temperature changes. Thus, techniques for the delivery of miniature seismometer packages and attaching them on the surface of a PHO to provide good seismic coupling between the asteroid and the sensor package must be developed and tested in the laboratory. SwRI can develop a miniaturized system for delivery and operation of seismometers on a Near-Earth Asteroid (NEA) in order to determine its internal properties and structure. They will design and test techniques for delivery of a sensor package and techniques for assuring good seismic contact with the body using existing facilities at SwRI. Results from the investigations will provide the information necessary for understanding how to anchor instruments to the surfaces of asteroids and make acoustic contact for effective seismic exploration. This in turn will allow determination of properties such as the velocity of seismic waves, material strengths, shock dissipation, elastic properties, etc., which will provide the information necessary to determine the conditions under which hyper-velocity impacts and explosives interact with a PHO.

In addition small science packages may be added to each missile or interceptor. The science package would contain instrumentation, for example cameras, so that we may learn more about these objects before they are nudged or blown up.

A database for geophysical and geological properties of asteroids and comet nuclei was initiated at a conference held here at Erice in 2001: Material properties in the database are divided into two parts: (1) Observed data from ground observations and from space missions to asteroids and comet nuclei and (2) simulated data approximating material properties of asteroids and comet nuclei. The observed and simulated data are further divided into static data for slow push approaches and dynamic data for impulsive approaches. Dynamic data are needed primarily for aggressive countermeasures, while the static data are useful for most missions (some deflection methods, such as the gravity tractor, require almost no geophysical or geological data). The data categories were determined at the conference organized by us for mitigation procedures. Some data, together with other relevant information, have already been collected and entered in the database, which can be found at http://neodata.space.swri.edu.

While several suitable deflection technologies have been proposed, none of them can be effective if we are taken by surprise. We must be prepared to launch on short notice. The "Citadel" program, as proposed in the Russian Lavochkin report of 2005, plans to launch PHO-intercepting spacecraft or missiles from space. We plan to investigate launches from the ground because the launch vehicles are easier to maintain. Ground-based launches also afford more flexibility (e.g., size and type of explosives and the choice of hyper-velocity impactors), and ground-based launches are politically less threatening and therefore more acceptable.

A small companion spacecraft may be launched to gather information about the effectiveness of the countermeasures. This may also provide valuable geophysical and geological data about a PHO. Instrumentation may include spectroscopes for chemical composition, and cameras to record the number of fragments, expansion velocities, etc, or the effects of nudging the PHO.

Work with conventional explosives and hyper-velocity impacts overlaps and requires coordination. We will also coordinate with European, Russian, and Japanese efforts. This will be done through the auspices of the Permanent Monitoring Panel for Cosmic Objects. The core of this panel is already set up and meets annually in August.

To develop an effective system, flexibility in the choice of countermeasures and readiness to launch will be key issues. Launch capabilities from the ground have many advantages over launches from space. A ground-based system affords continuous maintenance and readiness to launch. It also provides for last minute options for the type of countermeasures, such as hyper-velocity impactors, conventional explosives, nuclear explosives, size and number of explosives, etc. We also note that several countries, including the USA and Russia, have experience in maintaining rockets on alert.

Finally, the possibility exists that we may want to make use of existing missile silos that are scheduled to be closed. If it turns out that missile silos can be used as a means of readiness to launch against a short-warning time PHO, then this would be a powerful tool. However, modifications to the launch vehicles (possibly the Minotaur IV) are most likely necessary, which requires the involvement of the aerospace industry and requires further coordination and investigation.

International Support Teams have been established. Four task groups of the International Astronomical Union (IAU)—Commission for the Physical Study of Comets and Minor Planets (asteroids) are already at work in supporting areas for albedo—size determinations of asteroids and comet nuclei:

1. Task Group on Asteroid Magnitudes: (Team Leaders USA & Argentina).
2. Task Group for Albedo—Polarimetric Albedo Calibration: (Team Leaders Italy & Argentina).
3. Task Group on Comet Magnitudes: (Team Leaders Uruguay & Japan).
4. Task Group on Geophysical and Geological Properties of Asteroids and Comet Nuclei: (Team Leaders: Finland, Argentina & Japan).

INTERNATIONAL PARTICIPATION

In the area of international participation, we note many common goals between the USA, European, and Russian approaches to the asteroid—comet nucleus hazard: (1) The need for international agreements about countermeasures, (2) international coordination of the warning system and the system for countermeasures, and (3) the use of multiple spacecraft, one for reconnaissance, i.e., physical characterization of the PHO (if time permits), at least one for the application of the appropriate countermeasures, and one (perhaps the same as the reconnaissance spacecraft) for observation and determination of the effects of the countermeasures.

The building of a planetary defense system against asteroid and comet nucleus impact hazards will require adequate international legal instruments that may take the form of an international treaty. This treaty should contain principles governing the establishment and use financial fund with the aim of supporting the planning, development, management, and modernization of the system.

The United Nations might be the oversight organization to coordinate planetary defense. This may involve an international planetary defense center with regional offices. International legal instruments will have to be created to provide for the Earth's defense against asteroid and comet nucleus impact hazards. Political agreement between countries concerning countermeasures against PHOs is essential. It should define the procedures for informing the population about the collision hazard with a specific body and the extent of this hazard, who has responsibility for implementation of counteractions, what the responsibility of a country or a group of countries is to initiate the countermeasures against a hazardous object, and how to avoid misuse, failure, or a tragic mistake. These questions become especially important during phase 2 of the Spaceguard survey when the probability of discovery of a dangerous body on a collision course with the Earth increases many-fold. The Committee on NEO of the Association of Space Explorers and its International Panel on Asteroid Threat Mitigation were guided by similar considerations when preparing the report Asteroid Threats: A Call for Global Response, which was submitted to the UN Committee on Peaceful Uses of Outer Space at the end of last year.

There is good reason to hope that, given the legal status of the asteroid comet nucleus hazard and the consideration of the problem on an international level within the UN, the necessity to provide some specific obligations for participating in collective actions will result in raising attention to the PHO problem on the side of governments and secure financial support for development of effective countermeasures.

Zaitsev and Klapovsky[12] provide further details about content of an international agreement to support participation, planning, management, and use of countermeasures including the use of nuclear explosives in space. One of the most important problems will be testing all types of countermeasures in space.

CONCLUSIONS

The major objectives for countermeasures against asteroid and comet nucleus impacts with Earth can be met. Technical difficulties, including countermeasures against short warning-time objects in the 100 m size range, need to be investigated further. Simulation calculations and laboratory tests need to be conducted. Long-period comet threats need to be solved in the long run, but an important start has been suggested. International participation is building and progressing. International agreements need to be concluded.

REFERENCES

1. Harris, A.W. (2010) "Estimating the NEO population and impact risk: Past, present, and future." In *Proceedings of International Conference on Asteroid-Comet Hazard—2009*. In press.
2. Finkelstein, A.M., Medvedev, Yu. D., Shor, V.A. (2010) "Our view of objects of ACH-2009." In *Proceedings of International Conference on Asteroid-Comet Hazard—2009*. In press.
3. Kovtunenko, V.M., Zaitsev, A. V. (1995) "Protecting Earth from asteroid hazards is a real task for the world space states." *Space Bulletin* **2**:25-28.
4. Zaitsev, A.V. (2000) *Planetary Defense System `Citadel', the Conceptual Project*. Moscow, 80 pp.
5. Zaitsev, A., Pitchkhadze, K., and Tchesnokov, A. (2008) Possible Approaches to Implementation of `Citadel-1' International Planetary Defense System Project. Report on Forty-fourth Session of the Scientific and Technical Subcommittee of the Committee on the Peaceful Uses of Outer Space (12-23 February 2008).
6. Zaitsev, A., Koroteev, A., Liaschuk, B., and Popov, S. (2010) "The level of rapid response reaction of the planetary defense system." In *Proceedings of International Conference on Asteroid-Comet Hazard—2009*. In press.
7. Huebner, W.F., Johnson, L.N., Boice, D.C., Bradley, P., Chocron, S., Ghosh, A., Giguere. P.T., Goldstein, R., Guzik, J.A., Keady, J.J., Mukherjee, J., Patrick, W., Plesko, C., Walker, J.D., and Wohletz, K.A (2009) "Comprehensive program for countermeasures against Potentially Hazardous Objects (PHOs)." *Solar Syst. Res.* **43**:348-356.
8. Huebner, W.F., Boice, D.C., Bradley, P., Chocron, S., Clement, R., Ghosh, A., Giguere, P.T., Goldstein, R., Guzik, J.A., Johnson, L.N., Keady, J.J., Mukherjee, J., Patrick, W., Plesko, C., Tapley, M., Walker, J.D., Weaver, R.P., and Wohletz, K. (2010) "The engagement space for countermeasures against Potentially Hazardous Objects (PHOs)." In *Proceedings of International Conference on Asteroid-Comet Hazard—2009*. In press.
9. Boslough, M. (2008) "New Simulations suggest that Tunguska yield has been overestimated." In Conference: *100 Years since Tunguska Phenomenon: Past, Present and Future*. Also: Tunguska.sai.msu.ru/files/pr_boslough.zip.
10. Mastrodemos, N., Kubitschek, D.G., and Synnott, S.P. (2005) "Autonomous navigation for the Deep Impact Mission encounter with Comet Tempel 1." *Space Sci. Rev.* **118**:95.
11. Kubitschek, D.G. et al. (2008) "The challenges of Deep Impact autonomous navigation." *J. Field Robotics* **24**:339.
12. Zaitsev, A., and Klapovsky, (2010) "A. On the approach to form international legal foundations to ensure planetary defense." In *Proceedings of International Conference on Asteroid-Comet Hazard—2009*. In press.

MOMENTUM ENHANCEMENT FROM KINETIC IMPACTORS AND CONVENTIONAL EXPLOSIVES

JAMES D. WALKER AND SIDNEY CHOCRON
Southwest Research Institute, Engineering Dynamics Department
San Antonio, USA

ABSTRACT

Given the large quantity of objects in space near Earth, there is the possibility that one of these will impact the Earth's surface. The most mature technologies for deflecting such an object so that it does not strike the Earth are the conventional means of hypervelocity impactors or chemical explosives. These methods turn out to be very effective because of momentum enhancement, which is the increase in momentum transferred to a target due to ejecta material being thrown backwards during crater formation, i.e., the objects themselves supplying some of the material that leads to the change in momentum. Quantitative estimates of momentum enhancement are of interest in determining the effectiveness of hypervelocity impactors and chemical explosives in deflecting potentially hazardous cosmic objects. This paper explores momentum enhancement both experimentally and computationally, showing that enhancements on the order of 10 are a possibility.

INTRODUCTION

As this article is written at the end of June 2010, there are 7133 known near-Earth objects (NEOs), which are objects with closest approach to the Sun (perihelion) less than 1.3 AU units (1 AU is the distance of the Earth to the Sun). In the first six months of 2010, 401 NEOs were discovered, which is on average 2.2 objects discovered per night. Also at the end of June 2010 there are 1,132 known potentially hazardous objects (PHOs), which are objects that pass within 0.05 AU of Earth (7.5 million kilometers). In the first six months of 2010, 47 PHOs were discovered, which is roughly a rate of one every four nights. Of these 47, three are estimated to have an extent larger than 1 kilometer. In the first six moths of 2010, one near-Earth comet was discovered, bringing the near-Earth comet total to 85. There are many objects in our vicinity in the solar system. The sky-search programs are ongoing in an effort to better characterize our near-Earth space and to find potentially hazardous objects.

If an object appears to be on a collision course with Earth, we will want to do something about it. In 2005 NASA was tasked by the U.S. Congress to provide a report discussing NEOs in the context of their threat to the Earth. The report addressed various alternatives for the sky survey and an analysis of possible alternatives for diverting an object on a likely collision course. The required report was delivered in March 2007[1] (also see the detailed study containing the pre-decisional material[2]). The mitigation options focused on deflection of the asteroid or comet and fell into two general categories, either impulsive loads or slow-push loads. The former have high levels of technological readiness (we already know how to do it) while the latter are more speculative. In this paper we will discuss two of the techniques which are considered impulsive: hypervelocity impactors and conventional explosives. Both techniques can fall into the slow-push category if the application is lots of smaller impactors or lots of smaller explosive blasts employed over an extended time, rather than all at once. The technology for a hypervelocity impactor has been exhibited in two different scenarios: the direct impact of the comet nucleus of Temple 1 by the Deep Impact spacecraft on July 4, 2005 at 10.2 km/s[3] and by ballistic missile defense work which regularly performs impacts in the 7 km/s velocity regime.

A deflection mission is portrayed in Figure 1. Either a spacecraft would rendezvous, flyby, or directly strike an asteroid or comet nucleus. The momentum change to the asteroid or comet nucleus occurs either from impacts or explosive charges. This interaction could occur over an extended time (many impacts/blasts) or all at once (a single impulse). If possible, it would be ideal to place a transponder on the asteroid or comet nucleus surface for accurate orbit determination, to evaluate the effectiveness of the deflection operation and determine whether additional mitigation steps are required.

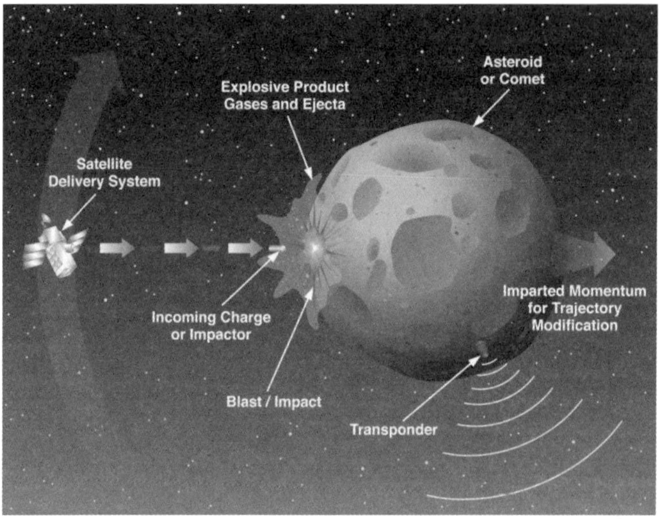

Fig. 1: A mission to deflect an asteroid or comet nucleus.

MOMENTUM ENHANCEMENT

To change the path of an asteroid or comet, momentum must be imparted to the object. Due to the expense and difficulty of delivering mass to the cosmic body from Earth, in an ideal scenario the asteroid or comet will assist in the momentum change by providing some of the reaction mass. When the object's mass helps provide for some of the momentum change, the effect is referred to as *momentum enhancement*.

To be specific, when an object is impacted at hypervelocities (speeds above 2 km/s), more momentum is imparted to the object than the original momentum of the impactor. This additional momentum is due to material from the impacted object being liberated and thrown backwards during crater formation. The amount of ejecta from a crater and its velocity depend on the impacted object's material and structure as well as the details of the impactor. In any event, however, this ejecta material provides an enhanced momentum transfer, above and beyond the initial momentum of the impactor (which is all transferred to the asteroid or comet nucleus). In a similar way, when explosives are detonated on the surface, in addition to momentum imparted by the explosive product gases, momentum is imparted to the body by the crater ejecta, and thus the amount of momentum transferred to the body depends on the object's material properties and structure.

Though this article does not address nuclear munitions, the same principle applies. In this case, nearly all the momentum change is due to mass liberated from the object itself. The mass is liberated either through an underground explosion producing a crater and throwing debris, or a standoff explosion producing radiation that subsequently vaporizes material beneath the surface, throwing off the surface layers back towards the explosive.

The topic of momentum enhancement shows why the impact and explosive scenarios for deflecting an object appear to be so beneficial: they use the objects own mass to assist in the deflection, rather than needing to transport all the mass to the object. The slow push alternatives that use transported fuels are unable to provide enhanced momentum transfer, as all the transfer comes from the thruster fuel, whether chemical or electric.

In general terms, it turns out that for the conventional momentum transfer techniques (impactors and chemical explosives), and it is likely true for nuclear explosives as well, that the momentum enhancement depends most strongly on the mechanical strength of the object, both in compression and tension. The momentum enhancement is also strongly dependent on the object's porosity—fully consolidated materials produce higher momentum transfers, while porous materials produce less enhancement. As a matter of notation, β will be used to denote the momentum transferred to the impacted body divided by the initial momentum of the impactor. Thus, there is momentum enhancement when $\beta > 1$. Given a lack of data, the NASA report to Congress[1] used bounding values of 1 and 10 for representative event evaluations.

EXPERIMENTAL DATA ON MOMENTUM ENHANCEMENT

There are very limited data on momentum enhancement. Though some work has been done with dust and extremely small impactors at hypervelocities, our interest is in impactors in at least the gram regime. Such experiments require relatively thick targets and a ballistic pendulum. Since most engineering work in hypervelocity impact has been shielding studies, where spaced plates provide better protection than thick targets on a weight basis, there are limited tests on thick targets where momentum enhancement data were collected.

Work was performed during the U. S. space program in the 1960s by Denardo and Nysmith.[4,5] In this work, the target was aluminum, which is not similar to our celestial bodies, but the results still have interesting implications. In impacts of aluminum spheres striking aluminum plates[5], there was a surprising scale effect, as is shown in a summary of their results striking 2024-T4 aluminum in Figure 2. It will be noted that as the radius of the spheres increases, so does the momentum enhancement, and that in all cases $\beta < 2$. The diameters of the spheres used were 0.15875, 0.3175, 0.635, and 1.27 cm (each scale differed by a factor of 2), with corresponding masses 0.006, 0.0468, 0.373, and 3 grams. It was possible to launch the smaller spheres to higher velocities. One is struck (as the original researchers were) by the apparent scale factor in these results, which they report went with approximately $D^{1/6}$, where D is the projectile diameter. This scale factor is of interest, in that if we have laboratory scale impactors of 1 cm and application scale impactors of 1 meter, then the full scale application may have $100^{1/6} = 2.15$ times the momentum enhancement that would be predicted from laboratory scale tests. As seen below for momentum enhancements of 7 to 8, this would result in β around 15, a substantial amount of enhancement.

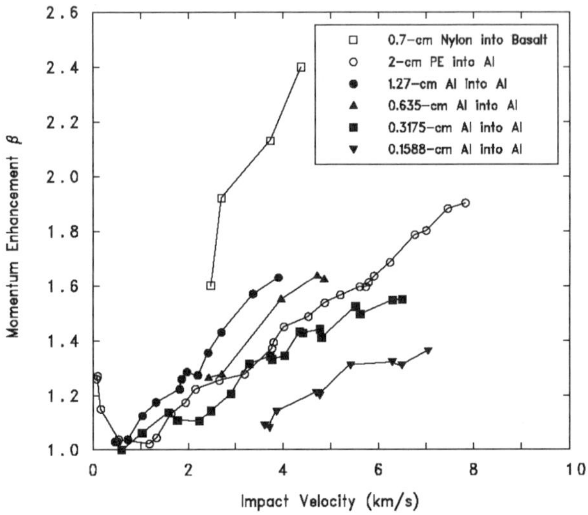

Fig. 2: Experimental results for momentum enhancement β for nylon spheres striking basalt, polyethylene cylinders striking 2024-T4 aluminum, and aluminum spheres striking 2024-T4 aluminum targets.

Large-scale numerical simulations with CTH (more description of the technique is in a later section) were performed with a 1-cm diameter spherical impactor with mass 1.46 grams. CTH was being used to explore the role of tensile fracture stress, and for fracture stresses of 10, 100 and 427 MPa the computations were within the spread of experimental results covered by the various scales in the experiments. However, the only item that could introduce a scale in the computations is the strain rate effect for the flow stress, and this has been shown to have a minimal effect in impact computations. Thus, we do not know the origin of the scale effect in the tests, and we do not expect to be able to replicate it in the computations. However, given these limitations, the computations with various tensile fracture stresses are on par with these experimental results.

The effect of polyethylene cylinders with a rounded nose striking the 2024-T4 aluminum as reported by Denardo[4] is shown in the figure. The cylinders are 2.032 cm in diameter and have a 1.524-cm radius spherical cap (thus, the side wall of the cylinder has height 1.148 cm). These cylinders massed from 4.1 to 4.2 g. Thus, they are both heavier and physically larger than the largest aluminum sphere used in impact. However, the momentum transfer from the polyethylene is seen to be less than that transferred by the aluminum. Yanagisawa and Hasegawa experiments with nylon spheres showed $\beta = 1.2$–1.5 impacting soft aluminum at 4 km/s[6], in agreement with Denardo.

Yanagisawa and Hasegawa performed experiments of 0.213-g nylon spheres ($D = 0.7$ cm) into basalt yielding larger βs: $\beta = 1.6$, 1.92, 2.13 and 2.40 (on average ±0.04) for impacts of 2.46, 2.68, 3.71 and 4.36 km/s, respectively (two with slight obliquity). These results are also shown on the Figure 2. The authors state the basalt has similar density and sound speed to aluminum, but the tensile properties are not stated. These data finally show experimental $\beta > 2$, for a material with presumably less tensile strength. A linear extrapolation of these points to 10 km/s gives $\beta = 4.8$.

MASS AND VELOCITY OF EJECTA

Holsapple and Housen[7] present experimental data and computational results on the mass and velocity distribution of crater ejecta material due to impacts. The impacts are into both hard targets and porous targets. The form of the results is presented in mass of ejecta traveling faster than a given velocity, and the data has been normalized by the impactor mass m and speed V. Thus,

$$\frac{M(>v/V)}{m} = f\left(\frac{v}{V}\right) = a\left(\frac{v}{V}\right)^b, \tag{1}$$

where $a > 0$ (mass is positive) and the functional form is suggested by a log-log plot of the results. We expect $b < 0$ since the ejecta mass with speed exceeding the stated velocity is a decreasing function of velocity. This form cannot be valid all the way down to zero ejecta velocity since for $b < 0$ it would predict infinite mass in the ejecta, thus it can only hold over a given velocity range (at zero velocity we have all the ejecta, hence the total mass of the material ejected from the crater $m_c = M(>0)$). In examining the plot in Holsapple[7] as well as his analysis, the following regions with their values are

$$\begin{aligned}
&\mathit{Nonporous}: && \frac{M(>v/V)}{m} = 0.06\left(\frac{v}{V}\right)^{-1.65}, && 3\times10^{-3} < v/V < 1.5,\\
&\mathit{Nonporous}: && \frac{M(>v/V)}{m} = 0.15\left(\frac{v}{V}\right)^{-1.4}, && 2\times10^{-4} < v/V < 3\times10^{-3},\\
&\mathit{Porous}: && \frac{M(>v/V)}{m} = 0.005\left(\frac{v}{V}\right)^{-1.4}, && 3\times10^{-4} < v/V < 1.
\end{aligned} \tag{2}$$

It will be noted that the consolidated material curve (nonporous, such as ice or rock) is not continuous, but then there is a fair amount of scatter in the presented results. The porous material was primarily sand and powders. Of particular interest in these expressions are the exponents *b*. Not only is $b < 0$, but $b < -1$. This velocity distribution has implications for momentum enhancement. The amount of mass in a given velocity range v to $v + dv$ is given by

$$\frac{dM}{m} = -ab\left(\frac{v}{V}\right)^{b-1}\frac{dv}{V}. \tag{3}$$

The momentum *P* due to ejecta velocities ranging from v_1 to v_2 is (assuming all the ejecta is moving in the same direction)

$$\frac{P}{mV} = \frac{1}{mV}\int_{v_1}^{v_2} v\,dM = -ab\int_{v_1/V}^{v_2/V}\left(\frac{v}{V}\right)^{b}\frac{dv}{V} = \frac{ab}{b+1}\left\{\left(\frac{v_1}{V}\right)^{b+1} - \left(\frac{v_2}{V}\right)^{b+1}\right\}. \tag{4}$$

Since $b+1<0$, the momentum is dominated by the lower velocity limit since the term is 1/(something small). Hence, the momentum enhancement is due to the lower "tail" of the crater debris – the high velocity components are not very massive, and the majority of the ejecta are at lower velocities. For example, use of this expression to evaluate P/mV for the ejecta velocity range of $0.1 < v/V < 1.5$ yields $P/mV = 0.56$, evaluating for the ejecta velocity range of $0.01 < v/V < 1.5$ yields $P/mV = 2.92$, and evaluating for the ejecta velocity range of $0.001 < v/V < 1.5$ yields $P/mV = 9.5$. If the integral is performed over the full range of definition $2\times10^{-4} < v/V < 1.5$, then $P/mV = 17$. Since it was assumed the ejecta all moved in the same direction, this value could be viewed as an upper bound to the momentum enhancement (plus 1 for the initial momentum of the impactor), based on these results, supposing the slowest ejecta had velocity $v/V = 2\times10^{-4}$.

The lower ends of these mass-velocity distributions involve large craters. For the consolidated material, $v/V = 2\times10^{-4}$ corresponds to $m_c/m = 2\times10^{4}$, and for the porous material, $v/V = 3\times10^{-4}$ corresponds to $m_c/m = 430$. Thus, a large amount of ejecta material is being created in these impacts. In the impact of Deep Impact into the nucleus of comet Tempel 1, it appears that large amounts of ejecta material were liberated.[3]

THE LOWER BOUND ON EJECTA VELOCITY

The previous section clearly shows that the lowest ejecta velocity drives the momentum enhancement. To obtain some idea of what this velocity might be, let us suppose we had an elastic bar of material of length *L* that had an initial linear velocity distribution from at rest at one end to a velocity v at the other end. Roughly, the bar stretches with uniform increasing (in time) strain until it comes to rest. If the corresponding uniform uniaxial stress state when it comes to rest is the fracture stress σ_f, then assuming the bar just fractures as it comes to rest leads to the following energy balance between initial kinetic energy (left) and final strain energy (right),

$$\frac{A}{2}\int_0^L \rho\left(\frac{vx}{L}\right)^2 dx = \frac{1}{6}AL\rho v^2 = \frac{A}{2}\int_0^L \frac{\sigma_f^2}{E}dx = \frac{1}{2}AL\frac{\sigma_f^2}{E}. \tag{5}$$

Thus, the lowest velocity that will break the bar is given by

$$v = \frac{\sqrt{3}\sigma_f}{\rho c}, \tag{6}$$

where c is a representative sound speed. For rock on an asteroid, 433 Eros' density is 2.7 g/cm^3 a representative sound speed might be 3 km/s, and a tensile strength of rock might be on the order of 10 MPa. When these numbers are used, a lower bound of 2 m/s ensues, which for a 10 km/s impact would yield $v/V = 2\times10^{-4}$, which is the lower end of the ejecta mass distribution results. For ice, with a density of 0.91 g/cm^3, a sound speed of 2.9 km/s, and a tensile strength of 1 MPa, the lower bound velocity of 0.7 m/s results, or for a 10 km/s impact $v/V = 7\times10^{-5}$, which is beyond the velocity range reported above. If we extrapolate with the equation beyond the results, we arrive at a value of $P/mV = 25$.

For soils the velocity is even less, as the lunar regolith is estimated to have strength 1 kPa. However, these materials' ejecta behavior falls on the porous curve developed for sands. For example, for the lunar regolith, representative values are a density of 1.5 g/cm^3, a sound speed of 100 m/s, and a strength of 1 kPa, which gives a lower velocity cutoff of 1 cm/s. If we extrapolate using Eq. (2) to this lower value of $v/V = 1\times10^{-6}$ for a 10 km/s, the resulting P/mV = 4.4. The value is less than the consolidated material due to the sand-like nature of the regolith (hence the reason we use the curve for porosity, which is based on sand and glass beads).

These equations also provide a potential way to investigate the role of scale. Suppose the failure stress depended on scale—for example, if there were random flaws distributed in a body, then the larger the body the more likely there would be available flaws. These flaws would lower the tensile failure stress. Thus, the cutoff velocity decreases as scale goes up. Inserting that fact into Eq. (4) then gives an increase in P/mV as scale goes up due to decreasing lower cutoff velocity v. This argument, however, seems weak in explaining the aluminum data, since aluminum is typically quite homogenous.

Of course, this distribution of kinetic energy in the material is the most efficient distribution there can be for breaking the bar, and we do not expect such a perfect distribution of energy to occur in an impact. Thus, again the estimates of momentum enhancement would be upper bounds.

On asteroids and comet nuclei, there is another lower velocity cutoff in the escape velocity—if the material cannot escape the body, it cannot lead to momentum change. The escape velocity is given by

$$v_e = \sqrt{\frac{2GM}{(D/2)}} = D\sqrt{\frac{2\pi}{3}G\rho}, \tag{7}$$

where G is the gravitational constant, M is the mass and D is the diameter of a spherical asteroid or comet nucleus. When the density 2.7 g/cm^3 is placed in this expression, a 100 m diameter asteroid has v_e = 6 cm/s and a 1 km diameter asteroid has v_e = 0.6 m/s. These velocities are in the vicinity of the strength based cutoff velocities.

MASS EJECTED FROM THE CRATER

Given that the majority of the momentum comes from the low velocity particles, and given that that high velocity particles have even higher relative energy, it is possible to find a bound on the momentum transfer assuming a uniform low velocity for the crater ejecta. In this case, with m being the impactor mass and V its velocity, M being the asteroid or comet mass with subsequent velocity v_a, and the crater ejecta mass being m_c with velocity v_c (assumed positive), momentum and energy conservation yield

$$\begin{aligned} mV &= -(m+m_c)v_c + (M-m_c)v_a, \\ (1/2)mV^2 &= (1/2)(m+m_c)v_c^2 + (1/2)(M-m_c)v_a^2. \end{aligned} \tag{8}$$

Manipulation of these equations and use of $\beta = (M-m_c)v_a/mV$ yields

$$(\beta-1)^2 + \left(\frac{m+m_c}{M-m_c}\right)\beta^2 = 1 + \frac{m_c}{m} \quad \Rightarrow \quad \beta = \frac{M-m_c}{M+m}\left\{1+\sqrt{1+\frac{m_c}{m}\left(\frac{M+m}{M-m_c}\right)}\right\}. \tag{9}$$

For a likely impact scenario, an asteroid or comet nucleus may be on the order of 50 meters across, and the impactor up to 2 meters across. Assuming a similar density for the impactor and asteroid or comet nucleus (a good choice, by the way), the ratio of M/m is over 15,000, and a good approximation is

$$\beta = 1 + \sqrt{1+\frac{m_c}{m}}, \tag{10}$$

or for large craters the momentum enhancement is proportional to the square root of the volume of the crater material ejecta for a consolidated material (for a porous material, a crater is formed by compressing the voids out of the material, so that material does not need to be ejected to open up the crater). Figure 3 shows this result for values of m_c/m from 1 to 1000 where M/m = 15000. The size of the crater can be much larger than the actual mass liberated, as material is displaced radially in the crater formation process, not just ejected. In hypervelocity impacts into strong metals, as typified by the impacts into 2024-T4 data described above, only 20–40% of the crater volume is liberated from the target[5] (their data of 47 experiments has only three values above 40%, the largest at 47%). For soft, ductile metals, the amount of liberated mass is even less: for each of nine shots into 1100-O aluminum, the liberated volume was less than 10% of the corresponding mass in the crater. Though the craters were much larger in the soft aluminum than in the stronger 2024 material, the corresponding momentum enhancement was less. Of course, metals typically are ductile. The upper value of $m_c/m = 1000$ corresponds to the crater diameter being at least 10 times the diameter of the impactor, for example, if all the mass were liberated.

Also shown in Figure 3 are the momentum enhancements as a function of ejecta mass from the relationships in Section 4 between ejecta mass and ejecta velocity. In particular, $P/mV + 1$ from Eq. (4) is plotted for the top line of the nonporous ejecta mass–ejecta velocity relationship (Eq. 2) and for the same relations for porous material. It is seen that these predicted momentum enhancements are considerably less than the bound. Also on this plot are results from four large-scale numerical simulations to be presented in the next section (Table 1). These results agree with Eqs. (2) and (4) for the consolidated material curve, which shows that the computations for aluminum with various tensile fracture stress are agreeing with integrated results from the ejecta mass ejecta velocity relationship, which in the controlling region is primarily based on aluminum striking basalt data.

The crater sizes seen in cratering experiments on Earth with explosives and impactors have a wide range of values. On bodies with large gravity, there is a distinction between strength dominated cratering and gravity dominated cratering, for mechanisms similar to those of cutoff velocities discussed in the previous section.

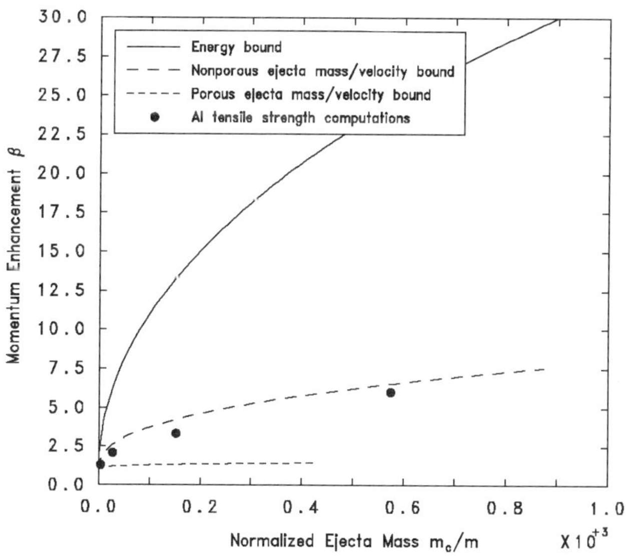

Fig. 3: Momentum enhancement vs. normalized ejecta mass.

LARGE-SCALE NUMERICAL SIMULATIONS

We also explore momentum enhancement through numerical simulations. Both impacts and conventional explosives were employed as the deflecting technology. High pressure and high temperature material states and material strength must be taken into account for both the impactor and impacted object in the energy and momentum transfer calculations of deflection or disruption of an object by a kinetic impactor. When explosives are involved, detailed equations of state for the explosive product gases are required. Spherical impactors of aluminum striking ice, rock, and aluminum with various tensile failure stresses were explored computationally, as were small explosives sitting on the surface. Related computations with other impactor materials, geometries, and impact speeds can be found elsewhere.[8,9]

Since momentum enhancement is related to cratering, the extensive difficulty of performing cratering computations comes into play. Highly resolved large-scale numerical simulations are required because the controlling event is the tensile fracture of the target material and the target material's subsequent launch in the direction opposite of impact. As is well known, fracture computations are extremely mesh sensitive, and our computations for this work showed that 50 to 100 cells across the diameter of a spherical impactor are required to produce computations that appear to be converging. The large number of cells across the extent of the impactor translates into even larger meshes since the crater at hypervelocity is typically much larger than the impactor. The crater formation time (during which the transferred momentum continues to increase) can be long for low strength materials.

Computations were performed with the impact physics code CTH[10] of 1-cm diameter spheres (volume 0.524 cm^3 and a mass of 1.46 g) of aluminum striking various targets at 10 km/s. Aluminum was modeled with a Mie-Grüneisen equation of state and with the Steinberg-Guinan constitutive model for Al 2024-T4 from the library. To model fracture, the CTH tensile failure model was used: when the tensile stress reaches a certain value void is inserted. The linking of void by the interface tracker leads to material separation and thus the formation of ejecta from within the crater. Rock was used as an impacted material as it is thought to be the prevalent material in asteroids, and ice was used as it is thought to be the prevalent material in comet nuclei. For rock, tensile fracture stress was the lone failure criteria, while for ice there was also a strain to failure model with 50% strain. The specific rock data was based on the estimated value of asteroid 433 Eros's density (2.7 g/cm^3)[11] and lunar rock properties.[12] It is seen that in density the asteroid rock matches aluminum. A material property of particular importance is the tensile fracture stress; for the rock the value used is 50 MPa, and for the ice the value used is 1 MPa. Figure 4 shows results of the spherical aluminum impactors striking rock and ice at 10 km/s. As is seen, the rock approaches a momentum enhancement β of 1.91. For the ice, as time continues, it appears the momentum enhancement is approaching a β of 8, and perhaps exceeding that value. The longer time that crater material is being ejected for the ice is seen in this plot, a result of the lower fracture stress. A large tensile fracture stress for the rock limits the amount of ejecta material created, leading to a lower value of β.

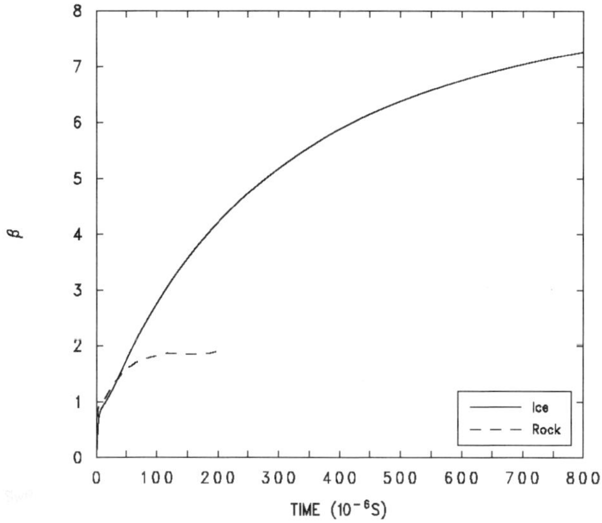

Fig. 4: Target momentum after 1-cm diameter spherical impactors strike rock and ice at 10 km/s.

When an explosive is placed on the surface of an asteroid or comet body and detonated, the expanding explosive product gases provide a momentum transfer to the body. In addition, a crater is formed and surface material is thrown outward, enhancing the momentum transfer. Consolidated, weak (in tension) materials have a larger momentum transfer than stronger materials. Computations were performed with CTH with various materials in contact with a 100 gram C-4 charge sitting on the surface in vacuum.[9] The explosive was detonated and the momentum transfer to the body was computed. No gravity was included in the computations. The computed momentum transfer was then normalized by the mass of explosive to calculate $\Delta\overline{v} = \Delta p / m_e$, where Δp is the momentum transferred to the NEO and m_e is the mass of explosive used. Explosives sitting on both rock and ice have been examined. The C-4 explosive was modelled with a JWL equation of state. Computations for the rock material led to a momentum transferred $\Delta\overline{v}$ of 1.04 km/s, and for the explosive sitting on ice the value was 2.22 km/s. Thus, the surface material can lead to an enhancement of momentum by over a factor of 2. The values of $\Delta\overline{v}$ can be compared to rocket fuel efficiencies by dividing by the acceleration of gravity at the surface of the Earth; essentially sitting the explosive on the top of the surface has a specific impulse I_{sp} of 106 s for rock and 227 s for ice. This latter value falls in the range of solid rocket fuels. The best rocket fuel used (liquid oxygen and liquid hydrogen) has an I_{sp} slightly above 400. Unfortunately it is difficult to compare these values directly to the values for the impactor, as the comparison in the long run (for both) is really based on how much mass it takes to deliver the momentum change to the object from Earth (the entire mass of the rockets delivering the momentum change).

For explosives sitting on the surface, a 1 km/s value is a lower bound on how much momentum transfer may occur. There will be some variation for different high explosives. These computations had the explosive sitting on the surface and there is increased momentum transfer when the explosive is either partially or fully embedded in the surface. These increases are due to both focusing the explosive product gases so that they move preferentially away from the asteroid surface as well as the formation of additional ejecta. In particular[13,14], show that when a right circular cylinder is half embedded or embedded to its top in strong rock, the momentum transfer increases by 31% and 38%, respectively, thus increasing $\Delta\bar{v}_e$ to 1.38 km/s for the fully embedded case. Fully embedded means the top of the explosive is flush with the surface. For strong materials, when the explosive is buried more deeply the momentum transfer can decrease, as an interior cavity can be opened which takes energy that does not translate into a momentum change for the asteroid or comet nucleus. However, for weak materials there may be gains by deeper embedment.

It might be thought that impacting at hypervelocities with an explosive could lead to higher momentum transfers. Our computations indicate that for rock as the surface material, higher momentum transfers are achieved till about 5 km/s, at which point the fact that the explosive is detonating does not appear to increase the amount of momentum transferred.[9]

One of the main findings of our research is the strong dependence of the momentum enhancement on the tensile fracture strength of the material. The momentum enhancement is very sensitive to the tensile strength of target. To examine this effect, the fracture stress was varied over a large range to study the effect on momentum enhancement. The tensile fracture stresses for impacts at 10 km/s were Y_t's of 1 MPa, 10 MPa, 100 MPa, and 1 GPa. Inspection of the curves (Figure 5) shows that for the higher strengths, the computed momentum enhancement factor β is reached after a relatively short amount of time. However, for Y_t = 1 MPa, the computations took an extended period to perform. What is observed is the strong dependence of the momentum enhancement on tensile fracture strength. Table 1 summarizes the results of these computations, where for the lower strength an extrapolation is listed as an estimate. As the aluminum material in the model becomes weaker and weaker in tension, it is possible to fracture and eject more and more target material. Also shown are values of mass ejected from the crater according to the computations. As impact velocity increases, more ejecta material leaves the crater, transferring momentum to the target, and the momentum enhancement increases.

Table 1. Momentum enhancement β for a 1-cm aluminum sphere striking aluminum targets of varying tensile fracture strength at 10 km/s.

Tensile Fracture Stress	β	m_c / m
1000 MPa	1.29	3.02
100 MPa	2.07	26.5
10 MPa	3.32	151
1 MPa	(6.0)	572

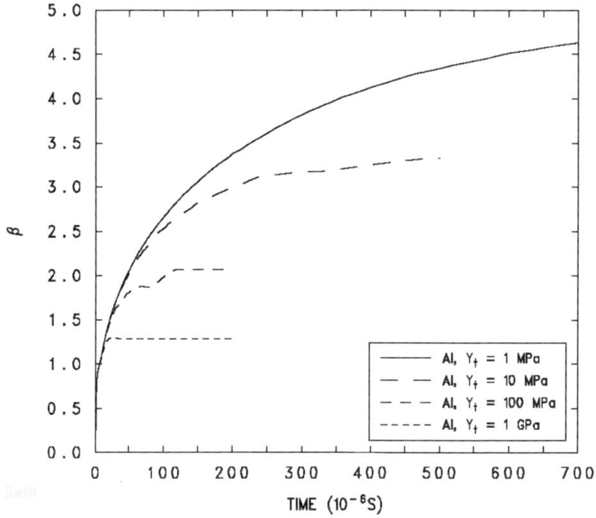

Fig. 5: Momentum enhancement β vs. time for a 1-cm aluminum sphere striking aluminum targets of varying tensile fracture strength at 10 km/s.

Strong aluminum alloys have a fracture stress on the order of 400 MPa (427 MPa was used in computations to compare with the above presented data). It is clear that the tensile strength of the material being impacted is of prime consideration. This conclusion is especially important in the context of striking celestial objects. It is likely that the surfaces of asteroids or comet nuclei are low cohesion materials. At the surface of the Moon, lunar regolith (soil) has irregular shapes, especially in the highlands which may be the most similar to asteroids. Cohesive strengths were measured during the Apollo flights (Apollo 16, e.g.) for very shallow depths (10-20 cm) to be on the order of 1 to 10 kPa. Those values appear to increase fairly rapidly with depth.[12] These values are even lower than what has been considered here. Offsetting this effect is the fact that dust and sand are also relatively porous materials, and the ability of the crater to grow in size by compressing material rather than moving it leads to less ejecta and hence less momentum enhancement. Computations were performed to explore the influence of target porosity. The assumed material properties of the target were those of granite, and the P-α model was used to represent the porosity. The granite strength was modeled with a Drucker-Prager yield surface with cutoff, and the zero-pressure intercept, the cutoff, and the tensile strength were viewed as a decreasing linear function of the porosity; for example, $\sigma_f = \sigma_{f0}(1-\phi)$, where ϕ is the porosity. The general result of porosity is that it decreases momentum enhancement. The craters are deeper and narrower, and less material is thrown back. Table 2 shows the results, demonstrating the decrease in momentum enhancement as porosity increases. More computations can be found in Walker and Chocron, 2010.[8]

For standoff nuclear explosives, ejecta is the primary source of the momentum transfer and computing how much energy is transferred to the NEO through radiation, how the NEO material vaporizes and fragments, and the amount and speed of that ejecta is extremely complex, which is why computations of the momentum transfer from nuclear explosives is very difficult. Nuclear explosive methods require the most complicated modeling techniques: these models include radiation transfer, opacities, and equations of state and neutron absorption cross sections. Opacity plays an important role in energy deposition and subsequent ablation of the NEO's surface materials.

Table 2. Momentum enhancement as a function of porosity for a spherical aluminum impactor striking granite at 10 km/s.

Porosity	Initial Density (g/cm^3)	Tensile Fracture Stress	β	m_c / m
0	2.765	50.0 MPa	1.88	41.6
10 %	2.489	45.0 MPa	1.58	23.2*
20 %	2.212	40.0 MPa	1.46	31.1
30 %	1.936	35.0 MPa	1.28	18.9

*Unlike the other computations, this one had a substantial crater lip remaining.

FUTURE WORK

Beyond computational modeling, there is a need to perform more impact tests determining momentum enhancement at hypervelocities. These tests can be performed with two-stage light-gas guns or other means. For example, SwRI operates an inhibited-shaped-charge-based hypervelocity test facility that launches 1-gram metal (such as aluminum) fragments into an evacuated chamber at speeds in excess of 11 km/sec.[15] Speeds that are relevant for PHO deflection can be simulated using this facility. These velocities and masses would give us a quantitative insight into the amount of momentum enhancement that could be obtained by hypervelocity impacts onto rock, ice, and other materials of interest for surfaces of asteroids and comet nuclei. The launcher, which employs a vacuum flight and target chamber, was developed originally to assess the risks to the International Space Station from impacts by meteoroids and space debris, and the current shielding on the Station was characterized at SwRI in the mid 1990s using this facility.

CONCLUSION

When an object is impacted at hypervelocities, more momentum is imparted to the object than the original momentum of the impactor. This fact, referred to as momentum enhancement, is due to target material being thrown backwards during crater formation. The amount of debris and ejecta from a crater and its velocity depend on the target material and structure as well as the impactor. We have shown results from simple arguments as well as large-scale numerical simulations that place the potential momentum enhancement on the order of $\beta = 10$ for some surface materials. There is still a great need to experimentally explore the effect above 10 km/s, and of course there is a great need to characterize asteroid and comet nuclei surfaces to determine their actual properties in order to be able to accurately determine what the momentum enhancement would be. In the work here, aluminium, rock, and ice were considered, with material models based on the best available data.

ACKNOWLEDGMENTS

We thank Walter Huebner for his encouragement and support of this work. The work was supported by Southwest Research Institute.

REFERENCES

1. NASA. Near-earth object survey and deflection: analysis of alternatives. Report to Congress, March 2007, available at http://www.nasa.gov/pdf/171331main_NEO_report_march07.pdf and http://neo.jpl.nasa.gov/neo/report2007.html.
2. NASA. 2006 near-Earth object survey and deflection study (draft pre-decisional material), available at http://www.hq.nasa.gov/office/pao/FOIA/NEO_Analysis_Doc.pdf and http://neo.jpl.nasa.gov/neo/report2007.html.
3. A'Hearn, M.F., Belton, M.J.S., Delamere, W.A., Kissel, J., Klaasen, K.P., McFadden, L.A., Meech, K.J., Melosh, H.J., Schultz, P.H., Sunshine, J.M., Thomas, P.C., Veverka, J., Yeomans, D.K., Baca, M.W., Busko, I., Crockett, C.J., Collins, S.M., Desnoyer, M., Eberhardy, C.A., Ernst, C.M., Farnham, T.L., Feaga, L., Groussin, O., Hampton, D., Ipatov, S.I., Li, J-Y, Lindler, D., Lisse, C.M., Mastrodemos, N., Owen, Jr. W.M., Richardson, J.E., Wellnitz, D.D., White, R.L. (2005) "Deep Impact: Excavating Comet Tempel 1." *Science;* **310**:258-264.
4. Denardo, B.P. (1962) Measurements of momentum transfer from plastic projectiles to massive aluminum targets at speeds up to 25600 ft/s. NASA Technical Note D-1210.
5. Denardo, B.P., Nysmith, C.R. Momentum transfer and cratering phenomena associated with the impact of aluminum spheres into thick aluminum targets at velocities to 24000 ft/s. In: *Proceeding of the AGARD-NATO Specialists*, Vol. 1: The Fluid Dynamic Aspects of Space Flight, Gordon and Breach Science Publishers, New York, 1964, p. 389-402.
6. Yanagisawa, M., Hasegawa, S. (2000) "Momentum transfer in oblique impacts: implications for asteroid rotations." *Icarus*; **146**:270-288.
7. Holsapple, K.A. About deflecting asteroids and comets. In: Belton MJS, Morgan TH, Samarasinha NH, Yeomans DK., editors, *Mitigation of Hazardous Comets and Asteroids*, Cambridge University Press, Cambridge, 2004, p. 113-140.
8. Walker, J.D., and Chocron, S. Momentum enhancement in hypervelocity impact. To appear in the proceedings of the 2010 Hypervelocity Impact Symposium, Freiburg, Germany, April 2010.
9. Walker, J.D., and Chocron, S. (2008) "Near-Earth Object Deflection using Conventional Explosives." *Int. J. Impact Engng,* **35:**1473-1477.
10. McGlaun, J.M., Thompson, S.L., and Elrick, M.G. (1990) "CTH: A three-dimensional shock wave physics code." *Int. J. Impact Engng*, **10:**351-360.

11. Veverka, J., Robinson, M., Thomas, P., Murchie, S., Bell, III J.F., Izenberg, N., Chapman, C., Harch, A., Bell, M., Carcich, B., Cheng, A., Clark, B., Domingue, D., Dunham, D., Farquhar, R., Gaffey, M.J., Hawkins, E., Joseph, J., Kirk, R., Li, H., Lucey, P., Malin, M., Martin, P., McFadden, L., Merline, W.J., Miller, J.K., Owen, Jr. W.M., Peterson, C., Prockter, L., Warren, J., Wellnitz, D., Williams, B.G., Yeomans, D.K. (2000) "NEAR at Eros: imaging and spectral results." *Science*; **289:**2088-2097.
12. Mitchell, J.K., Carrier, W.D., Houston, W.N., Scott, R.F., Bromwell, L.G., Durgunoglu, H.T., Hovland, H.J., Treadwell, D.D., Costes, N.C. Soil Mechanics. In: Apollo 16 Preliminary Science Report, NASA Manned Spacecraft Center, U.S. Govt. Printing Office, Washington, DC, 1972, p. 8-1–8-29.
13. Walker, J.D. (2003) "Loading sources for seismological investigations of asteroids and comets." *Int. J. Impact Engng,* **29**:757-769.
14. Walker, J.D., and Huebner, W.F. Seismological investigation of asteroid and comet interiors. In: Belton, M.J.S., Morgan, T.H., Samarasinha, N.H., Yeomans, D.K., editors, *Mitigation of Hazardous Comets and Asteroids*, Cambridge University Press, Cambridge, 2004, p. 234-265.
15. Walker, J.D., Grosch, D.J., and Mullin, S.A. (1995) "Experimental impacts above 10 km/s." *J. Impact Engng,* **17:**903-914.

LOOKING BEFORE WE LEAP: AN ONGOING, QUANTITATIVE INVESTIGATION OF ASTEROID AND COMET IMPACT HAZARD MITIGATION

CATHERINE S. PLESKO
Applied Physics Division, Los Alamos National Laboratory
Los Alamos, USA

ROBERT P. WEAVER
Applied Physics Theoretical Design, Los Alamos National Laboratory
Los Alamos, USA

PAUL A. BRADLEY
Applied Physics Theoretical Design, Los Alamos National Laboratory
Los Alamos, USA

WALTER F. HUEBNER
Theoretical Physics, Los Alamos National Laboratory
Los Alamos, USA

ABSTRACT

There are many outstanding questions about the correct response to an asteroid or comet impact threat on Earth. Nuclear munitions are currently thought to be the most efficient method of delivering an impact-preventing impulse to a potentially hazardous object (PHO). However, there are major uncertainties about the response of PHOs to a nuclear burst, and the most appropriate ways to use nuclear munitions for hazard mitigation.

INTRODUCTION

Asteroids and comet nuclei have collided with planets throughout the history of our planetary system. The evidence is clear from the impact craters on the Moon, Mars, Mercury, other planets and moons in our solar system, as well as from over 160 identified impact craters on Earth. Impact craters were first documented by the cartographer Thomas Harriot in August of 1609 and confirmed in 1610 by Galileo, who used the Greek root for 'bowl' to describe circular depressions on the Moon without commenting on their origin. At the time these observations were significant for settling the contemporary debate between Averroe's interpretation of Aristotle's cosmology of geometric purity and the hypothesis originally by Plato, Plutarch, and others that the Moon was Earth-like in having a surface shaped by dynamic and ongoing processes.[1] Meteor showers and falls were known from prehistory, but meteorite origins were not scientifically accepted until the 1880s.

Asteroids were not discovered until 1801 when Giuseppe Piazzi discovered Ceres. It took another hundred years of improvement in telescope observations and geological field work until G. K. Gilbert and D. M. Barringer proposed an impact origin for lunar and terrestrial craters.[2] Gilbert's hypothesis was not widely accepted for many years. The chief criticism of the impact hypothesis was that nearly all observed impact craters are circular, whereas low-velocity impact experiments could only produce circular craters from nearly vertical impacts—suggesting the improbable result that all natural impacts were vertical. It took many more years until the mechanics of high-velocity impacts and their parallels to explosion cratering were better understood.[3] Two independent lines of research, the study of large explosions and high strain-rate geophysical responses to them during World Wars I and II, and the Cold War[4,5], and the observational and experimental work on impact cratering by Shoemaker, Gault, Ahrens, and others began to converge in the later half of the 20th century. A comprehensive understanding of impact processes and their context in the solar system was aided by observations and sample returns from Apollo and other space missions.

Impacts have occurred throughout history of our planetary system and indeed still occur now. The Tunguska event[6], the near miss of a similarly sized object in March 2009[7], collision of Comet Shoemaker-Levy 9 with Jupiter in 1994, and the August 2009 impact of a 500-m-diameter object on Jupiter[8] are reminders and warning signals that we should take seriously. The extinction of the dinosaurs has been attributed to the impact of a large asteroid or comet nucleus on Earth. Zaitsev[9] has listed six objects hurtling between Earth and the Moon since 1991. Two large asteroids, which are each several hundred meters in diameter (99942 Apophis and 2004 VD17), will approach the Earth on 19 March 2029 and 4 May 2102, respectively. Besides the Tunguska event, there were several other notable events in the last hundred years: on 13 August 1930 in Curuçá, Amazonas, Brazil; on 12 February 1947 in Sikhote-Aligne, Russia; on 24 September 2002 in Vitim, Bodaybo, Russia; and the Carancas event on 15 September 2007 in Alta Plana, Peru.[10]

PHOs strike Earth with a frequency (commonly quoted as a function of object diameter) that is inversely correlated to their mass. There are more small objects so they strike more often according to a predictable size-frequency distribution. Objects below a threshold diameter of 10 m have minimal consequences on the ground (similar to the Carancas impact of 2007), and an impact frequency of 1:10 years globally[11] or 1:500 years in an urban area. The energy of the impact is simply the kinetic energy,

$$E = \frac{1}{2} m v^2.$$

The smallest events are seen as shooting stars. Several hundred metric tons of these small particles burn up in Earth's atmosphere every day.[12] Larger objects, up to perhaps 50 m in diameter, depending on composition, may burn up as they transit Earth's atmosphere. Some objects in this size range may strike the ground as fragmented debris, as happened in the Carancas event in 2007, and some air bursts may have severe consequences on the surface, as was the case for the Tunguska event. Consequences of larger impacts are described in Collins et al. (2005)[13] and are analogous to the damage caused by the fireball from a nuclear detonation with much less ionizing radiation, and with the potential to be of much higher explosive yields.

PHOs are defined as small solar system objects that meet two criteria: (1) they are asteroids or cometary nuclei with diameters larger than 150 m, and (2) approach within 0.05 astronomical units (5% of the Earth-Sun distance, or 7.5 million km) of Earth's orbit.[14] The minimum diameter is under debate. It was originally set to be comparable to the object that caused the Tunguska airburst[15], but recent calculations by Boslough and Crawford[16] indicate that the Tunguska impactor may have been as small as 30 m in diameter.

Asteroids are a diverse group of objects whose chemical composition varies between carbonaceous objects that make up about 75% of the population of small inner solar system objects, stony basaltic objects that make up another 17%, and with the remainder made up of nickel-iron objects.[17] Small asteroids may be solid objects, indicated by a rotation rate that generate centripetal forces larger than the gravitational forces generated by a mass of that size, although this hypothesis is under debate.[18] Larger asteroids have been shattered repeatedly by collisions with other objects and so tend to be unconsolidated piles of gravitationally bound rubble.

There are a variety of current and ongoing surveys and missions that have the potential to significantly improve the planetary science community's understanding of near-earth and main belt objects. At the time of this writing, the Japan Aerospace Exploration Agency (JAXA) Hayabusa mission[19] sample return canister has been retrieved, but the results have not been released. It is hoped that this canister contains the first samples of a near-Earth asteroid collected in situ and returned to Earth. If so, it will provide ground truth for the comparison of asteroidal and meteoritic compositions, which will allow for better matches between the chemical compositions of laboratory samples and spectral measurements of basaltic asteroids. The NASA Dawn mission[20] is currently en route to 2011 and 2010 rendezvous with asteroids Vesta and Ceres where it will study the chemical properties of their surfaces and the bulk densities of the objects' interiors. The wide-field infrared survey explorer (WISE) infrared space telescope mission is just over halfway through its planned mission. During this mission it is expected to discover 100,000 previously unseen main-belt asteroids and 300 previously undiscovered near-Earth objects during its mission.[21] Earth-based surveys for PHOs continue as well. The PanSTARRS PS1 telescope has recently started its science mission, which will map the sky down to 24th magnitude, four magnitudes fainter than the current best data from the Catalina Sky Survey, and over a wider area of the sky. It is expected to significantly improve our catalog of small solar system objects, including increasing the number of known Kuiper belt objects by two orders of magnitude.[22]

Much less is known about comet nuclei. Determining the chemical composition of a comet nucleus is a complicated problem. Remote observations of the coma of a comet, formed by outgassing from the icy component of its nucleus under the influence of solar heat, yields only indirect information, (i.e., mother molecules of the ices are contaminated by molecular radicals and ions produced by solar ultraviolet radiation and chemical reactions in the coma). In addition, abundances in the coma are *not* the same as in the comet nucleus because the mixing ratio of chemicals (e.g., CO/H_2O) changes with heliocentric distance.

Space missions to comets have given the most reliable information. Among these missions are ESA's Giotto mission, the Russian's Vega 1 and Vega 2 missions, and Japan's Sakigake and Suisei missions, all to Comet 1P/Halley; NASA's Deep Space 1 mission to Comet 19P/Borelly; NASA's Stardust mission to Comet 81P/Wild 2; and NASA's Deep Impact mission to Comet 9P/Tempel 1. Combining data from these missions and from ground based observations of comets 1P/Halley, C/1995 O1 Hale-

Bopp, and Hyakutake resulted in the best estimates for the composition and density of comet nuclei. More than 30 neutral cometary molecules have been identified that also exist in the interstellar medium. Scientists found that most chemical elements in comets are present in amounts consistent with solar system abundances, except for nitrogen, which is under abundant, and hydrogen, which is present only in amounts consistent with condensable species that can bind hydrogen (i.e., hydrogen molecules do not condense as ice because the temperature in comet nuclei is too high). ESA's Rosetta mission is the most likely to explore the interior of a comet. It was launched on 2 March 2004. After entering an orbit around Comet 67P/Churyumov-Gerasimenko in 2014, the spacecraft will release a small lander onto the nucleus and then spend the next two years orbiting the comet as it heads towards the Sun.

The most commonly quoted "impossible" hazard mitigation scenario is that of a previously unknown Oort cloud object. This object would have a diameter greater than 1 km, would be on a parabolic orbit with a high angle of inclination relative to the plane of the ecliptic (orbital plane of the solar system), would be discovered only after its first and only orbit brought it around the sun, and as a result we would have only months or weeks of response time remained before an impact. A comprehensive survey of the Kuiper belt and Oort cloud are unlikely to be completed for many years, but such surveys would both significantly add to our understanding of the solar system and minimize the statistical uncertainty on the threat of large impactors.

Fig. 1: A common hypothetical impact hazard scenario that would be technologically difficult to respond to would consist of a previously undiscovered Oort cloud object with a diameter larger than 1 km. This object would approach the inner solar system on a parabolic or hyperbolic orbit that is highly inclined to the plane of the ecliptic, and would intersect Earth's orbit in such a way that it would not be detectable until a few months before impact.

Minimizing the risk of damage to the Earth from a PHO or its fragments is known as hazard mitigation. It is accomplished through the disruption of a PHO and the dispersal of its fragments, or through modification of the PHO's orbit so that the threat to Earth is reduced or eliminated. Orbit modification can be accomplished in a variety of ways that fall into either "fast push" or "slow push" categories. Slow push methods include gravity tractors, albedo modification, and mass drivers[23], which are intended to provide a small change in velocity ($\Delta\mathbf{v}$) over time scales of years to decades and would require a substantial amount of time from the time of deployment to the time of potential impact in order for the process to work. On top of this delay, all of these proposed methods are untested and well beyond modern spacecraft technical readiness levels for deployment in the case that a hazard is discovered with a time to Earth impact of less than a century.

Fast push methods rely on well-tested technologies such as kinetic impactors[24], high explosives[25], and nuclear bursts[26] to launch, deploy, and provide the impulse but are limited by modern launch mass limitations and uncertainties in the coupling of momentum from the explosion to the object. Modern launch capabilities are limited to at most a 10,000 kg payload for an escape trajectory and only an additional 5,000 kg to low earth orbit. This limitation places severe restrictions on the explosive yield available for a fast push deflection and means that nuclear explosives would be required to provide the yield necessary to deflect very massive PHOs or those requiring a larger change in velocity because of a short lead time. Exploration of momentum coupling from the explosion to the PHO is part of the work presented here.

METHODS

Before we can say with certainty that an explosive yield Y at height of burst $h < 0$ or depth of burial $h < 0$ will produce a change in velocity or dispersion of an object, we need to quantify how and where energy is deposited into the complicated media that may make up the object. We then need to understand how shock waves propagate through the system, what causes them to disrupt, and how long the gravitationally bound fragments of a disrupted system take to recombine.

We begin with a careful look at the coupling of energy from the mitigation technique to the PHO. Both explosive mitigation techniques and impacts can be thought of as inelastic collisions. The energy imparted to the PHO by the mitigation technique is known. For an impactor, this variable is the kinetic energy of the impactor. For a buried explosion, this variable is the yield of the explosion. For a stand-off explosion, it is the fraction of the explosion energy that intersects the surface of the PHO, which is simply the fraction of the yield that passes through the solid angle subtended by the PHO,

$$E_{in} = \frac{Y}{4\pi}\iint_S \frac{\hat{\mathbf{n}}\cdot d\mathbf{a}}{r^2},$$

where $\hat{\mathbf{n}}$ is a unit vector from the origin (usually the center of the explosion), $d\mathbf{a}$ is the differential area of the patch of surface area subtended by the PHO, and r is the distance from the origin to the patch.

However, kinetic energy is not conserved in any of these scenarios. Momentum imparted to the system of particles that make up the PHO may be estimated; however, the coefficient of restitution that would permit us to analytically calculate the velocity of the object or objects left after the mitigation impulse is a complicated function of the chemical and physical composition of the PHO system. We can model this response using numerical methods called hydrocodes to explore the effects of properties like equation of state, porosity, strength, and PHO shape on the system's response to an impulse. These models allow us to estimate the potential range of the coupling constant for different objects, and by extension, the change in velocities imparted to the components of a PHO system by a proposed mitigation technique, along with the range of uncertainties on those velocities, and the dynamical evolution of the post-mitigation system.

Various computational models separately track energy deposition from x-rays, gamma rays, or neutrons into different materials based on experimentally determined absorption cross-sections. These energy deposition processes are independent, so a piecemeal approach is physically reasonable. The well-known Monte Carlo particle transport simulation packages GEANT4[27] and MCNP[28] are used to estimate neutron or gamma-ray deposition. Once the location and amount of deposited energy is known, it can be sourced into the initial conditions of a radiation hydrocode model.

A hydrocode is a computer modeling framework that uses the equations of fluid motion to study the response of different materials and objects to rates of strain and pressure wave propagation that are large relative to the object's properties (e.g., viscosity, strength, sound speed). Hydrocodes are widely used in planetary science to explore impact[29] and volcanic processes.[30] A radiation hydrocode further couples a model of radiation transport to the equations of fluid motion in order to more accurately model problems where a large amount of the energy in the system is carried by light. Deflection or disruption of a PHO by nuclear burst is just such a problem. According to Glasstone and Dolan[4], about half of the energy released from a nuclear explosion is in the form of thermal radiation. The actual percentage is a complicated function of yield, design, and environment. This fraction makes thermal radiation a very important part of the problem and means that hydrocodes without radiation transport are insufficient to the task of modeling this method of deflection.

Here we use the Radiation Grid Eulerian (RAGE) hydrocode developed by Los Alamos National Laboratory (LANL) in collaboration with SAIC. RAGE is an Eulerian hydrocode with continuous adaptive mesh refinement (CAMR). RAGE uses a 'gray' diffusion model for radiative transfer using flux-limited nonequilibrium (two-temperature) diffusion, and tabular opacities. A variety of equations of state (EOS) are available to RAGE. Of these EOSs, the most appropriate for the materials encountered in the hazard mitigation problem is the SESAME library. SESAME is a temperature-based, tabular EOS library maintained by the Mechanics of Materials and Equations of State group at LANL. RAGE and SESAME have been through extensive verification and validation tests at every stage of their development.

Verification is the process of confirming that the physical models that the authors intend to include in the code are accurately implemented. It is done through comparisons of code results with analytical models of specific idealized physical processes. Validation is the process of comparing code model results to experiments of similar scenarios to establish that the physics implemented in the code is sufficient to model the scenario in question. RAGE model results have been compared with nearly 100 separate experiments and analytical models. Gittings et al. (2008)[31] document RAGE's performance on six separate verification problems, including spherical adiabatic compression, the Sod shock tube, Sedov blast wave, and the Marshak problem. Further detailed verification work is reported in Gisler (2005)[32], Kamm et al. (2002, 2000).[33,34]

Validation work has been conducted for many different physical situations including turbulent fluid flow experiments by Baltrusaitis et al. (1996)[35] and Zoldi (2002)[36], inertial confinement fusion (ICF) explored by Holmes et al. (1999)[37], Goldman et al. (2000)[38], Schappert et al. (2001)[39], Foster et al. (2002)[40], Wilde et al. (2000)[41], Parker et al. (2004)[42], Foster et al. (2005)[43], Lanier et al. (2006)[44], and Lanier et al. (2007).[45] Validation tests for physics of particular interest to hazard mitigation include Plesko et al. (submitted)[46] models of the Nakazawa et al. (2002)[47] study of shock propagation through basalt, comparisons of RAGE impact crater models with impact scaling relations from Holsapple (1993)[48] in Plesko (2009)[49], and comparisons of hydrocode models with laboratory-scale impact experiments in Pierazzo et al. (2008).[50] Further validation work is ongoing at this time.

Constraint of energy coupling will draw heavily on experimental shock propagation work in relevant media, such as Housen and Holsapple (2003)[51] on impacts into porous media. It will draw on previous numerical studies of collisions of small solar system bodies and the energy required to disrupt them, Q^*_D, such as Benz and Asphaug (1999)[29] and Housen et al. (1999)[52], and on numerical techniques to approximate the relevant physics within current computational abilities, such as damage models[53], rigid body dynamics codes[54] and particle transport codes.

Energetic subsurface bursts are another method under consideration for impact-hazard mitigation. This technique has been popularized in the media, but still faces significant technical difficulties in the emplacement of the explosive device. There are two potential scenarios for the use of subsurface explosions to mitigate PHOs. First, if the source explosion were emplaced near the surface of the object (h~-10 m to -50 m) then the resultant shocks would preferentially eject material from the surface near the explosion and by conservation of momentum the remainder of the body will be given a significant force in the opposite direction. The goal of this type of intercept would be to impart a large enough velocity to the remaining object/fragments so that they would miss the Earth's orbit by a significant margin. The second subsurface method of mitigation would be to emplace the explosive source near the center of the object and independent of the composition of the PHO, the explosion would have enough power to significantly disrupt the entire body, leading to radial ejection velocities well above the escape velocity. Given a large enough explosion, here we consider energies of 1–10 megatons (Mt) TNT equivalent, the PHO would be fractured into smaller fragments with sufficient velocity to again miss the Earth's orbit by a significant margin. To begin we consider the second option, a centrally located explosion. We also build our computational models from simple to more complex by first considering uniform composition objects and then non-uniform, or "rubble pile" initial geometries. These rubble piles can have a very large range of actual internal compositions for which we have no actual data. We consider various rubble pile geometries as shown below. In this work we do not consider any political or engineering questions that might be involved in achieving the initial conditions assumed in these model hydrocode simulations.

RESULTS

We have explored energy deposition from stand-off bursts ($h > 0$) onto realistically shaped PHOs and the effects of internal object structure on the response of a PHO to a buried burst ($h < 0$). We find that our model results are physically reasonable and challenge previous assumptions about the response of PHOs to nuclear bursts.

In order to confirm our ability to study hazard mitigation with the tools available to us, we began with simplified models of nuclear deflection stand-off bursts at near Earth asteroid 25143 Itokawa. We chose to model deflection scenarios for asteroid Itokawa because the asteroid is so well characterized. 25143 Itokawa[19] is a member of the Apollo asteroid dynamical family. It has an S-type reflectance spectrum, indicating a basaltic composition, and an ellipsoidal, potato-like shape with dimensions of 535 m by 294 m by 209 m. (See Figure 2.) Spacecraft observations indicate that it is a gravitationally bound rubble pile composed of fragments that vary in size over many orders of magnitude, and approximately 40% void space. It rotates in the plane of its long axis with a rotation period of 12 hours, well below a rate that would require strength to hold it together.[18] So its rotation rate does not give a lower limit to its cohesive or tensile strength. Surface slopes in the range of 35–50 degrees indicate that the surface regolith does have some cohesive strength, of order or greater than granular materials observed on Earth. Its orbit crosses both that of Earth and Mars, although it does not currently pose an impact hazard to either planet. The asteroid was visited by the JAXA Hayabusa mission in September 2005. A sample collection was attempted and the capsule has been returned to Earth. The Hayabusa mission took extensive photographic, spectral, and LIDAR data in addition to the attempted sample return. This substantial set of observations makes 25143 Itokawa the best characterized subkilometer near Earth asteroid at this time.

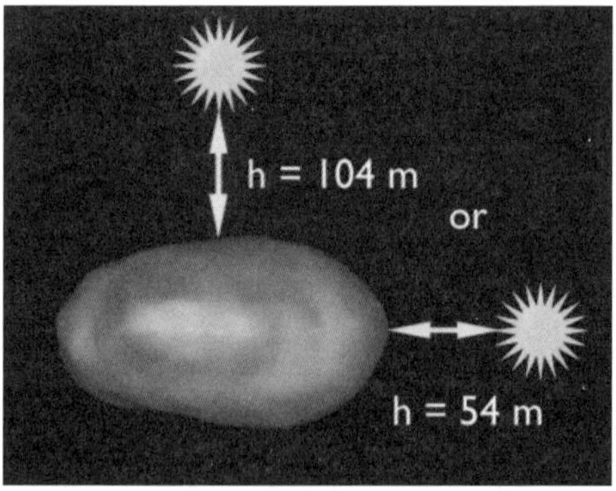

Fig. 2: Initial conditions for two different hypothetical nuclear deflection bursts at a target shaped similar to near-Earth asteroid 25143 Itokawa using the geometric optimum heights of burst, h = 0.4r.

We begin with a set of two two-dimensional (2-D) axisymmetric RAGE models of nuclear stand-off bursts. Ahrens and Harris (1994)[55] estimate that a total yield of order 100 kt would be required to impart a change in velocity ($\delta\mathbf{v}$) of 1 cm/s to a 1-km-diameter asteroid, given a geometrically optimal stand off distance, h/r = 0.414. We take their estimates as a starting point and present the initial results of a model of a 10-kt burst, 52 m away from the object on a line perpendicular to the plane of the shorter axis. This model, and one with a 10-kt burst, 104 m away from the object along a line perpendicular to its long axis. We use the Ostro et al. (2004)[56] shape model of 25143 Itokawa obtained from Goldstone radio telescope data with 20 m resolution. The object is modeled with a minimum mesh resolution of 25 cm as porous SESAME Nevada Alluvium with an initial compressive strength of 0.2 bar, up to an assumed 1-kbar pressure to crush it completely, after which the material strength is set to zero (see Figure 3).

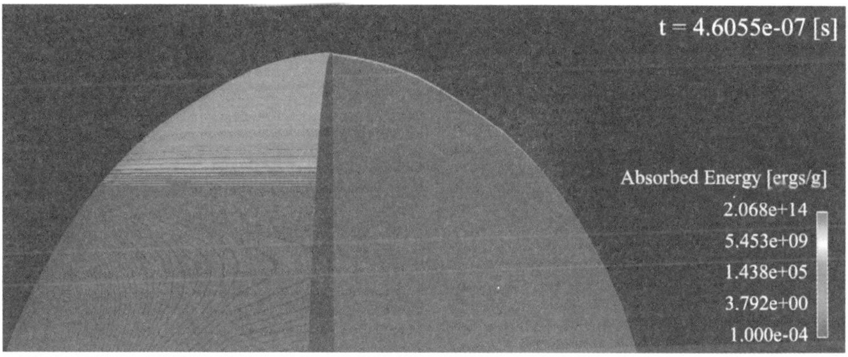

Fig. 3: Cut-away image of the Itokawa-like PHO target model, colored by absorbed energy after a 10-kt stand-off burst, 54 m off of the long axis of the object.

Figure 4 demonstrates that approximately $8\text{x}10^{19}$ ergs are deposited into the target's near-surface by x-rays. $3\text{x}10^{19}$ grams of material are heated above the vaporization temperature of the target material. In this model, we see surface temperatures that are an order of magnitude higher than Ahrens and Harris anticipated from a burst yield that is an order of magnitude smaller than they recommended. This emphasizes the importance in validating the assumptions used in planning for any deflection scenario.

After the energy from the burst is deposited, the vaporized region of the target is expected to explosively decompress, pushing against the remaining solid portions of the PHO. This decompression sends a shock through the object. We are currently exploring the effects of the material properties of the target on both the initial deposition of energy, and on the post-deposition propagation of shocks through highly porous and heterogeneous media.

(a)

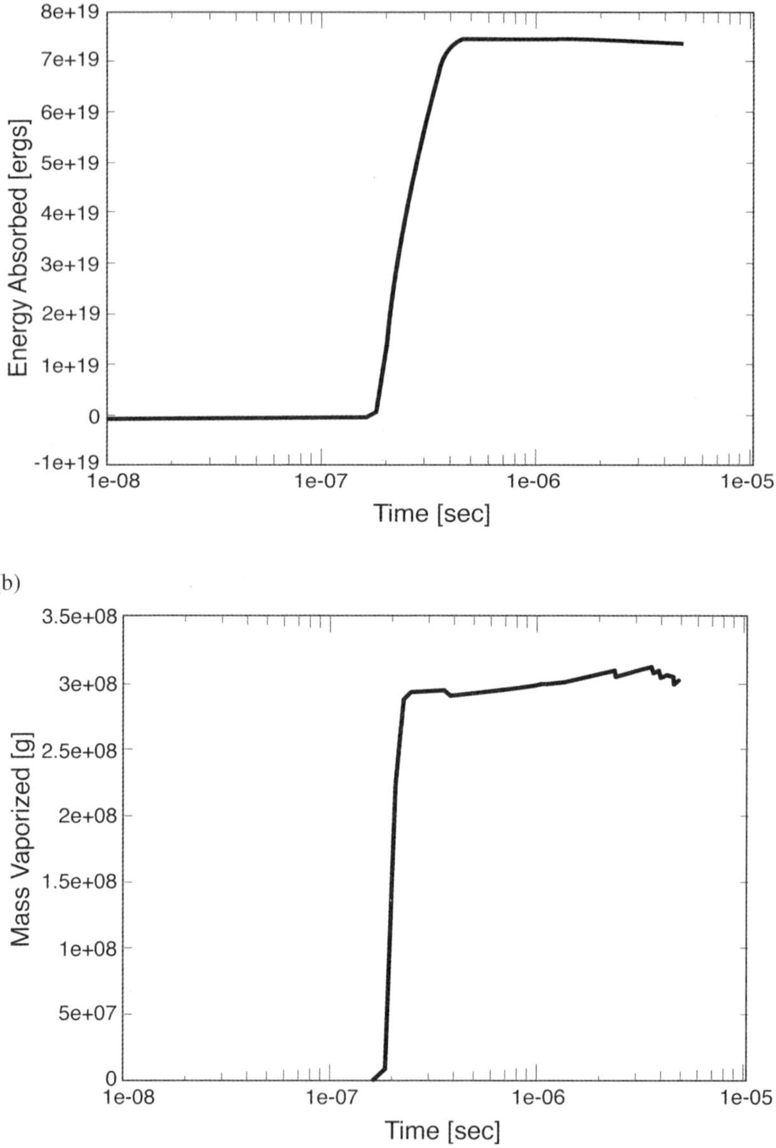

Fig. 4: (a) Modeled energy deposition by x-rays onto the target PHO and (b) the amount of PHO mass vaporized as a result.

In a parallel effort, we used the same RADAR shape model of asteroid Itokawa in RAGE hydrocode models of the shock-generated disruption of PHOs by subsurface, centrally located energetic bursts. We use known specific asteroid shape models in order to assess the consequences of various shapes and look for optimal emplacement of the source explosion. We will explore 2D and 3D models for the disruption by a large energy source at the center of such PHO models (1-10 Mt TNT equivalent), specifically for Itowaka (see Figure 2), with future work on the Mars-crossing asteroid 6489 Golevka. We have conducted three sets of parameter studies to test the sensitivity of our models to the values of specific variables; over the plausible range of values those variables might take. First, we examined the response to explosive yield, Y, in the range 1 Mt $\leq Y \leq$ 10 Mt. Second, we explored the sensitivity to various parameters used by the Steinberg-Guinan strength model used for these buried burst models. Third, we explored the effects of an inhomogeneous rubble pile-like composition on PHO response.

We are interested in assessing the optimum depth of burial and energy required to essentially disrupt the PHO into much smaller objects many of which will not recollect and therefore mitigate the hazard. This work starts with a uniform composition model as shown in Figure 5.

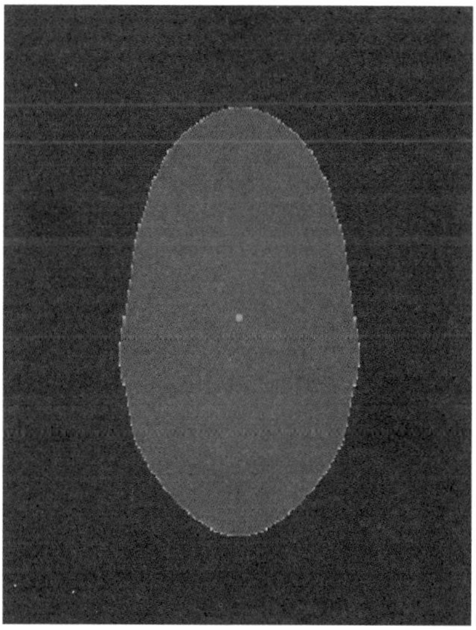

Fig. 5: An example of the initial conditions for a RAGE calculation of the Itowaka asteroid with a uniform iron composition and a central explosion.

In this example simulation we use a 1 Mt energy source with cylindrically symmetric geometry (2D). The RAGE hydrocode has been extensively validated for this type of strong shock, uniform composition model with detailed material properties, as described above. The result of this simulation is shown in Figure 6 as a series of images from the time-dependant simulation.

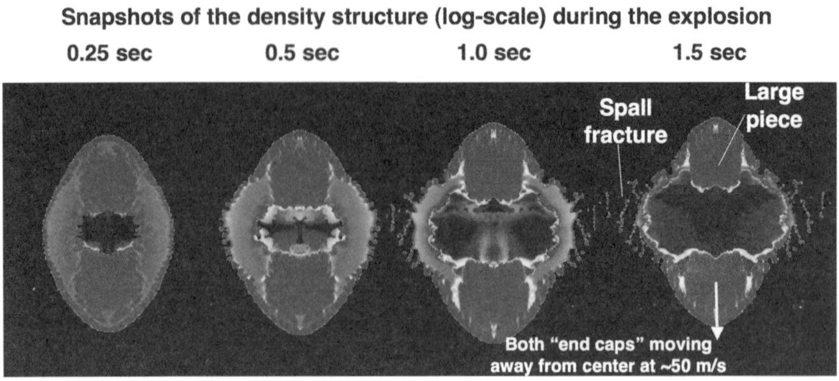

Fig. 6: An example of the RAGE calculation of the internal disruption of a model of the Itowaka asteroid by a massive explosion out to a time of 1.5 sec after the explosion. The model shows significant disruption of the modeled PHO.

We follow the time history of this simulation. The initial source energy is so large that it creates extremely high pressure surrounding the explosion. This high pressure leads to the formation of a strong shock wave that propagates spherically outwards leading to fracture of the surface of the object closest to the explosion. The high internal pressure continues to expand to the surface of the object, creating fractures planes and significant disruption of the object. Finally, we see that the high internal pressure results in the ejection of the large end-cap fragments with very high velocity. With an ejection velocity of 50 cm/s, the majority of the PHO mass would move two Earth-diameters away from the center of mass of the system after as little as one week. Given proper timing and orbital dynamics, we believe that this type of disruption scenario could effectively mitigate a PHO hazard. Further work is needed to confirm the accuracy of the mass-velocity distribution.

Next, we consider that same shape model of asteroid Itokawa but fill the shape with a rubble pile of solid rocks within a uniform background alluvium material. Each rock is composed of granite with appropriate strength models for both the rocks and the alluvium. The evolution of this RAGE simulation is shown in Figure 7. In this simulation we have used a $Y = 10$ Mt TNT equivalent energy for the source. This larger yield leads to faster and more complete disruption of the object. The initial setup ($t = 0$ sec) image shows the rubble pile composition of the PHO model filled with uniform 5 m radius rocks. The evolution of this disruption model is qualitatively different from that of the uniform composition object. Here the high pressure blast wave is able to shock more of the (weaker) PHO target material to a higher pressure, which results in vaporization of more material, and breakup of the non-vaporized mass into much smaller fragments.

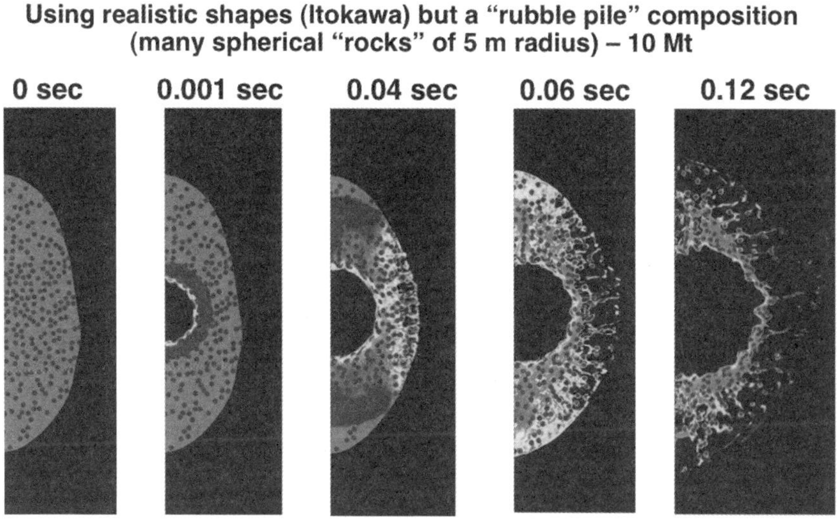

Fig. 7: An example of the RAGE calculation of the disruption of a non-uniform composition model of asteroid Itowaka by a massive explosion. Here we have used uniform size 5 m radius "rocks" to fill the shape contour. The model shows significant disruption of the "asteroid".

These simulations are examples of the variety of simulations we are pursuing in both 2D and 3D. Future models will include more realistic material physics and a variety of internal compositions.

REFERENCES

1. Artemieva, N. (2002) "Tektite Origin in Oblique Impacts: Numerical Modeling of the Initial Stage," in *Impacts in Precambrian Shields*, eds. J. Plado and L. J. Pesonen, Springer.
2. Abramov, O. and D.A. Kring (2004) "Numerical modeling of an impact-induced hydrothermal system at the Sudbury crater," *J. Geophys. Res.*, **109** (E10).
3. Melosh, H.J. (1989) *Impact Cratering: A Geologic Process*, Oxford University Press.
4. Glasstone, S. and P.J. Dolan (1977) *The Effects of Nuclear Weapons*, U.S. Government Printing Office.
5. Greeley, R. (1994) *Planetary Landscapes*, Chapman and Hall.
6. Boslough, M.B.E. and D.A. Crawford (2008) "Low-altitude airbursts and the impact threat. " *Int. J. Impact Eng.*, **35** (12), 1441-1448.
7. Herald, D. and e. al. (2009) "2009 DD45." *Minor Planets Electronic Circular* **2009**(F01).
8. Martinez, C. (2009) "New NASA Images Indicate Object Hits Jupiter," *JPL News*

& Features, http://www.jpl.nasa.gov/news/news.cfm?release=2009-112.

9. Zaitsev, A. (2006) "Near Misses," presented at International Seminars on Nuclear War and Planetary Emergencies (Erice, Italy).
10. Brown, P., D.O. ReVelle, et al. (2008) "Analysis of a crater-forming meteorite impact in Peru." *J. Geophys. Res.*, **113** (E9), E09007.
11. Brown, P., R.E. Spalding, et al., "The flux of small near-Earth objects colliding with the Earth," *Nature*, **420** (6913), 294–296 (2002).
12. Ceplecha, Z. (1992) "Influx of interplanetary bodies onto Earth," *Astron. Astrophys.*, **263** (1-2), 361–366.
13. Collins, G.S., et al. (2005) "Earth Impact Effects Program: A Web-based computer program for calculating the regional environmental consequences of a meteoroid impact on Earth," *Meteorit. Planet. Sci.*, **40** (6), 817–840.
14. Yeomans, D., "NEO Basics," *Near Earth Object Program*, http://neo.jpl.nasa.gov/faq/ June 28, 2010.
15. Atkinson, H., Crispin Tickell, and David Williams (2000) Report of the Task Force on potentially hazardous Near Earth Objects. London, British National Space Centre.
16. Boslough, M.B.E. and D.A. Crawford (2008) "Low-altitude airbursts and the impact threat. " *Int. J. Impact Eng.*, **35** (12), 1441-1448 (2008).
17. Norton, R.O. (2002) *The Cambridge Encyclopedia of Meteorites*, Cambridge University Press.
18. Holsapple, K.A. (2007) "Spin limits of Solar System bodies: From the small fast-rotators to 2003 EL61," *Icarus*, **187** (2), 500–509.
19. Fujiwara, A., J. Kawaguchi, et al. (2006) "The Rubble-Pile Asteroid Itokawa as Observed by Hayabusa," *Science*, **312** (5778), 1330–1334.
20. Rayman, M.D., T.C. Fraschetti, et al. (2006). "Dawn: A mission in development for exploration of main belt asteroids Vesta and Ceres, " *Acta Astronautica*, **58** (11), 605-616.
21. McMillan, R.S.M., A.K. Walker, R.G. Wright, et al. (2009) "NEOWISE: Proposed Discovery of Near-Earth Objects in the Infrared by the WISE Mission," *AAS* (American Astronomical Society Meeting, 2-8 January, Long Beach, California).
22. Veres, P., R. Jedicke, et al. (2009) "Detection of Earth-impacting asteroids with the next generation all-sky surveys," *Icarus,* **203** (2), 472–485.
23. Melosh, H.J. (1994) "Nonnuclear Strategies for Deflecting Comets and Asteroids," in *Hazards Due to Comets and Asteroids*, Ed. T. Gehrels, U. Arizona Press, 1111–1132.
24. A'Hearn, M., M. Belton, et al. (2005) "Deep Impact: A Large-Scale Active Experiment on a Cometary Nucleus," *Space Science Reviews*, **117** (1), 1-21.
25. Walker, J.D. and S. Chocron (2008) "Near-Earth object deflection using conventional explosives," *Int. J. Impact Eng.*, **35** (12), 1473–1477.
26. Berkhouse, L., S.E. Davis, et al. (1983) "Operation Dominic 1 1962", *Defense Nuclear Agency*: 440.
27. Agostinelli, S. and e. al. (2003) "G4a simulation toolkit." *Nuclear Instruments and Methods in Physics Research Section A: Accelerators, Spectrometers, Detectors and Associated Equipment*: 506.

28. Booth, T.E., F.B. Brown, et al. (2008) MCNP5 1.50 Release Notes.
29. Benz, W. and E. Asphaug (1999) "Catastrophic Disruptions Revisited." Icarus **142**:5-20.
30. Ogden, D.E., G.A. Glatzmaier, et al. (2008) "Effects of vent overpressure on buoyant eruption columns: Implications for plume stability." Earth and Planetary Science Letters **268**(3-4):283-292.
31. Gittings, M.L., et al. (2008) "The RAGE radiation-hydrodynamic code," *Comput. Sci. Discovery*, **1:**(1).
32. Gisler, G.R. (2005) "Two-Dimensional Convergence Study of the Noh and Sedov Problems with RAGE: Uniform and Adaptive Grids," Los Alamos National Laboratory report LA-UR-05-2809.
33. Kamm, J.R., W.J. Rider, et al. (2002) "Space and Time Convergence Analysis of a Crestone Project Hydrodynamics Algorithm," Los Alamos National Laboratory report LA-UR-02-5962.
34. Kamm, J.R. (2000) "Investigation of the Reinicke and Meyer-ter-Vehn Equations: 1. The Strong Conduction Case," Los Alamos National Laboratory report LA-UR-00-4304.
35. Baltrusaitis, R.M., M.L. Gittings, et al. (1996) "Simulation of shock-generated instabilities," *Physics of Fluids*, **8** (9) 2471–2483.
36. Zoldi, C.A. (2002) "A numerical and experimental study of a shock-accelerated heavy gas cylinder," Ph.D. thesis, SUNY Stony Brook.
37. Holmes, R.L., G. Dimonte, et al. (1999). "Richtmyer Meshkov instability growth: experiment, simulation and theory," *J. Fluid Mech.*, **389:**55–79.
38. Goldman, S.R., C.W. Barnes, et al. (2000) "Production of enhanced pressure regions due to inhomogeneities in inertial confinement fusion targets," in *Proc. 41st Annual Meeting of the Division of Plasma Physics of the American Physical Society* (DATE, Seattle, Washington), *AIP*, **7**:2007–2013.
39. Schappert, G.T., S.H. Batha, et al. (2001) "Rayleigh-Taylor spike evaporation," *Phys. Plasmas*, **8** (9), 4156–4162.
40. Foster, J.M., B.H. Wilde, et al. (2002) "Supersonic jet and shock interactions," in *Review, Tutorial and Invited Papers from the 43rd Annual Meeting of the APS Division of Plasma Physics* (11-15 November, Long Beach, California), *AIP*, **9**: 2251–2263.
41. Wilde, B.H., P.A. Rosen, et al. (2000) "Simulations of Supersonic Jet and Shock Interaction Experiments at OMEGA," *APS Meeting Abstracts*.
42. Parker, K., C. J. Horsfield, et al. (2004) "Observation and simulation of plasma mix after reshock in a convergent geometry," in *Proc. 45th Annual Meeting of the APS Division of Plasma Physics* (DATE, Albuquerque, New Mexico), *AIP*, **11:** 2696–2701.
43. Foster, J.M., B.H. Wilde, et al. (2005) "High-Energy-Density Laboratory Astrophysics Studies of Jets and Bow Shocks," *Astrophys. J. Letts.*, **634** (1), L77–L80.
44. Lanier, N.E., G.R. Magelssen, et al. (2006) "Validation of the radiation hydrocode RAGE against defect-driven mix experiments in a compressible, convergent, and miscible plasma system," *Phys. Plasmas*, **13** (4), 042703.
45. Lanier, N.E., J. Workman, et al., (2007) "Highly resolved measurements of defect

evolution under heated-and-shocked conditions," *Phys. Plasmas*, **14** (5), 056314.

46. Plesko, C.S., E. Asphaug, et al. (submitted) "Hydrocode Validation against Laboratory-scale Impacts into Basalt." *Meteoritics and Planetary Science*.
47. Nakazawa, S., S. Watanabe, et al. (2002) "Experimental Investigation of Shock Wave Attenuation in Basalt." *Icarus* **156**:539-550(12).
48. Holsapple, K.A. (1993) "The Scaling of Impact Processes in Planetary Sciences." *Ann. Rev. Earth Planet. Sci.* **21**:333-73.
49. Plesko, C.S. (2009) Automated Feature Detection and Hydrocode Modeling of Impact-Related Structures on Mars, U. C. Santa Cruz.
50. Pierazzo, E. and e. al. (2008) "The Impact Hydrocode Benchmark and Validation Project." *Earth and Planetary Science Letters*.
51. Housen, K.R. and K.A. Holsapple (2003) "Impact cratering on porous asteroids." *Icarus* **163**:102-119.
52. Housen, K.R. and K.A. Holsapple (1999) "Scale Effects in Strength-Dominated Collisions of Rocky Asteroids." Icarus **142**:21-33.
53. Johnson, G.R. and T.J. Holmquist (1994) "An improved computational constitutive model for brittle materials." *High-pressure science and technology* 1993. S.C. Schmidt, J. W. Shaner, G.A. Samara and M. Ross. Colorado Springs, Colorado (USA), *AIP*. **309:**981-984.
54. Korycansky, D.G. and E. Asphaug (2003) "Impact evolution of asteroid shapes 1. Random mass redistribution." *Icarus* **163**:374-388.
55. Ahrens, T.J. and A.W. Harris (1994). "Deflection and fragmentation of NEAs." *Hazards Due to Comets and Asteroids*.
56. Ostro, S.J., L.A.M. Benner, et al. (2004) "Radar observations of asteroid 25143 Itokawa (1998 SF36)." *Meteoritics and Planetary Science* **39**:407-424.

DEFENSE AGAINST ASTEROID-COMET NUCLEUS DANGERS—A REAL TASK OF TODAY

ANATOLY V. ZAITSEV
NPP "Planetary Defense Center
Moscow, Russia

INTRODUCTION

Since it is well recognized and agreed that there exists a real danger to Earth and its biosphere from asteroid and comet nucleus impacts, we do not need to elaborate these circumstances. However, because the asteroid-comet threat is a serious risk factor for our civilization, development of measures for impact prevention should become one of the most important tasks that should be solved by mankind in the twenty-first century.[1-9] Evidence for the possibility to find a solution is based on the technologies developed in the former USSR, in the USA, and in other countries in which components for planetary defense systems (PDS) were developed, including space and missile systems, nuclear weapons, communications facilities, navigation equipment and controls, etc. Thus, there exists a unique possibility at this time to apply them not for destruction but for protection of all mankind.

These tools provide the main methods for creation of an international planetary defense system, "Citadel,"[10-12] to defend the Earth from the asteroid-comet threat. One of the main tasks is the official justification and approval for the creation of the system for defense of the Earth against the asteroid and comet nucleus impact danger, based on the existing technologies in the next five to seven years,

The "Citadel" project is based on technologies available in Russia and CIS, however, during creation of the system it is planned to use all the best achievements of mankind worldwide.

PRINCIPLES OF ARCHITECTURE OF THE "CITADEL" PLANETARY DEFENSE SYSTEM

It is obvious that it is practically impossible to create and keep in a state of readiness a system of defense against large asteroids and comet nuclei that may collide with Earth on average in intervals ranging from hundred thousands to tens of million of years.

At the same time, the organization of defense against small bodies that collide with the Earth sufficiently frequently is not only possible, but extremely necessary. For this purpose the echelon (level) of short-term (operational) reaction (ESTR) of the PDS,[10,13-16] which will be in constant readiness and intended for defense against relatively small (diameters from tens to hundreds of meters) asteroids and nuclei of comets, and which may be detected in time periods of only a few days, weeks, or months prior to the collision shall be created. As this takes place, these objects comprise 99.5% of the total number of near-Earth objects (NEOs).

If necessary, it may be possible to form sufficiently quickly an echelon (level) of long-term reaction (ELTR) on the basis of ESTR, intended for reacting to the more dangerous celestial bodies (DCBs), through mobilization of the potential of all countries with missile and space capabilities and nuclear explosives. That is, the ELTR will exist in the "virtual" form as an international project, allowing mobilization of necessary methods: launch vehicles (LV), spacecraft (S/C), spacecraft launch sites, etc., only in the case if a dangerous situation arises.

Both these PDS levels shall meet the following main requirements.[17-20]

- Provide a defense against any DCBs moving along the trajectories leading to impact on Earth with an approach time in the range from few days to tens of years.
- Guarantee the use of PDS for the defense of any country or object, excluding a possibility arising from the "non-use dilemma," which may consist, for some reasons, in refusing to defend a certain territory.
- Guarantee the responsiveness and reliability of warning about space danger, excluding a possibility arising from the "announcement dilemma," consisting in the intentional delay or even misinformation of data about a DCB.
- Ensure nonuse the PDS means for military purposes (exclude the "use dilemma").
- Provide the detection and interception of a DCB at the maximum possible range from the Earth or another protected object.
- Bring minimal damage to the Earth, near-Earth space, etc., from DCB fragments and consequences of using methods of countermeasures on a DCB, including nuclear explosives.
- Maintain and enhance capabilities using replacements of old technologies by new ones.

Many other constraints must be taken into account for PDS development and deployment that follow from its specificity. The feasibility of these requirements represents a sufficiently complicated problem. Nevertheless, the experience of development of the complicated rocket and spatial combat and other systems provides the confidence in the possibility of their fulfillment.

Based on these requirements it is possible to distinguish the following as the most important principles on which the PDS architecture shall be constructed:

- The PDS shall be based on the permanently acting level of short-term (operational) response and service of global control of outer space.
- The service of the outer space control shall include the ground-based and space-borne methods united in the common global international network.
- The maintenance of DCB interception technology shall be ground-based and include a few regional segments, developed on the basis of missile, space, and nuclear capabilities of Russia, USA, and a number of other countries.
- The developmental testing of the PDS methods shall be combined with fulfillment of different science and technological programs and, in particular, space related methods.

Let us consider the possible configurations of the PDS architecture.

The scheme of construction and operation of the STR "Citadel-1"

The STR has to incorporate the international ground- and space-based surveillance service, two regional segments of the reconnaissance and interception service—the European-Asian, "East," and the American, "West," and, accordingly, two regional centers of planetary defense (CPD) (Figure 1).[10, 11, 21]

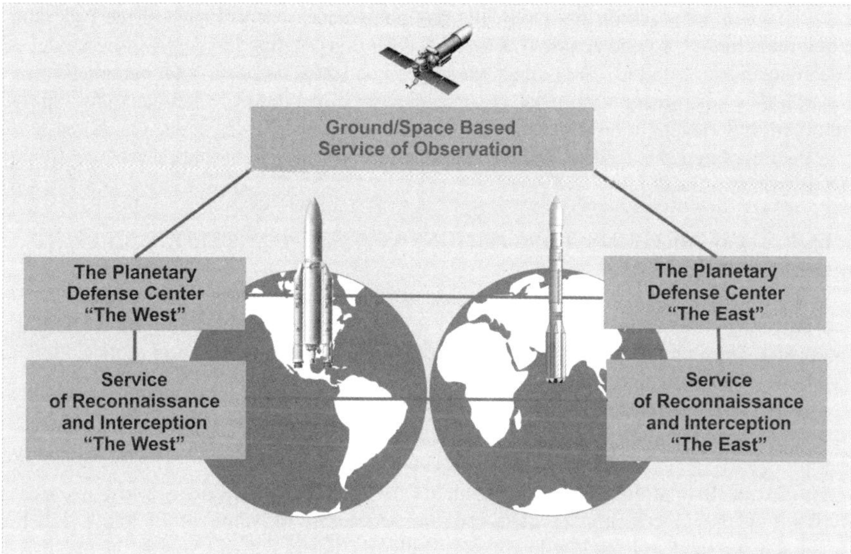

Fig. 1: Composition of the PDS "Citadel-1".

The basis for the ground- and space-based surveillance service will be formed by the space surveillance segment (SSS). As one of possible variants of construction of the operative SSS, the SSS "Cone"[22] can be used.

The regional segments of the ground- and space-based reconnaissance and interception service are meant for a refinement of the DCB characteristics with the reconnaissance spacecraft and for deflecting the DCB from its impact trajectory or destruction of the DCB by the intercepting spacecraft.

The ground- and space-based reconnaissance and interception service must contain:

- Rocket-space complexes with a reconnaissance spacecraft and an intercepting spacecraft;
- Ground-based infrastructure incorporating the launch preparation hardware of the spacecraft, the launch itself, and its controls.

The STR of the PDS will operate as follows:

After identification of a DCB all ground- and space-based systems available in the world will start to observe it. The Planetary Defense Centers using information acquired from the ground- and space-based systems will evaluate the degree of hazard and methods for its prevention will be developed. Upon coordination of a plan of the methods at the international level, instructions to launch the reconnaissance spacecraft and then to the intercepting spacecraft will be issued.

When approaching the DCB, the reconnaissance spacecraft using the patrolling equipment has to provide acquired data needed for constructing the engineering model of the dangerous asteroid or comet nucleus. For this purpose, the reconnaissance spacecraft's instrumentation must incorporate: high-resolution video cameras, an IR multispectral camera, a laser or radar finder, and some other instruments. On the basis of the data acquired the method for the interception and the countermeasures against the DCB will be refined.

CHOICE OF METHODS FOR ACTION AGAINST DANGEROUS CELESTIAL BODIES

To prevent asteroid-comet hazards, many variants of methods for action on a DCB are proposed. They can be broken down into two groups differing by duration of the action:

- The methods of short-term (impulse) action, and
- The methods of gentle long-time action.

To the first group belong the methods of sudden impact (kinetic) and explosive types. For the kinetic impact methods, the spacecraft or some of its parts, kinetic impactors, or small celestial bodies ("space billiards") can be used. Among the explosive methods are nuclear devices.

The long-term action group can be divided into the methods of direct and distant action. The direct-action methods are rocket propulsion systems of various types (using chemical propellant; electric or nuclear engines, using substances from celestial bodies, and so on), methods for changing the DCB's albedo, etc. The distant-action means are laser facilities, solar concentrators, "gravitational tractor," and so on.

The first-group of methods are more flexible, they can be used both for deflection of a DCB and for its destruction.

The second-group of methods can be used only for the deflection of a DCB. Taking into account its low thrust-to-weight ratio because of the big masses of a DCB, these methods must be able to operate for many months and even years, in order to impart the necessary impulse to deflect the DCB in its orbit. Therefore, their operative use is impossible for practical purposes. Thus, only the impulsive-action methods—kinetic impacts and explosives—can be applied in the short-term reactions.

It has been estimated[23] that stony asteroids with a diameter between 50 to 150 m can be destroyed by kinetic impactors with a mass of about 20 to 50 t. At small collision velocities they become inefficient, whereas nuclear devices (ND) can be applied over the whole range of possible DCB parameters. The efficiency of the ND application to the defense against DCBs was presented by Wood et al.[24], Shubin et al. [25], Huebner et al.[26] and others. The high efficiency of NDs, and their characteristics such as compactness, reliability, and convenience to service, makes the ND a main defense system in the STR against DCBs.

This paper is concerned with a variant of the ND application to a sub-surface explosion as the most efficient means of counteraction. An ND design for a single application of ~2 Mt of energy[27] was taken as a basis. It allows crushing a DCB of 200 to 250 m size, i.e., objects similar to asteroid Apophis. In the case of a group application of NDs[28] the efficiency of action on a DCB may be increased significantly.

THE INTERCEPTION OF A DCB BY SHORT-TERM REACTION METHODS

Strict requirements for the effectiveness of a launch of the spacecraft, used in the STR method, dictate using launch vehicles (LV) with a minimum time for launch preparation. For this purpose the LV Zenit, whose launch preparation period amounts to 1.5 hours total, is most promising.

This LV makes it possible to realize an acceleration of the reconnaissance and intercepting spacecraft up to asymptotic velocities of departure from Earth as high as 12 and 3.3 km/s, respectively. The masses of these spacecraft comprise approximately 120 and 4000 kg.

As an example, let us take one of the most critical options as an initial precondition, when the time-period from the moment of the asteroid detection up to its impact with the Earth does not exceed 3 days. Let us assume that the asteroid's velocity is equal to 50 km/s, which is evidently close to the extreme velocities of such objects.

The events are illustrated in the sequences shown in Figure 2:

T = 0. Asteroid detection. Warning. Beginning of preparation of the rocket/space facilities to the launch. Involvement of the additional observation facilities. Updating of the asteroid parameters. Making a decision on the S/C launch.
T ≈ 12 hours. First intercept S/C launch.
T ≈ 18 hours. First reconnaissance S/C launch.
T ≈ 24 hours. Second (reserve) reconnaissance S/C launch.
T ≈ 48 hours. Second (reserve) intercept S/C launch.

It is clear, that such rate of launches can be provided only at the conditions of a number of possible LV launch azimuths and at launches from different Space Launch Sites.

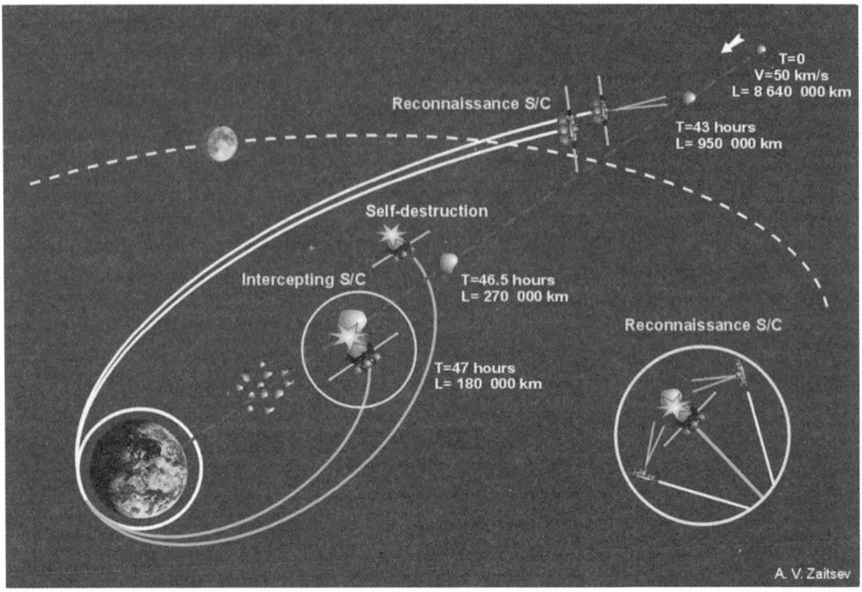

Fig. 2: The Scheme of the operative interception.

The first intercept S/C is launched with the aim to encounter the asteroid at a maximum distance from the Earth. The reconnaissance S/C, due to its small mass, will be accelerated up to considerably greater velocity and, by outstripping the first S/C, it will be the first to meet the asteroid.

On the basis of the data acquired from these S/C, the Centers of Planetary Defense generate the model of this body that is utilized for preparation of all necessary data, which are to be transmitted on-board of the intercept S/C. All these operations last about 8 hours. Then the intercept S/C performs the maneuver of approaching to the asteroid. At their collision the nuclear warhead is exploded and the asteroid is distracted or deflected from the Earth impact trajectory.

A possible overall appearance of the reconnaissance spacecraft is shown in Figure 3. When built, the most recent accomplishments in the world's astronautics technology are conceived to be used. This would make it possible to develop the spacecraft with the minimum overall dimensions and mass characteristics (with a mass of about 100 kg).

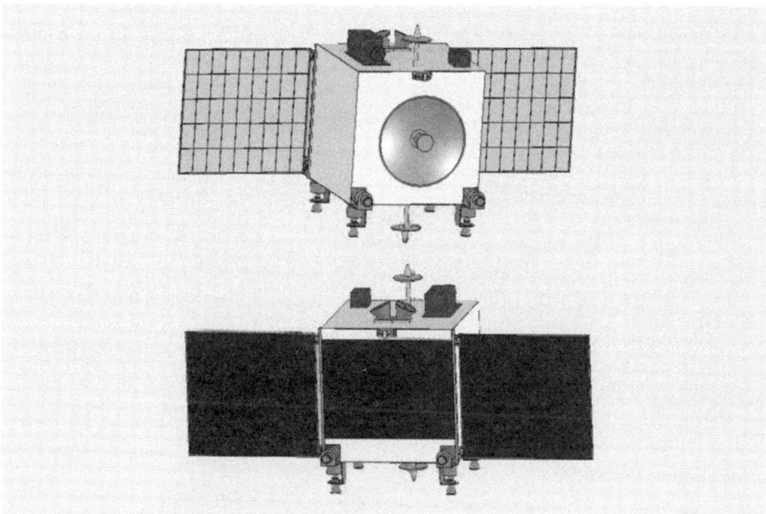

Fig. 3: The overall appearance of the S/C- reconnaissance.

The S/C-interceptor is intended to deliver the payload to deflect the dangerous celestial body. A general view of S/C-interceptors is shown in Figure 4. On board is the module for interception with a nuclear device or hypervelocity impactor.

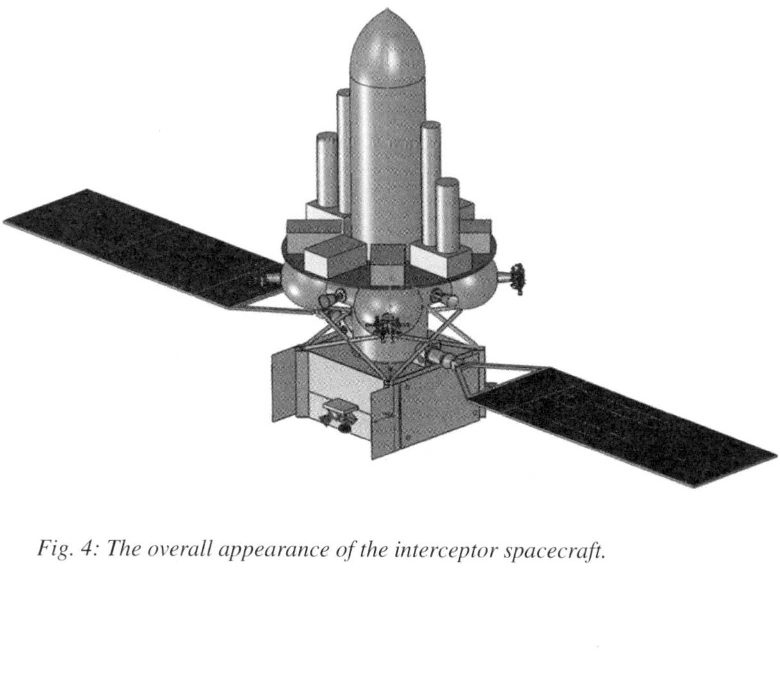

Fig. 4: The overall appearance of the interceptor spacecraft.

S/C mass.........................- 4000 kg
Including:
The module of interception - 3500 kg
(Energy of nuclear device.....- 1 - 2 Mt)

It should be designed and built by space and other agencies of the world.

Simulations of the DCB interception process showed the velocity requirements for the interception spacecraft to rendezvous and the needed accuracy of the DCB ephemeredes.[29] For instance, the magnitude of the interception delta-v will be about 600 m/s with the relative velocity of the intercepting spacecraft and the DCB as high as 60 km/s.

The scenario of a DCB interception is shown in Figure 5. During approach to the DCB the interception module separates from the spacecraft and heads for the DCB with the aid of low-thrust engines. The orbital module continues its motion on the flyby path and follows the interception process implementing the reconnaissance. In the case of a miss or the need for additional action on the DCB another intercepting spacecraft is used.

Fig. 5: A scenario of a DCB interception.

In forming the operative interception scheme it is obviously necessary that it be governed by "the minimum damage criterion" such as the undesirability of explosions of the ND in the Earth magnetosphere and possible damage to objects on Earth and in near space by DCB fragments.

If necessary, the ND power in the intercepting spacecraft may be increased. For this purpose it is required to perform the docking of two units in near-earth orbit or to use, with an available time reserve, the high power LV Proton.

In both cases the ND mass can be nearly doubled and the ND power can attain 10 Mt.[30] In this case, or with the method of groups of NDs, it will be possible to destroy a 500 m stony asteroid or deflect an even larger DCB from an impact trajectory.[28]

The concept of the proposed PDS suggests deflection as the main objective of defense against DCBs. In some events, however, in operating the STR a situation may arise when the DCB would be unintentionally destroyed. This may happen when the energy necessary to deflect a DCB exceeds the energy for destroying it. In this case the DCB fragments can be hazardous. To prevent it, the rocket-space defense system may be applied to intercept the fragments before their entry into the Earth atmosphere.[31]

Thus, having combined a few observational, reconnaissances, and intercepting spacecraft into "operational patrol", we shall be able to ensure the defense against 99.5% of Earth threatening asteroids.

Progress in observation and detection techniques promises that the rest (0.5%) of asteroids the size over 0.5 to 1.0 km would be disclosed years and even decades in advance of a potential collision. In some cases both methods for LTR and STR can be used. Specifically, it has been demonstrated in the case of Apophis[32] that if a kinetic impactor is used before 2029, the impact mass may between a few hundred of kilogram to several tons subject to timing.

Therefore, for the deflection of large asteroids the intercepting spacecraft at the level of rapid response reaction may be used. In such a case, the interception scheme will be similar to that of standard missions to Solar System objects.

INTERNATIONAL LEGAL BASES FOR ENSURING THE PLANETARY DEFENSE

Evidently, as the first step in building the PDS, we should prepare and approve an international treaty that will serve as a foundation for creating the PDS and financial support for it. It may be executed in the format of an international treaty "On the principles of ensuring the Earth's defense against asteroid and comet impact hazards"[20] (hereafter referred to as the Treaty).

We outline general principles that will be needed:

- When drafting the Treaty it is desirable to have a concept for the PDS, drafted by the international community. "Citadel"[10] is one of the options to be considered.
- We need to attribute international status to the legal instruments governing the PDS formation, use, and execution.
- The legal instruments should be of a lawfully binding nature for the countries involved in the establishment of the PDS.
- The legal instruments may be structured according to the basic areas to be encountered (scientific, technical, organizational, financial etc.), and the stages of building of the PDS etc.

Furthermore, the treaty should contain a provision that prohibits countries united in a single military and political association to be the sole or independent creators and developers of the PDS. This is necessary in order to avoid a so-called "no-use dilemma,"[17] which allows for the possibility for the association to refuse to defend other countries or exert influence on them with a view to change a geopolitical situation or even to destroy them.

It is necessary to include into the treaty general principles for allocating and using a special international fund that may be called a security fund for humanity[10], in order to guarantee financial support for PDS development, use, management, and modernization.

CONCLUSION

The research results make it possible to conclude:

1. Russian and foreign technologies allow for practical stages of development of the International Planetary Defense System against the asteroid-comet hazard.
2. The System shall be based on the level of short-term (operative) reaction for protection from suddenly appearing dangerous celestial bodies the size from tens to hundreds of meters, which can be discovered on timescales as short as a few days, weeks or months before their collision with Earth.
3. If necessary, it will be possible to expand the level of short-term reaction to a level of long-term reaction to counteract larger, dangerous celestial bodies, which would allow completely solving the problem of Earth defense against asteroids and, in part, the comet nucleus hazard (for comet nuclei 0.5 to 1 km in size).
4. Tackling the issue of defense against asteroid and comet impact hazards requires developing international legal instruments that could take the form of an international Treaty "On the principles ensuring the Earth's defense against asteroid and comet impact hazards."
5. For providing financial security of development, maintenance, and modernization of the system there is a need to establish a special international fund, that may be given the name the Insurance Fund of Mankind.
6. The Organization of the United Nations, more specifically, the international Center of Planetary Defense, specially created for these purposes under its aegis, and its regional offices, could coordinate the PDS formation, use, and management.

So "Citadel" PDS project can become the first global project of mankind in the third millennium that would turn the Earth into unassailable fortress protecting us from space threat.

ACKNOWLEDGEMENTS

I thank Dr. W. Huebner for his helpful comments, valuable advices and updating of translation on this manuscript, my colleagues E.G. Chistov, S.V. Ivanov and B.D. Yakovlev for assistance in preparing this manuscript.

REFERENCES

1. Zaitsev, A.V. Proposals on Development of the System of Prevention of the Earth Collision with Asteroids and Comets (Re-orientation of Works Carrying Out in the Framework of the SDI into Peaceful Objectives). Report to the General Secretary of the Central Committee of the Communist Party of the Soviet Union, N 629203 from 10.20.86, Babakin SRC, 1986. (17 pp.)
2. Zaitsev, A.V. Some Principles of Construction of the System of Prevention of the Earth Collision with Asteroids and Comets. Proceedings of 23rd readings of K.E. Tsiolkovsky. Section: Problems of the rocket/space technology. Moscow. IHST of AS of the USSR, p. 141-147, 1989.
3. Yeomans, D.K. (1991) "Killer Rocks and the Celestial Police." *The Planetary Report*. Vol. XI, No. 6. pp. 4-7.
4. The Spaceguard Survey: Report of the NASA International Near-Earth-Object Detection Workshop. (Ed. D. Morrison). JPL/CIT, Pasadena, CA, Jan. 25, 1992.
5. Resolution 1080 (1996) "On the Detection of asteroids and comets potentially dangerous to mankind." Parliamentary Assembly of the Council of Europe, 20 March 1996.
6. U.S. Congressional Hearings on Near Earth Objects and Planetary defense, 1998.
7. Review of the implementation of the recommendations of the Third United Nations Conference on the Exploration and Peaceful Uses of Outer Space (UNISPACE III). Note by the Secretary-General. Vienna. 1999. (141 pp).
8. Report of the Task Force on Potentially Hazardous Near Earth Objects. UK. 2000. (56 pp).
9. Organization for Economic Co-operation and Development (OECD). Global Science Forum. Workshop on Near Earth Objects: Risk, Policies and Actions. Final Report. January 20-22, 2003. Frascaty, Italy. (13 pp).
10. Zaitsev, A.V. Planetary Defense System "Citadel". The conceptual project. Moscow, 2000. (70 pp).
11. Bashilov, A.S., Volk, I.P., Zaitsev, A.V., Konyukhov, S.N., Pichkhadze, K.M., Pobedonostsev, K.A. The Planetary Defense System "Citadel", Proposals. Planetary Defense Center. 2002. (23 pp).
12. Zaitsev, A., Koroteev, A., Liaschuk, B., Popov, S. Rocket-Space Means of Echelon of Short-Term Reaction of the Planetary Defense System. International Conference "Asteroid-Comet Hazard-2009" (ACH-2009), St.-Petersburg, Russia. Book of Abstracts. 2009. p. 40.
13. Kovtunenko, V.M., Zaitsev, A.V. (1995) "Protecting Earth from Asteroid Hazards is a Real Task for the World Space States." *Space Bulletin*, Vol. 2, N4, 25-27.
14. Kovtunenko, V.M., Zaitsev, A.V., et al. *Principles of Construction of the System of Prevention of the Earth Collision with Asteroids and Comets*. Lavochkin Association, Babakin SRC. 1995. (69 pp).
15. Kovtunenko, V.M., Simonenko, V.A., Rogovsky, G.N., Papkov, O.V., Bojor, J.A., Zaitsev, A.V., Kotin, V.A., Feshin, I.V., Maglinov, I.D. Opportunity to Create the System For Space Protection of the Earth Against Asteroids and Comets on the Base of Modern Tecnology. Proceedings of the Planetary Defence

Workshop. Lawrence Livermore National Laboratory. Livermore, California, May 22-26, 1995, 453-463.
16. Huebner, W.F., Patrick, W.C., Walker, J.D. Countering Cosmic Collisions. Even relatively small space objects can cause damage to Earth. SwRI 2009, *Technology Today*.
17. Zaitsev, A.V. (1997) "Some of Problems and Sequences of Development of the Planetary Defense System." Chelyabinsk Scientific Center News, Special Issue *Space Protection of the Earth*. (pp. 243-246).
18. Zaitsev, A.V. Safety Problems While Nuclear Devices Using in Planetary Defense System. In: "Nuclear Safety: Social and humanitarian structures". Moscow. 1998. 102-104.
19. Zaitsev, A.V., Klapovsky, A.A., Koulik, S.V. Organizational and legal aspects of Planetary Defense System creation and application. All-Russian conference "Asteroid-Comet Hazard – 2005" (ACH – 2005), 3-7 October 2005, St. Petersburg. Materials of the Conference, 148-150.
20. Zaitsev, A.V., Klapovsky, A.A. About the Approach to Formation of International-Legal Bases of Ensuring Planetary Defense. International Conference "Asteroid-Comet Hazard-2009" (ACH-2009), St.-Petersburg, Russia. Book of Abstracts. 2009. 188.
21. Zaitsev, A., Pitchkhadze, K., Tchesnokov, A. Possible Approaches to Implementation of "Citadel-1" International Planetary Defence System Project Report on Forty-fourth session of the Scientific and Technical Subcommittee of the Committee on the Peaceful Uses of Outer Space (12-23 February 2007).
22. Dobrov, A.V., Zaitsev, A.V., Maglinov, I.D., et al. Possible Approaches to Forming of Space Observation Service for Asteroids and Comets. International Conference "Asteroid Hazard-96". Program and abstracts. July 15-19, 1996. St. Petersburg. 49-50.
23. Tedeschi, W.J. Mitigation of the NEO Impact Hazard Using Kinetic Energy. Proceeding of the Planetary Defense Workshop Lawrence Livermore National Laboratory, Livermore, CA, May 22-26, 1995. 313-323.
24. Wood, L., Hyde, R., Ishikawa, M., Ledebuhr, A. Cosmic Bombardment IV: Averting Catastrophe in the Here-And-Now LLNL Doc. No Phys. .Brief 94-029. International Conference "SPE-94", Snezhinsk, Sept. 26-30, 1994.
25. Shubin, O.N., Nechai, V.Z., Nogin, V.N., et al. Nuclear Explosion Near Surface of Asteroids and Comets. Common Description of the Phenomenon. Proceeding of the Planetary Defense Workshop Lawrence Livermore National Laboratory, Livermore, CA, May 22-26, 1995, 383—396.
26. Huebner, W.F., Patrick, W.C., Walker, J.D. Countering Cosmic Collisions. Even relatively small space objects can cause damage to Earth. SwRI 2009, *Technology Today*.
27. Rodionov, V.V., Vozovikov, V.V., Zadvornov, Yu. A., et al. (1997) "Nuclear means for affecting near-Earth-objects. Composition and scheme of operation." Chelyabinsk Scientific Center News, Special Issue *Space Protection of the Earth*, 111-117.

28. Simonenko, V.A. Space impacts mitigation: deflection and dispersion based on nuclear explosions. All-Russian conference "Asteroid-Comet Hazard-2005". Materials of the conference. 3-7 October 2005. St. Petersburg. 2005. 295.
29. Zaitsev, A.V., Dobrov, A.V., Kotin, V.A., Simonov, I.V. Impact experiment for project Space Patrol. International Journal of Impact Engineering, V. 20. Proceedings of the 1996 Hypervelocity Impact Symposium, 839-848.
30. Koblov, P.I., Simonenko, V.A., Ivanov, Yu. A., et al. (1997) "Main requirements to nuclear explosive device (NED) to be used in system of Earth protection against asteroids and comets." Chelyabinsk Scientific Center News, Special Issue *Space Protection of the Earth*, 107-110.
31. Arsenyev, G.N., Semyonov, B.I., Torgovkin, S.N., Treckin, V.V. (2000) "The possibility of two-fold use of missile-space defense systems for solving the problem of asteroid-comet hazard." *Information-measuring and Control Systems.* V. 4, No 5. 5-12.
32. Ivashkin, V.V, Stikhno, C.A. A study of the orbit correction of the asteroid Apophis. International Conference Asteroid-Comet Hazard–2009. Russia, St. Petersburg, September 21-25, 2009. Book of Abstracts. P. 171.

SECTION 9

SESSION 9

MITIGATION OF TERRORIST ACTS

FOCUS: SCIENTIFIC CONTRIBUTIONS TO BIOSECURITY AND MITIGATION OF TERRORISM

LESSONS FROM THE 2009 SWINE FLU PANDEMIC, AVIAN FLU, AND THEIR CONTRIBUTION TO THE CONQUEST OF INDUCED AND NATURAL PANDEMICS

RICHARD L. GARWIN
Thomas J. Watson Research Center, IBM Research Division
Yorktown Heights, USA

The PMP for Mitigation of Terrorist Acts, now co-chaired by Sally Leivesley and Alan Leigh Moore, has long studied the mitigation and even the prevention of pandemic disease from terrorist acts. In this effort we have benefited from experience and motivation of natural epidemics such as smallpox, SARS, and seasonal and specific influenza disease.

For fighting disease:

- **limit occasion for infection—personal protective measures,**
- **vaccine to build immune response to prevent infection**
- **pharmaceuticals to prevent symptoms or to kill bacteria or microbes in the body**

Emphasizing personal protective measures, the 2006 Report[1] of that PMP (sub-panel at that time) reads:

Since 2002, the Mitigation Subgroup has considered measures to reduce the impact of potential terrorist use of biological agents against humans; included have been improved air filtration, vaccination, and pharmaceutical intervention against the agent causing the disease or, in some cases, the toxins responsible for severe illness or death. Over the past year, much public attention has focused on pandemic influenza, such as might arise from reassortment of the Type A (H5N1) avian flu that has been spreading from Southeast Asia, but expert consensus is stronger that a flu pandemic is likely than is the judgment that it will derive from H5N1. Even a recurrence or an image of the 1917-18 H1N1 "Spanish flu" that killed some 50 million people world wide would be a disaster in the modern age of specialization and globalization, and such a pandemic that occurred in the next few years could not be much eased by available stocks of vaccine or antiviral drugs.

Because mitigation of a natural flu pandemic can teach us much about defense against some bioterror agents, because there is a receptive audience to measures against pandemic, and because we have some novel and important perceptions to counter this serious and likely threat to health, life, and society, in the attached paper, "Conquering Pandemic Flu by Practical Measures," we present our analysis and recommendations for countering pandemic flu by nonpharmaceutical means.

---

[1] Permanent Monitoring Panel On Terrorism (PMPT), Fourth Meeting, Erice, 18-22 May 2006, Final Report.

We speak of an epidemic in terms of a single reproductive factor R_0 ("R-naught") and a serial interval ν ("nu"). For the SARS epidemic, R_0 is about 3 and ν about 8 days. For smallpox, R_0 is about 3 and $\nu \approx 14$ days. And for influenza, $R_0 \approx 1.7$-2.4 and $\nu \approx 4$ days. In our paper we take for flu $R_0 = 2$, although we recognize that it will vary from society to society and in various groups within society. Unchecked, an epidemic that begins with N "index cases" would give rise $\nu \approx 4$ days later to 2N additional cases, ν days later to 4N more, 8N more, and so on, so that after M serial intervals there will be $N (1 + R_0 + R_0^{\ 2} + R_0^{\ 3} + ... + R_0^{\ M})$ cases altogether, until the susceptible population is exhausted and a substantial fraction of the population is resistant or even dead.
[emphasis added in bold font]

The paper proposes a set of practical Personal Protective Measures—PPM--that, in principle and if practiced by almost everyone in the population of a large group or society, could essentially eliminate the damages of pandemic flu in that group or society, by reducing the reproduction factor below 1.0 and thus limiting the number of flu infections per index case to the finite sum of the above series—namely to $1/ (1- R_0)$ cases. If such measures reduce an initial $R_0 = 2$ to $R_0 = 2/3$, a 3-fold reduction, each index case would then correspond to a total of 3 infections—not to 2000 in 40 days and doubling every 4 days thereafter. To have any prospect of such benefits, the great majority of individuals must practice modest protective measures; if all complied perfectly except 33%, the number of cases from index case would no longer be finite.

Even if the efficacy of PPM were demonstrated in a test evaluation, which is not yet the case, it is clear that relatively few in society would be ready to use such measures if the necessity presented itself, in the absence of sound information as to what to do and how to do it. The information does not move itself; major staff resources and effort would need to be expended to evaluate the effectiveness not only of the PPM but of the means of communication and persuasion that could be used via schools, place of employment, clubs, pharmacies, and faith-based organizations to persuade people that they should equip themselves and practice the PPM.

Pandemic influenza is unlikely to be a terrorist weapon, both because of its potential for worldwide destruction, even of the society for which a potential bioterrorist some esteem, but also because it is likely to spread worldwide in any case. Avian flu itself, in the form that does not spread from human to human, is a more likely weapon of terrorism, which might be used to do serious damage to the commercial poultry sector of a target country.

Thus, a successful counter to human pandemic influenza does not eliminate a bioterror weapon of concern, but the measures that individuals and groups take to protect against pandemic flu are themselves directly relevant to real bioterror threats such as the smallpox epidemic that would result from human-to-human transmission in a population in which vaccination ceased in 1972 when the elimination of smallpox as a naturally occurring disease was in sight. Furthermore, ***the availability to individuals and groups of practical measures such as hand washing, disposable and improvised masks, elbow bumps instead of handshakes, and improved air filtration, and the frequent practice of such measures to prepare for an eventual flu pandemic, can provide effective tools for reducing the incidence of sickness and death from the more commonly expected bioterror weapons such as anthrax or tularemia.*** *The direct empowerment of individuals and groups, would be an important supplement to measures that may be taken by government, such as an alert that bioterror germs may have been liberated into the atmosphere or building lobby*

There are major differences between protection against pandemic flu and against an anthrax bioterror attack on a city. Protective measures against diseases without significant human-to-human transmission do not have the prospect for reducing casualties by factors of 1000 or more, but only by the protective factor that might be 3 or 5. Against bioterror attack, warning is of greater importance, and those close to the atmospheric release point may not be able to protect themselves against breathing the agent. Still, for an attack on a large city, a factor 3 reduction in casualties might save hundreds of thousands of people. Those who care about mitigating the potential damage from bioterror attacks should welcome the benefit that would come from a capability in-being to use personal protective measures against pandemic influenza.

"It's probably not going to be practical to contain a potential pandemic by merely trying to limit contact between people (such as by travel restrictions, quarantine or even closing schools), but we find that these measures are useful in buying time to produce and distribute sufficient quantities of vaccine and antiviral drugs," said Germann of Los Alamos' Applied Physics Division.

"Based on our results, combinations of mitigation strategies such as stockpiling vaccines or antiviral agents, along with social distancing measures could be particularly effective in slowing pandemic flu spread in the U.S.," added Longini.

(Timothy C. Germann, Kai Kadau, Ira M. Longini, Jr., and Catherine A. Macken, Mitigation strategies for pandemic influenza in the United States, *PNAS 2006 103:5935-5940; published online before print April 3, 2006)*

Day 50

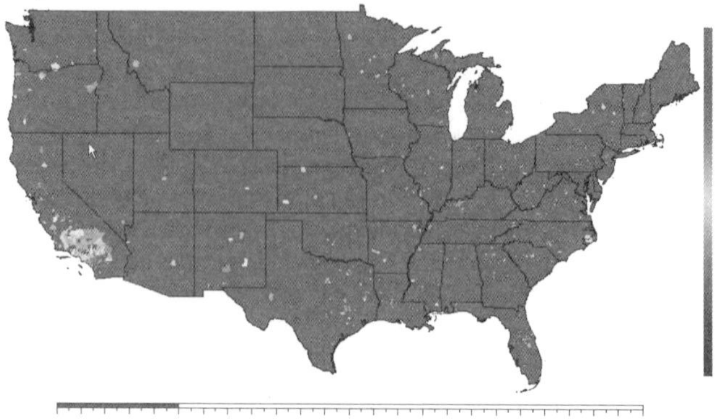

Day 90

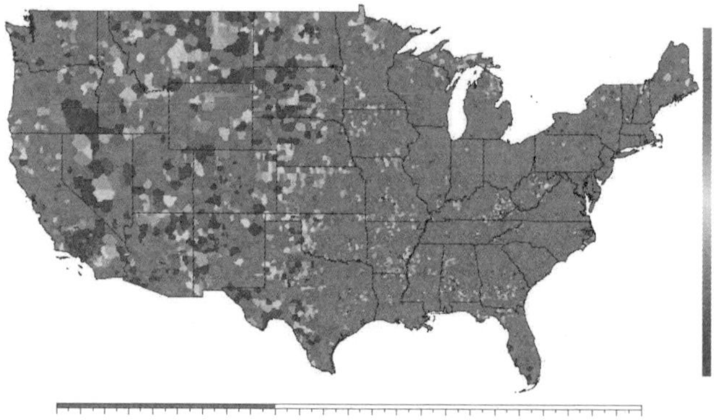

Day 130

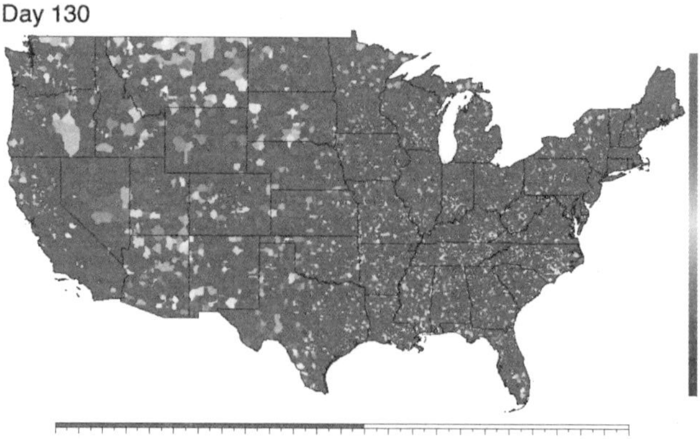

[Type-A Influenza Viruses]

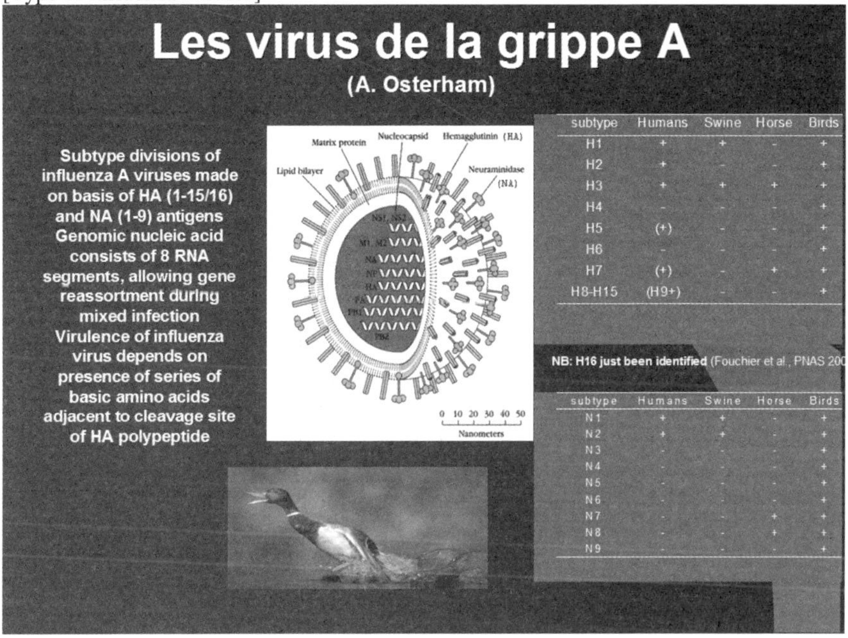

WHAT HAVE WE LEARNED?

- Emerging infections
- Influenza H5N1 2003, high mortality, low transmissibility (among humans)
  - Result: global preparedness
- H1N1, Mexico, April 2009: rapid spread, detected early, could not be stopped, Peaked in Oct-Nov 09
  - Mortality low, mild symptoms, affected healthy young people
  - "30 July 2010—As of 25 July 2010, worldwide more than 214 countries and overseas territories or communities have reported laboratory confirmed cases of pandemic influenza H1N1 2009, including over 18398 deaths." [Annual deaths from normal "seasonal flu" about 250,000 worldwide—so 7%]

<u>Lessons Learned</u>

- Detection mechanisms worked
- Response was national; information reliably available through WHO
- Vaccine development went well
  - Vaccine distribution and vaccination varied among and within countries

Perceived Failures

- International Health Regulations (IHR) and the WHO
  - Moving from Phase 5 to Phase 6 pandemic
  - Undue influence of laboratories and pharmaceutical manufacturers?
- Resulting overemphasis on antivirals and vaccine, vs. personal protection

Successes

- Detection and monitoring
- Technological advances
  - Vaccine developed and tested in 3 months
  - Inhalable vaccine
- International commitment and collaboration
- Absence of panic in public response

Lessons for the Future

- Technical developments
  - Early diagnostics, pre-symptomatic assay of classrooms, offices…
  - Cell-based vaccine production to delay commitment to full stock
  - Develop vaccines against conserved elements of flu viron
  - Emphasize learning details of transmission of particular strain
  - Develop and tune models reflecting detailed transmission means—e.g., via droplets, aerosols, fomites (surfaces to hand to mouth, nose, etc)
- Plan for interaction of security and health structures in case of infection spread by terrorists
- Evaluate potential for personal protective measures—masks, bleach, hand sanitizer, elbow bumps [instead of handshake].
- Change criteria for declaration of Phase 6 pandemic—severity as well as geographic spread.
- Recognize non-identity of public health goals and pharmaceutical company profits.

I thank Diego Buriot for his contributions to this report.

THE ECOLOGY OF INTERNATIONAL SECURITY

CHARLES D. FERGUSON
President, Federation of American Scientists
Washington DC, USA

How much security does more than $1 trillion buy? It may buy very little or even make matters worse. According to the U.S. Congressional Research Service, the United States has spent to date this amount of money on the War on Terror, especially about $784 billion on the Iraq war and $321 billion on the military operations in Afghanistan.[1] Has this massive expenditure made the United States and other countries more secure? From the perspectives of tens of thousands of casualties—including numerous civilian deaths—and the continuing instabilities in Afghanistan and Iraq and growing discontent in the wider region, many could argue that the United States has purchased insecurity. But from the perspective of disrupting terrorist training camps in Afghanistan and the safe haven for al Qaeda in that country, others could argue that the United States has bought some greater security against castastrophic terrorism because planning for and carrying out such acts requires a safe haven. However, recent disclosures by WikiLeaks of more than 92,000 pages of U.S military and intelligence documents indicate that the Taliban and al Qaeda have remained potent in this region.[2] But these leaks may jeopardize the lives of Afghan and coalition forces.

Was this massive expenditure with so little positive change and considerable adverse consequences inevitable? That is, could the U.S. government and, in particular, the Bush administration, have been persuaded not to commit such blood and treasure particularly to the war in Iraq? Even if that particular government was determined to head down the warpath and especially widen the armed conflict to include Iraq, what lessons should we in the scientific community learn in contributing to a more secure world and specifically in informing, interacting with, and influencing policymakers, politicians, and diplomats?

As identified in a recent report by the Royal Society and the American Association for the Advancement of Science (AAAS), science and diplomacy can interact in three conceptual areas:

- "Science in diplomacy": Science can provide advice to inform and support foreign policy objectives.
- "Diplomacy for science": Diplomacy can facilitate international scientific cooperation.
- "Science for diplomacy": Scientific cooperation can improve international relations.[3]

---

1 Stephen Daggett, "Costs of Major U.S. Wars," Congressional Research Service report, June 29, 2010.
2 C.J. Chivers et al. "View is Bleaker than Official Portrayal of War in Afghanistan," *New York Times*, July 25, 2010.
3 The Royal Society and the American Association for the Advancement of Science, *New Frontiers in Science Diplomacy*, January 2010.

While that report discusses many valuable ways in which science and scientists should take part in these issue areas, it does not specifically address the subject of science's role in mitigating terrorism and improving nations' security. This essay will focus on that topic and will apply the conceptual frame outlined in the Royal Society-AAAS study.

The hypothesis here is that a science-informed and science-infused foreign policy will tend to produce greater security for all nations. Unfortunately, foreign policy does not lend itself to reproducible situations such as a biological or physical scientist could create in a laboratory. Each political situation is unique. Thus, this hypothesis is not testable in a strict scientific sense. Nonetheless, scientific analysis can and should reveal and elucidate the systemic structure of international security. For example, instead of a perspective based on one side's politics, a system's view should open analysts' minds to take into account as many sides' political perspectives as identifiable and feasible. Moreover, instead of analysis based on simple cause and effect relationships, a system's view would identify complex feedback loops influencing political perceptions and showing how a multiplicity of factors such as government policies, energy availability, water and food resources, economic development, and cultural issues interact and affect security.

SCIENCE IN DIPLOMACY

What can science offer diplomacy? Science provides a clear way of thinking about vexing problems. The scientific habit of mind asks probing questions that should challenge conventional wisdom. Answers to these questions should be based on solid evidence. For example, in the lead up to the Iraq war, relatively few independent U.S. scientists questioned the Bush administration's view that Saddam Hussein had definitely resurrected weapons of mass destruction programs and that he and his government had been supporting al Qaeda. The U.S. government had hidden behind the cloak of secrecy giving the impression that classified assessments led to these conclusions. Nonetheless, even when the international team of inspectors led by Hans Blix did not find evidence to support these claims, the U.S. government still clung to these beliefs. In this case, independent scientists and analysts should have called for sufficient disclosures of evidence by the U.S. government so as not to reveal information that would have jeopardized sources and methods but would have at a minimum allowed for peer review of the government's assessment. In addition, independent experts should have performed a system's analysis of the potential consequences of an attack on Iraq and the toppling of Saddam's regime. Greater transparency of information (taking into account protecting sources and methods) and rigorous analysis will help promote better policy. The analysis should assess the efficacy of different policy options.

The advice provided by scientific analysis, however, will not be monolithic. Scientists do not have one point of view. The national citizenship of a particular scientist will have influence on his or her viewpoint. Also, not all scientists of the same nationality will have the same perspective. This is why it is fundamentally important to involve as many different scientists as feasible in the analysis. While scientists are national citizens, science fosters a sense of global citizenship. At their best, scientists can transcend national boundaries. In sum, science offers the values of rationality, transparency, and universality.

Scientific advice will be essential in shaping government responses and in helping to prevent overreaction to catastrophic terrorism. Involving a multidisciplinary team of scientists (biological, physical, political, psychological, and social) is essential to give policymakers and decision makers integrated analysis. For example, deciding how to respond to a nuclear attack demands assessing a complex mix of data. The decision maker will need to know the uncertainties in the attribution of the origin of the fissile material used in the improvised nuclear device or intact nuclear weapon from a nation's arsenal. Chemists and physicists will therefore be essential members of that analytic team. In addition, psychologists and terrorist analysts who understand the motivations and command-and-control structure of the terrorist group are needed to assess the likely counter-response of the terrorists.

DIPLOMACY FOR SCIENCE

How can greater scientific cooperation be fostered in the mitigation of terrorism acts? The Erice conference serves as one important example of how a multinational group of scientists of diverse areas of expertise can come together to assess and advise. Governments can also formally link scientists who are government employees to share information and analysis on security threats. For example, the lab-to-lab program has connected American and former Soviet scientists. International government agencies such as those under the auspices of the United Nations have in addition created organizations and forums for scientists to work together. In particular, the International Atomic Energy Agency includes experts from many countries teaming up to improve nuclear safety, safeguards, and security. The latter area is especially relevant for prevention and mitigation of nuclear and radiological terrorism. Scientists in government employment can often feel constrained in offering advice that goes against government policy. Faulty policy may remain unless there are those who can provide credible voices on alternative policies. Consequently, independent scientists can play an important function in being these credible voices.

But what determines credibility? Certainly having the appropriate educational experience is essential. But this may not be sufficient. Scientists with previous government experience or with security clearances to get access to sensitive information can lend additional credibility to scientific groups assessing and advising on security issues. However, these scientists may also have constraints on what they can share with others who do not have such clearances. One group that has had significant influence on U.S. government policy is the JASONS, which includes senior scientists who have access to classified information as appropriate for their analysis. With the JASONS, government has facilitated bringing together the group: a prime example for diplomacy for science. Even when a group of scientists does not possess security clearances, it may still be able to provide highly credible assessments of how to achieve greater security by applying known scientific principles to open source information. Such networking of scientists may foster greater international security.

SCIENCE FOR DIPLOMACY

Environmental stresses, economic development, political tensions, and international security are linked. As eloquently described by renowned ornithologist and environmental scientist Jared Diamond:

> First ask some ivory-tower academic ecologist [from a "First World" country] who knows a lot about the environment but never reads a newspaper and has no interest in politics to list the overseas countries facing some of the worst problems of environmental stress, overpopulation, or both. The ecologist would answer, "That's a no-brainer, it's obvious. Your list of environmentally stressed or overpopulated countries should surely include Afghanistan, Bangladesh, Burundi, Haiti, Indonesia, Iraq, Nepal, Pakistan, the Philippines, Rwanda, the Solomon Islands, and Somalia, plus others." Then ask a First World politician who knows nothing, and cares less, about the environment and population problems to list the world's worst trouble spots: countries where state government has already been overwhelmed and has collapsed, or is now at risk of collapsing, or has been wracked by recent civil wars; and countries that, as a result of their problems, are also creating problems for us rich First World countries, which may be deluged by illegal immigrants, or have to provide foreign aid to those countries, or may decide to provide them with military assistance to deal with rebellions and terrorists, or may even (God forbid) have to send our own troops. The politician would answer, "That's a no-brainer, it's obvious. Your list of political trouble spots should surely include Afghanistan, Bangladesh, Burundi, Haiti, Indonesia, Iraq, Nepal, Pakistan, the Philippines, Rwanda, the Solomon Islands, and Somalia, plus others."[4]

Given this observation, it follows that a promising path to reduce security threats to countries is to improve their economies, their access to clean energy sources, their management of water and agricultural resources, etc. All of these sectors would benefit from scientific collaborations. One approach is to offer technical assistance from the developed to the developing world. This is more of a one-way conduit that often has the underlying assumption that developed world scientists know best and just need to instruct their counterparts in the developing world.

A far more collegial approach is to bring these two groups of scientists together in a real partnership in which each side can learn from the other. Developed world scientists will likely discover that their counterparts have practices that would benefit the developed world and vice versa. The author has recently begun such an endeavor called the International Science Partnership. The pilot project will be in Yemen and will likely focus on the water-energy nexus.[5] One reason Yemen was chosen was to find out if scientists from the U.S. and Yemen can work together to improve the infrastructure in a country that has a very challenging set of problems, including the presence of al Qaeda, a weak central government, tribal unrest, fast growing population, exhausting water

---

[4] Jared Diamond, "The Last Americans," *Harper's*, June 2003.

[5] For more on this project, see www.fas.org.

supplies, significant electricity generation shortfalls, about one quarter of the population unemployed or severely underemployed.

In addition to linking scientists to work on physical projects like water management, scientists and other experts can connect virtually to form extended networks. For example, using the inherent networking capability of the Internet, these professionals can form a virtual community devoted to work on specific subjects such as bio-security, nuclear security, and counter-terrorism. The Federation of American Scientists has recently developed a virtual bio-security center that will use a hub Website to link together many partner organizations. This Website will act as the one-stop-shop for information, news, and analysis on bio-security. Moreover, it will include dual-use case studies and tutorials about the subject from renowned biological scientists and experts in biological weapons.

The next steps for science for diplomacy can be for scientists to take an even more active role in networking to offer peer review and advice for governments. The Erice movement may benefit from an improved Web-hub that would include assessments of options for mitigation of terrorism. This Web portal could also include a database of scientific experts that governments and the news media could contact. The news media especially needs to be able to call on credible, independent scientists to provide the public with non-hysterical advice and analysis immediately following a terrorist attack. One issue that must be addressed, though, is how to assess the credibility of the scientists in the database. Would a panel of experts need to vet those scientists in the database? Who would select the panel of experts? What criteria would determine credibility? These and other concerns would have to be addressed. Nonetheless, there may be significant value to better organize networks of scientists to address mitigation of terrorism, in particular, and improved international security, in general.

ASSESSMENT OF RDD EVENT MEDICAL RESPONSE, RECOVERY, AND MITIGATION IN A WORLD OF ONE SCIENCE

ANNETTE SOBEL, M.D., M.S.
University of Missouri, Vice President
Columbia, USA

PURPOSE OF STUDY AND BACKGROUND

The purpose of this study is to perform an objective assessment of current medical operational capabilities, gaps, and shortfalls pertaining to RDD response in the context of specific representative study parameters (Cs-137, Am-241, and Co-60). Devices such as blood and research irradiators and teletherapy machines containing these radionuclides are considered of greatest relative risk for malevolent acquisition and are in greatest quantity and distribution throughout the U.S. according to the NRC Interim Database (RDD Global Protection Architecture, L. Connell and L. Trost, May 2006). Unlike improvised nuclear devices, RDDs are more likely to result in large-scale medical effects secondary to contamination and psychological effects rather than the wide scale death and destruction due to fission devices (personal communication, Dr. Roger Hagengruber, Director, Institute of Public Policy and the Office of Policy, Security, and Technology, University of New Mexico). Although the global threat of nuclear devices is clearly a high priority for our Nation, the ubiquitous nature of sources for RDDs, namely, Cobalt-60, Americium-241, and Cesium-137 are of great national concern due to the lower security measures employed to safeguard, track, and account for these materials, and due to the pervasive commercial and industrial usage.

The persistence of these materials in the environment makes them particularly attractive as weapons of terror. The half-lives of Cesium-137 and Cobalt-60 are thirty years and five years, respectively. Cesium-137 is found throughout the world, and is characterized as an external and internal hazard due to its biogeochemical properties, enabling absorption into the food chain (ICRP, 1979-1988; UNSCEAR, 1969). It is used extensively in blood and research irradiators, but also in gamma radiography and produces beta and gamma irradiation. Cobalt-60 is used in teletherapy devices and research and industrial irradiators, and is both a beta and gamma emitter.

The special case of Americium-241 deserves discussion. Although representing a low-probability event, there is a historical precedent for terrorist use of an Americium-241 RDD, with thirteen individuals having been apprehended in the United Kingdom during August 2004 on Terrorist Act charges: "Conspiracy to commit a public nuisance by use of radioactive materials, toxic gases, chemicals, and/or explosives." Although Americium is a fissile material, approximately 80 kilograms are needed to produce a bomb; the worldwide production is estimated at a few kilograms, hence making the production of such a device highly unlikely. However, the psychology of possessing such a finished device with the intent of producing mass contamination versus human casualties should not be ignored. (www.cbwinfo.com/radiological/radmat/am241.shtml).

Americium oxide would be the form of Americium in a RDD. Decay occurs primarily by alpha emission to Neptunium-237 and the half-life is 432 years. Decay products are detectable as a 59.5 keV gamma and in the urine of exposed individuals.

Classic radiation sickness is unlikely except with extremely large total exposure (greater than 1 gram from all routes of exposure). The primary commercially available source of Americium is in smoke detectors (the average home detector contains 1 micro Curie as compared with industrial-use detectors which contain up to 50 micro Curies), and hence more of a concern that Americium may be used as a weapon of terror. Management of individuals exposed to Americium would primarily consist of mechanical decontamination and through use of chelating agents (such as DTPA; Emergency Response Guidebook Number 161, 2004) to prevent binding in bone and liver (primary routes of absorption are through inhalation and ingestion). In summary, threat awareness dictates a low likelihood of Americium use as an RDD to generate large numbers of casualties, however, similar to the other radionuclides previously described, and this material could be used quite effectively as a weapon of terror through a misinformation campaign.

Radiological terrorism due to RDDs is generally believed to be easily achievable, pragmatic and inexpensive, requires less technical sophistication, and requires far less source material and lead time for weaponization than is necessary for classical military warfare agents and platforms. The classical military warfare agents are generally recognized as chemical and biological warfare agents, however this definition has recently been expanded more broadly to include a wider spectrum of novel and bioengineered agents such as toxins or genomes coding for specific proteins that induce specific physiologic and toxic effects.

The three main objectives of RDD attacks are: contamination of food/water supplies; area denial; and direct exposure (Len Connell and Lawrence Trost, RDD Global Protection Architecture, May 2006). Extrapolation of these themes leads to additional objectives of inducing fear and uncertainty, and undermining confidence in government. A review of radiological dispersal accidents lessons-learned underscored a major emphasis on the consideration of psychosocial effects in the consequence management, clean-up standards, population relocation guidelines, and anticipatory management (both policy and operational) of such events (Len Connell, Lawrence Trost, Presentation to The DTRA US/Singapore RDD Workshop). Public opinion phone and internet surveys performed by the George H. W. Bush School of Government and Public Service, Texas A&M University, demonstrate public support for enhanced public awareness training and support for additional U.S. government efforts to deter radiological and nuclear terrorism in the homeland. (Sandia Report SAND2006-0753P, March 2006).

As previously indicated, terrorist motivations include relative ease of weaponization, and deployment, and paralysis and fear in the targeted population. These characteristics combined with the high efficacy of RDDs to create a long-term contamination problem further justify the purpose and methodology of this study to understand schisms in medical situation understanding, response and mitigation capability.

Ultimately, recommendations in this study are focused on prevention and countermeasure strategies to minimize casualty generation across all sectors, to include responders, and to minimize environmental impact. Information sources such as technical reference documents, Subject Matter Expert (SME) interviews, and national U.S. policy and doctrine, form the basis of the assessment.

STANDARDS OF MEDICAL MANAGEMENT

A comprehensive set of recommendations for population protection was published by the National Council on Radiation Protection and Measurement, and may be found in a report entitled "Management of Terrorist Events Involving Radioactive Material" (NCRP Report No. 138, October 24, 2001). Although some of the material included in this document is clearly outdated, the basic principles remain solid. Differentiation of a RDD event from one due to a conventional weapon is particularly helpful when attempting to characterize the optimum medical response. Similar to other unconventional weapons such as biological agents, consequences of RDD events may be characterized as complex systems, and are not readily predictable or neatly bounded by space and time. Medical consequences such as psychological and physical trauma may occur to persons not wounded by the immediate or localized attack, and the end-effects may be disproportionate to the physical effects. Although the primary zone of psychosocial and external contamination effects will be local, a wider and broader slice of society will be affected over space and time.

Bennett (1999) published a basic synopsis of understanding of responses to risk in *Risk Communication and Public Health* (Bennett, P. and Calman, K., Eds., Oxford University Press, New York). In this document and others, Bennett refers to the involuntary and irrational public response to unfamiliar threats. These threats are described as those posing hidden/irreversible damage, the psychological overlay of "dread" of impending illness or death, and potential special or enhanced dangers to children and pregnant women. RDD and biological agent effects are consistent with this category and reinforce the need for effective pre-event planning at federal, state and local levels, education and training, and supportive technologies to support and mitigate the effects of such events.

Response to any domestic nuclear or radiological accident is generally partitioned into the initial (urgent) and secondary (less urgent) or follow-on phase of the response. Criteria for action are further specified by the International Committee on Radiation Protection and the International Atomic Energy Agency. Although underlying response philosophies may vary, adding complexity to translating raw data from the field into action, generally, high- level medical response activities are straightforward and separated into two primary phases. Response to RDD incidents should be differentiated from more traditional medical responses and associated response criteria due to the potential crime scene element of the event that is geographically separated from the event, and significant psychological overlay (immediate and delayed).

The initial medical response phase to a RDD event includes the immediate emergency management procedures necessary to contain the threat and mitigate further loss or consequences to health, environment or property. Primary tasks involved in the initial phase include: incident command and control to include information flow; intelligence, surveillance, and reconnaissance; search and rescue; fire suppression; casualty triage and initial management to include hazard mitigation. Primary tasks involved in the second phase of the response include: employment and coordination of Federal assets to include specialized quick response teams from Department of Defense (including National Guard and Defense Threat Reduction Agency assets), Department of Energy/NNSA (Radiological Assistance Program), Department of Homeland Security

(FEMA), and FBI. These teams provide additional capability for site surveillance and reconnaissance, and forensics analysis and threat/vulnerability characterization.

The relationships between the aforementioned agencies are evolving, and the intergovernmental coordination is complex, however, these relationships have incrementally improved within the new National Response Plan and National Incident Management System (NIMS) framework. One area in this relationship-building that needs continued emphasis and improvement is information-sharing, particularly between non-medical and medical assets of the response. All too often, these entities do not share a common operational picture and do not have access to relevant, vital information such as threat, actionable intelligence, and vulnerability analysis information that is essential to an optimized response. Advanced information systems and knowledge discovery and dissemination approaches using new data-mining and post-analytic fusion engines can help enormously.

A number of information-sharing initiatives are underway to assist in alleviating problems that have been identified through training and exercises and real-world homeland security and defense operations. The 9/11 (WMD) Commission Report helped to focus resources on information-sharing gaps. One solution includes the establishment of an office responsible for information sharing under the Office of the Secretary of Defense (Ms. Debra Filippi). This office works closely with civilian and military agencies involved in the entire spectrum of homeland security and defense response and recovery. The author of this study was also requested to address the JASONS on this topic, thus indicating significant White House interest. On the operations level, a number of interagency MOUs/MOAs exist to facilitate information-sharing at the national, state, local and regional level. Perhaps most relevant to the operational picture are inter-state emergency management assistance agreements that specify sharing of resources, assets, and underpinning critical information during a homeland security/homeland defense incident. All fifty-four states and territories have implemented inter-state such emergency management assistance agreements.

HISTORICAL MEDICAL CONSEQUENCE MANAGEMENT LESSONS-LEARNED

Our global medical experience pertaining to the management of medical consequences of RDD exposure largely stems from experiences in Japan, Chernobyl and Goiania. Jacobs et al. published data on the correlation of white blood cell depression with mortality in 1963 (Military Medicine, 128, 732-739, 1963). Post-Chernobyl, Cesium-137 was determined to be the key residual environmental contaminant (NCRP, SC 64-23, April 1986). The Chernobyl experience documented a total of 134 Acute Radiation Syndrome (ARS) of the total of 400 documented worldwide (V.K. Ivanov, et al. Medical Radiological Consequences of the Chernobyl Catastrophe in Russia and Consequences of the Chernobyl Catastrophe on Human Health, Nova Science, New York, 1999). The Cesium-137 radiological accident in Goiania, Brazil in 1987 resulted in a total of 249 contaminated patients.

Acute Radiation Syndrome (ARS) comprises the comprehensive set of pre-clinical/clinical and biological effects arising from whole body penetrating radiation exposure. End-organ effects can lead to death within hours or weeks. (NCRP Commentary No.19, 31 December 2005)

There is much to be learned from the clinical experience in Goiania, Brazil. Of the 249 contaminated patients (age range 4 to 38 years), 48 acute radiation syndrome patients were triaged and treated with Prussian Blue (ferric hexacyanoferrate), selected as the agent of choice for enhancing excretion of cesium. This population had a documented exposure of 1-7 Gy (100-700 rad) as indicated by bone marrow suppression and supportive cytogenetic testing. Subsequent statistical analysis yielded a Poisson distribution of cells with chromosomal aberrations. This finding supported the management hypothesis that some patients had received non-uniform exposure to radiation. (The Radiological Accident in Goiania, IAEA, Vienna, 1988; Richmond, CR. Accelerating the Turnover of Internally Deposited Radiocesium. In: Kornberg HA and Norwood WD, Eds. *Diagnosis and Treatment of Deposited Radionuclides*. Excerpta Media Foundation, Amsterdam, 1983; 315-327.)

Based on these experiences, the range of 10 mGy (1 rad) to 100 mGy (10 rad) has been determined to correlate most strongly with a 1/1000 risk of cancer and observable cytogenetic chromosomal aberrations. As a corollary to this finding, radiosensitivity is directly related to cell turnover rate, and the most sensitive early biological marker is lymphocyte count. Cytogenetic techniques are extremely helpful in determining exposure as stated above. When combined with blood measurements and clinical findings, an accurate picture of radiation exposure can be determined. In summary, observable end-organ effects correlate with radiation dose and other host susceptibility factors. White blood cells/lymphocytes are the most susceptible organ system to RDD exposure and nerve/brain tissue is the most resistant. This information has practical applications in triage. Additional tools in the tool kit are derived from Chernobyl lessons-learned. These aggregate findings form the basis for triage and differential diagnosis by symptoms and blood counts, and the LD/50 guidelines (lethal dose for 50% of the population) of 6.5 Gy (650 rad) with skin injuries and 6-8 Gy (600-800 rad)without skin injuries. This assumes medical care and surge capacity for care.

In summary, Chernobyl and Goiania have taught us much regarding effective triage and management of radiation casualties. To this end, development of rapid triage capability employing biological markers which may be integrated with environmental sensing and plume modeling data would greatly support the first responder end goal of rapid situation awareness/common operational picture generation. To this end, the Chernobyl and Goiania experiences support a multi-modal system for early evaluation and triage of casualties, which may provide additional insight that is useful in development of policy decisions on allowable radiation exposure in responders and health care workers with repetitive exposure to contamination.

CURRENT BASELINE OF MEDICAL MANAGEMENT OF RDD EVENTS

A number of myths exist regarding exposure to RDDs. First, it is important to recognize that contamination to a health-endangering level is unlikely in the event of a RDD event. This observation applies to both responders and caregivers. In addition, contamination of responders via ingestion, inhalation and absorption through the skin is also very unlikely. Additionally, similar guidelines will apply to exposure to improvised nuclear devices (INDs).

The initial radiological survey is an essential element of the initial, baseline assessment of exposure to radioactivity and should be performed by health physicists or other similarly trained personnel. The accuracy and timeliness of this assessment is critical to effective consequence management. An integrated, common operating picture employing a distributed, on-site dosimetry monitoring system is highly desirable and represents an effective management tool for consequence management. Such a dosimetric system would map exposures to geolocation and result in a highly accurate and predictive map.

Radiation exposure continues long after the initial event, hence requiring an essential baseline understanding of chronic effects and exposure levels considered as "medical risk" triggering events for protective action. Generally, the "trigger" implies a .5% cancer risk and no acute effects (corresponds to about 0.1 gray or 0.1 sievert, absorbed dose versus protective quantity). (DHS Working Group on RDD Preparedness, May 2003, and personal communication with Dr. Fred Mettler, 2006) This quantitative, science-based approach is much more meaningful than an ALARA approach, although an ALARA approach may endure due to public pressure (see below), since clear-cut medical consequence management and mitigation/preventive actions correspond to dose and anticipated relative hazard and physiologic effects. Although, prevention of exposure is clearly a more viable approach than secondary mitigation and consequence management, guidance for post-exposure medical management is certainly necessary in a tiered approach to medical management. Tiered approach implies layers to management to include time/distance/shielding and targeted treatment of end-organs. Evaluation of Chernobyl workers reinforced this approach, which yielded a dose to the medical staff of approximately 10 mGy or 10 mSv (DHS Working Group on RDD Preparedness, May 2003). Historical data refutes the logic for use of Prussian Blue in RDD incidents, although considered the standard of care in management of patients in Goiania.

Once an understanding of the event and resulting/potential exposures to radioactivity has been achieved, the issues of triage, decontamination, contaminated medical waste disposal, and mortuary affairs are issues that require further attention. First and foremost, treatment of life-threatening injuries supercedes decontamination procedures. However, it is important to recognize that treatment facilities and highly skilled personnel are "renewable resources" and should be considered as such when managing contaminated or partially decontaminated patients. For those ambulatory and minimally-affected individuals, advance guidance for "sheltering-in-place" and home care is an important adjunct to medical treatment facility based care and should be considered a proactive component of a public, health care provider, and media education campaign.

Revised interdisciplinary training for medical staff which better integrates a balanced, situation-awareness approach to medical management of radiological casualties is needed. Interviews with physician SMEs (see Katona, Kadlec, Gray, and Smock) indicate that retention and time dedicated to maintaining currency with respect to management of these patients is limited and receives limited emphasis in continuing medical education programs. One continued theme is maximizing the efficiency and use of current processes and protocols in disaster management crossing the spectrum of all threats and hazards. For example, more efficient management of triage systems, contingency facilities, hospital discharges of non-critical patients, and specialized equipment mapped to personnel and facility preparedness to handle special incidents. Individual knowledge and leadership is essential to build, implement, and enforce standards and protocols. This knowledge and leadership must transcend across all levels of care providers, pre-hospital to definitive care. Some cities, such as New York, have invested in specialized RDD teams with embedded medical capability that are analogous to military teams described below.

Critical enablers of disaster management are communications, transportation infrastructure, information sharing, and implementation of medical standards of care with operational relevance and pragmatism. These systems and practices must be exercised regularly to ensure inter-facility and inter-agency interoperability and uninterrupted transportation of such items as medical treatment fluids, water (potable and for decontamination), and stockpiled materials. A number of cities have established specialized communications networks for use in disasters (i.e., Readinet in LA County). A number of information sharing systems exist. The Terrorism Early Warning System is one such system used to network first responders, intelligence professionals, and law enforcement during a terrorism event.

Across the spectrum of operations, the concept of health protection incorporating radiation dose limits is one that forms a singular point of significant contention among agencies. For example, the EPA has not endorsed the Nuclear Regulatory Commission's (NRC) rule for site clean-up and has issued its own (conflicting) guidance. This publicly visible conflict of established standards openly undermines the confidence of the general public and critical sectors of the responder and receiver communities. (Background information on "Ionizing Radiation-Safety Standards For The General Public", Position statement of the Health Physics Society, reaffirmed March 2001). Specifically relevant to this study is the necessity for a "jumping-off" point from baseline standards, enabling optimized outcomes-based medical management in the setting of an act of radiological terrorism.

The EPA guidance (Comprehensive Environmental Response Compensation and Liability Act [CERCLA or Superfund], Resource Conservation and Recovery Act [RCRA], and Toxic Substances Control Act [TSCA]) is derived from a source constraint of 0.15 mSv in any year that assumes lifetime cancer risk should not exceed a goal of approximately 10^{-4}. This premise is in striking contrast to the ICRP (International Commission on Radiological Protection, ICRP Pub. 60, Ann. ICRP 21 (1-3) and NCRP (National Council on Radiation Protection and Measurements, Report No. 116; 1993) recommendations which specify an allowable dose limit and derived reductions in dose based on the ALARA principle. The Health Physics Society opposes the quantification of ALARA and supports its primary use as "...a philosophy for striving for excellence in the practice of health physics" ("Occupational Radiation-Safety Standards and Regulations are Sound: Position Statement of the Health Physics Society," Mclean, VA, March 2000). In the context of a RDD or other unconventional mass care scenario, it is generally unacceptable to apply criteria like CERCLA or extrapolate guidelines from occupational health and safety derived settings. (Also, see comments from Jarrett below). Historically, in the event of a large accident, they have never gone as low as CERCLA, because it would be impractical.

MILITARY MEDICAL CAPABILITY PERSPECTIVE

Interviews with military medical SMEs included those tasked with supporting and developing procedures for domestically-focused National Guard quick response team leadership (CERF-Ps, CBRNE Enhanced Response Force Packages and CSTs, Civil Support Teams), stressed the need for consistency among modeling and simulation systems. Specifically, predictive analysis of anticipated bio-effects (to include psychological) and contamination (environmental and personnel) due to RDDs in an urban environment is lacking standardization and consistency. The ability to rapidly determine the extent of individual exposure and potential long-term health consequences through a wearable device is viewed as a current gap. Real-time epidemiologic surveillance of contamination needs improvement, and would assist in optimizing outcome, medical responder performance, and mitigate the medical consequences

Regarding the issue of decontamination protocols, current protocols emphasize decontamination in parallel with essential life-saving measures. This process is dynamic, with a recognized need for adaptability and real-time adjustment of priorities, depending upon real-world events. However, current thinking is that facilities will be considered contaminated post-receipt of casualties, and a number of contingency locations will need to be pre-designated. Suggestion of "clean versus dirty facility" pre-designation suggested by Dr. Fred Mettler is viewed as unrealistic by many military medical responders (many of whom work or have worked in civilian medical facilities) from political, business operations, and implementation perspectives.

CIVILIAN FIRST RESPONDER COMMUNITY PERSPECTIVE

Prior studies (Musolino and Harper, 2006) have indicated a number of significant concerns voiced by key end-user populations spanning responders and senior policy and decision-makers across government, academic, and private sectors impacting measurable health consequences. The radiological accident in Goiania in 1985 underscored these themes and also reinforced the importance of radiological source security and regulatory/compliance processes. Goiania emphasized the need for collaborative international databases such as those which currently exist and continue to function as an important element of sharing lessons-learned and mitigation of future loss in accidental or terrorist events.

To further elaborate, these **concerns** included, but were not limited to:

- Triage and evacuation protocols.
- Indications for sheltering in-place (quarantine and containment).
- Facility environmental management.
- Unambiguous operational "triggers" to initiate and terminate protocols.
- Sustainable medical management of mass casualty events, triage of the "worried well" and management of post-event or delayed psychological trauma patients.

A number of communities have taken a proactive approach to addressing these concerns. For example, Louisville, Kentucky has established a Joint Emergency Services

Unit (ESU) consisting of team members deputized by the U.S. Marshalls and with expertise across multiple domains such as Law Enforcement, Health Care (physicians and nurses), and Health Physics. This unit establishes medical response protocols for a wide range of RDD incidents, conducts regular training with the National Guard Civil Support Teams, (CSTs), shares actionable information with the CSTs, and is highly functional due to being embedded within the emergency response mechanism of the State of Kentucky (Smock and Gray, personal communications). The Joint ESU is authorized to respond to any incident, regardless of local, state, or federal jurisdiction, and is especially prepared to respond when there is a need for law enforcement capability in a hot zone or there is a known or suspected victim/human biological vector. This Unit may serve as a model for other states and regions wishing to build an integrated response capability with a horizontal organizational structure.

The concerns listed continue despite the U.S. EPA 1992 guidelines which clearly enunciate evacuation criteria in the acute explosive RDD scenario. These concerns appear to reflect the greater umbrella of concerns of ambiguity regarding science-based standards for radiation based decisions or "triggers" for allowable radiation exposure. This lack of standardization is partially due to the focus on planning and personal protection to achieve optimal near-term survival (the bar is set at twenty per cent survival for nuclear attacks) and to mitigate other health and environmental near-term consequences.

Unfortunately, a somewhat ambiguous operational medicine strategy has resulted in a less-than optimal approach to population survivability and morbidity. The net health effects from chronic and geographically separated RDD events have been described in NCRP forums (NCRP Symposium on "The Control of Exposure of the Public to Ionizing Radiation in the Event of an Accidental Attack", Reston, VA, 15 May 1982).

SUMMARY OF SUBJECT MATTER EXPERT OBSERVATIONS TO ACHIEVE ENHANCED MEDICAL OUTCOMES

In an effort to achieve a more balanced and effects-based approach to near-, mid-, and long-term medical consequences, the following observations and recommendations have been collated from SMEs:

- Education and Training

  DTRA subject matter experts strongly recommend that members of specialized response teams receive focused technical and health physics training.

  Fred Mettler (39th Annual Meeting of the NCRP; published in Health Physics "Medical Perspective on Ways to Improve Radiation Protection Standards") summarized the frustration of the medical community at-large by the following statement: "...radiation protection in medicine is not likely to be solved with the addition of more standards but will require a combined approach with the medical and educational communities." The previously described integrated Joint ESU conducts combined technical, medical management, and health physics training and has begun to address the challenges inherent to integration of new technologies into existing response and medical management protocols. Such strategies should be considered as an important pillar to all-hazards response capability building across the spectrum of threats.

- Need for implementation of situation-dependent emergency dose and dose-rate guidelines: next-generation ALARA

  ALARA (as low as reasonably achievable) techniques are viewed by the response community are difficult to interpret and comply with. As a whole, generic parameters are unpredictably responsive to large uncertainties, and do not reflect the most credible assessments of levels of contamination (NCRP Report No. 138). For example, in an urban setting, wind and microclimate conditions may be highly variable and unpredictable, thus adding to the uncertainties of response.

  In addition, the inconsistencies and inappropriateness of equally applying ALARA guidelines to occupational exposure settings and terrorism events has been noted by Mettler (personal communication, July 2006) and cries for operational exposure guidance from the perspective of the incident commander. These inconsistencies may be summarized by differentiating the settings of occupational exposures from accidental or deliberate radiation exposures. The situations demand different personal or force protection methodologies emphasizing what is "correct" for the given situation vice what is "minimal". In an attempt to address many of these concerns, Dr. Craig Conklin of DHS heads a working group nearing completion to establish ongoing exposure limits across operational response settings.

  In 1991, the International Commission on Radiological Protection (ICRP) referred to the multi-factorial concept of risk as equaling dose X risk factor (per unit dose), with risk including the variables of age, weight, sex, and dietary habits. The ICRP continued to emphasize that risk must be based on scientific rigor, credible assumptions, appropriate methods, and actionable decisions based on known site contamination parameters.

- Sensors and associated response protocol "triggers" represent an applied technology gap

  Current sensor networks and protocols do not effectively test for alpha and beta-emitters in field environments and are not adequately integrated into automated "triggers" for protocols. This observation identifies a large gap in current sensing protocols. A summary of suggested instruments to be considered during the planning for radiological terrorist attacks is collated in the NCRP Report No. 138. This sensor data would merit integration with other data sources previously described to achieve a field-deployable common operational picture.

- Suboptimal integration of the media in RDD medical response, education, and training

  The media can be a strong ally for information dissemination and synchronization during a disaster. Federal, state, and local authorities should consider employing the media to assist with self-triage, management of the "worried-well", and updated evacuation procedures and route usage.

- Inconsistent standards for data analysis, need for unambiguous "triggers" for decision-making and action, and application to field operations.

Over the years, policy guidance regarding radiation-based allowable exposure has become increasingly restrictive ((NCRP, 15 May 1982). Although response philosophies may depend on such dose modifying factors as: occupancy factor, geographic distribution (mean maximum dose rates) and time distribution factors, as well as immediate implementation of focused medical management, history has taught that pre-determined, scalable standards-of-care result in optimized effects- based operations, especially in disasters with mass casualties. The basic medical tenets for management and mitigation of radiological casualties still hinge on the cornerstones of time, distance, and shielding (Jarrett, personal communication, 2006). Increasingly restrictive values for data interpretation have resulted in early trigger criteria for intervention and action and have substantial economic and social burdens which negatively influence the establishment of rigorous scientifically-based and standardized criteria. In addition, highly restrictive criteria unrealistically assume an infinite resource of skilled responder and receiver personnel and also do not account for the need to protect and retrieve critical infrastructure.

The bottom-line is ambiguity of medical protocols and response criteria, and the resulting confusion regarding medical standards of care during RDD events. For example, patients at "ground zero" of a RDD event are not "salvageable" and should be considered "non-rescue-able". In this scenario, precious resources and personnel should not be deployed and unnecessarily expended or jeopardized (Jarrett, personal communication, 2006). This would depend upon the scenario. GZ may be too contaminated or may not. This underscores the need for fast, accurate surveys of contamination. Adoption of early triage and management criteria which focus on data sources such as biomarkers and environmental data appears to be a scientifically sound and politically robust strategy worthy of consideration. Data models are an integral component of medical decision-making and strategic planning and should be treated as such through integration into the larger "situation awareness" picture. A "train as we respond" approach which includes such information tools within the accreditation and credentialing/continuing education requirements would be integral to improving this process.

Such a comprehensive information management approach would address a number of concerns voiced by SMEs, and would enable first responders and physicians alike to use unambiguous indicators and "triggers" for specific, pre-determined community-wide actions. Although a notional set of solutions was presented in this study, such bio-informatics-based operations will support the end goal of optimized medical management of RDD events.

SUMMARY AND RECOMMENDATIONS

Globally, the U.S. Department of Energy/National Nuclear Security Administration is viewed as a premier authority on the physical effects, science and technology and weaponization potential of radionuclides to be incorporated into weapons of radiological terror. To this end, medical consequences represent a significant area of national concern that could substantially benefit from the DOE/NNSA leadership role. For example, this leadership naturally ensures close coupling of the scientific understanding of such devices, the potential health and environmental effects, appropriate methodologies for prediction, modeling, simulation, detection, and mitigation. This comprehensive understanding impacts ability to influence public policy on medical consequence management.

This study identified a number of shortcomings and opportunities for initiatives to address these shortcomings. The DOE/NNSA could play a lead role at the national and international levels to address these shortcomings.

The identified shortcomings are as follows:

- Decentralized leadership and chain-of-command;
- Limited public awareness campaign;
- Lack of integrated, standardized modeling and simulation capability;
- Separate sensor and health care databases resulting in lack of unified situation awareness from the field to definitive care facilities;
- Lack of interdisciplinary training;
- Limited international partnership networks focusing on RDD medical management at all levels from public health policy to implementation of care;
- Lack of interoperable sensor platforms;
- Lack of multi-threat countermeasures with focus on common pathways of injury and disease;
- Limited international technical information exchanges.

The potential initiatives to address the identified shortcomings are:

- Champion establishment of a single point-of–contact across government agencies that will serve as the lead in response and preparedness across the spectrum of threats at Federal, State, and Local levels (Conklin);
- Champion the public information and awareness of the global threat of RDDs and individual and community protective measures, and invest in a program which targets the specific needs of health care providers/responders and senior executives. Some high priority topics include: vulnerability of existing safeguard, tracking, and accountability systems of radionuclide sources for such devices and awareness of medical systems/components that employ radionuclides;
- Support integration of existing modeling and simulation capability into a national system for incident and medical consequence management (National Incident Management System); natural locations for this integrated capability include the DHS Homeland Security Operations Center, NORTHCOM, and FEMA operations center. Intelligence fusion centers throughout the United States would benefit from such a common operational picture and associated education and training initiatives;
- Support integration of responder and health care provider (first responder, first receiver and medical treatment facility personnel) detection device

data/information into a Common Relevant Operational Picture (CROP) to achieve rapid situation awareness of the medical epidemiology of RDD incidents. Such a system would help achieve appropriate triage, customized, isotope-based management, appropriate resource allocation, and requests for information (RFIs) which are medical outcomes-driven within the scope of a rapidly evolving event or series of events at the local, state, regional, national, and transnational levels. Interagency partners include DOE/NNSA, DoD, DHS and DHHS;

- Champion integrated and collaborative training emphasizing response and preparedness of health physicists, rapid response team members, and health care providers. Such training would optimally promote more comprehensive (environmental, operational, and physiologic/psychological) exposure guidelines emphasizing pragmatic and operationally-relevant standards to include clean-up and aftermath environments. Such guidelines would replace ALARA, would provide operational experience frameworks, and a much-needed tool in the toolkit of health care providers confronted with the patient management challenges due to radiological terrorism;
- Consider a "way-ahead" that builds an international network of partnerships between National Laboratories, strategic DoD educational centers (such as the George C. Marshall Center) and health/public policy academic centers (such as Texas A&M Health Sciences Center and the George Bush School of Government and Public Service); the focus would be promotion of senior executive education and training in CBRNE terrorism and mitigation/ prevention of health consequences to include acute and chronic environmental effects; leverage and expand the use of phone/internet surveys to include medical consequence management of RDD/nuclear terrorism and to track performance metrics of new community-based education and training initiatives (as performed by the George H.W. Bush School of Government and Public Service, Texas A&M University and the University of New Mexico Institute for Public Policy under contract to Sandia National Laboratories);
- Further invest in research and development of field-deployable, portable, interoperable platforms that integrate biomarkers and other minimally invasive physiologic indicators (acute and chronic) of radiation effects. R&D should focus on a broad range of applications of medical decision-making paradigms, to include tactical (intra-operative, triage and decontamination) and strategic (policy for mass decontamination and countermeasures use, population containment, etc.);
- Encourage further R & D of multi-threat responsive, field-deployable radiological countermeasures and pre-treatments that support a CONOP of environmental robustness (for example, implantable or encapsulated system delivery platforms) and protection of perishable human-capital limited personnel assets within the first responder and first receiver communities. For example, little research has been performed near-term in the very promising field of cytokine work relevant to this threat environment (personal communication, COL Dave Jarrett and Roberto Rebeil, Ph.D.);

- Leverage current international programs within DHS and DoD to standardize an events-based information-sharing network; the core focus of this network would be medical consequence management (preparedness and response) of all threats; RDD medical consequence management protocols and training would leverage "lessons-learned" from recent Asia-Pacific regional and country-specific workshops, and would serve as a baseline for technical investment and information exchanges in high priority geographic regions.

CONVERSIONS OF CONVENTIONAL AND INTERNATIONAL SYSTEM OF DOSIMETRIC EQUIVALENTS

The NCRP (1985) supports the use of the International System of Units, in which absorbed dose is given in gray (Gy) and other protection quantities are given in Sievert (Sv). This is in contrast to the U.S. convention of using rad for absorbed dose (1 rad equals 10 mGy), and rem for other protection quantities such as equivalent dose and effective dose (1 rem equals 10 mSv). However, in the context of international terrorism, the use of the newer system would be more useful. For radiation survey instruments, 1 Roentgen equals .01 Gy or .01 Sv. (NCRP Report No. 138, Management of Terrorist Events Involving Radioactive Material, October 24, 2001).

SUBJECT MATTER EXPERTS (SMES) INTERVIEWED

Brooke Robert Buddemeier, CHP, WMD ER/CM Program Manager, Emergency Preparedness & Response, Plans, Programs, and Budgets, U.S. Department of Homeland Security, Science and Technology, Washington, D.C.

LTC Julie A. Bentz, Ph.D., DATSD (NCB) Nuclear Matters, Pentagon, and former Director for Nuclear Defense Policy, Homeland Security Council, The White House.

CDR Charles Blue, M.S., M.B.A., Deputy Chief of Staff, Office of the Chief Medical Officer, Department of Homeland Security, Washington, D.C.

LTC Daniel Bochicchio, M.D., National Guard Bureau, Joint Surgeon's Office, Arlington, VA.

Paul K. Carlton, M.D., LTG (USAF, Ret.), Director of Homeland Security, Texas A& M Health Sciences Center, College Station, Texas.

Julie A. Canepa, Ph.D., Technical Staff Member, Defense and Homeland Security Programs, Associate Directorate for Threat Reduction, Los Alamos National Laboratory.

C. Norman Coleman, M.D., Department of Health and Human Services, U.S. Public Health Services, Office of Public Health Emergency Preparedness, Washington, D.C.

Craig Conklin, Ph.D., Acting Director, Chemical and Nuclear Preparedness and Protection Division Office of Infrastructure Protection Preparedness Directorate, Department of Homeland Security, Washington, D.C.

Honorable Jack David, Office of the Secretary of Defense, WMD Countermeasures, Pentagon.

James D. DeZetter, CHP, Manager, Radiation Safety, University of New Mexico, Safety, Health & Environmental Affairs (Metropolitan Medical Response System, Radiologic Advisory Group), Albuquerque, NM.

Dectra Dixon, Epidemiologist, City of Albuquerque, Environmental Health Department (Metropolitan Medical Response System, Radiologic Advisory Group), Albuquerque, NM.

Ms. Debra Filippi, SES, NII/DoD-CIO, Arlington, VA.

Aubrey Godwin, Arizona Radiation Regulatory Agency, Phoenix, AZ.

Amy Sue Goodin, Associate Director of Research, University of New Mexico, Institute for Public Policy, Albuquerque, NM.

Kent Gray, M.D., Centers for Disease Control, Atlanta, GA.

Roger Hagengruber, Ph.D., Director, University of New Mexico, Institute for Public Policy and former Director, Office of Policy, Security, and Technology, Albuquerque, NM.

Frederick T. Harper, Ph.D., High Consequence Assessment & Technology Program, Sandia National Laboratories, NM.

Robert Huelskamp, Nuclear Material Monitoring and Advanced Technologies, Sandia National Laboratories, NM.

David G. Jarrett, COL, MC, USA, OSD-ATL (CBD & CDP), Pentagon and formerly, Armed Forces Radiobiology Research Institute (AFRRI).

Joseph A. Jones, P.E., Radiological Consequence Management and Response Technologies, Sandia National Laboratories, NM.

Robert P. Kadlec, M.D., Director, PRTM, Washington, D.C.

Peter Katona, M.D., Associate Professor, Department of Internal Medicine, UCLA Medical Center, and Emergency Management Agency for LA County, National Bioterrorism Hospital Preparedness Program.

Carol Linden, Ph.D., Office of Public Health Emergency Preparedness and the Office of Public Health Emergency Medical Countermeasures, Department of Health and Human Services, Washington, D.C.

Ruth McBurney, President, Health Physics Society, Austin, Texas.

Fred A. Mettler Jr., M.D., M.P.H, Department of Radiology, New Mexico Federal Regional Medical Center, Albuquerque, NM.

Honorable Paul McHale, Assistant Secretary of Defense for Homeland Defense.

National Guard Bureau, J3, CERF-P (CBRNE Enhanced Response Force Package) and CST (Civil Support Team) SMEs (Lt Col Davis, OPR), Arlington, VA

Gerald Parker, M.D., Principal Deputy Assistant Secretary, Department of Health and Human Services, U.S. Public Health Service, Office of Public Health Emergency Preparedness, Washington, D.C.

Arthur Payne, Ph.D., National Security Studies, Sandia National Laboratories, NM.

Roberto Rebeil, Ph.D., Sandia National Laboratories, NM.

Amelia Rouse, Ph.D., Deputy Director, University of New Mexico Institute for Public Policy, Albuquerque, NM.

Tammy P. Taylor, Ph.D., P.E., Group Leader, Safeguards Systems, Nuclear Nonproliferation, Los Alamos National Laboratory.

James Tritten, Ph.D., Director, Defense Threat Reduction University, DTRA, Kirtland AFB, NM.

Alfredo Robert Santistevan, Director, City of Albuquerque, Environmental Health Department, Albuquerque, NM.

Frederick M. Scudiery, Health Physicist, Defense Nuclear Weapons School, Kirtland AFB, NM.

F. Joseph Schelling, Ph.D., Radiological Consequence Management and Response Technologies, Sandia National Laboratories, NM.

Sara C. Scott, Ph.D., Program Director, Civilian Nuclear Programs, Science, Technology and Engineering Directorate, Los Alamos National Laboratory.

Greg Scharrer, Manager, Energetic Systems Research Department, Sandia National Laboratories, NM.

Jose H. Saloio, Jr., Manager, International Science and Technology Development, Sandia National Laboratories, NM.

Kasey Smith-Alexander, Epidemiologist, City of Albuquerque, Environmental Health Department (Metropolitan Medical Response System, Radiologic Advisory Group), Albuquerque, NM.

Bill Smock, M.D., Police Surgeon, Louisville Metro Police Department, Professor, Division of Protective Medicine, Department of Emergency Medicine, University of Louisville Hospital, Louisville, KY.

Peter A. Terrill CWO, USN, Master EOD Technician, WMD Training Specialist, Defense Threat Reduction Agency, Kirtland AFB, NM.

Ruben (Val) A. Valencia SFC, USA, Exercise Branch NCOIC, Defense Threat Reduction Agency, Kirtland AFB, NM.

RADM Craig Vanderwagen, M.D., Assistant Secretary, Department of Health and Human Services, U.S. Public Health Services, Office of Public Health Emergency Preparedness, Washington, D.C.

ACKNOWLEDGEMENTS

The author of this study would like to acknowledge the National Security Studies Department, Sandia National Laboratories, the DOE/NNSA sponsor of this effort, and the Subject Matter Experts who so kindly provided their time and expertise.

The views and opinions expressed herein do not necessarily state or reflect those of the United States Government, any agency thereof, or any of their contractors.

DEFEATING RELIGIOUS TERRORISM—WHAT WILL IT TAKE?

PERVEZ HOODBHOY
Physics Department, Quaid-i-Azam University
Islamabad, Pakistan.

By posing this mother of all questions, I hope to stimulate thinking rather than to merely hazard answers or prescriptions. Indeed, no putative answer can possibly be satisfactory. Irrespective of the efforts and technical tools that can be brought to bear on this problem, those using violent means to impose their ideological goals can, at best, only be weakened. Mathematicians, physicists, and engineers have surely developed wonderful tools, but these are not the right ones for this job. As the pursuer uses ever better tracking tools, those pursued will also develop better means of evasion. So, while various technical fixes and mitigation attempts to be discussed at this meeting must not be belittled, serious thinking has to go far beyond these.

Some may object to this paper's exclusive focus on religious terrorism, and even more particularly to the terrorism perpetrated by Islamic groups. This is done to limit its scope although, arguably, other forms of terrorism may be equally or more important. If one defines terrorism as the deliberate targeting of non-combatant civilians with the intent to kill, harm, or terrify, then nation-states have shown a far greater capacity to exercise terror. The nuking of Hiroshima and Nagasaki, the fire-bombing of Dresden, the wholesale slaughter of ordinary Bengalis by the Pakistani Army in 1971, the aerial bombardment of occupied apartment buildings in Lebanon by Israeli aircraft in 1982 and 2006, and the gassing of 100,000 Kurds in Halabja, are just a few examples. But these belong to a domain that lies away from the present focus.

In the following, I shall pose certain critical questions and suggest answers that may help provide an understanding of terror perpetrated by pan-Islamic groups around the world.

1. Is Islam The Problem?
2. Who Are The Islamists?
3. What Are They Fighting For?
4. How Did Islamists Become So Powerful?
5. Combating Terror: What Should The West Do?
6. Towards Eliminating Terror: What Should Muslims Do?

1. IS ISLAM THE PROBLEM?

An overwhelming majority of Muslims insists that Islam is only about peace and that it has been hijacked by extremists. They choose to view the present epoch as an aberration rather than the norm. On the other hand, a common Western belief is that Islam is an inherently violent religion. "*Islam has bloody borders*[1]", says Samuel Huntington, and "*This is particularly true along the boundaries of the crescent-shaped Islamic bloc of*

[1] *The Clash of Civilizations and the Remaking of World Order* by Samuel P. Huntington, New York, Simon & Schuster, 1996, ISBN 0-684-84441-9.

nations from the bulge of Africa to central Asia. Violence also occurs between Muslims, on the one hand, and Orthodox Serbs in the Balkans, Jews in Israel, Hindus in India, Buddhists in Burma and Catholics in the Philippines."

Both views are wrong; no religion is about peace or war. Every religion is about absolute belief in its own superiority and the divine right to impose its version of truth upon others. But in any given historical period, it might have either a peaceful temper that allows it to coexist with others. Or it could be possessed with an aggressive temper that seeks to conquer and subdue.

Whether Islam today has a peaceful or warlike temper depends upon what one chooses to identify as the "real Islam". Since there is no *mufti* endowed with central authority similar to the Pope in Catholicism, an enormous diversity of Islamic beliefs cohabits under a single canopy. If Islam is taken to mean Sufi Islam—or the vastly different forms of folk Islam that have become indistinguishable from tradition—then Islam is indeed a peaceful religion. There is no such a thing as a jihadist Sufi.

These are not marginal groups; Sufis were responsible for much of Islam's rapid spread after its initial military conquests in the 7th and 8th centuries. Converts from caste-ridden Hinduism were attracted towards Islam because it was viewed as a people-friendly religion. The Kurds of Iraq, Iran, and Turkey were also converted in this way and still maintain their respective brands of folk Islam. For Sufis, the Qur'an has to be interpreted allegorically. Harsh instructions for punishment were smoothed away, just as divine Quranic miracles were smoothed away by Muslim rationalists like Syed Ahmad Khan. Charismatic Sufi masters like Mansur al-Hallaj and Jalaludin Rumi invested in the concept of subjugating the self (*jihad bi nafsihi*) to the service of the Creator and His creation. Allah, they argued, must be worshiped not out of duty or fear but because he loved his creation and was loveable. Many dedicated their lives to the service of the weak and needy. In searching for that divine love, Sufi Muslims pray at shrines, venerate local saints, sing, and dance themselves into ecstatic oblivion. In India, Sufi saints continue to be revered by Muslims and Hindus alike.

But this kind of "soft Islam" has a mortal enemy in "hard Islam", which vociferously claims to be the real Islam. The conflict between them has a long history. *Wahabism*, which originated in the 18th century in Arabia, started as a reaction to Shia'ism and Sufism. In its early years, it succeeded in destroying all shrines, together with priceless historical monuments and relics from the early days of Islam. This process has continued. In June 2010, the widely venerated shrine of Data Darbar in Lahore was targeted by two suicide bombers who killed over 50 worshippers.

The *Salafis*—who seek the "purification" of Islam by returning to the pure form practiced in the time of Prophet Muhammad and his companions—are also prone to violent extremism. Among the most extreme manifestation of *Salafism* is *Takfir-wal-Hijra*. In 1996 the group is said to have plotted to assassinate Osama bin Laden for being too lax a Muslim. Muslims of the *Deobandi-Salafi-Wahabi* persuasion fiercely decry the syncretism of popular Islam, claiming that it arises from ignorance of Qura'nic teachings. Mosques in the U.S. and Europe are increasingly dominated by *Wahabi* imams. Some are imported from Arab countries, others are local. They insist on a literalist interpretation of the Qur'an. Salaries are paid for by donations from oil-rich countries or raised locally.

Steadily, the culture of the mosque is defeating the culture of the shrine. The support provided by Saudi Arabia has been critical to this. Petrodollars have promoted the work of leaders of the *Ikhwan al-Muslimeen* (the Muslim Brotherhood) such as Hasan al-Banna and Sayyid Qutb, and financed Maulana Abdul Maudoodi's Jamaat-i-Islami in South Asia. Bosnians, who had welcomed Saudi-financed Wahabist fighters in their desperate efforts to avoid being ethnically cleansed by Serbs, soon learned that this came at a high price. To their dismay, the imported "saviors" soon set about uprooting "corrupted" traditional Bosnian Sunni Islam and asserted their purified brand as the only correct one.

The bottom line: although the overwhelming majority of Muslims remain peaceful and opposed to terrorism, the centre of gravity is moving away from soft Islam towards jihad-oriented hard Islam. Dangerous times lie ahead.

2. WHO ARE THE ISLAMISTS?

It is conventional to understand violent Islamic fundamentalism as a response to poverty, unemployment, poor access to justice, lack of educational opportunities, corruption, loss of faith in the political system, or the sufferings of the working class. As partial truths they are indisputable. Those condemned to living a life with little hope and happiness are indeed terribly vulnerable to calls from religious demagogues who, in exchange for unquestioning obedience, offer a happy hereafter. Orphans and impoverished madrassa-educated lads in Pakistan can be readily turned into suicide bombers. But they are mere pawns used by those with better education and skills, whose distorted ideals and visions owe to factors other than poverty. Consciousness and world-view are not straightforward consequences of material conditions; its determinants include intangibles and are psychologically rooted. In fact, terrorists who have planned or carried out attacks inside Western countries are sometimes homegrown, relatively well off, and quite a few are university graduates. Many have had a technical education.

There are countless examples: Mohammed Atta, the 9/11 mastermind who piloted one of the two planes into the World Trade Center, studied architecture at Cairo University and then continued on at the Technical University of Hamburg. Among doctor terrorists, Dr. Ayman Al-Zawahiri, described to be the real brains behind Al-Qaida, stands out. There are, of course, less well-known doctors too. Of the eight people arrested in Britain after two bungled car bombings in 2007, seven men were physicians and the lone woman, the wife of one of them, was a medical technician. Hafiz Saeed, the founder of the *Lashkar-e-Tayyaba* which executed the Mumbai attacks in 2008, was a professor at the University of Engineering Technology in Lahore. Syed Bashiruddin Mahmood, who was arrested after meeting with Osama bin Laden in Afghanistan, is a nuclear engineer with a degree from Manchester University and received a patent for a probe that detects reactor leaks.[2] The list goes on.

Researchers at Oxford University have recently studied the sociology of those involved in extremist groups or in violent terrorism.[3] In a paper entitled "Engineers of Jihad", they arrive at an interesting conclusion: "*We find that graduates from subjects*

[2] Mahmood and the author had a rather vitriolic exchange about 20 years ago about whether djinns could be used to solve Pakistan's energy problems. See the appendix of "Islam and Science—Religious Orthodoxy and the Battle for Rationality" by Pervez Hoodbhoy, Zed Books, London 1990.

[3] *Engineers of Jihad* by Diego Gambetta and Steffen Hertog, Department of Sociology, University of Oxford, Manor Road, Oxford OX1 3UQ, www.sociology.ox.ac.uk/swp.html.

such as science, engineering, and medicine are strongly overrepresented among Islamist movements in the Muslim world, though not among the extremist Islamic groups which have emerged in Western countries more recently. We also find that engineers alone are strongly over-represented among graduates in violent groups in both realms. This is all the more puzzling for engineers are virtually absent from left-wing violent extremists and only present rather than over-represented among right-wing extremists."

What explains this phenomenon? The authors consider four hypotheses[4]: "*Is the engineers' prominence among violent Islamists an accident of history amplified through network links, or do their technical skills make them attractive recruits? Do engineers have a 'mindset' that makes them a particularly good match for Islamism, or is their vigorous radicalization explained by the social conditions they endured in Islamic countries?*"

While these hypotheses appear to be well-framed, and the answers found within this valuable study are interesting, it does not really help us understand what motivated these men. This is because the authors shy away from exploring the link between various Western invasions of Muslim countries and the anger they subsequently generated. But for a scientific understanding of motivations, it is essential to listen to what some violent Islamists are saying.

3. WHAT ARE THEY FIGHTING FOR?

Faisal Shahzad, the failed Times Square bomber with a degree from the University of Bridgeport, is a terrorist with a clear mind. Calling himself a "Muslim soldier", he pleaded guilty on June 21, 2010 and read a prepared statement: "It's a war...I'm going to plead guilty a hundred times over because until the hour the U.S. pulls its forces from Iraq and Afghanistan and stops the drone strikes...we will be attacking the U.S. And I plead guilty to that."

Shahzad was clearly grandstanding, but his words inspired some resonant sympathy even among those Muslims who condemned his tactics. Indeed, I would hazard that Western foreign policy over the last several decades stands as the number one cause for Islamic radicalism. There can be little doubt that the illegitimate war waged by the United States in Iraq provided an extraordinary impetus to radicalism and terrorism.

Religious radicalism can be produced by carefully cultivating grievances, particularly those grievances inflicted by "the other" rather than one's own side. The vodka-drinking Muslim Uighur from the Xinjiang province in China who has been discriminated against by the Han, the Chechen from bombed-out Grozny, and the secular Palestinian whose orchard of olive trees has just been destroyed by Israeli bulldozers, are victims of some oppressive order. Although initially they may not be driven by some theologically purist fantasy of the world caliphate, they can fall prey to religious ideologues and believers in violent solutions.

Internal decline also generates radicalism, which competes in importance with that produced by Western military invasions. Faced with internal failure, manifest decline from a peak of greatness many centuries ago, and afflicted by cultural dislocation in the age of globalisation, many Muslim societies have turned inwards. Muslims have little

---

[4] ibid

presence in today's world affairs, in science, or in culture. This has led to diminished self-esteem, as well as increasing recourse to political Islam. Some dream of a new global caliphate.

Although Islamic radicalism may be bad news for Washington, Tel Aviv, and London, it is much worse for Muslims. It pits Muslims against Muslims, as well as against the world at large. Only peripherally directed against the excesses of corrupt Muslim ruling establishments, and taking opportunistic advantage of existing injustices and inequities, the primary targets of extremists today are other Muslims living within Muslim countries. Some religious fanatics terrorise and kill others who belong to the "wrong" Muslim sect. The Shia-Sunni conflict accounts for the majority of those killed in Iraq. In Bangladesh, religious fanatics set off 400 bombs in public places in just one day. These fanatics accuse "modernised Muslims" of being vectors of hellish sinfulness—the so-called *jahiliya*—deserving the full wrath of God. The greatest ire among them is aroused by the simplest of things, such as women being allowed to walk around bare-faced, being educated, or the very notion that they could be considered the equal of men.

Contrary to its claims, Islamic radicalism is indifferent to the suffering of Muslims; the victims of suicide bombings are often those praying mosques or at funerals. On the other hand, fundamentalist fury explodes when the Faith is seen to be maligned. For example, mobs set afire embassies and buildings around the world for an act of blasphemy committed in Denmark; others violently protested the knighthood of Salman Rushdie. Even as Muslim populations become more orthodox, there is a curious, almost fatalistic, disconnection with the real world. This suggests that fellow Muslims do not matter any more—only the Faith does.

To conclude: Islamists have gained enormous strength from the general anger generated by numerous Western invasions of Muslim countries. But their wider agenda is to radically change society in the direction of 7th century Arabia. Even if the rest of the world was to mend its ways, or even to miraculously disappear, millenarian inspired zealotry will remain.

4. HOW DID ISLAMISTS BECOME SO POWERFUL?

Every religion, including Islam, has its crazed fanatics. Few in numbers and small in strength, they can properly be assigned to the "loony" section. This was true for Islam as well until 1979, the year of the Soviet invasion of Afghanistan. Indeed, there may well have been no 9/11 but for this game-changer. Since Western analyses often assign this to a mere footnote, I feel that this needs some substantive discussion here.

From the first day onwards, American diplomatic strategy was to mobilize world opinion against the Soviets. American ire was aroused not out of sympathy for the particular victim but by the act of aggression itself and what it portended for the future. Afghanistan was declared to be a domino and the "warm waters of the Persian Gulf" was declared to be the Soviet's real goal. With Ronald Reagan as his rival presidential candidate, Jimmy Carter could not afford to appear soft on the Soviets. Officials like Richard Perle, Assistant Secretary of Defense, immediately saw Afghanistan not as the locale of a harsh and dangerous conflict to be ended but as a place to teach the Russians a lesson. Such "bleeders" became the most influential people in Washington.[5]

Given the highly conservative nature of Afghan society and the spontaneous resistance to the Afghan communist resistance, it did not need a genius to suggest that Islamic international solidarity could be used as a powerful weapon. The task of creating such solidarity fell upon Saudi Arabia, together with other conservative Arab monarchies. This duty was accepted readily and they quickly made the Afghan Jihad their central cause. It was a natural course of action to take. First, they felt genuinely threatened by the Soviets. Second, it shielded their patron and ally, the United States, whose direct confrontation with the Soviets would have been dangerous and unwise in a nuclear-armed world. But still more importantly, to go heart and soul for jihad was crucial at a time when Saudi legitimacy as the guardians of Islam was under strong challenge by Iran, which pointed to the continued occupation of Palestine by America's partner, Israel. An increasing number of Saudis were becoming disaffected by the House of Saud—its corruption, self-indulgence, repression, and closeness to the U.S. Therefore, the Jihad in Afghanistan provided an excellent outlet for the growing number of militant Sunni activists in Saudi Arabia, and a way to deal with the daily taunts of the Iranian clergy.

Support for the Mujahideen also fitted perfectly with the Reagan Doctrine—a global package of widely publicized covert aid for anti-Communist guerrillas fighting the established governments in Nicaragua, Angola, Kampuchea, and Afghanistan. Now the United States decided to play in the global game of guerrilla politics and to do what the Soviets had done in the sixties and seventies when they had encouraged wars of national liberation. The U.S. would henceforth do the same by sponsoring right-wing guerrilla movements in the eighties.

The bleeders soon organized and armed the Great Global Jihad, funded by Saudi Arabia, and executed by Pakistan. A powerful magnet for militant Sunni activists was created by the U.S. The most hardened and ideologically dedicated men were sought on the logic that they would be the best fighters. Advertisements, paid for from CIA funds, were placed in newspapers and newsletters around the world offering inducements and motivations to join the Jihad.

American universities produced books for Afghan children that extolled the virtues of jihad and of killing communists. Readers browsing through book bazaars in Rawalpindi and Peshawar can, even today, sometimes find textbooks produced as part of the series underwritten by a U.S. AID $50 million grant to the University of Nebraska in the 1980's.[6] These textbooks sought to counterbalance Marxism through creating enthusiasm in Islamic militancy. They exhorted Afghan children to "pluck out the eyes of the Soviet enemy and cut off his legs". Years after the books were first printed they were approved by the Taliban for use in madrassas—a stamp of their ideological correctness and they are still widely available in both Afghanistan and Pakistan.

At the international level, Radical Islam went into overdrive as its superpower ally, the United States, funneled support to the mujahideen. Ronald Reagan feted jihadist

---

5 "Bloody Games", Eqbal Ahmad and Richard J. Barnet, The New Yorker, 11 April 1988, and, "Confronting Empire", Eqbal Ahmad, interviews with David Barsamian, South End Press 2000.

6 See, for example, Craig Davis in World Policy Journal, Spring 2000. The author, who was a doctoral candidate at Indiana University, conducted fieldwork in Afghan education in Afghanistan and Pakistan in 1999-2000. The examples quoted in the present essay are from his work.

leaders on the White House lawn, and the U.S. press lionized them. When Soviet troops withdrew from Afghanistan in the face of the U.S.-Pakistani-Saudi-Egyptian alliance in 1988, a chapter of history seemed complete. But, as things turned out, it was a pyrrhic victory. Even in the mid 1990s—long before the 9/11 attack on the U.S.—it was clear that the victorious alliance had unwittingly created a monster now beyond its control.

5. COMBATING TERROR: WHAT SHOULD THE WEST DO?

From the early 1950s, following the era of decolonisation, a sense of grievance and frustration had produced a multitude of Islamist movements spreading from Algeria to Indonesia. But they were inconsequential. Had the U.S. not cultivated them as allies against communism during the Cold War, history could have been very different.

All this is history—and unchangeable. Today, relations between Islam and the West are worse than ever before. A civilizational clash it may not yet be, but it is probably around the corner. How can it be avoided? Focussing upon the U.S., imagine for a moment that it had a sudden change of heart, realised the error of its ways, and wanted to bury the hatchet with Muslims. How could the U.S. atone for its past? Here are five key steps:

First, as demanded by both by Muslims and non-Muslims across the globe, the U.S. needs an attitudinal change. Indeed, welcome steps in this direction have been taken by President Barack Obama, as in his Cairo address of 2009, in seeking to repudiate grand imperial designs as well as U.S. claims to being an exception among nations. These reflect the growing realization within the U.S. that economic realities will eventually force it to relinquish its earlier goal of gaining total planetary control (as in PNAC, Program for the New American Century) and, instead, accept the more modest goal of being just one of the world's big players. Now that America's wheels are coming off in Afghanistan, this seems increasingly more likely.

Second, the creation of a Palestinian state must not be further postponed. The dispossession of Palestinians has been appropriated as a Muslim cause with huge symbolic significance. The U.S. had given Israel carte blanche for military action against the Palestinians, as in the invasions of Lebanon in 1982 and 2006. So far there is little reason for optimism here. A powerful pro-Israel lobby has resulted in even popular U.S. presidents being forced to bow to Israeli interests. The point-blank refusal of Israel to halt its settlement expansion in spite of Obama's urging is clear proof.

Third, the U.S. must take seriously the impact of collateral damage on civilian populations. The heavy use of airpower in Iraq and Afghanistan inevitably led to large numbers of non-combatant casualties. While drones have greater ability to discriminate, the rules of engagement need to be continuously upgraded to reach the goal of zero civilian casualties.

Fourth, the U.S. must stop threatening Iran with a nuclear holocaust for trying to develop nuclear weapons, while rewarding, to various degrees, other countries—Israel, India, Pakistan, and North Korea—that have developed such weapons surreptitiously. It would, of course, be highly preferable if Iran could be dissuaded by peaceful means, including sanctions, from making a bomb. But if it does make the bomb in spite of everything, then it must be accepted as one more nasty fact of life. There is no strong moral argument available to the U.S. against Iran's nuclear ambitions, given both its own nuclear stance and the fact that Iran's initial nuclear capability was provided by the U.S. during the Shah's rule. Attacking Iran will unleash dynamics over which the U.S. will be powerless, and perhaps exceed the global impact of the Iraq war.

Fifth, the West must seize opportunities that project it as generous, rather than aggressive. Providing disaster relief (including following the 2004 tsunami and the 2005 Kashmir earthquake) did much to build a positive image. Soft power is critical. Draining the swamps where extremism breeds will require increasing foreign aid to poor Muslim countries, creating economic and employment opportunities there, and desisting from policies that reward only the elites of the recipient societies.

6. TOWARDS ELIMINATING TERROR: WHAT SHOULD MUSLIMS DO?

A paradigm shift is essential. Muslims must realise that the awesome strength of Western civilisation—which also made possible its predatory imperialism—springs from accepting the premises of science and logic, respecting democratic institutions, allowing value systems to evolve, and boldly challenging dogma without being condemned for blasphemy. They must connect the West's success with personal freedom and liberty, superior work ethics, artistic and scientific creativity, and the compulsive urge to innovate and experiment.

Muslims, if they are to be a part of mainstream civilisation, will have to adapt to a new universal cultural climate, one that accepts human rights as defined by the United Nations Charter and the Universal Declaration of Human Rights, including the equality of men and women. On the part of Muslim minorities and immigrants to non-Muslim countries, this means acceptance of different behavioural norms, and a move away from the current tendency of ghettoisation and towards greater integration into the larger society.

Meanwhile, Muslims themselves must stop believing convoluted conspiracy theories that purport to explain their states of weakness. For example, it is widely held that today's sectarian warfare is a consequence of some cunningly remote manipulations by enemies of Islam. Or, as in Pakistan, that suicide bombers are Blackwater recruits.

Muslims must also stop dreaming of theocracy and *sharia* law as solutions to their predicaments. This means acknowledging the sovereignty of the people rather than the rule of Allah, the latter by way of a self-appointed priesthood, such as *vilayat-e-faqih* and *khilafat-e-arz*. These are essentially prescriptions for a theocracy run by mullahs. It is simply impossible to run modern states while remaining shackled to medieval religious laws. Economic development, an expansion of individual liberties, democracy, an explosive growth in scientific knowledge and technological capabilities—these and a host of other benefits will forever remain distant dreams without the modernisation of thought.

The only way by which Muslim societies can become democratic, pluralistic and free from violent extremism is by going through their own internal struggles. Indigenous reform is difficult but possible. Islam is certainly as immutable as the Quran, but values held by Muslims have changed over the centuries.

CONCLUSION

Looking at planet Earth from above, one can see a bloody battlefield, where civilizations are poised to clash with each other. America and the West cannot win. Nor can the Islamists. This calls for reasonable compromise. Not all Islamist demands are ridiculous. Some are, and must be ignored. But some are rooted in objective reality, and must be negotiated with in the same spirit and manner as in other disputes.

Ultimately, the hope for a happier and more humane world lies in reason, education and economic justice. Secularism, universalistic ideas of human rights, and freedom of belief are non-negotiable and we must struggle for them. Biology, which endowed our ape ancestors with territoriality, is perhaps the ultimate problem. But biology may also be the solution because it has given us the human mind—the best survival tool known.

TERRORISM THREATS DUE TO WEAPONS OF MASS DISTURBANCE[1]

FRIEDRICH STEINHÄUSLER
Physics and Biophysics Division, University of Salzburg
Salzburg, Austria

ABSTRACT

Terrorists using a Weapon of Mass Disturbance (WMDi) have the capability to cause massive disarray in the administration in the targeted area and interrupt daily routines of society at a catastrophic level. The following major categories of WMDi and modes of their deployment have been identified:

WMDi categories:
Radiological Dispersal Device (RDD), covert radiation weapon, disruption of food- and drinking water supply, cyber threats, contamination of monetary supply, disruption of energy supply, disruption of telecommunication, disruption of IT infrastructure and banking sector, and disruption of mass transportation system.

Modes of WMDi attack:
Radioactive aerosol dispersion, dirty bomb, plane as guided missile, contamination of roads, radiation as a covert murder weapon, covert public irradiation, radiation blackmail; threats against personnel; threats against oil- and gas Installations; attack on the food supply at the farm level and storage site, during food processing, at the food distribution centre, at food services; cyber attacks via WLAN, forged messages, denial-of-service, physical vulnerability of stations; different modes of violent physical attacks on sectors of the national critical infrastructure (energy, communication, transport, financial sector).

DEFINITION OF WEAPONS OF MASS DISTURBANCE

A *Weapon of Mass Disturbance* (WMDi) impairs or even destroys the composure of the society under attack by damaging the order in the business community, or by causing disarray in the administration or severely interrupting daily routines of society: Thereby, the deployment of a WMDi can lead to a major disturbance in the population. This includes damages by a WMDi to the critical infrastructure, such as disruption of the financial sector, impediment of the mass transport and distribution of goods, unsettlement of the energy system or upsetting the IT infrastructure.

MOTIVATION FOR DEPLOYMENT OF A WMDI

Deployment of a WMDi may be attractive for terrorists for the following reasons:

- Increasing fear among many members of the public with a single unconventional attack, multiple synchronized attacks, or simply threatening to attack.
- Indicating to the public that the authorities can no longer ensure security for a large segment of society.

---

[1] This project is co-funded by the European Commission DG Energy & Industry (FP 7 (GA No. 218070-CAST; www.castproject.eu)) and the Austrian Federal Ministry for Science and Research. In view of its *EU Confidential* content only unclassified content is contained in this publication.

- Disturbing life for a *large* number of people in a significant manner at a relatively low cost due to an unconventional mode of terror attack rather than by inflicting mass casualties.
- Inflicting large scale economic damage resulting from such a terror attack.

A WMDi attack does not result in more casualties than a conventional terror attack but the circumstances of the WMDi attack increase the complexity of the aftermath as compared to the conventional attack. For example, in case of a *Dirty Bomb* the victims may not only have suffered injuries due to a bomb blast, but they will also be contaminated by radioactivity.[2] The inherent fear of many people of the health risks associated with radiation per se adds to the already prevailing feeling of helplessness. The extent of clean up procedures of the residual radioactivity in the environment (buildings, streets, parks, etc.) will depend significantly on the existing regulatory framework for environmental standards and radiation protection. Applying the low levels recommended in the internationally widely accepted IAEA Basic Safety Standards (BSS)[3] can result in excessive financial costs, as was shown in the clean-up of the incident at Goiania (Brazil, 1987) and Chernobyl (Ukraine, 1986).[4]

WMDi CATEGORIES

The main purpose of a WMDi is the severe disruption of society. In view of the relatively low resilience of many members of the public in industrialized countries against any reduction of its standard of living over an extended period, terrorist successfully interfering with the daily routine and its technical amenities will be able to inflict harm on the targeted society. WMDi can be grouped into the categories listed below.

Radiological Dispersal Device

A *Radiological Dispersal Device* (RDD) enables the terrorist to disperse radioactivity in the environment without the target population necessarily noticing the presence of radiation at the time of the attack. Terrorists have a relatively wide spectrum of possible radiation sources to chose from: (a) Cesium (Cs-137) sources used in medicine (activity: about 10E14 Becquerel) can contain caesium chloride, a highly dispersible powder; (b) Teletherapy cobalt (Co-60) sources contain typically 1,000 metal pellets (activity: 10E11 Becquerel each); (c) Industrial radiography sources use mostly metallic iridium (Ir-192), to a lesser extent cobalt (Co-60), selenium (Se-75), or ytterbium (Yt-169), with a typical activity of about 3 trillion Becquerel each; (d) Industrial gauges and other applications can contain cobalt (Co-60), cesium (Cs-137), or americium (Am-241) (typical activity: 0.1 to 50 x 10E9 Becquerel). Figure 1 shows the iso-dose curves (TEDE) after the detonation of a dirty bomb, based on Ra 226, in a non-urban environment.[5] Within the area marked with a red boundary (187 km²) a person would receive a TEDE of 10 mikro-Sievert for every *hour* of residence time. For

---

[2] A *Dirty Bomb* represents a device, consisting of conventional explosives combined with a suitable radioisotope.

[3] International Atomic Energy Agency (IAEA), International Basic Safety Standards for Protection against Ionizing Radiation and for the Safety of Radiation Sources, Safety Series No. 115 (1996).

[4] F. STEINHÄUSLER, Countering Radiological Terrorism: Consequences of the Radiation Exposure Incident in Goiania (Brazil), in: Social and Psychological Effects of Radiological Terrorism (Eds.: I. Khripunov et al.), NATO Science for Peace and Security Series, Vol. 29, IOS Press (2007).

[5] TEDE (= EDE*external* + CEDE*internal* 50-year dose).is the most complete expression of the combined dose from all applicable delivery pathways. The 50-year committed effective dose equivalents are calculated per one-hour residence time. Therefore, an individual would be committed to the output dose for each one hour at the output location.

comparison, the average *annual* radiation exposure from natural sources to an individual in the United States is about 3 000 mikro-Sievert.

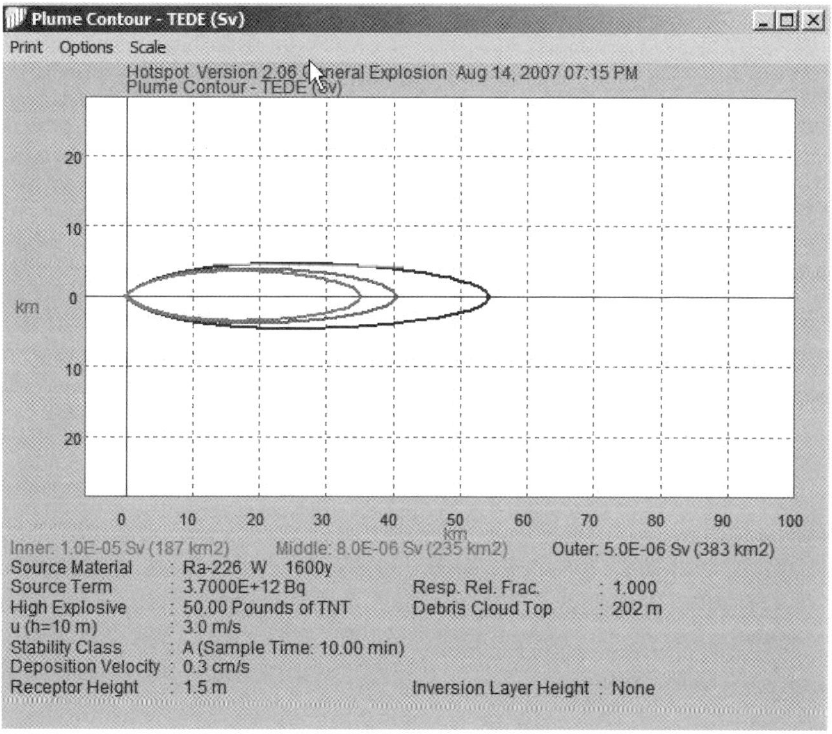

Fig. 1: Radium (Ra-226) based Dirty Bomb*; TEDE contour plot in a non-urban environment (Activity: 3.7 TBq; Assumption: release time 10 min to < 3 h, atmospheric transport max 8 h).*

Covert Radiation Weapon

Since radiation cannot be detected with our senses, terrorist can deploy a WMDi as a covert weapon. Radiation can be used to contaminate, injure or even kill a victim, depending on the absorbed dose. It can also be deployed to enhance the effects of a conventional terror attack, i.e., significantly complicating search-, rescue-operations by first responders having to fulfill their duties in a radioactive environment. Also, terrorists can use this approach to blackmail society by threatening to contaminate or irradiate, covertly irradiate a large number of persons, or kill targeted individuals.

Disruption of Food Supply

Food can be tampered with at: (a) At the storage areas of raw materials used for food production; (b) During transport from the storage area to the food processing factory; (c) At the food processing factory; (d) During transport from the processing factory to the distribution centre; (e) At the food distribution centre; (f) During transport from the distribution centre to the retail shop; (g) At the retail shop. Food contamination can be achieved by radioactive isotopes, biological contaminants (e.g., *E. coli* bacteria), or simply by using dyes.[6]

Contamination of Money Supply

All world reserve currencies, such as the EURO, US Dollar, or the Japanese Yen, are at risk of being contaminated intentionally, since they in circulation in large numbers worldwide. In case of the EURO, since 2002 about 43,767.61 million EURO banknotes have been produced.[7] For example, the radioactive contamination and international distribution of 10 000 popular € 5 notes would be sufficient to trigger wide-spread concern about *all* EURO denominations, once the media have been informed by the terrorists about the contamination of bank notes. This could lead to wide spread reluctance to accept any EURO notes at the national level and in international commercial transactions (e.g., tourism).

Disruption of Energy Supply

Since 2004 approximately 1,597 attacks of all terrorist events recorded have targeted the energy infrastructure (oil, gas, electricity production) worldwide, i.e., about 4%.[8] In the EU member states about 40 such attacks have occurred in the same period. Based on the experience gained in the past ten years, in the EU electricity production is at higher risk than locations housing oil- and gas facilities.[9]

Disruption of Telecommunication and IT Services

Modern society depends on a continuously functioning telecommunication system and access to comprehensive IT services. A major disruption of these services can be achieved by:

(a) Physical attack and subsequent destruction of a centre serving as a regional node;

(b) Physical attack and subsequent destruction of a first responder (FR) dispatcher system;

(c) Denial-of-service attack on the system installed at the FR dispatcher centre(s). In addition, several of the attack modes, listed in section on *Cyber Threats* below, are also applicable here.

Disruption of Mass Transportation System

Industrialized countries enjoy a multimodal mass transportation system, such as passenger train service, subway, rapid rail, street cars and trams, buses, ferries, and civil aviation. All of these components are part of an open system, used by a large number of persons for almost 24 hours every day. An attack on one of the components is likely to impact on a secondary level also on some of the other modes of mass transport, leading to major delays in the transport of persons and freight at significant costs to society.

Cyber Threats

Cyber attacks are increasing in number, degree of sophistication, severity, and cost. The overall number of threats to date has tripled between 2005 and 2007.[10] In 2009, 75% of

---

6 Radionuclides can be used for radioactive contamination of drinking water. Such an attack can occur on a public water supply system by contaminating the water supply to a level considered unsafe for chronic consumption. However, radiological contamination introduced prior to water entering the distribution system (e.g., contamination of wells or supply canals) is likely be detected by routine public water supply monitoring and mitigated in the water treatment process.

7 ECB Banknotes and coins production; http://www.ecb.int/stats/euro/production/html/index.en.html#Y2009 (last visited: 14 November 2009).

8 WITS Database (http://wits.nctc.gov/).

9 Brynjar Lia, The Terrorism Threat to Energy Supply Chains (2009): http://www.counteract.eu/Documents/Final%20Conference%20Brussels,%20March%202009/Energy/4%20EN%20Lia%202.pdf (last visited: 14 November 2009) .

10 Symantec Internet Security Threat Report Volume XI, March 2007.

businesses worldwide have experienced a cyber-attack at an average cost of € 1.5 million.[11] The United States was the top country of origin for attacks in 2009, accounting for 36 % of all detected attacks. The United States was also the top country of origin for attacks detected against targets globally in 2009, with 23% of the worldwide total.[12] Once subverted by terrorists, these computers can then be used to launch massive denial-of-service attacks, steal or corrupt sensitive information and disrupt network usage round the globe.

MODES OF WMDi ATTACK

Provided terrorists have access to the suitable radioactive or bio-chem material, explosives and weapons needed for a WMDi, as well as training, they can select the most suitable mode of attack among the examples listed below.

Terrorist Attacks with Radioactive Material

Once the radioactive material has been obtained by the terrorist, the material itself can be hidden in practically any type of containment, ranging from a soft-drink bottle to a briefcase or a parcel. The volume of the actual radioactive material in a radiation source without the shielding is rather small (typically a few mm^3 to about 100 cm^3 only). The ultimate size and weight of a radioactive material used for a terror attack depends on the shielding required: whilst a strong gamma radiation source needs heavy metal- or concrete shielding, an alpha radiation source can be shielded with aluminum foil. The subsequent intentional release of the material can occur from any type of land-, water-, or air-based vehicle. Intentional irradiation is possible indoors as well as outdoors.

Terrorists can disperse radioactive material into the environment by using an RDD, e.g., designed to generate radioactive aerosols over an extended period of time. This aerosol can be introduced into a ventilation system (e.g., subway, shopping mall, convention centre) in a covert manner. Alternatively, this could be achieved outdoors with an adapted crop dusting plane or commercially available mosquito-control flogging apparatus.

In case of a Dirty Bomb radioactive material is combined with conventional explosives (e.g., a truck bomb, loaded with radioactive material blanketed with explosives). Alternatively, a fully fuelled general aviation plane can be loaded with radioactive material and conventional explosives.[13] Provided the aircraft has been leased by a suicide commando, it could be crashed into a civilian target.

The detonation of a Dirty Bomb results in conventional damage due to the blast but also distributes radioactive material in the environment. Blast victims, first responders, and the area of the attack will be contaminated to a largely varying degree, depending on the prevailing weather conditions at the time of the detonation, and the degree of aerosolization.

Another possibility is the use of a radioactive solution for large scale environmental radioactive contamination, preferably in an urban area of high commercial value. This requires radionuclides to be dissolved in large containers, e.g., oil drums, to which a drain valve has been added. The drums are located inside a truck. At the end of a forecast rainy period, terrorists can cause widespread environmental contamination by spreading the solution, e.g., in a commercial district, with only a low probability of being detected.

Radioactive material intended to kill requires that a hidden source (preferably a strong gamma source) is positioned in such a manner as to expose the targeted person(s) to an excessively high dose rate for a defined period of time. Upon application of the desired dose

---

11 The Sunday Times, *Global cyber attacks on the rise: report*, 22 February 2010.
12 Symantec, Internet Security Threat Report, April 2010.
13 The plane could be adapted for criminal purposes at a remotely located minor airport or even private airfield with low operational security.

(exposure period: weeks to months, depending on the dose rate), the radiation source is removed. Subsequently the victim(s) can be informed about the irradiation to instill fear.

A strong radiation source in a public place (e.g., gamma source in a fire extinguisher, located in a shopping centre) with a high frequency of pedestrians is less likely to represent a major health hazard due to the presumably short exposure time of people passing by. Nevertheless, upon revelation of the fact by the terrorists to the media, largely independent of the magnitude of the actual radiation dose received, many members of the public will react in panic.

Terrorist could also be bluffing and only placing a few such radiation sources to remain credible. The actual (or fake) locations would only be revealed upon fulfillment of the demand by the terrorist.

Terrorist Attacks on Food Supply

Food contamination can be achieved by radioactive isotopes, biological contaminants (e.g., *E. coli* bacteria), or simply by using dyes. Since the objectives are to cause major disturbance and economic losses, the actual concentration of the contaminant are of lesser importance. Once the food has been contaminated, terrorists will inform the media.

Terrorist Attacks Contaminating Bank Notes

Bank notes represent the component of the money supply which is easiest to contaminate. Notes can contain routinely a wide range of contaminants, such as human or animal body fluids (urine, feces, vomit, and infectious blood), fine hazardous powders from detonated explosives, dye, or drugs.[14] Since most banknotes are made from cotton paper (weight: 80 to 90 g/m²), sometimes mixed with linen, abaca, or other textile fibres, it can be contaminated with a solution containing a radioactive isotope or biological contaminant.[15] Subsequently, persons handling these notes, as well as vending machines, cash dispensers, and counting-/sorting machines, will be contaminated.

Terrorist Attacks on Energy Supply Systems

Over the next twenty-five years oil and gas will provide about 60% of global energy, i.e., they will remain the prime sources of energy for industrialized and developing countries alike. Therefore, the oil- and gas industry represents a major component of every national critical infrastructure.

A pipeline system, basically protected by being buried between 1 m to 10 m underground, stretches over thousands of kilometres with little dedicated physical protection. It faces the following main security threats: (a) Placement of explosives at exposed sections of the pipeline; (b) Simultaneous detonation of explosives at several block valve- and compressor stations.

A special case is a terror attack against oil drilling operations. In case terrorists carry out simultaneous attacks on several such platforms operating in the same region, it is unlikely that the operators or the national authorities would be able to regain control for an extended period of time. At present in several countries platform security is inadequate, enabling even poorly armed gangs to take over temporary control on oil drilling platforms (e.g., Nigeria). The worst case scenario would be the hijacking of several oil platforms and feeder vessels by

---

[14] Trichur, Rita, *Bankers wipe out dirty money,* Toronto Star, 28 September 2007, http://www.thestar.com/Business/article/261441 (last visited 21 July 2010).

[15] Polymer banknotes (e.g., issued by Australia), made from polypropylene, may be slightly more resistant to contamination.

terrorists.[16] A coordinated attack on pipeline and associated stations could result in a significant environmental impact due to an uncontrolled oil spill, as well as reduced supply. In general, national energy supply systems are complex and largely open system, facing the following security threats: (a) Bombing of energy company offices, including attacks with truck-bombs; (b) Bombing of electrical transformers; (c) Arson attacks against electrical transformers; (d) Bombing or mechanical sabotage of electricity pylons, cutting of power lines; (e) Attacks against vital areas of power plants with rocket propelled grenades (RPG), mortars, or missiles; (f) Explosive-laden Unmanned Aerial Vehicle (UAV) attack on power plants; (g) Coordinated attacks (including diversion attacks), deploying mortar combined with truck bombs; (f) Commando raids by suicide terrorists.

Terrorist Attacks on Telecommunication and IT Infrastructure

Essential telecommunication- and IT services to society could be severely slowed down or even shut down by: (a) Denial-of-service attacks; (b) Attack on systems used by emergency call centres; (c) Attack on dedicated phone lines between government agencies; (d) Corruption of electronically stored vital information; (e) Attack on the computer systems used for broadcasting via television or radio or operating official websites, thereby impeding official use of these media or inserting erroneous messages; (f) Attack on the supervisory control and data acquisition system (SCADA).

Provided that terrorists can gain unauthorized access to vital IT components, this could result in: (a) Tampering with technical data; (b) Theft of strategic data; (c) Intrusion into a security-sensitive database; (d) Undetected hacking into the system and subsequent misuse; (e) Data snooping; (e) Introducing malware into the system.

Terror Attacks on Mass Transportation

Major disruption can be caused by one or more of the following modes of attack: (a) Small arms; (b) Rocket-propelled grenades; (c) Light anti-tank weapons; (d) Synchronized attacks with multiple improvised explosive devices (IEDs); (e) Improvised incendiary devices (IIDs); (f) (Sabotage of the Supervisory Control and Data Acquisition (SCADA) systems used for controlling and monitoring mass transit systems.

These modes of attack can be aggravated further by suicide attackers and/or insider support. Primary attacks may be followed by secondary attacks.

CONCLUSIONS

Terrorists attacking society with a WMDi have the potential to cause disarray in the population and damage to the national critical infrastructure. The number of different modes of attack deploying WMDi ranges from the use of radiation to cyber threats, to disruption of food- and energy supply, telecommunication, mass transport, IT services and the financial system. It is important to note that the primary impact (number of dead and injured) resulting from such an attack is likely to be similar to a conventional terror attack. However, the secondary (*Domino-*) effects, in particular at the economic and psychological level in the targeted population, can be devastating.

---

[16] A group of terrorists will gain access to 3 or more oil platforms and one feeder vessel collecting crude oil from the platforms. The operation will be prepared with insider assistance or as a covert attack by speed boats against the platform and vessel. All personnel will be held hostage in order to prevent any police or military rescue operation. Explosive charges will be placed on strategic points on the platform as well as at the critical underwater safety valves and drilling rods. Upon preparation of the explosive charges the platforms will be evacuated and destroyed.

CELL PHONES, TEXTING, POSITION REPORTING, AND SELF-ASSEMBLY IN EMERGENCY RESPONSE MANAGEMENT

ROBERT V. DUNCAN
University of Missouri
Columbia, USA

There is an increasing awareness of the advantages associated with the use of cell phones and Personal Data Assistants (PDAs) in the management of disasters. The Safe America Foundation has initiated a national campaign called ***Text First. Talk Second***, which is an attempt to encourage the public to rely on texting, not on placing calls to cell phones, to verify that their relatives are safe when an emergency situation arises. While cell phones are most frequently considered to be communications devices in an emergency, they also provide an important opportunity for use as sensors that can provide exceptionally important data to emergency managers as they mitigate the cascade of threats that arise in the immediate aftermath of a disaster. Here we will first consider the advantages for the use of cell phones primarily as distributed sensors, and then discuss how the management of cell phone communications, both broadly throughout society, and in more focused local networks that may be formed using portable cell service provider towers at the perimeter of the disaster area, may save lives through better medical triage and through more efficient evacuation strategies.

CELL PHONES AS A NETWORK OF SENSORS

Smart cell phones are used today in many applications to provide composite information from cell phone position tracking throughout the cell phone network to a central system that then synthesizes and reports this composite information back to all operators within the network, which results in robust situational awareness. For example, this is currently used to provide traffic delay indications in cities by various cell-phone network providers, where the rate of progress of each cell phone on the freeway is monitored and processed for robustness with hundreds of other such reports. This composite information is then fed back to all PDAs and cell phones on the network to inform the drivers of the level of congestion delays on all area roads and freeways. This information, when coupled with electronic maps and automatic route optimization software, then generates the optimal routing advice for the driver.

We propose a similar but more complex system to provide robust, unambiguous information to everyone in an emergency zone to assist them to safety, and to provide information for genuine data-driven management of the overall emergency response. This system will be useful in the management of a wide range of emergency situations, so it will effectively provide valuable support in any emergency where the local cellular networks remain available. Like in the traffic display system, the resulting information that is fed back to the operator will be robust, since it will be found to be consistent with many other operators who are also in the same situation in the emergency zone. This sort of emergency tool is designed to replace the single-point emergency operation hubs that are not robust against information errors or even intentional misinformation, and which often operate on assumptions regarding the distribution of operators within the emergency zone, and hence not on actual data. People caught in an emergency situation will naturally depend first upon their cell phones and PDAs, so the utilization of this system will be intuitive, in fact even reflexive, for those who are caught within the emergency zone. Conceptually, this entire system could be designed to self-assemble all available positioning reports over time to detect successful patterns of evacuation that have been discovered by some within the disaster zone, and to detect regions where the population in the emergency zone are not ambulatory. This data could be managed by an expert system and an experienced emergency manager to assure that only critical situational information is transferred back to those in the emergency, informing them to do only exactly what they must do to survive the situation. Grouping and tagging of position data from the various cell phones at the time when the disaster occurred, or at the latest during the start of the emergency management of the event, with each individuals' position data noted relative to the epicenter of the disaster, may be used to triage those fleeing the emergency zone for progressive critical care based upon each individual's level of initial exposure to the horrific event.

SYSTEM DESIGN

In an emergency situation anyone accessing the internet via their cell phone may be tentatively identified using automated phone number lookup systems. There are other existing systems that are based upon the GSM protocol (discussed in more detail below) that provide accurate position information at a level that varies with level of clutter in both standard urban environments.[1] These systems may, in the future, provide comprehensive location information automatically from all cell phones within the disaster region to a central disaster management system. If these cell phones are confirmed to be within the emergency region, and if adequate emergency preparedness software were installed within the phone, then their display screen may be set to divert to an emergency web site where this expert system could provide the phone's owner with only simple and succinct information and displays regarding what their next move, if any, should be. In more advanced applications, survivors may be asked by the expert system and/or emergency management staff for their assessment of damage, local fires and other threats, and the accessibility of roads, electricity, and other local infrastructure. Photos taken by the cell phones may be delivered over the wireless network using one of many possible protocols for this purpose as well. This information may be used with the phone's GPS location information, if available, and other relevant data (wind data, output of dispersive models or other modeling to predict the threat evolution with time) to update the situational awareness of the central system, which in turn determines the best evacuation or emergency management strategy for feedback to the cell phone and hence to the survivor. Clearly this basic sense-and-communicate network platform may be used in many ways, each centered on the specific disaster at hand, to provide the most relevant response possible. Collectively this system will provide the interface and data system layer to support a wide class of emergency response development tools, and the development of these specific tools could become a substantial effort. The main advantage of this platform is that it provides extensive real-time data for the emergency response managers, thus making the disaster response far more scientific and much less dependent on assumptions and course modeling that is used when actual data is not available.

EXAMPLE, SEPTEMBER 11, 2001

Imagine how the situation would have been different on September 11, 2001 if such a system were in place, and if GPS positioning and wide-bandwidth cellular phone communications were both available to those within the World Trade Towers in New York City. If those conditions were met, then such a system may have detected which stairwells were being successfully accessed for evacuation based simply on position and inferred velocity information from individuals within the buildings, with no direct intervention from the evacuees on the move. Similarly, trapped, wounded, or disabled occupants of the building could have been recognized by their lack of motion as observed by this positioning system, and strategies could have been developed to efficiently reach these people with emergency personnel. Most importantly, this information could be fed back to building occupants via their cell phones, both survivors and emergency personnel operating in the building, to extensively increase their situational awareness as they were being led efficiently only to the demonstrated evacuation routes, and as emergency personnel were being led directly to those who were no longer ambulatory. Furthermore, a related open-source system could be 'outward looking' for expert advice from a pre-defined network of trusted responders to glean urgently needed information that could be used immediately to develop a broader strategy for emergency response. This network may be pre-structured to assure that only those pre-vetted experts with the most relevant specific knowledge to assist with the specific situation are engaged in a well structure advisory network. For example, such an 'outward looking' disaster management component could have been well advised in real-time by civil engineers of the threat from thermal weakening of the aluminum-alloy exoskeletons of the buildings, and hence of the high likelihood of the towers' collapse. If organized in advance, there would have been ample time following the onset of this emergency to receive this information, which would have shifted strategies to immediate assisted evacuation, even on floors below the fires, instead of a 'stay in place' strategy for rescue once the fires were better controlled.

ROBUSTNESS

As discussed above, each cell phone will be located using the technology mentioned above to determine if they are in the immediate disaster zone. If so, then they will be added to the emergency management system server. This server will only download direct, simple text that instructs the survivor of their next best move. This disaster zone would be serviced first by the resident cellular towers is they survive the disaster, but later by a new set of mobile service towers that would use their portable high-gain antennas and concentrated transmitting power over the disaster region to form a local wireless intranet with external access permitted to only a very limited network of 'outward looking' communications. This will prevent the wild information overload that a standard commercial system would experience from the surge of concerned relatives world-wide who would be attempting to determine if their loved ones were safe. The technology necessary to establish such a limited and local wireless cell phone network has been developed by a number of companies such as Alto***. These systems are lightweight and portable, and they can be used to limit and manage communications within the disaster zone. For example, such a system may filter to permit only outside text messages to be delivered or transmitted to/from the phones within the disaster zone when the capacity of the local cellular system permits. This will assure that adequate bandwidth is reserved for the emergency managers to operate the system, and to assure that instructions from outsiders who are not part of the focused emergency response do not confuse the survivors inside the disaster zone. Tools may be developed to manage the situation in different event scenarios, and expert systems may be permitted to operate to determine movement throughout the disaster region and to feed this information back to the survivors with no direct operator intervention. Such a system would be designed to permit those within the disaster zone to self-organize through the increased awareness that this fed-back information would provide. This may permit the emergency response staff to actively plan for later phases of the disaster management while survivors are determining how to evacuate, if that was their instructions, from the egress that is being discovered by others within the disaster zone.

As an example, if the emergency zone was the immediate destruction zone from a nuclear blast, then the emergency coordinator may use the information assembled electronically from this network to group respondents based upon their location relative to the blast. This position tag of everyone who connects with this system may be used to design the type of medical care that they will receive once they escape the immediate blast zone. Wind, and other relevant environment data such as street maps, may be used to direct people fleeing the blast zone to areas so that they avoid advancing fires, and so that they evacuate to regions that are much less likely to experience severe fallout. The survivors may be instructed to go to muster stations on the basis of their likely initial exposure to the flux of gamma radiation from the blast, thereby meeting medical personnel who are prepared to triage their level of exposure. Strategies may be developed to airlift or to otherwise deploy essential potable water and supplies to those who have been instructed to stay in place and to await rescue, and to intercept those who are moving out from the emergency zone along effective escape routes.

In every case, composite information automatically detected from the fleeing operator's cell phones and PDAs will be used in favor of rare individual reports, assumptions, etc. This way the rescue and disaster management will be driven by real time data, using system tools that have been developed in advance.

HIGHER-LEVEL OPERATING SOFTWARE

This system described above will provide the base level communications for advanced situational awareness in a disaster. Additional layers will overlay this information on scalable maps, such as those that exist now for GPS navigation, and permit the emergency zone to be defined graphically. Fortunately much of this software is already in use to register GPS data on map sets on commercial handheld PDA and GPS devices. It will be desirable to develop three dimensional maps of cities to extend this capability into the mitigation of building emergencies, and to develop more advanced feedback systems that will parse the situational awareness data at the central emergency management location and send it out to each individual within the emergency zone. This system would provide to each cell phone a simple map that indicates their present position and their demonstrated route to safety, or an instruction to stay in place if no self-evacuation route is feasible. This system would ask each individual in the emergency zone to provide feedback through texting details if they see a reason why the recommended evacuation route is no longer feasible. Of course the demonstrated position information from each cell phone over time will provide such information on the basis of observed flow, but this more immediate and 'high-bandwidth' feedback will be useful for alternative routing considerations and for forward emergency responder deployment in a very efficient, data-driven manner. Other higher-level overlays may include external expert correspondence, and scenario-specific concerns. For instance, in agent release of nuclear detonations it will be useful to run models of the dispersion of the agent or fallout based upon atmospheric models that access overlays of the current weather, wind-speeds, etc. This entire system may be developed in a modular way to permit more scenario-specific software to overlay the basic system as the specific concern warrants, much like the toolbox applications in MATLAB.

THE SAFE AMERICA FOUNDATION'S TEXTING DRIVE

In general preparation for the use of cell phones in disaster management, the Safe America Foundation has recently started a national campaign to encourage users to text instead of placing calls in an emergency. This approach is motivated by the fact that the information content is a brief text message such as "IMOK" is much less than the information content in a one-minute telephone call that is intended to convey the same information. Texting has the additional advantage that it utilizes latency, meaning that the text message under most protocols will be stored until such time that the necessary network support becomes available to deliver the message. This is a huge advantage compared to the delays that are often inherent in securing a viable voice communication line when the system is overloaded.

To quantify the advantage of texting, consider that the message "IMOK" requires only four bytes of information to convey using the standard SMS protocols. There are certainly routing data and other metadata that are included with this information to route and place the text message properly, and the level of this 'overhead' data varies with the text service provider, as discussed below. But a typical message such as this simple one will require much less than 120 bytes to fully convey. The amount of digital information required to capture and convey voice varies considerably depending on compression techniques and specific protocols used by each service provider, but typically runs at an effective sampling rate of 16 kbaud, which produces 2,000 bytes of conveyed information per second. This would produce and convey 120,000 bytes of information in a one-minute long call. Hence the information transferred in a single emergency telephone call is likely to be one thousand times greater than the information that must be transferred in a simple text message. This, combined with the latency considerations listed above, makes texting the communication modality of choice. The actual bandwidth within the communication channel that is necessary to transmit the text message varies significantly depending on the protocols of the communication network, as discussed below, but clearly the single most significant way of protecting the capacity of our cellular communications infrastructure from becoming overwhelmed and saturated during a disaster is to encourage the public to text, not call, in an emergency situation.

Texting has increased rapidly over the last few years, with over 8.1 trillion SMS text messages sent world-wide in 2008 alone. Typically telecommunication companies obtain about \$0.11 (U.S.\$) per text message, which secures literally many billions of dollars per year in profits to telecommunication companies, since the marginal cost of text messaging is virtually zero. Texting, which first became possible in 2008 with the advent of SMS, remained virtually non-existent as a communication modality prior to the year 2000. The volume of texting has increased by a factor of three each year over the past decade. The SMS protocol originally ran using GSM, but now it has been expanded to run on CDMA and other competitive protocols. Please see the "Summary of Protocols" section below for a description of how these protocols and location capabilities have progressed through the four generations of mobile communications performance standards.

BRIEF SUMMARY OF PROTOCOLS

The SMS system (Short Message System) was developed in 1982, and subsequently modified to allow concatenation so that the system may support much larger text messaging. Originally SMS ran only on the GSM (Global System for Mobile communications) protocol, but now SMS has been expanded to work on most competitive protocols, including the CDMS protocol discussed below. The GSM protocol is used in 80% of all cell phone calls that are placed world-wide, and this protocol is used by over three billion cell phone owners world-wide. GSM is a second-generation (2G) protocol, but it has now been extended to third-generation speed (3G) through a new protocol called EDGE (Enhanced Data rates for GSM Evolution). More recent and advanced GSM protocols incorporate and support earlier cell phone communication protocols, such as TDSM (Time Division Multiple Access), which use time sharing of a given bandwidth allocation to permit multiple calls to access a single frequency assignment by switching rapidly between the multiple calls to provide data refresh from the last channel assignment position. GSM, and its extension EDGE, are used by T-Mobile and AT&T world-wide.

The competitor of the GSM / EDGE protocols is the CDMS (Code Division Multiple Access) protocol, which is utilized by the Sprint, Verizon Wireless, and Virgin Mobile service providers. CDMS is often referred to as 'spread spectrum' technologies, since a digital code is utilized to dynamically allocate bandwidth as necessary to support a much faster data transfer rate than was provided by GSM during the second-generation (2G) wireless capability.

CONCLUSIONS

The necessary telecommunications technology exists today to both provide superior directed communications during a disaster, and to use the cells phones within the disaster region as a sensor network to obtain actual data on the movements and actions taken by those in the disaster region. This data may be used to develop a scientific process of testing different hypothetical methods of disaster management, and to determining if certain universal emergent behaviors result. It is important that disasters and emergency situations occur on a continuous basis, with small emergencies occurring on a daily basis, and huge disasters occurring only rarely. This proposed sensor/communications network may be deployed by sub-setting the overall network on the relevant scale that is appropriate for the emergency at hand. This will permit the scaling of the response models and procedures that are developed scientifically against data to be determined. These opportunities depend on developing agreements with various service providers within the cellular communications service providers throughout the world, and with universities and national laboratories as appropriate. These agreements do not exist, and their establishment should become a priority, since, while little additional hardware investment will be required, extensive software and calibration efforts are required on a large scale.

REFERENCES

1. "Practical Metropolitan-Scale Positioning for GSM Phones" by Mike Y. Chen, Timothy Sohn, Dmitri Chmelev, Dirk Haehnel, Jeffrey Hightower, Jeff Hughes, Anthony LaMarca, Fred Potter, Ian Smith, and Alex Varshavsky, (P. Dourish and A. Friday (Eds.): Ubicomp 2006, LNCS 4206, pp. 225–242, 2006, © Springer-Verlag Berlin Heidelberg 2006).

SECTION 10

SPECIAL SESSION:

LECTIO MAGISTRALIS

WHY SCIENCE IS NEEDED IN EVERYDAY LIFE

IGNAZIO LA RUSSA
Minister of Defence of the Italian Republic
Rome, Italy

As Professor Zichichi said, I have been following your activities in Erice for almost half a century. Every year, thousands of scientists present their latest discoveries and inventions here, thus promoting a Science without secrets and without borders. But you do not limit yourselves to conducting studies and research that the general public might consider purely theorctical. You caught my attention with your study of Planetary Emergencies and with your worldwide implementation of projects aimed at proving that we can address and solve these Planetary Emergencies. You have realised a hundred projects in 50 countries in order to convince us politicians that where there is a political will, Planetary Emergencies can be addressed and solved.

As Professor Zichichi was saying earlier, I have tried in my actions to introduce the rigour that Science requires to address the solution of a problem. It is unfortunately true that when faced with issues of security and activities necessary for the defence of a nation, as part of a strategy in which too many variables are involved, the complexity of the problems increases tremendously, also because our actions must always be based on the values of Freedom and Democracy.

I have carefully read through the programme of your work which has led you, the scientific community, this year to focus your research on the world energy crisis, the "global monitoring" of our planet, the problems of pollution, cyber security, protection from cosmic objects and "mitigation of terroristic acts", all examples of the 63 Planetary Emergencies that your study groups were able to identify.

I would now like to offer some thoughts. If Galilei's scientific methods were adopted in everyday life, in the laws set down by the Parliaments of democratic countries and in their effective transformation into actions aimed at enforcing the laws themselves, the huge number of errors which hosts of innocent people are victims of would be drastically reduced. Speed and rigour are not at all incompatible. This is demonstrated by the enormous quantity of scientific truths discovered in just four centuries of Galilean Science. Before Galilei, for thousands and thousands of years, man had hardly understood anything about the way the world functions. None of the old truths have been able to withstand the rigour of a Galilean-style analysis. Thanks to Galilei, in just four centuries, your Science has brought us to the threshold of the Superworld. This is why Science should finally enter the structure of everyday life, by which I mean the Culture of the third millennium, which is the focus of this 43rd Session of the Seminars on Planetary Emergencies.

Let me now quote a few examples relative to Italy. The serious delays in our legal system could be eliminated by developing scientific principles to solve problems. The study of "a software" capable of addressing those problems, whose solution does not depend on words but on deeds, would be of great help.

In Medicine, Mathematics has so far been absent from the study of diseases. I read in your Proceedings that a group of scholars has set up a project called

"Mathematics and Medicine", precisely to establish what will one day be an area of great value to avoid errors on a large scale, as was the case with H1N1.

A topical issue is the so-called Global Warming. And here you have clearly stated that "Climate Science" is not first-level Galilean Science. The predictions are based on mathematical models with a huge number of free parameters, which can be modified to achieve what you want. The link between human activity and climate depends on the weather thermodynamic engine in which over 90% of the power is due to Nature and not to man. Many of you have told us that we cannot waste billions of dollars and euros by ignoring these truths.

The fact that it is possible to solve problems that at first sight may seem without a solution is proved by the countless technological inventions born in just four centuries of Galilean Science: cell phones, computers, radio, television, medical technology (TAC, PET, X-ray, Nuclear Magnetic Resonance, etc.), and the Internet would not exist if we were the children of chaos. Unfortunately, the Culture of our time, so-called modern culture, is actually pre-Aristotelian. Indeed, neither rigorous Logic nor Science belongs to it. We live as if Galilei had never been born. Yet today no one would be able to do without the technological inventions produced in just 400 years of Galilean Science. Many of these inventions were born at the heart of the true great Science, which is CERN, where three giants of Physics (Blackett, Bohr and Rabi), straight after the Second World War, pooled their efforts to create the infrastructure of what is now the largest Laboratory of Subnuclear Physics in the world. This is where the Fundamental Laws that govern everything that exists are studied: laws, not chaos.

But recent events in the news regarding justice (blatant errors), medicine (catastrophic predictions about H1N1), weather and climate forecasts (Global Warming and the coldest winter in recent decades), seem to prove that chaos is at the root of our existence. However, you tell us that chaos is not at the root of our existence.

Indeed, the greatest achievement of Science is the discovery that rigorous Logic rules the world, from its smallest structures, such as those of the subnuclear Universe, to the larger systems such as galaxies and the boundaries of the Cosmos. No one can produce this Logic from chaos. Therefore, there must be an Author. Atheistic culture says there is no need for one, but cannot prove it. If we were the children of chaos the scientific Laboratories in which the Book of Nature continues to be deciphered would not exist. These Laboratories contain the real engine of economic, civil and social progress. If Science should cease to continue deciphering what is written in the Book of Nature, we could say goodbye to new technological inventions.

The foundations of our existence are undisputedly based on rigorous Logic. This should thus prevent us from making glaring errors like the ones mentioned before.

Some claim there is no perfect form of activity as determined by Heisenberg's Uncertainty Principle and by Gödel's Theorem. These seem to be the limits of perfection beyond which Science (Heisenberg) and mathematical Logic (Gödel) will not allow us to go. These limits, however, have nothing to do with the blatant errors that the use of Science and rigorous Logic would enable us to avoid. This is not just an Italian problem, and indeed the WFS (World Federation of Scientists) has established several groups of specialists whose goal is to study the sixty-three Planetary Emergencies. These Emergencies are not the inevitable consequence of scientific and technological progress, as the dominant Culture often maintains. The sixty-three Planetary Emergencies are the result of pervasive political and economic violence which continues unabated.

I would like to end by quoting another very topical problem in Italian politics: bipolarism, which is the system by which the numerous parties line up behind two main ones, and fluctuating political alliances. The WFS has a group studying "Mathematics and Democracy", committed to introducing logical rigour into democratic life in order to understand how to link Democracy to the ability to govern, while respecting the freedom and dignity which are the sacrosanct right of each voter.

To deal with Planetary Emergencies we need a great alliance between Science and Politics. It is time for Science to enter all the activities which determine the quality of life that characterizes civil society, which means it has to become part of everyday life.

SECTION 11

SESSION 10

CLIMATE

FOCUS: DATA, MATHEMATICAL STRUCTURES AND PREDICTIONS

CARBON DIOXIDE, FRIEND OR FOE

WILLIAM HAPPER
Physics Department, Princeton University
Princeton, USA

DECARBONIZE THE WORLD!
Gretchenfragen

1. If CO_2 is held constant, will the climate stop changing?
2. What is the optimum CO_2 level?

We have heard a lot about CO_2 during the past few days, and especially in Europe, one can hardly escape reminders of the need to minimize our "carbon footprint." The United States Supreme Court, in a strange decision, has ruled that CO_2 is a "pollutant." Our exhalations contain about 40,000 ppm of CO_2, so it is a bit unnerving to realize that each of us is guilty of "spewing poison into the air." Perhaps the ultimate solution to "stop climate change" is to eliminate the human race. But before we agree to severe measures to fight CO_2, perhaps we should ask some fundamental questions, not unlike Gretchen's question to another persuasive scientist, Faust, before he betrayed her. Here are two such questions:

1. If CO_2 is held constant, will climate stop changing? Not likely, we have had periodic ice ages; in spite of the hockey-stick hysteria, there really was a medieval warm period around the year 1000 that was very similar to our current warm period, and there really was a little ice age. Perhaps, as argued by Nir Shaviv, the sun is having a big impact on climate, perhaps climate changes as the natural result of a chaotic system. So holding CO_2 constant will not stop climate change. Adding CO_2 will probably shift the mean temperature, and we have heard arguments this week that the shift will be small, about 1°C for CO_2 doubling. This will probably be a net benefit for mankind.
2. Assuming we can control the CO_2 level, what level is optimum?

"TOUT EST POUR LE MIEUX DANS LE MEILLEUR DES MONDES" *VOLTAIRE, CANDIDE*

What is the "best of worlds" with respect to CO_2?
150 ppm—plants stop growing?
280 ppm—preindustrial Garden of Eden?
390 ppm—present value?
1000—2000 ppm—phanerozoic norm?

It is hard to understand what drives the campaign to limit CO_2 in the atmosphere. What levels are we aiming for and why? We sometimes hear that the mean global temperature rise should be limited to 2°C. But this raises a related question, what global temperature is optimum? If we could do so, should we make the planet hotter or colder? From the point of view of geological history, we are currently in a CO_2 famine.

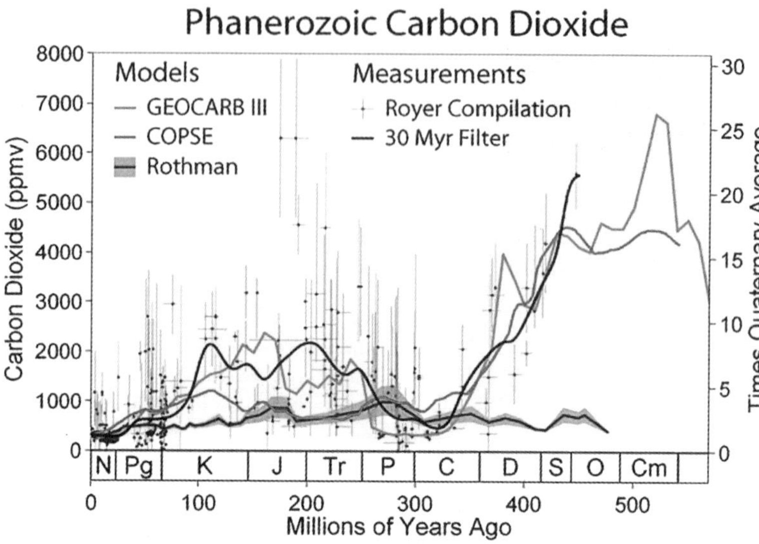

Increased levels of CO_2 will cause some warming, but there are good arguments that the warming will be modest, about 1°C for a doubling of CO_2. This increase in temperature will probably do more good than harm. Since the response of global temperature to CO_2 is logarithmic, you would have to quadruple CO_2 levels to get about 2°C warming. This would give a CO_2 level of more than 1000 ppm. There is no doubt that the increased CO_2 levels will be very good news for plants, as is illustrated by data in the next viewgraph.

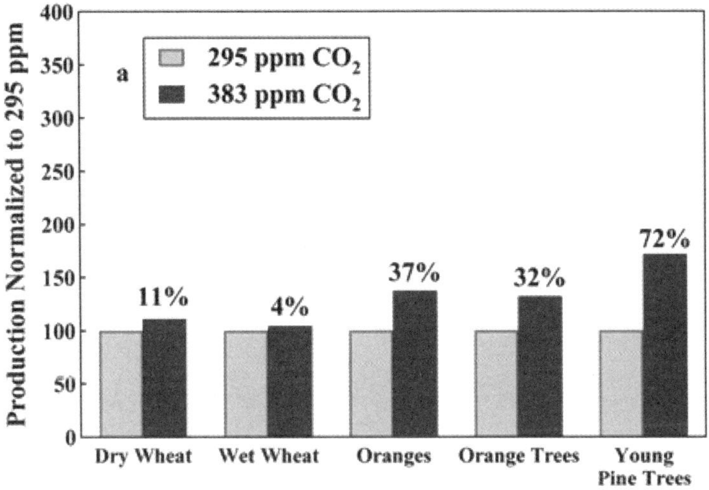

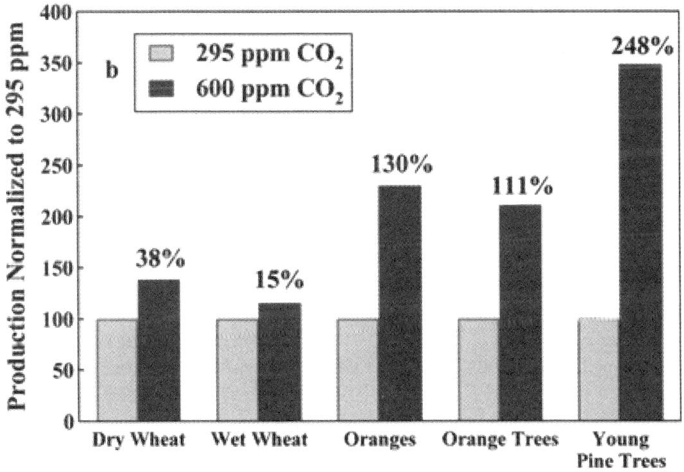

Enhanced agricultural yields from more CO_2

Here are data for the growth enhancement of various plants of economic importance. CO_2 is especially helpful for plants growing under dry conditions. At higher CO_2 levels, there will be more vegetation in dry areas that currently have very sparse vegetation. Deserts will shrink.

STOMATA

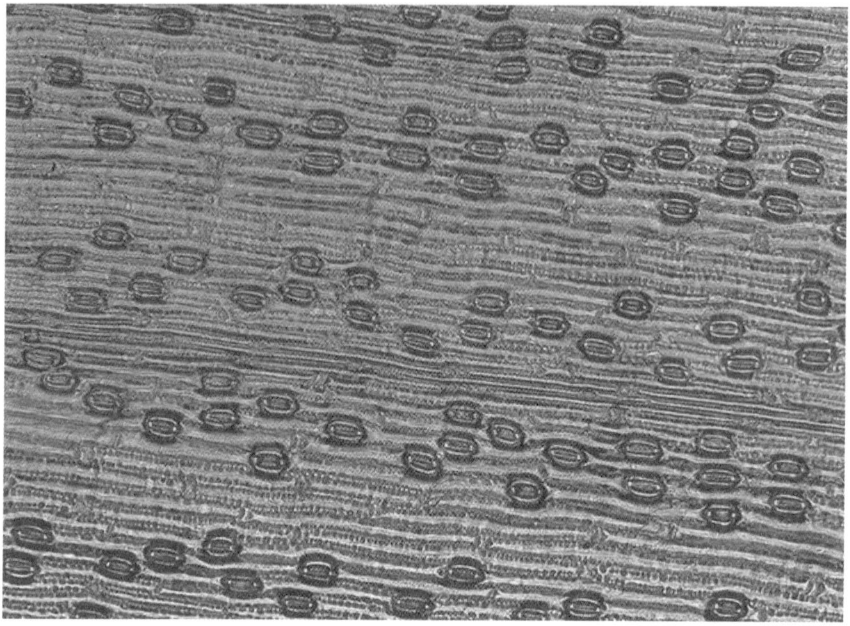

Leaves of grass at high magnification

Plants grow better with more CO_2 because they don't need so many stomata, the little openings in the surface of a plant leaf that allow CO_2 to diffuse into the leaf. But water vapor also diffuses out of the stomata and dries up the leaf. As Jan Veizer mentioned in his talk yesterday, at current CO_2 concentrations, about 1000 molecules of H_20 are transpired to the air for every CO_2 molecule that diffuses into the leaf. All plants have ancient feedback paths that cause them to reduce the number of stomata when the levels of CO_2 in the air increase. Plants with fewer stomata can cope more easily with arid conditions.

CO_2 ENRICHMENT OF GREENHOUSES

Typical sales pitch: "CO_2 enrichment at 2, 3, and 4 times natural concentration will cause plants to grow faster and improve plant quality. ... The **** generator automatically provides the carbon dioxide to meet maximum growing potential."

Cost < $1000 plus cost of propane. Not eligible for stimulus funds.

The benefit of higher CO_2 levels is well known in the commercial greenhouse community, if not to academics or policy makers. If you Google "CO_2 for greenhouses," you can find lots of advertisements like this one.

CAUTION ON THE BRAVE NEW DECARBONIZED WORLD

Jonathan Swift, *Gulliver's Travels*, Voyage to Laputa:

> "In these colleges the professors contrive new rules and methods of agriculture and building, and new instruments, and tools for all trades and manufactures; whereby, as they undertake, one man shall do the work of ten; a palace may be built in a week, of materials so durable as to last for ever without repairing. All the fruits of the earth shall come to maturity at whatever season we think fit to choose, and increase a hundred fold more than they do at present; with innumerable other happy proposals. The only inconvenience is that none of these

projects are yet brought to perfection: and in the mean time the whole country lies miserably waste, the houses in ruins, and the people without food or clothes. By all of which, instead of being discouraged, they are fifty times more violently bent upon prosecuting their schemes, driven equally on by hope and despair."

CLIMATE SENSITIVITY: VARIOUS APPROACHES

RICHARD S. LINDZEN
Earth, Atmospheric & Planetary Sciences, Massachusetts Institute of Technology, Cambridge, USA

1. INTRODUCTION

Climate sensitivity has come to be defined as the equilibrated average response of the surface temperature to the radiative forcing produced by a doubling of CO_2. Since at least 1967 (Manabe and Wetherald, 1967, Manabe, 1971) the question of climate sensitivity has been the major question in the issue of CO_2 induced climate change. Nonetheless, one would hardly guess that this is a central issue from reading the reports of the IPCC. Only a few pages are devoted to this in the text, and hardly anything in the summaries.

It was early understood that doubling CO_2, alone, would lead to a sensitivity of about 1°C. Moreover, there has been no data since 1967 that explicitly points to any sensitivity greater than this. This is a very strange situation for an allegedly incontrovertible and incomparable threat to life on earth. However, in the 1960's, Manabe and Wetherald (1967) suggested that if relative humidity would remain constant in the presence of climate change, this would provide a positive feedback that would double the sensitivity. The basis for their suggestion was, and remains, meager. The basis in 1967 was a generally unavailable 1957 U.S. Airforce report by Julius London. Remember that this was before satellites. RH was measured by essentially non-functioning sensors on balloons. Cirrus was from ground observations. Manabe and Wetherald argued that the similarity between Winter and Summer in Figure 1 suggested constant RH.

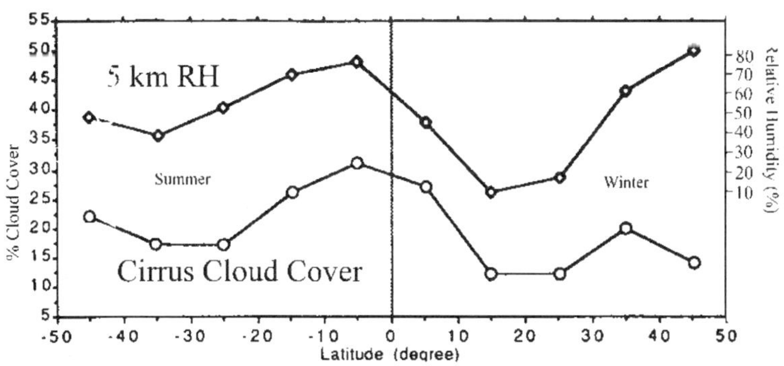

Fig. 1: Cloud cover and relative humidity at 5 km as functions of latitude.

Even Figure 1, taken from Manabe and Wetherald (1967) does not suggest constancy. Nonetheless, constant RH has become a dogma—especially since models behave 'as though' RH were approximately constant.

One might have thought that pinning down climate sensitivity would be a high priority for climate science. Indeed, when in the 90's the Climate Research Board of the U.S. National Research Council was preparing its five year plan, it did assign high priority to this question, but Jerry Mahlman, then director of the NOAA Geophysical Fluid Dynamics Laboratory at Princeton, strongly objected. He argued that this would be a difficult issue that should be deferred to a later date. The final five year plan still mentioned this issue but no longer emphasized it.

It proves helpful to use the common feedback formalism for sensitivity.

$$\Delta T = \frac{\Delta T_0}{1-\sum_i f_i} = \frac{G_0 \Delta F}{1-\sum_i f_i}, \qquad (1$$

(Climate sensitivity is simply the ΔT corresponding to ΔF=3.7 Wm^{-2}; ΔT_0=1.2C.) Manabe and Wetherald had found that with constant RH, f_0 was already 0.5. This meant that the addition of smaller feedbacks from clouds could readily push sensitivity to very large values.

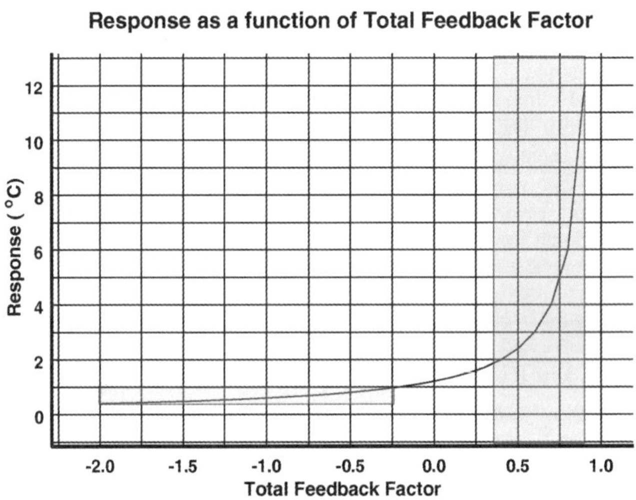

Fig. 2: Response to a doubling of CO_2 as a function of feedback factor.

This is clear in Figure 2. For positive feedbacks, relatively small variations in feedback lead to large changes in response (shown in pink shaded region). For negative feedbacks, large variations in the feedback lead to only small changes in response (as shown in green shaded region). It is the positive feedbacks in the models that lead to the alleged uncertainty in model results.

2. EARLY ALTERNATIVE ESTIMATES OF CLIMATE SENSITIVITY

By the end of the 70's there were quite a few papers looking further into this. One by Hummel and Kuhn (1981) suggested that sensitivity might actually be about 0.8C. Several other papers by Idso (1980) and by Newell (1979) argued that evaporation would be sufficient to render sensitivity very small. Although I think Idso and Newell were technically wrong, they were on to something as we will see in Section 4.

What is notable, however, was the viciousness with which Idso in particular was attacked. At a meeting in May 1980 of the NRC National Climate Board (later to be called Climate Research Board), Walter Orr Roberts (head of UCAR) and Joseph Smagorinsky (head of GFDL) asked the Board to formally censure Idso for questioning the results of Manabe and Wetherald. Sylvan Wittwer pointed out that the Board didn't do that sort of thing. Money was already flowing toward NCAR and GFDL for further studies of the effect of CO_2, and the interest depended crucially on large sensitivity.

It is interesting that it was Idso rather than Newell who was attacked. Already, the argument was that Idso (who was not a climatologist but a physicist by training) had no right to criticize climatologists. Newell, however, was a climate scientist at MIT.

The reason I doubted the arguments of Idso and Newell appeared in a paper we published in 1982 (Lindzen et al.1982) where we were primarily concerned with the impact of improved convective representation on climate sensitivity.

Turbulent transfers are usually parameterized by the linearized bulk aerodynamic formulas:

$$E = C_D u_* \left[q_s(T_*) - q(T(0)) \right] \quad (2a$$

$$F_s = C_D u_* C_p [T_s - T(0)] \quad (2b$$

where E is the evaporation, F_s the sensible heat flux, C_D the aerodynamic drag coefficient, u_* a "gustiness" factor, T(0) the air temperature at the ground, and T_* the ground temperature. The water vapor mixing ratios q and q_s are respectively the saturated humidity at the surface temperature and the specific humidity of air at z=0 (=RH x $q_s(T(0))$).

What we found was that the turbulent exchange at the surface was largely balanced by changes in the temperature jump so as to keep sensitivity largely unchanged. We will return to this matter in Section 4.

3. THE 'HOT SPOT' AND THE MOIST ADIABAT

In Lindzen et al. (1982), where we investigated the role of convective parameterization on the response to doubling CO_2, we found that the proper inclusion of moist convection—either by the explicit use of a convective model or by the use of the moist adiabatic convective adjustment—led to the result shown in Figure 3. We see that with moist physics included, the response in the upper troposphere is more than twice as large as the surface response. The upper tropospheric maximum has come to be known as the 'hot spot.'

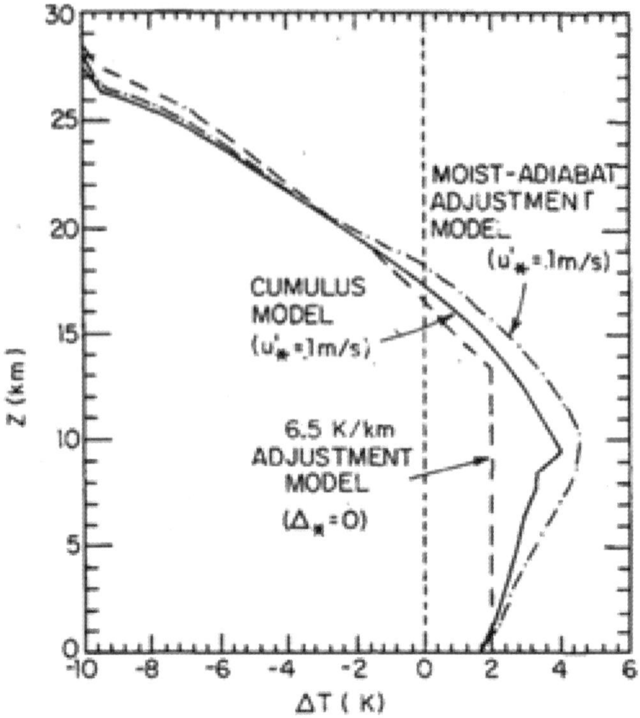

Fig. 3: The tropical response to a doubling of CO_2 for different choices of cumulus parameterization. From Lindzen et al. 1982.

As we see in Figure 4, from Lee et al. (2008), the hot spot appears in current climate GCMs as well. However, as Figure 5 shows, this is not found in current analyses of the data. Since the hot spot arises from fundamental physics (namely the fact that the tropics is dominated by moist convection), something must be wrong with the analyzed data. Given the many problems with the data and its analysis, this should not be surprising. Supposing that the upper tropospheric data is not greatly in error (and it would have to be greatly in error to eliminate the hot spot), and given that independent balloon and satellite data agree in this region, it is reasonable to suppose that the error resides in the surface analysis. Thus, to be consistent with the hot spot, the observed analysis giving point *a* at the surface would have to be moved to point *b*—a value about a third of that given by *a*. This would eliminate most of the surface temperature rise since 1979, and thereby suggest a much smaller climate sensitivity than current models suggest.

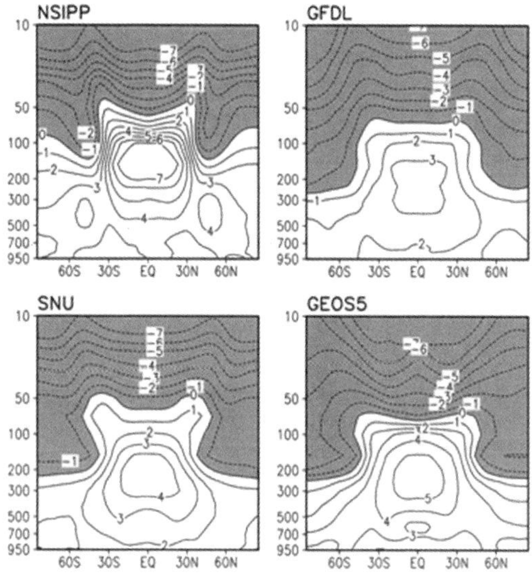

Fig. 4: Meridional distribution of response to a doubling of CO_2 in typical GCMs. From Lee et al. 2008.

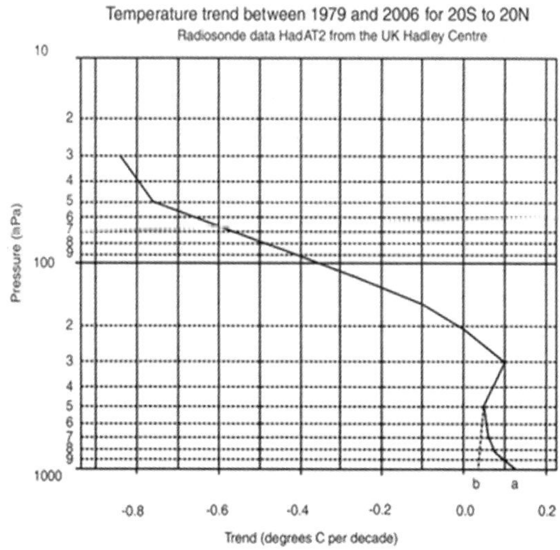

Fig. 5: Observed temperature trends as a function of altitude. From Lindzen, 2007.

4. SENSITIVITY DERIVED FROM OBSERVATIONS OF SURFACE EVAPORATION

The feedback formalism that we discussed in Section 1 refers to the relation between surface temperature change and something referred to as radiative forcing. The radiative forcing is generally taken to be a radiative imbalance between net incoming solar radiation, and thermal radiation to space. The radiative forcing is a flux of energy corresponding to this imbalance. As can be seen from Lindzen et al. (1982), the surface of the earth is coupled to the radiative emission level by convection, and the flux at the surface is dominated by evaporation and sensible heat transfer rather than by radiation. Evaporation dominates over most of the earth. The gist of the previous examples is that the surface flux is nonetheless approximately the same as the radiative forcing. This was discussed early on in Dickinson (1981) and is essentially what was found in Lindzen et al. (1982) (viz Section 2). Thus, the change in evaporation associated with change in the surface temperature is essentially a ratio of energy flux change to temperature change; ie, the inverse of the climate sensitivity.

Recently, Wentz et al. (2007), who seem unaware of this relation, studied with space based data, the changes in evaporation associated with changes in temperature. Quoting from their paper

"The GCMs indicate that E should increase about 1 to 3% K^{-1} of surface warming. However, according to the bulk formula (eq. S1) (8), evaporation increases similarly to C-C (Clausius-Clapeyron) as the surface temperature warms, assuming that the other terms remain constant. For example, a global increase of 1 K in the surface air temperature produces a 5.7% increase in E (8). To obtain the muted response of 1 to 3% K^{-1}, other variables in the bulk formula need to change with time. The air-sea temperature difference and the near surface relative humidity are expected to remain nearly constant (8), and this leaves t_0 as the one variable that can reduce evaporation to the magnitude required to balance the radiation budget in the models. To bring the bulk formula into agreement with the radiative cooling constraint, $t_0^{-1/2}$ would need to decrease by about 4% K^{-1}. Thus, a muted response of precipitation to global warming requires a decrease in global winds (2, 3, and 7)." (The numbers refer to references in Wentz et al. 2007.)

Reference 8 refers to supplementary material that seems to have no explanation for assertion concerning near surface RH and air-sea temperature difference. Thus, there really is no basis for assuming that the discrepancy is due to a decrease in global winds, and, indeed, no such decrease is found.

In point of fact, a 1% change in evaporation corresponds to about 0.8 watts m^{-2}, and a doubling of CO_2 changes flux by about 3.7 watts m^{-2}. Thus, the model range of 1-3% per degree corresponds to a climate sensitivity of from about 1.5-4.5C, which is, indeed, the range of model sensitivities. The observed 5.7% per degree, however, corresponds to a sensitivity of only about 0.8C.

Thus, Wentz et al. have probably provided the simplest observational estimate of climate sensitivity—albeit inadvertently. Given the fact that Wentz et al. explain in their supplemental material that they have thrown away measurements that are 'too far' from the models, it is reasonable to conclude that the observed sensitivity is an upper bound.

5. OTHER ESTIMATES OF CLIMATE SENSITIVITY

In view of what we have just noted, climate sensitivity is a ratio of ΔT to ΔF at the surface. Given that the surface is mostly ocean with a substantial heat capacity, high sensitivity means that the ocean will take longer to change temperature since a given ΔT is associated with a smaller ΔF. This was noted by Hansen et al. (1985) (though with an oversimplified model), and is the basis for what is known as the 'delayed response.' The calculated response times for reasonable ocean parameters range from a few years for low sensitivities to many decades for higher sensitivities.

In Lindzen (1994) and Lindzen and Giannitsis (1998), an attempt was made to use this behavior to estimate climate sensitivity from the response to volcanic eruptions. It was found that the response to a single volcano did not effectively distinguish different sensitivities, but that there was a discernible difference in the response to a sequence of large volcanoes that would, for high sensitivities, lead to secular cooling, but for low sensitivities would simply produce transient 1-2 year dips in temperature (viz Figure 6). The record (Figure 7) seems to favor dips.

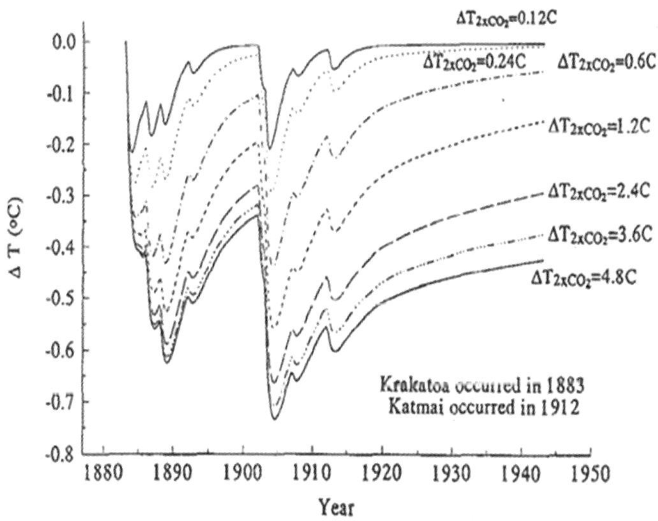

Fig. 6: Calculated response to a sequence of volcanos for different choices of climate sensitivity. From Lindzen and Giannitsis, 1998.

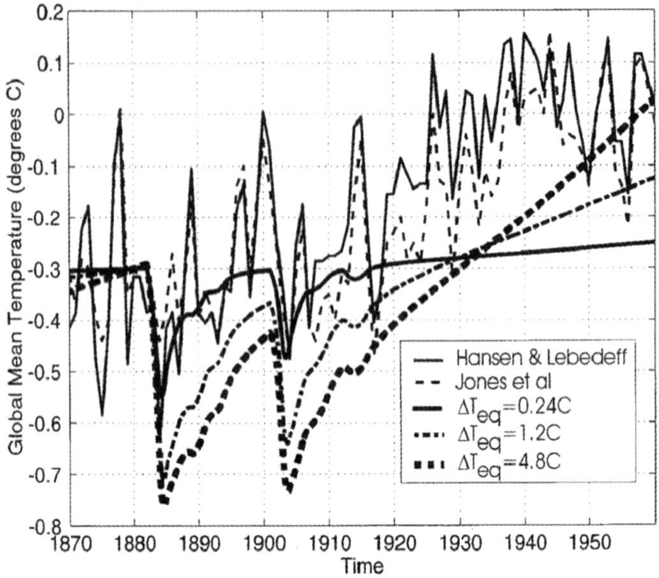

Fig. 7: Observed temperature and calculated temperature including response to series of volcanos using different values for climate sensitivity. From Lindzen and Giannitsis, 1998.

6. CONCLUDING REMARKS

So far, we actually have three profoundly different empirical estimates of climate sensitivity. They all point to sensitivities requiring modest negative feedback.

There are several other approaches that all lead to the same conclusion. Chris Monckton (in this volume) described some of them—including the recent work by myself and Yong-Sang Choi. Time does not permit describing them. The only arguments to the contrary have been the alleged use of paleoclimate. However, such studies are bogus since they assume that we have almost perfect knowledge of such things as clouds and dust loading in climates of 18 thousand years ago. Lacking this information, the authors can get any answer they want. The same is true for the IPCC 'consistency' studies. In these studies, one uses essentially unknown aerosol and solar forcing to 'correct' the excessive warming produced by current models. Schwartz et al. (2010) showed this in detail. That said, if one simply assumes that the warming over the past century is due to CO_2 (and introduces no fudge factors), then one again comes up with a low sensitivity.

A perhaps more important point is that the problem of determining climate sensitivity, while not trivial, is eminently approachable, and there really is only one excuse for not doing so. Namely, the answer has the ability to end this issue—especially since most legitimate attempts (including Shaviv's in this volume) do lead to low sensitivity.

ACKNOWLEDGEMENT

The preparation of this paper was supported by DOE grant DE-FG02-01ER63257.

REFERENCES

1. Dickinson, R.E. (1981) "Convergence Rate and Stability of Ocean -Atmosphere Coupling Schemes with a Zero-Dimensional Climate Model." *J. Atmos. Sci.*, **38**:2112-2120.
2. Hansen, J., G. Russell, A. Lacis, I. Fung, and D. Rind (1985) "Climate response times: Dependence on climate sensitivity and ocean mixing." *Science*, **229**:857-859.
3. Hummel, J.R. and W.R. Kuhn (1981) "An atmospheric radiative-convective model with interactive water vapor transport and cloud development." *Tellus*, **33**:372-381.
4. Idso, S.B. (1980) "The climatological significance of a doubling of earth's atmospheric carbon dioxide concentration." *Science*, **207**:1462–1463
5. Lee, M.-I., M.J. Suarez, I.-S. Kang, I.M. Held, and D. Kim (2008) "A Moist Benchmark Calculation for Atmospheric General Circulation Models." *J. Clim.*, **21**:4934-4954.
6. Lindzen, R.S. (1994) Constraining possibilities versus signal detection. pp 182-186 in *Natural Climate Variability on Decade-to-Century Time Scales*, Ed. D.G. Martinson, National Academy Press, Washington, DC, 630pp.
7. Lindzen, R.S. (2007) "Taking greenhouse warming seriously." *Energy & Environment*, **18**:937-950.
8. Lindzen, R.S., A.Y. Hou and B.F. Farrell (1982). "The role of convective model choice in calculating the climate impact of doubling CO_2." *J. Atmos. Sci.*, **39**:1189-1205.
9. Lindzen, R.S. and C. Giannitsis (1998) "On the climatic implications of volcanic cooling." *J. Geophys. Res.*, **103**:5929-5941.
10. Manabe, S. (1971) Estimate of future changes of climate due to man's increase of carbon dioxide concentration in the air. *Man's Impact on Climate*, W.H. Matthews, W.W. Kellogg and G.D. Robinson, Eds., The MIT Press, 594 pp
11. Manabe, S. and R.T. Wetherald (1967) "Thermal equilibrium of the atmosphere with a given distribution of relative humidity." *J. Atmos. Sci.*, **24**:241-259.
12. Newell, R.E. (1979) "Climate and the Oceans." *American Scientist*, **67**:405-416.
13. Schwartz, S.E., R.J. Charlson, R.A. Kahn, J.A. Ogren, and H. Rhode (2010) "Why Hasn't Earth Warmed as Much as Expected?" *J.Clim.* **23**:2453-2464.
14. Wentz, F.J., L. Riccaiardulli, K. Hilburn and C. Mears (2007) "How much more rain will global warming bring?" *ScienceExpress*, **31** May 2007.

CLIMATEGATE AND THE INQUIRIES

STEPHEN MCINTYRE
University of Guelph
Toronto, Ontario, Canada

CLIMATEGATE AND THE INQUIRIES

The biggest news event in climate during the past year, for better or worse, was Climategate. It was widely covered in the media, with even the noted climate scientist, Donald Trump, expressing an opinion. It was an even bigger internet event—at one point, Climategate ranked higher in Google hits than Tiger Woods. As a long-standing critic of the scientists involved, I was pretty much at Ground Zero for Climategate.

Editorials at both *Nature*[1] and *Science*[2] have worried about the impact of the affair on public trust in climate science.

Three books have been written about the events: The Hockey Stick Illusion[3]; The Climate Files[4] and Climategate: the CRUTape Letters.[5]

Four inquiries have been completed. The UK Parliamentary Science and Technology Committee's brief consideration[6] was interrupted by the recent UK elections. The Muir Russell[7] and Lord Oxburgh[8] inquiries were commissioned by the University of East Anglia—the Oxburgh panel having a unique letterhead combining the House of Lords logo with the email address of East Anglia registrar's office. Penn State had a two-stage academic misconduct inquiry.[9,10] A panel on IPCC procedures[11] has yet to report, but its terms of reference preclude it from examining past conduct by IPCC scientists.

As of mid-August, things have quieted down, but there is a remarkable difference between the attitude of the public and official climate science to the results. Fred Pearce, a long-time and very green environmental reporter, concluded in The Climate Files[12] that, while the Climategate scientists were not guilty of "outright fraud", they were guilty of pretty much everything else—cutting corners, playing down uncertainties, and then covering their tracks and suppressing dissent, and that there was a systemic problem. Clive Crook of the Financial Times said[13] that the inquiries' failure to discharge their responsibilities actually made things worse, bemoaning the inability of the climate science establishment to either understand or repair their problems.

On the other hand, climate scientists, such as Mann's coauthor Raymond Bradley, seized on the inquiries as a form of exoneration and demanded apologies from media for even covering the story.[14]

INQUIRIES

In finalizing my remarks on the Climategate inquiries, I was impressed by Dr Isaacs's description yesterday of the procedures employed by successful outreach programs to obtain public confidence. Every one of his rules was broken by the climate inquiries.

No CRU critics whatever were interviewed by either of the two inquiries commissioned by East Anglia. The Muir Russell report[8] justified this strange procedure as a form of "natural justice" saying that "natural justice requires that those in respect of whom findings will be made should have an opportunity to be heard: this does not apply to the authors of submissions and other parties, in respect of whom the Review has made no findings." Reasonable people can disagree as to whether this is in fact a form of "natural justice", but it is a singularly ineffective way of getting to the bottom of dispute. Nor did the Muir Russell panel even have much actual contact with the CRU scientists themselves. After the press conference in February announcing the composition of the panel, Muir Russell never met with Phil Jones or any other CRU scientist. Nor did the majority of his panel. There were only two interviews with CRU scientists at which questions were asked—each attended by only a couple of panellists. The interviews were not recorded nor were transcripts taken. The perfunctory minutes show that major topics assigned to the inquiry were not addressed.

The Oxburgh inquiry[7] was even worse. It was announced on March 22; the report appears to have been written on the afternoon of April 6 and released on April 14. It was only 5 pages long. It did not interview CRU critics nor did it take submissions. It had no written terms of reference. Its report said that it considered 11 supposedly representative CRU articles selected on the "advice of the Royal Society". This was untrue. The 11 articles were not in fact representative, nor were they selected by the Royal Society. CRU chose the articles to be examined; IPCC conduct was not included. Most people being audited would like a similar right - to be able to choose the tax years that they were being audited on.

The university had told the Parliamentary Committee that the Oxburgh panel would reappraise CRU's science.[15] After the publication of the report, I heard from a reliable source that Jones had told the Oxburgh panel that it was "probably impossible" to do 1000-year reconstructions with any accuracy—the very point in question.[16] This was not mentioned or referred to in the Oxburgh report. I emailed Oxburgh, asking him to confirm this statement and, if correct, issue an addendum. Oxburgh refused to confirm or deny the statement or to issue an addendum, stating in no uncertain terms that "the science was **not** the subject of the inquiry". Key members of the UK Parliamentary Committee were furious with the Oxburgh panel and the university. Former chairman Phil Willis called the panel's conduct a "sleight of hand".[17]

The Penn State inquiry was described nicely by Clive Crook of the Financial Times as being "difficult to parody".[13] At the preliminary stage, they dismissed most of the charges without talking to any critics and without carrying out an investigation. In their final report, they stated that Mann's very success in obtaining grants was all the evidence that they needed.

TWO KEY EMAILS

The failure of the inquiries to carry out their assigned responsibilities is vividly exemplified in their handling of the two most notorious emails.

TRICK TO HIDE THE DECLINE

In perhaps the most notorious email[18], Phil Jones wrote to colleagues saying that he had just used Mike Mann's Nature trick to hide the decline. This email has prompted much public ridicule, with a satirical video entitled Hide the Decline receiving over 600,000 views.[19] The email did not concern temperature, but the proxy reconstruction diagram that had featured prominently in the IPCC's Third Assessment Report—shown in the background behind John Houghton at the press conference releasing the Third Assessment Report.[20] Figure 1 (left) is a plot of data available for the first time in the Climategate emails, showing the Briffa et al 2001[21] proxy reconstruction to 1994; the official archive[22] deleted values after 1960. Briffa had expected tree ring widths and density to increase linearly with temperature in the last half of the 20th century; instead, they went down. People in the field call this the divergence problem and didn't like to talk about it very much. The IPCC 2001 proxy reconstruction[23] diagram in Figure 1 (middle) gives a strong rhetorical impression of consistency; the divergence problem has disappeared. On the right, the blowup of the IPCC diagram shows why—the data for the Briffa graph after 1960—the data shown in red in the diagram at the left—had been deleted.

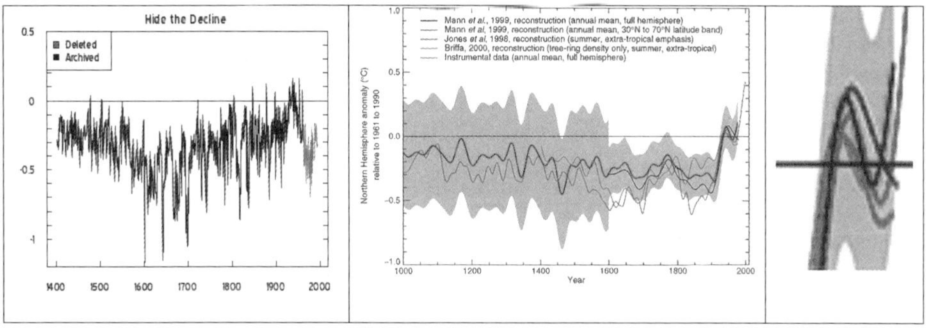

Fig. 1: Left—actual Briffa data, deleted portion in red; Middle—IPCC 2001 Figure 2.21; right—blowup of Figure 2.21.

The Climategate emails indicate that the data had been deleted to produce a diagram more palatable to senior IPCC authors.[24] A different version of the proxy diagram in which data had not been deleted had been presented at a lead authors' meeting leading with disappointing results. IPCC Coordinating Lead Author Folland tells the others that a proxy diagram was wanted for the policy-makers, but that the Briffa reconstruction was perceived as "diluting the message rather significantly". Briffa worried about the pressure to present a statement about 'apparent unprecedented warming in a thousand years or more' given the unexpected changes in proxy response. Mann, who was Lead Author of this section, noted that the Briffa reconstruction had been perceived by everyone in the room at the IPCC meeting as a "problem and a potential distraction/detraction from the reasonably consensus viewpoint we'd like to show". This problem was "solved" by simply deleting the adverse data in the IPCC diagram. As predicted in the Climategate exchange, the proxy diagram was indeed given a prominent place in the Third Assessment Report and at the announcement press conference.

The issue recurred in the Fourth Assessment Review. I was a reviewer and, by this time, had identified the trick. As a reviewer, I asked that the deleted data be displayed and justified as best they could. This time Briffa was Lead Author. He refused, saying that it would be "inappropriate" to show the actual data.[25]

None of the inquiries gave a remotely coherent account of the trick to hide the decline.

At the preliminary inquiry[9], Penn State stated that the trick was a "legitimate statistical technique" that was supposedly widely used but did not provide any references or citations, nor can any be supplied. Oxburgh[7] did not agree that the method was legitimate—on the contrary, they said that it was 'regrettable", but blamed it on IPCC authors "neglecting" to show the data. They didn't mention that the IPCC authors were all Climategate correspondents—or for that matter that the deletion had been intentional. Muir Russell[8] dodged the IPCC graphic, saying that splicing or deletion wasn't per se wrong—without addressing whether it was wrong in the specifics of this incident.

The American comedian Jon Stewart had a more insightful analysis than any of the inquiries, Stewart[26] mugging for the TV audience:

> It means nothing…He's just using a trick to…hide the decline. It's just scientist-speak for a standard statistical technique for calibrating data in order to…trick you…into not knowing about...the decline.

DELETE ANY EMAILS

The second most notorious email[27] came in May 2008, following a freedom of information request to CRU for any IPCC review comments not included in the official archive as IPCC policies required. Phil Jones asked Michael Mann to delete any emails that he had had with Keith Briffa concerning the Fourth Assessment Report and to ask Eugene Wahl to do likewise. Jones told Mann that he'd be contacting Briffa and Caspar Ammann to do likewise.

The email prompted something even more corrosive than criticism: satire—such as the cartoon shown here—in which the CRU scientist is drawn in the image of Richard Nixon.[28]

Fig. 2: Cartoon satirizing CRU scientists.

The conduct even raised questions of whether a criminal offence had been committed under the UK Freedom of Information Act. The UK Information Commissioner stated that a more cogent prima facie example of an offence was "impossible to contemplate".[6] However, the Commissioner observed that the law had a six-month statute of limitations from the time of the offence, rather than the more typical language of six months since the offence was brought to the attention of the prosecutor. In this case, the incident had not been discovered for eighteen months and the Information Commissioner was unable to bring charges. He recommended that the law be changed.

The UK Parliamentary Committee asked for a thorough investigation.[6] However, the Muir Russell panel didn't ask Jones a single question about the deletion of emails. In its final report, the Muir Russell inquiry made findings that simply flew in the face of simple facts. Muir Russell stated that there was no governing Freedom of Information request at the time of Jones' delete email and thus no offence. This finding was made despite the fact that the previous day's email, saying that Briffa should deny the existence of any correspondence, bore the subject line of FOI 08-23, the identification number of the governing FOI request and the email in question bore a subject line of "FOIA and IPCC". These facts that were known to tens of thousands of people who were following the dispute on the internet, making the inquiry simply look foolish.[29]

DATA OBSTRUCTION

The Climategate emails showed many examples of IPCC scientists refusing to provide data to people regarded as potential critics, while providing the same data to insiders. There had been evidence of this for some years, such as Jones' well-known statement that they had "25 years invested" in this so why he should he provided data to a potential critic who was only interested in finding something wrong.

In another email[30], Michael Mann supplied residuals from his reconstruction to CRU in England with the caveat that this data was his "dirty laundry", supplied to CRU only because they were "trusted colleagues", emphasizing to them the vital importance of ensuring that this dirty laundry didn't fall into the "wrong" hands. Later that year, Ross McKitrick and I asked Mann for the same residuals, copying our request to the U.S. National Science Foundation since other requests had also been refused. In a remarkable letter, the NSF said that Mann had no obligations; that we were entitled to our view of science and Mann to his. This delayed our analysis by many months.

The public is very unsympathetic to data hoarding in any area where public funding has been obtained, but they wouldn't be bothered by it in areas where public policy wasn't engaged. However, in climate science, where data has not only been obtained with public funding, but the public are also being asked to accept possibly far-reaching policies based on the data, the public has no sympathy whatever for any supposed intellectual property rights of IPCC scientists.

PEER REVIEW

Comments in the emails about peer review attracted wide attention. In another case, CRU correspondents worry that a statistical criticism of their tree ring climate reconstructions could cause damage, complaining that getting the article killed will be difficult because it appears to be correct mathematically.[31] In another email, Phil Jones says that he "went to town" against two articles criticizing the handling of urban heat island issues in Siberia—an important and relevant issue in assessing the modern temperature record.[32] Neither article ever appeared.

David Rutledge of Caltech, who edited over 10,000 IEEE pages, said that the peer review conduct evidenced in these and other incidents was totally outside his experience.[33] Muir Russell did not report on these or numerous other incidents.

CONCLUSION

The matter has subsided but has not ended. Any person encountering modern climate science is startled by its angriness and animosity. One of the greatest shames of CRU and the Climategate correspondents has been their role in spreading this culture. The public has now seen this and doesn't like it. Unfortunately, rather than Climategate being regarded as a wake-up call to the climate science community to change this culture, perhaps even adopting Erice cheerfulness, if anything, the inquiries have reinforced this culture and the community is angrier and more sullen than ever—a culture that badly needs to change if climate scientists are to successfully communicate with a public that is desperate not simply for information, but for information from people that it trusts.

REFERENCES

1. *Nature*, 2010. 466:7.
2. *Science*, 2010. 327:50.
3. A.W. Montford, 2010. *The Hockey Stick Illusion* (Interlink Publishing Group Inc.).
4. Fred Pearce, 2010. *The Climate Files* (Random House UK).
5. Steven Mosher and Thomas Fuller, 2010. *Climategate: the CRUtape letters* (CreateSpace).
6. House of Commons Science and Technology Committee. 2010. The disclosure of climate data from the Climatic Research Unit at the University of East Anglia. http://www.publications.parliament.uk/pa/cm200910/cmselect/cmsctech/387/387i .pdf .
7. Lord Oxburgh et al. 2010. Report of the International Panel set up by the University of East Anglia to examine the research of the Climatic Research Unit. http://www.uea.ac.uk/mac/comm/media/press/CRUstatements/SAP.
8. Muir Russell et al., 2010. The Independent Climate Change Emails Review. http://www.cce-review.org/pdf/FINAL%20REPORT.pdf .
9. Foley, Henry, Alan Scaroni, and Candace Yekel. 2010. RA-10 Inquiry Report: Concerning the Allegations of Research Misconduct Against Dr Michael E. Mann, Department of Meterology, College of Earth and Mineral Sciences, The Pennsylvania State University. http://www.research.psu.edu/orp/Findings_Mann_Inquiry.pdf.
10. Final Investigation Report Involving Dr. Michael E, Mann, The Pennsylvania State University, 2010. http://live.psu.edu/fullimg/userpics/10026/Final_Investigation_Report.pdf .

11. Shapiro, Harold et al., 2010. Climate Change Assessments: Review of the Processes and Procedures of the IPCC. http://reviewipcc.interacademycouncil.net/report/Climate%20Change%20Assessments,%20Review%20of%20the%20Processes%20&%20Procedures%20of%20the%20IPCC.pdf .
12. Pearce, 2010: "The evidence of scientists cutting corners, playing down uncertainties in their calculations and then covering their tracks by being secretive with data and suppressing dissent suggests a systemic problem of scientific sloppiness, collusion and endemic conflicts of interest, but not of outright fraud."
13. Crook, Clive. 2010. Climategate and the big green lie. http://blogs.ft.com/crookblog/2010/07/climategate-and-the-big-green-lie/. "I had hoped, not very confidently, that the various Climategate inquiries would be thorough. This would have been a first step towards restoring confidence in the scientific consensus. But no, the reports make things worse, by failing to take seriously the charges that competent critics were actually making…The climate-science establishment…seems entirely incapable of understanding, let alone repairing, the harm it has done to its own cause."
14. Braley, Raymond, 2010. Climategate review: verdict. http://www.guardian.co.uk/commentisfree/cif-green/2010/jul/07/climategate-review-expert-verdict "[The CRU researchers] are honest, hardworking scientists whose reputations have been unjustifiably smeared by allegations of unscrupulous behaviour…This was a shameful chapter in the history of news reporting."
15. Memorandum submitted by the University of East Anglia to the Science and Technology Committee, 2010. http://www.uea.ac.uk/mac/comm/media/press/CRUstatements/Memorandum.
16. http://climateaudit.org/2010/07/01/oxburgh-and-the-jones-admission/
17. http://news.bbc.co.uk/today/hi/today/newsid_8795000/8795643.stm
18. http://www.eastangliaemails.com/emails.php?eid=154&filename=942777075.txt
19. http://vodpod.com/watch/2589937-hide-the-decline-climategate
20. http://climateaudit.files.wordpress.com/2005/03/houghtonhockeystick.jpg
21. Briffa, K.R., T.J. Osborn, F.H. Schweingruber, I.C. Harris, P.D. Jones, S.G. Shiyatov, and E.A. Vaganov, 2001. Low-frequency temperature variations from a northern tree ring density network. *Journal of Geophysical Research* 106, 2929-2941.
22. http://www.ncdc.noaa.gov/paleo/pubs/briffa2001/briffa2001.html
23. Houghton, John et al. 2001. *IPCC Third Assessment Report. Climate Change 2001: The Scientific Basis.*
24. See discussion and references at http://climateaudit.org/2009/12/10/ipcc-and-the-trick/
25. See discussion and references at http://climateaudit.org/2009/11/26/ipcc-reviewer-show-the-decline/.
26. http://www.disclose.tv/action/viewvideo/34415/Jon_Stewart_on_Climate_Gate/
27. http://www.eastangliaemails.com/emails.php?eid=893&filename=1212073451.txt
28. http://farm3.static.flickr.com/2711/4176810875_5515980d71_m.jpg
29. See discussion at http://climateaudit.org/2010/07/22/blatant-misrepresentation-by-muir-russell-panel/

30. http://www.eastangliaemails.com/emails.php?eid=345&filename=1059664704.txt
31. http://www.eastangliaemails.com/emails.php?eid=321&filename=1054756929.txt. Discussed at http://climateaudit.org/2009/12/16/climategatekeeping/
32. http://www.eastangliaemails.com/emails.php?eid=407&filename=1080742144.txt. Discussed at http://climateaudit.org/2009/12/16/climategatekeeping/
33. http://rutledge.caltech.edu/Caltech%20talk%202010/videos.html

ENERGY LEGISLATION IN THE USA

RICHARD WILSON
Department of Physics, Harvard University
Cambridge, USA

SOME HISTORY

The United States is blessed with abundant sources of energy of all sorts, and the legislation about it has been minimal and were mostly price control, in the states, until 1945.

The first important legislation was the Atomic Energy act of 1946. It was recognized that civilian government control could prevent excessive military control. The act established the United States Atomic Energy Commission (AEC), a civilian agency with sweeping powers over all military and civilian aspects of nuclear energy. Manhattan Project assets transferred to the Atomic Energy Commission at midnight, December 31, 1946. The AEC exercised governmental control over military, regulatory, and developmental aspects of the atom until 1975 when the agency was disestablished. In its place, Congress created the Nuclear Regulatory Commission to oversee the nuclear power industry and other civilian uses and the Energy Research and Development Administration (ERDA) to coordinate energy development including nuclear power. The AEC's weapons program was folded into ERDA. In 1977, ERDA and the energy programs from a number of other agencies were brought into the new Department of Energy. From 1946, the Congress exercised oversight with the Joint Committee of Atomic Energy (JCAE) on which sat members from both the house of representatives and the Senate. As far as I know JCAE was unique.

Under this legislation nuclear power became a viable energy source. There is a lot of talk about subsidies in this period, but they were primarily indirect. Support for nuclear engineering departments throughout the USA with many research reactors. Encouragement in general for building up an infrastructure and generally building up optimism for nuclear energy among the American people.

During the Presidencies of President Eisenhower, President Kennedy and President Johnson, the theme was to provide plentiful cheap energy. Admiral Strauss expressed this in a very over optimistic manner when referring to nuclear fusion. 'Electricity will be too cheap to meter" Many others have argued the importance of having affordable energy including Benjamin Franklin over 200 years ago (his words are on the on the energypmp website). In the USA and the developed world we can afford other incentives. But many people would oppose putting coping with global warming ahead of raising the living standards of the poor in China and India.

This began to change in 1973. It became clear in spring 1973 that the American demand for oil was outstripping supplies in the western hemisphere, and this gave economic power to the rapidly developing countries of the middle east. They exercised that power in oil embargo in late 1973 and America was shocked. President Nixon reacted by proposing an energy bill with the aim of "energy independence" to provide emphasis on domestic resources. Interestingly, as expressively stated by Michael Lubell of the American Physical Society, congress received this and other energy legislation until 1990 with a giant yawn. Simultaneously the environmental movement got under way starting on earth day April 22, 1970.

By steady rhetoric which a few of us struggled to counter, the U.S. environmental movement had basically stopped nuclear energy by 1975 after which no nuclear power plants were proposed that were completed. Nuclear energy was a logical candidate for assisting energy independence but there was no agreement on this either in the administration, congress or the environmental movement and the infrastructure built up and nurtured by Chairman Glenn Seaborg in particular under Kennedy and Johnson was largely dismantled

But in about 1990 another issue arose in the public consciousness. This issue was and still is largely pushed by non-American sources. The International Program on Climate Change was created and the first report of note was IPCC3. This came to a dramatic head in the Kyoto conference in December 1997. The academic world had been slowly pushing the idea that carbon burning would be a source of global warming. Originally a "pork barrel" project of the University of California at Davis the National Institute for Global Environmental Change (NIGEC) was started in 1990 with 4 centers at Davis, Tulane, Harvard and Indiana and I had the honor of directing the NE center. Still Congress was slow and continued the bipartisan yawn and ignored the executive branch proposals on energy whether democrat or republican till about 5 years ago

Ever since King Croesus invented money, it has been a very effective way of understanding commerce, and control of money gives guidance to millions of individual decisions. The energy system of the USA is big enough that there is not enough money to change it by subsidizing what we want, it has to be by banning or taxing what we do not want. But to achieve efficiency it is necessary to have an educated populace. Microeconomists, particularly Jorgensen of Harvard and Manne of Stanford had calculated that if there was a tax on carbon with the funds going in to the general treasury as reduction of other taxes, it would be stimulating to the economy. I invert that argument and argue that if we try to take funds out of general taxation to subsidize non-carbon technologies, it would dampen the economy. This was done to a small extent in the Clinton administration but fortunately only to a small extent.

INDIRECT LEGISLATION AND CONTROLS

Electricity came into wide use a century ago electric utility companies used marketing tactics to encourage its use. In UK in 1930, it was correctly assumed that everyone would switch from gas light to electric light as soon as possible. The charge for electricity for lighting was high, but the charge for other uses such as heating was 1/3 as much! Although modern legislation stops this, it seems unwise to assume that an electricity company will try to stop people using its product. It has only partially worked.

Encouraging efficient use by legislation follows two paths. The first adopted 35 years ago by Massachusetts and California alike was to legislate that information be available. A customer must be able to know how much the electricity to operate an appliance will cost him. California went further and has led the country in mandating efficiency standards for all products sold there. I have argued with Art Rosenfeld about this both in Erice and elsewhere over the years. Art estimates that even with information people will normally only make decisions based on 2 year running costs. But this remains a matter for study.

Electric utilities historically encouraged increased consumption and sales. Public information and legislation to encourage efficient use may well be necessary. Interestingly the report by Revelle, Singer and Starr in a much maligned paper called for such strategies in 1991. I like Carmen Difiglio's discussion and particularly like his change of title. A "no regrets strategy" implies that someone (undefined) will do the regretting.

A number of indirect legislative and legal actions, both at the national level and the state level have undone the encouragement and infrastructure put in for nuclear power in the time of Glenn Seaborg. In nuclear waste, for example, the proposal in 1970 to put high level nuclear waste in a salt bed in Lyons, Kansas, was stopped by state action, and the present proposal at Carlsbad New Mexico is much hampered and delayed. Even low level waste sites have been stopped at the state level in spite of supportive reports by the National Academy of Sciences. But the recent actions at the federal level are probably more damaging. Starting in the late 1970s the federal proposed a national site. In the 1990s the congress chose a site in Nevada. Although Nevada had originally suggested that their state would have such a site, they now changed their minds and it was opposed in the U.S. Supreme court who insisted that more study be done to determine if it would be safe for a million years. $10,000,000 was spent on such studies and remarkably scientists can say something sensible even at that time. But now President Obama has pulled out the rug, presumably to help Senator Harry reid. But I am reliably informed that both the secretary of energy (Dr. Steven Chu) and the President's Science Adviser, (Dr. John Holdren), informed him that there is no scientific reason for rejecting the site and the crafty words were used that "it is not feasible". These attitudes must change for a nuclear revival to take place.

Another indirect approach is mandatory changes such as fuel efficiency standards for automobiles/ Again this is, in a sense, a way of persuading the public to do something that they will not do by price alone. The arguments against them were described in the first version of Ginzton's NAS report in 1977, which he scrapped because he realized that they did not correspond to the nation's will. To achieve a high fuel efficiency means the car will cost more in other ways (usually a capital cost) and someone who drives only 5000 miles a year is relatively penalized.

"Washington should make greater use of a powerful policy tool that it has largely neglected: public sector procurement. Procurements by the U.S. Department of Defense (DOD) and other agencies were the foundation for major waves of innovation after the Second World War, including those in jet propulsion, Earth orbiting satellites and so on" (*Nature*, Vol. 466, 15 July 2010).

I note that 30 years ago the largest cogeneration plants were U.S. military bases.

SMALL LOCAL DECISIONS

The only energy legislation of any importance has been a number of small, local decisions, often empty decisions because the local government has no effective authority. The most obvious are the "nuclear free zones" ostensibly for stopping military activities, but of which the only operational significance has been the hindrance of the development of nuclear energy in a rational way. An obvious example is the ban in New London, CT. This ban has no effect on the nuclear submarines in the base across the river but stopped nuclear fuel, both fresh and spent fuel, coming from Brookhaven National Laboratory by the ferry from Orient point.

There are many other local decisions, that may have no legal impact but are "road blocks" for nuclear power. These include, for example, the decision to oppose the Shoreham Nuclear power plant after its' NRC license was issued by the Governor of NY State, the abandonment of the Yucca Mountain nuclear waste site, the decision of the Bureau of Indian Affairs (BIA) and the Bureau of Land management to block the temporary waste site on the Goshute reservation in Utah just as NRC was prepared to issue a license and opposition to wind turbines off Cape Cod. These mostly involved with nuclear power and increase the costs of these non carbon alternatives.

But no local community has enacted legislation to restrict the use of oil for transportation, or for reducing the hating and cooling costs in stores and public buildings. Yet these are places where there are huge inefficiencies. So on balance I argue that the myriad of local regulations have done little so far to address the basic problems and cannot make a major improvement but as noted below a careful attention to local conditions can make inroads.

THE STERN REPORT

Meanwhile Europeans were pushing ahead. I refer in particular to the report by Sir Nicholas Stern,(now Lord Stern) reporting to the U.S. Treasury. The executive summary is on the energy pmp website. Although there were of course other reports this seems to have had a major influence on the European Union. In it Stern accepted that global warming was happening. He estimated the economic consequences thereof and argued that some action was immediately necessary. As adopted there were several aspects.

Carbon dioxide emissions were to be controlled in some but not all users of carbon. There was to be a cap and trade system with each industry setting a historical cap to start with. There were to be incentives for some, but not all, low carbon energy technologies. But in spite of specific recommendations by the WFS energy PMP and a number of other organizations carbon sequestration and nuclear energy were not included.

The problem with a historical cap is that the industry sets its cap as high as they can justify which is often higher than is the truth. This occurred in Europe and made the system, at least initially, a failure. The problem with setting the cap on emissions, rather than on carbon as it comes out of the ground is two fold, as discussed at Erice many times. Firstly we do not know the emissions. But we do know the amount of carbon as it comes out of the ground. After all, that is what the customer actually pays for so it is known and reliable. Secondly each industry is treated separately and it a a centralized command and control system that must decide which industry bears the brunt. Again Stern ignored the resolutions from the energy PMP which to my certain knowledge were sent to him.

THE WAXMAN-MARKEY BILL

This bill, all 1300 pages, is on the Energy PMP website. It was passed by the U.S. House of Representatives in 2009 but not picked up by the U.S. Senate. It also proposes a cap and trade system with a historical cap, on emissions. It therefore has all the disadvantages of the Stern report plus a few more. To make it "acceptable" to powerful interests especial subsidies were proposed. The basic criticism is, therefore that it is a "pork barrel" bill with a slight veneer of climate change A leading climate researcher and advocate for strong action is Dr. James Hansen of NASA and adjunct professor at Columbia University. He has written (also on the Energy PMP website) that the bill is so bad that it should be abandoned and start over again. I agree with Jim. So it seems do many in the U.S. Senate. During this discussion, an energy company proposed 20+ new coal fired power plants in Texas, without carbon sequestration or gasification to make it more carbon efficient. The environmental community objected and the company finally, in a compromise agreed to only build six. But to those of us examining the situation carefully six was all they actually wanted to build! Such deliberate over statements followed by a "compromise, is clearly what one can expect from any regulatory scheme for Carbon with a historical cap.

THE CANTRELL-COLLINS BILL

This bill of a modest 49 pages, also available on the Energy PMP website, concentrates solely on the issue of global warming and makes no attempt to legislate what the new energy structure of the U.S. is going to be. Senators Maria Cantrell and Collins proposes to control carbon upstream at the mine, well, gas field or port of entry as the enrgy pmp had proposed. He also has a "cap and dividend" approach: a tax on the carbon that is repaid as a dividend to taxpayers in just the way Jorgensen and Manne had suggested. When this was submitted in March 2010, I was wildly enthusiastic. But it was too good to be true. It seems to be forgotten already in the senate

THE KERRY-LIEBERMAN BILL

I have also placed the details of this bill on the Energy PMP website. It seems to be a compromise. It is a draft for discussion. It has a lot of detail (perhaps a synonym for pork barrel) in its 987 pages. A main reason for the complexity is that it tries valiantly to address three issues at once:

a. Energy independence
b. Energy shortages
c. Climate change

As I read it it Kerry and Lieberman propose what seems to be a command and control system which I believe will be much more expensive than needed. But maybe this is a useful point to stop for discussion. It is unlikely that the bill will come up in 2010. Th oil spill in the Gulf of Mexico is far more politically important although it has killed no one in the public. I do note, however that everyone seems to agree that the well must be sealed and abandoned. As a physicist and professor I note that my former graduate student, Dr. Arthurr Kuckes of Vector Magnetics is using his knowledge of electromagnetism to guide the drilling of the relief wells.

SUGGESTED REASONS FOR FAILURE OF THE KERRY LIEBERMAN BILL

A search of the internet suggests the following:

1. http://energytopic.nationaljournal.com/2010/06/epa kerry lieberman.php
2. http://www.brookings.edu/opinions/2010/0517_climate_bill_muro.aspx
3. http://www.greencarcongress.com/2010/05/kerry lieberman 20100521.html
4. http://www.glgroup.com/News/The Effects of the Kerry Lieberman Bill on Employment 48361.html
5. http://news.yahoo.com/s/usnw/20100721/pl_usnw/DC38585
6. http://www.ucimc.org/content/billions new nuke giveaways kerry lieberman bill exposed
7. http://energytopic.nationaljournal.com/2010/06/epa kerry lieberman.php

None of the first 5 seem sensible. The last two worry about any encouragement about nuclear power. But Michael Schellenberger and Ned Norhaus suggest that the cost gapis a more serious reasons in the next session. http://www.thebreakthrough.org.

MY LIST OF PROBLEMS WITH EXISTING CAP AND TRADE PROPOSALS

1. The major scientific technical problem in discussing these bills is that there is a big gap between the cost of fossil fuels and the cost of most proposed

alternatives. Moreover the portability and flexibility of using oil is hard to equal. I note that of the alternates nuclear power WAS cheaper than alternates in 1973. None of the legislative proposals address this clearly. I believe that cost is an issue we should discuss at Erice. As I have emphasized for 35 years nuclear power4ed electricity has been cheaper than coal and could be again. It is unclear whether that will happen. But it remains the only candidate in my view for a large scale substitution of fossil fuel by other sources.

2. Too many pork barrel issues attached to the Waxman Markey and Kerry Lieberman bills.
3. Cap and Trade with a historical cap is an initial give away to Industry.
4. Setting caps separately on each sector is inefficient and reduces choise to the consumer.
5. The control is too late in the carbon cycle.

Numbers 2-5 would be avoided with a carbon tax such as in the Cantrell bill. If we had a carbon tax, this could be tested.

SOME CHEAPER PROPOSALS

I believe that there is, as now, no good simple solution (other than nuclear without obstacles) for all locations. There should be much more emphasis on choosing a technology suitable for the location.

1. Solar water heating is cost effective in the front range of Colorado with clear weather and high UV content. But it is NOT in Massachusetts.
2. Small hydro, abandoned by big electric utilities 80 years ago may now be cost effective in some locations.

SECTION 12

SESSION 11

WFS GENERAL MEETING

PMP REPORTS—DEBATE AND CONCLUSIONS

PERMANENT MONITORING PANEL ON MOTIVATIONS FOR TERRORISM

THE LORD ALDERDICE FRCPsych
Chairman, PMP on Motivations for Terrorism
House of Lords, London UK

In my report to last year's AGM of the World Federation of Scientists I set out the history and background to the work of the two Permanent Monitoring Panels on Terrorism with particular reference to the PMP on Motivations for Terrorism. I will not go over that ground again for it is available in the excellent volume of the 42nd (2009) session of the WFS International Seminar on Nuclear War and Planetary Emergencies.

I pointed out last year that one of the obstacles to our scientific work was that, because those who undertake terrorist attacks are consciously breaking the law, the world's most powerful countries are relatively united in seeing terrorist behaviour as criminal. As the terrorist threat has not diminished, despite the enormous resources devoted to the 'War on Terror', states (and the United States in particular) have hardened and extended their security operations. One of the results has been that meetings with those who promote or engage in terrorism are no longer regarded as legitimate activities. This direction of flow has continued over the past year. Our attempts to enable interlocutors to travel for meetings and data gathering have become increasingly difficult because of visa and entry/exit problems, and most recently and significantly the U.S. Supreme Court, in a decision in June of this year, has made the position of such activities even more difficult for U.S. citizens—a matter addressed by two of our members, Professors Scott Atran and Bob Axelrod in a joint op-ed in the New York Times on 29 June 2010. The Supreme Court had the previous week in *Holder vs. Humanitarian Law Project* ruled that any "material support" of a foreign terrorist group, including talking to terrorists or the communication of expert knowledge and scientific information, helps lend "legitimacy" to the organization. As Atran and Axelrod noted, while sometimes, undoubtedly, that is the case, American law has to find a way to make a clear distinction between illegal material support and legal actions that involve talking with terrorists privately in the hopes of reducing global terrorism and promoting national security. I will return to the implications of this situation later in my report, suffice to say that while it makes some of our scientific work much more difficult it is notable that this judgement is delivered at the same time that there are increasing calls for engagement with some of these very same groups in order to make progress in the Middle East Peace Process. The U.S. at the most senior levels has a track record of congratulating successful peacemakers who have done exactly what is now being declared illegal. On the other hand the MEPP in which the U.S. has invested significant political capital is regarded by many as hopelessly moribund, especially in the absence of participation by those groups who are regarded most disturbing the peace. This gloomy assessment persists despite the announcement a few days ago that direct Israeli-Palestinian talks will shortly re-commence at leadership level.

Unfortunately the last year has also seen various extensions of terrorism.

Its geographical extension involves the progression westwards in Saharan and Sub-saharan Africa and the increasingly daring and sophisticated operations of international piracy off the coast of Somalia. As the U.S. and its allies withdraw from Iraq and set down timetables for withdrawal from Afghanistan, the insurgencies there and the instability in Pakistan threaten to get worse and some other states such as Yemen teeter on the edge of being unmanageable. I would emphasize that despite the preoccupation of the populations in the West about their own safety, the overwhelming majority of terrorist attacks and the death and destruction associated with them are to be found in this band of unstable countries stretching from West Africa to Indonesia, not in the developed Western world.

The problems are not only of instability within the countries concerned but the likelihood that they will become bases for relatively organized terrorist operations as Afghanistan had become for Al Qaeda prior to the U.S.-led invasion. Additionally instability has a tendency to leak across into neighbouring countries. Pakistan is an example of both problems developing. As well as the Afghanistan/Pakistan border problems, the deterioration of the situation in Kashmir, whatever the causes, will likely see increasing terrorism across the Line of Control, and the Mumbai bombings demonstrate that Pakistan is also a base for terrorist attacks on India at arms-length. India is also suffering from the home-grown terrorism of the Naxalite/Maoist groups throughout the country and this only seems to get worse from year to year.

There is a further extension of terrorism which is of concern to us, and here again Pakistan is one of the significant countries. For some years after the end of the Cold War there was a lower level of concern about the prospect of conflict involving nuclear weapons, however nuclear concerns are again on the rise. There is increasing instability in Pakistan, which is a long-time nuclear state, and the recent dreadful floods will not help at all in terms of political stability in the country. The commencement of the fuelling of Iran's first nuclear reactor reminds us that Iran is almost certainly moving irrevocably in the direction of becoming a nuclear state too, raising the stakes for other states in the region. The worsening tension between the U.S. and Iran specifically on the nuclear question does not help in this regard. Setting aside any concern about state-sponsored nuclear terrorism (and this seems unlikely in respect of either Pakistan or Iran) there is an increasing risk of expertise or materiel coming into the hands of non-state actors—an eventuality that would add a new dimension to the 'terror' in terrorism. How far other rogue players such as North Korea, may be relevant in this regard is difficult to ascertain, but concern is certainly rising.

In addition to the increasing significance of the internet in the rise of fundamentalism, radicalism and terrorism, the developments I have already identified in the global picture of terrorism have accompanied some other changes in focus. Homeland security remains a matter of profound concern especially in the U.S., the UK and France, but with very few successful 'homeland' attacks from AQ and AQ-sympathetic operatives from outside these countries increasingly attention is being paid to the emergent problems of home-grown radicalization. While these concerns are well based and the potential for trouble is significant, the actual number of successful attacks in the West has been very low, but there is little sense that this represents a lesser radicalization amongst young Muslims especially in Europe, and the longer the economic crisis lasts and its effects are felt in unemployment and economic disadvantage, the greater the risk of alienation amongst young men in minority communities. Anxiety is not lessened by the knowledge that the U.S.-led military interventions in Iraq and Afghanistan have not succeeded in getting rid of the global terrorist threat (on the contrary), and that the U.S. and its allies are progressively withdrawing from those military involvements, possibly in short order and without an obvious Plan B for dealing with terrorism. There is a danger that Western populations will negatively withdraw their interests into addressing immediate homeland threats rather than recognizing that globalization does not really give us this option.

As I pointed out last year, even treating terrorism as a 'phenomenon' to be studied is often misunderstood and criticized. In the Motivation Panel however we have continued our psychological and social anthropological studies and strengthened our network of cooperating colleagues. All of this work has continued to give evidence for the view that terrorism is primarily a phenomenon of group rather than of individual psychology. Groups should be thought of as a form of organism each with its own history and memories, its own identifications and forms of functioning, its own evolution and development and its own possibility of dysfunction, dissolution and death. Perceptions of humiliation and shame can be transmitted down through successive generations as well as across groups or communities and a sense of injustice, unfairness and disrespect may persist for generations after the social and economic disadvantages have been addressed. If non-violent routes to the resolution of these feelings of communal shaming and injustice are continually blocked, a rage is generated in the group that can spread with contagion to others who identify with the group, even when they have not themselves suffered directly.

It is not religious faith, or even religious fundamentalism itself that leads to terrorism, but the radicalizing effects of the moral outrage felt at certain social and political developments. These groups, like individuals do not operate rationally in their own best interests and their actions are inevitably damaging to their own community, but when what we have called 'sacred values' are involved, they seem to trump other values. Such values are not sacred in the sense of being religious, but transcend short-term self interests such as comfort, shelter, health and general well-being. Things such as the life of my child, a sense of justice for my people, and the importance of self and group esteem, have a value which is not negotiable in the usual economic way, is not dependent on the prospect of success and may provoke violent reactions.

<u>How did we take the work forward in the past year?</u>

1. We have published our work in a number of books and scientific journals, as well as in the more public media and press. Examples include a series of articles by Carlo Strenger, the most recent in the 19 July 2010 issue of Israel's Ha'aretz newspaper) and upcoming in October 2010, Scott Atran's latest book 'Talking to the Enemy' will be published containing much useful material and some important conclusions. It will be of substantial interest to those of you who follow our work.
2. Our Lounsbery Foundation (USA) grant of U.S.$100,000 enabled us to address the relevance, albeit indirect, of Water, Energy and the Environment in the wider Middle East, to combating the spread of terrorism in that region. We continued our WFS study visits to the region and hope to make a final visit under this grant and complete our report in the very near future. We have also worked with the Strategic Foresight Group in Mumbai on these issues

and the results have been most encouraging. The Swiss, Swedish and Turkish Governments have sponsored and facilitated two conferences on our proposals in the past year (in Turkey and Switzerland) and a further meeting will be held next month (September 2010) in Turkey. These meetings have persuaded the Governments of Iraq, Jordan, Lebanon and Turkey to explore the development of formal structures for cooperation on Water, and we continue to encourage the Syrian Government to engage formally. Our colleague Munther Haddadin (Jordan) has also been working on these issues with Prince Hassan of Jordan and SFG has been liaising with him. Separately we have been encouraged in our meetings with the Israeli and Palestinian officials who deal with the problems of Water in their jurisdictions and we have arrangements in place to take this further in the autumn of this year. We have also had some very encouraging indications from other European Governments about assisting our projects.

3. ARTIS Research and Risk Modelling, the companies set up by some of the members of the PMP to support our research work, have continued to make some modest progress and this can be followed on our website. (www.artisresearch.com).
4. The group which has developed the International Dialogue Initiative working through Bahçeşehir University in Istanbul is continuing its regular meetings—in Belfast in September 2009 and in Ankara in February 2010—and the next meeting in September 2010 in Stockbridge, Massachusetts will continue our exploration of the motivational components in the relationship between Iran and the West, as well as the rapidly changing situation in Turkey itself. I should report that our meetings in Turkey have generated considerable public interest because of our work on Kurdish terrorists, with a number of television and front page newspaper stories and meetings at the level of the President and Foreign Minister. There are also plans for a web-site for this group.
5. Members of the PMP continued to meet extensively in the Middle East to develop our work, not least because of the problems of bringing cooperating partners together in U.S. and European centres.

I would emphasise again as I did last year that almost all of our initiatives are undertaken in cooperation with bodies outside of WFS itself, enabling us to inject our scientific perspective into the work undertaken by others, but also acquiring information, contacts and resources that would not be available to us on our own as a PMP. In addition to those colleagues noted already this also includes the Critical Incident Analysis Group (CIAG) at the University of Virginia in Charlottesville, CRISMART at the Swedish Defence Academy and the S. Rajaratnam School of International Studies (RSIS) at Nanyang Technological University in Singapore.

How do we hope to take the work forward in the next year?

1. We will continue to publish our work in books and scientific journals, and in the public media and to legislators, government officials and others who can affect the attitudes of our societies. We will also coordinate our work closely with the PMP on Mitigation of Terrorist Acts and are currently exploring how

to jointly progress a 'Road Map' for our work over the next few years. This includes funding proposals.

2. We will shortly complete the work under the Lounsbery Foundation Grant (USA) looking at the contribution of problems of Water, Energy and the Environment in the wider Middle East to the spread of terrorism in that region but hope that the other support we have identified will enable us to continue to move forward on this project with Sundeep Waslekar and the Strategic Foresight Group.
3. Professors Scott Atran and Bob Axelrod, Dr. Marc Sageman, Richard Davis and I will continue to pursue work through ARTIS Research and Risk Modelling (www.artisresearch.com) on direct field research on the motivations for terrorism.
4. The International Dialogue Initiative working with Professor Vamik Volkan and Dr. Robi Friedman will continue, though it seems likely that we will find it beneficial to move to a new academic partner in Istanbul. This will be discussed at the meeting in Stockbridge, Massachusetts in September 2010.
5. Members of the PMP will continued to travel to the Middle East to develop our work but will also continue to try to pursue meetings with cooperating partners in centres within or close to Europe for practical reasons.

We had hoped last year that the election of the new U.S. President would create more openings both for the funding of projects and for attention to our findings, but the initial promise no longer burns so brightly in light of the decisions of the U.S. courts, the pressures in the Congress and the reactions of the executive branch. It is not easy to be optimistic either that this will change for the better in the short-term, or that our findings however convincing will be espoused, but we will nevertheless strive to continue as we have planned.

I must emphasize that in this short report I cannot do justice to an enormous and growing agenda and a number of key areas, for example the use of the internet, to which we have continued to pay significant attention do not figure heavily in what I have said. This is inevitable in a brief document which cannot hope to be exhaustive, but is I hope at least informative.

I would again thank the various individuals and institutions who have assisted our work enabling us make real scientific progress during a difficult year in the field, and especially to thank Professor Nino Zichichi, Mr. Claude Manoli and the staff of WFS as well as our other WFS colleagues, especially in the PMP on Mitigation of Terrorist Acts for their continuing and much appreciated support and encouragement during 2009/2010.

2010 PROGRESS REPORT OF THE MCD-2/7 PROJECT AND 2011 RESEARCH PROJECT. EAST-AFRICA AIDS RESEARCH CENTER AT THE UGANDA VIRUS RESEARCH INSTITUTE (UVRI), ENTEBBE UGANDA

DR. FRANCO M. BUONAGURO
Istituto Nazionale dei Tumori, "Fondazione G. Pascale"
Napoli, Italy

In the year 2010 several activities have been conducted at the Centre in synergism with the program of *"Global Support to the National Plan for HIV/AIDS control in Uganda"*. Furthermore research activities, directly related to the Uganda HIV epidemic and to the fight against it, have been conducted at the INT-Naples within the context of 4 subprojects:

1. Epidemiology of HIV-1 in Uganda, within the frame of the Ugandan National Plan for HIV/AIDS Vaccine Development (MCD-2/7/1):
2. Molecular Characterization of HIV Strains from Subjects exposed to HIV Infections, who either fail to seroconvert or do not develop AIDS (MCD-2/7/2);
3. Etiopathogenesis of AIDS associated Neoplasias (MCD-2/7/3);
4. Development of HIV vaccine (MCD-2/7/4).

Studies performed on the molecular characterization of HIV HMA, nucleotide sequencing and phylogenetic analysis show that >60% of the samples cluster within the A clade, ~ 25% in the D clade, <20% does not show a clear cut result suggesting a divergence greater than 30%.

Within the theme of HIV characterization other studies have been performed on:

- HIV characterization in Iran (clade A);
- HIV characterization in Russia (clade A);
- HIV characterization from Ukrainian subjects (clade A);
- HIV characterization from Nigerian subjects (HIV 1 and 2).

Furthermore we are contributing with Saladin Osmanov (Director of the WHO HIV Vaccine program) to the re-establishment of the pan-European HIV monitoring system, for predicting the efficacy of new vaccine approaches.

The studies conducted on the etiopathogenesis of AIDS associated cancers have focused on: Kaposi's sarcoma, with the continuation of sample collection and virus characterization; Genital Cancer, with virus genotyping and vaccine design; Conjunctival carcinoma, with analysis of genetic susceptibility and virus characterization; Liver Cancer, with characterization of genes involved in progression from chronic hepatitis to cirrhosis and liver cancer; lymphoproliferative diseases, with characterization of genes involved in the progression from mixed cryoglobulinemia.

Within the theme of Virus-associated cancers other studies have been performed on:

- HPV-association to genital cancers in Ecuador (in collaboration with Claudio Maldonado, President of the Latin-American section of the American Society for Colposcopy and Cervical Pathology);
- HPV-role in genital lesions in Colombia (Dr. Alvaro Mauricio Flórez, Universidad de Santander;
- HPV role in esophageal cancer in Italy (Dr. A. Monaco, Hospital Cardarelli, Naples-IT);
- HCV association to Liver Cancer within the context of the ILCS (international Liver Cancer Study coordinated by the IARC-Lyon).

Moreover in 2010 a Plant-based Vaccine Monograph has been published on Expert Review of Vaccine, focused on the strategy to develop in plants vaccines targeted to pathogens associated with chronic infections at high risk of cancer progression. Finally FMB is leading a COST grant application of a large Network of Scientists for Pathogen-associated Cancers. The application on Nov. 19, 2010 has been selected for full project submission. If approved in March 2011, the 5-year grant will finance activities/workshops for standardization of diagnostic and therapeutic approaches within Europe, to be widely disseminated following accurate statistical evaluation.

P53 ROLE IN CANCER PATHOGENESIS: TWO DISTINCT MODELS

A relevant result has been the definition of *the role of P53 polymorphism at codon 72 in cancer pathogenesis*, whose preliminary data have presented in a plenary presentation at the African Society of Human Genetics, Yaoundé—Cameroon, 13-15 March 2009, and a more complete version is going to be presented at the AfSHG Meeting, Cape Town-South Africa, 6-9 March 2011.

The study has been conducted (as shown in Figure 1) using two sets of specific primers in order to selectively amplify either the Arginine (Arg) or the Proline (Pro) allele, with the amplification of 141 or 177 bp fragment, respectively.

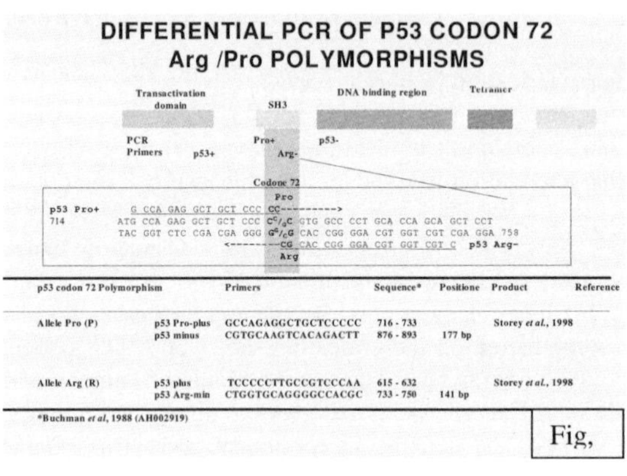

Fig,

Fig. 1:

The amplified products, where subsequently detected by gel electro-phoresis as shown in Figure 2, where homozygous P (Pro) sample is in lane 2 and 3; heterozygous P/R (Pro/Arg) is in lane 4 and 5; homozygous R (Arg) is in lane 6 and 7.

The study has been performed on three different mucosal cancers and in two populations: African subjects enrolled at the East Africa AIDS Research Center, with low prevalence of Arg homozygosity, and Caucasian subjects at the Natl Cancer Institute in Naples-Italy, with high prevalence of Arg homozygosity.

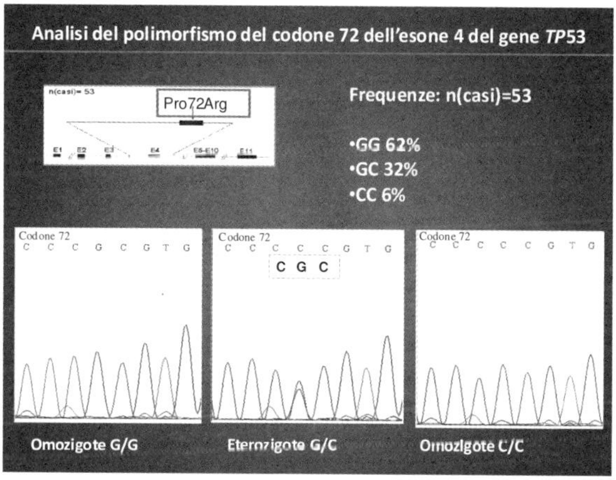

Fig. 2:

- In **Conjunctival Carcinoma**, which has been associated with HIV-associated immunodeficiency and UV-exposure, p53 polymorphism is playing a role [subjects homozygous for Arg at codon 72 are at high risk of progression].

- In **Penile Squamous Carcinoma**, which has been associated with oncogenic HPV types (in particular HPV16), and in **Liver Cancer**, associated with HCV, the p53 polymorphism does not play a role.

This data suggest that polymorphisms of the p53 onco-suppressor gene, which plays a major role in DNA repair pathways, are relevant for the susceptibility to cancers associated with DNA damages, but are not relevant for cancers associated with the chronic infections, analysed so far.

DEVELOPMENT OF HIV VACCINE

Within the theme of VLP-based Vaccines

- Specific studies have been financed by:
  - EU within the FP7 to characterize the relevance of the humoral immunity in protection against HIV/AIDS: our NGIN project received a 5-year support for a total 12-million €;

- NIH within the SVEU program for a pre-clinical primate evaluation;
- A plant-based model of HIV vaccine has been developed;
- A HIV vaccine patent has been submitted;
- Our group contributed to the Bob Gallo's annual Conference, which has been held in Tropea-Italy, the first time outside of USA.

TRAINING AND EXCHANGE PROGRAMS

Within the context of the Training and Exchange Programs several activities have been pursued:

- Direct support to the GuluNap project the newly established School of Medicine in Northern Uganda with teaching Missions and exchange programs;

Independent journals' Editor profiles

Franco M. Buonaguro
The Natl Cancer Institute "Fondazione Pascale", Naples - Italy.

INFECTIOUS AGENTS AND CANCER

Website: http://www.infectagentscancer.com/

Infectious Agents and Cancer is an open access, peer-reviewed online journal that encompasses all aspects of basic, clinical and translational research providing an insight into the association between chronic infections and cancer. There are established links between specific chronic infections and cancer. Recent advances in genomic and proteomic technologies have accelerated the discovery of novel pathogens, cancer antigens and the molecular pathways involved in disease progression. Several new fields are developing very rapidly, in particular the application of immunogenomics and pharmacogenomics to fight specific cancers, targeting immuno-tolerated cellular/viral molecules by immuno and drug therapy.

Editor-in-Chief of *Infectious Agents and Cancer*

Franco M. Buonaguro PhD
The Natl Cancer Institute "Fondazione Pascale", Naples - Italy.

"Cancers related to infectious diseases represent a very significant component of the cancer burden in humans spanning from 20% of all malignancies in industrialized countries to 40% in developing countries. Studies on pathogen-related cancers allow the identification of mechanisms relevant to oncogenesis as well as molecular markers for diagnostic/prognostic methods and targets for therapeutic strategies. Furthermore their relevance is due to the concrete possibility of preventive strategies, which are always cost-effective and can usually be implemented with relatively modest investments. The open access policy of Infectious Agents and Cancer represents a key factor to foster an even participation and contribution by Western as well as Eastern/Southern Countries' colleagues"

A short biography
Currently the Director of Molecular Biology and Viral Oncology Unit, in the Department of Experimental Oncology at the Natl Cancer Institute "Fondazione Pascale", Naples - Italy.

Dr Buonaguro has worked on various aspects of cancer pathogenesis including chemical, physical and viral carcinogenesis. His current research area is mainly focused on pathogen-associated cancers, including HPV and genital cancers, HHV-8 and Kaposi's sarcoma, HBV/HCV and liver cancer.

Dr Buonaguro gained his doctoral degree from Frederick II University in Naples and has subsequently worked in Chicago, at Argonne National Laboratory, and in Seattle, at the Fred Hutchinson Cancer Research Center. He was a recipient of WHO Fellowship in 1983; and has served as WHO Virology Consultant of WHO/ UNAIDS Network for HIV Isolation and Characterization (since 1996) and WHO HPV DNA International Collaborative Study Group (2003).

He is also currently Director of the East Africa AIDS Research Center in Uganda, established by the World Laboratory, Geneva.

To see profiles of some of the other Editors who have chosen to bring their journal to BioMed Central please see the independent journals' Editor profiles.

Furthermore an International PhD in immunology has been established at the University of Milan, available also to foreign students, supported by a tuition fellowship.

Finally a Training program has been arranged to host two Iranian post-docs in our Unit to train them in molecular biology and virology. This approach is continuously pursued to contribute to post-graduate educational programs in developing countries. To this goal is finalized also the Infectious Agents and Cancer online journal, which this year is going to receive the first Impact Factor following the ISI-Thomson.

The relevance of the journal is due to its focus on a significant proportion of Cancers (particularly in developing countries), the worldwide open free accessibility, and the no-fees policies for authors from low income countries, as reported in the Journal cover page at http://www.biomedcentral.com/info/publishingservices/profiles/101161:

> *"Cancers related to infectious diseases represent a very significant component of the cancer burden in humans spanning from 20% of all malignancies in industrialized countries to 40% in developing countries. Studies on pathogen-related cancers allow the identification of mechanisms relevant to oncogenesis as well as molecular markers for diagnostic/prognostic methods and targets for therapeutic strategies. Furthermore their relevance is due to the concrete possibility of preventive strategies, which are always cost-effective and can usually be implemented with relatively modest investments. The open access policy of Infectious Agents and Cancer represents a key factor to foster an even participation and contribution by Western as well as Eastern/Southern Countries' colleagues."*

SEPT 2009–OCT 2010 PUBLICATIONS ON PUBMED-TRACKED JOURNALS

1. Monaco, A., Marincola, F.M., Sabatino, M., Pos, Z., Tornesello, M.L., Stroncek, D.F., Wang, E., Lewis, G.K., Buonaguro, F.M., Buonaguro, L. "Molecular immune signatures of HIV-1 vaccines in human PBMCs." *FEBS Lett.* 2009 Sep 17; 583(18):3004-8.
2. Buonaguro, F.M., Tornesello, M.L., Buonaguro, L. "Virus-Like Particle Vaccines and Adjuvants: the HPV paradigm." *Expert Review of Vaccines*, 2009 Oct; 8(10):1379-98.
3. De Giorgi, V., Monaco, A., Worchech, A.,Tornesello, M.L., Izzo, F., Buonaguro, L., Marincola, F.M., Wang, E., Buonaguro, F.M. "Gene profiling, biomarkers and pathways characterizing HCV-related hepatocellular carcinoma." *J. Translational Medicine* 2009 Oct 12;7:85.
4. Tornesello, M.L., Biryahwaho, B., Downing, R., Hatzakis, A., Alessi, E., Cusini, M., Ruocco, V., Katongole-Mbidde, E., Buonaguro, L., Buonaguro, F.M. "Tp53 Codon 72 Polymorphism in Classic, Endemic and Epidemic Kaposi's Sarcoma in African and Caucasian Patients." *Oncology* 77:328-334;
5. Tornesello, M.L., Biryahwaho, B., Downing, R., Hatzakis, A., Alessi, E., Cusini, M., Ruocco, V., Katongole-Mbidde, E., Loquercio, G., Buonaguro, L., Buonaguro, F.M. "Human herpesvirus type 8 variants circulating in Europe, Africa and North America in classic, endemic and epidemic Kaposi's sarcoma

lesions during pre-AIDS and AIDS era." *Virology*. 2010 Jan 13. [Epub ahead of print] PubMed PMID: 20079510.

6. Buonaguro, L., Petrizzo, A.., Tornesello, M.L., Napoletano, M., Castello, G., Benedice, G., De Renzo, A., Perrella, O., Dolcetti, R., Buonaguro, F.M. "Immune signatures in human PBMCs of idiotypic vaccine for HCV-related HCV-lymphoproliferative disorders." *J Transl Med*. 2010 Feb 19;8:18.
7. Buonaguro, L., Tornesello, M.L., Buonaguro, F.M. Virus-like particles as particulate vaccines. In the Special Issue on "Novel Vaccines and Adjuvants" (Issue Editor L. Buonaguro). Curr HIV Res. 2010 Jun;8(4):273. PMID: 20462398.
8. Puzelli, S., Buonaguro, F.M., Facchini, M., Palmieri, A., Calzoletti, L., De Marco, M.A., Arace, P., de Campora, E., Esposito, C., Cassone, A., Rezza, G., Donatelli, I. Surveillance Group for Pandemic H1N1 2009 Influenza Virus in Italy; Campania H1N1 Task Force. Cardiac tamponade and heart failure due to myopericarditis as a presentation of infection with the pandemic H1N1 2009 influenza A virus. *J Clin Microbiol*. 2010 Jun;48(6):2298-300. PMID: 20392922.
9. Tornesello, M.L., Loquercio, G., Tagliamonte, M., Rossano, F., Buonaguro, L., Buonaguro, F.M. "Human Papillomavirus Infection in Urine Samples from Male Renal Transplant Patients," *J Med Virol* 2010 Jul;82(7):1179-85. PMID: 20513081).
10. Giorgi Rossi, P., Bisanzi, S., Paganini, I., Carillo, G., Angeloni, C., Scalisi, A., Macis, R., Chini, F., Carozzi, F.M. and the HPV Prevalence Italian Working Group. "Prevalence of HPV high and low risk types in cervical samples of the Italian general population." *BMC infectious Diseases* 2010 Jul; 10(7): 214.
11. Buonaguro, F.M., Campadelli-Fiume, G., De Giuli Morghen, C., Palù, G., "Updates and Achievements in Virology" *Expert Rev Vacc* 2010-Jul; 9 (7): 703-706.
12. Buonaguro, F.M., Butler-Ransohoff, J.E., "PharmaPlant the new frontier in Vaccines." *Expert Rev Vacc* 2010-Aug; 9 (8):805-807.
13. Tagliamonte, M., Visciano, M.L., Tornesello, M.L., De Stradis, A., Buonaguro, F.M., Buonaguro, L. "Constitutive expression of HIV-VLPs in stably transfected insect cell line for efficient delivery system." *Vaccine*. 2010 Sept 28(39): 6417-6424.
14. Scotti, N., Buonaguro, L., Tornesello, M.L., Cardi, T., Buonaguro, F.M. "Plant-based anti-HIV-1 strategies: vaccine molecules and antiviral approaches." *Expert Rev Vacc* 2010-Aug; 9 (8): 925-936.
15. Sirugo, G., Williams, S.M., Royal, C.D.M., Newport, M.J., Hennig, B.J., Mariani-Costantini, R., Buonaguro, F.M., Velez Edwards, D.R., Ibrahim, M., Soodyall, H., Wonkam, A., Ramesar, R. and Rotimi, C.N. Report on the 6th African Society of Human Genetics (AfSHG) meeting, 12th-15th of March 2009, Yaoundé, Cameroon. American Journal of Tropical Medicine and Hygiene. 2010 Aug;83(2):226-9.
16. Carozzi, F.M., Tornesello, M.L., Burroni, E., Loquercio, G., Carillo, G., Angeloni, C., Scalisi, A., Macis, R., Chini, F.. Buonaguro,F.M., Giorgi Rossi, P. and the HPV Prevalence Italian Working Group. Prevalence of HPV types in high

grade cervical intraepithelial neoplasia and cancer in Italy. *Cancer Epid, Bio & Preven* 2010-Sept; 19(9): 2389-2400.

17. Tornesello, M.L., Buonaguro, L., Cristillo, M., Biryahwaho, B., Downing, R., Hatzakis, A., Alessi, E., Cusini, M., Ruocco, V., Viviano, E., Romano, N., Katongole-Mbidde, E., Buonaguro, F.M. MDM2 and CDKN1A gene polymorphisms and risk of Kaposi's sarcoma in African and Caucasian patients. Biomarkers. 2010 Oct 28. [Epub ahead of print].

MOTHER AND CHILD HEALTH PMP: MATERNAL AND CHILD MORTALITY IS A PLANETARY EMERGENCY, REPORT OF 2010 ACTIVITIES

DR. NATHALIE CHARPAK, DR J.G. RUIZ, DR S. DE LEON-MENDOZA
Kangaroo Foundation
Bogotá, Colombia

FIRST PART

Background: Mission of the Mother and Child PMP

1. To recognize new and monitor existing tools that decrease maternal and infant mortality and morbidity.
2. To highlight the impact of the other planetary emergencies on Maternal and Infant mortality and morbidity and therefore contribute to the enhancement of the quality of the present life and future generations.

Progress Report on globalization of the KANGAROO MOTHER CARE (KMC) Intervention

1. Diffusion (dissemination)
   Kangaroo Mother Care is our specialty. KMC is a method to decrease the mortality and to improve the quality of survival of the premature and Low Birth Weight infant in all levels of care. Twenty millions of newborns are candidates for KMC per year.

   KMC has 3 components: 1. the Kangaroo Position (carrying 24h/24h the premature baby in direct skin to skin contact on the chest of the mother) 2. the Kangaroo nutrition (based on breastfeeding the premature infant) and 3. Kangaroo discharge policies (early discharge in Kangaroo position with strict follow up:

   - The Colombian experience 2009-2010: A public health edict published by Health authorities in 2009, imposed an obligation to implement and promote KMC in all the institutions working with mother and infant. We redact for the health ministry the technical guidelines to implement a KMC program with the tool to evaluate this implementation of a KMC program and the criteria to give the KMC label to a newly implemented program. Our hope: Evaluation of the impact of KMC at a country scale (Colombia) in 2012?
   - Practical Kangaroo book for parent: 2010

- Special issue on KMC in "Current Women's Health Reviews" to complement WHO KMC practical guide. Spearheaded by the Kangaroo Foundation, Colombia It is a International Collaborative effort of KMC experts (Mali, Bangladesh, South Africa, Canada, Spain, Colombia, Sweden, USA, Philippines) We think in a comprehensive updated guide on KMC with 9 chapters:
  - Editorial: Kangaroo Mother Care: past, present and future.
  - KMC, Concepts, definitions and praxis: what elements are applicable in what settings in which local circumstances?
  - Evidence-based review of physiologic effects of Kangaroo Care
  - Prematurity and morbidity: could KMC reverse the process?
  - Human Milk and Kangaroo Mother Care
  - "Early" or timely discharge in kangaroo mother care: evidence and experience
  - Kangaroo Mother Care: Optimal support of preterm infants' transition to extra-uterine life in the high tech NICU environment
  - The Newborn Individualized Developmental Care and Assessment Program (NIDCAP): An Encompassing Approach for the Care of Preterm Infants
  - Implementation of a Kangaroo Mother Care program to decrease the morbid-mortality of the premature infants born in the CHU—Gabriel Touré of Bamako, Mali
  - Experiences with Community Kangaroo Mother Care in Very Low-Income Settings

- Participation in the organization of the VIII International conference on KMC in Quebec
  32 countries were present; more than 100 lectures were presented, including new perspectives as: **"Kangaroo Mother Care from a Neuroscience Perspective"** or **"the economic Impact of KMC"**

2. Research:
   A collaborative and multi-faceted Québec-Bogotá follow-up of 15 year-old preterm children having experienced the Kangaroo Mother Care was conducted in Colombia during this year.

   Participant institutions:
   - Kangaroo research team (Fundación Madre Canguro, Bogota),
   - Neuroscience and Pediatrics Unit (University of Laval—Quebec City),
   - Imagine Grupo de Informática Grafica & Procesamiento de Imágenes (Universidad de los Andes),
   - Departamento de Imágenes Diagnósticas (Hospital Universitario—Fundación Santa Fe de Bogotá),
   - Unidad de Epidemiologia Clínica & Bioestadística (Javeriana University, Bogota).

   Summary:
   - There are fundamental gaps in the understanding of the relationship between cognitive, motor, sensory and psychosocial capacities of the infant with prematurity and/or low birth weight at birth.
   - An excess of behavioral problems among preterm and very LBW infants when they reach the school level (5 years of age to late adolescence) has been clearly documented.
   - These gaps are limiting our ability to accurately tackle psychosocial misadaptation of children born preterm which may result from subtle cognitive and sensory-motor impairments due to known immature brain malfunctioning.
   - Beyond a fragmented knowledge in which the integrated impairment is not quantified, a multi-disciplinary and multi-centre approach could represent a really new relevant avenue to address the inter-relationship and dependence of basic behavioral functions.

   Objective:
   - To explore cognitive, behavioral neuroimaging and neurophysiological variables and school performance in a sample of 30 subjects who participated in a RCT comparing KMC and "traditional" care in incubators early in life together with 10 normal term (control) subjects, when reaching 14-15 years of age.

Design:
Cross sectional study (observational) for generating and refining hypotheses.

Case study (Alexis):

- 21 year-old, single mother, very low income, secondary education
- Good prenatal follow-through with noted pre-eclampsia and fetal growth retardation
- Baby delivered prematurely by C-section at 31-weeks, small for age, 1.2kg
- Nursery stay 30 days with diseases of prematurity and discharged 1.3 kg.
- Enrolled in KMC upon discharge and kept in kangaroo position at home 39 days until 41 weeks, 2.1kg
  - 1-year follow-up:
    - Wt 7.52kg, ht 65cm, head 45cm
    - Eye exam: normal
    - Diagnosed Cerebral palsy
    - Griffith's score low at 6mos., borderline 1yr
    - Exclusive breastfeeding 3mos
    - Accumulated re-admission 4 times (13days total) for bronchopneumonia
  - At 15 years:
    - Right handed
    - Anthropometric data: weight 37,5 kg, height 146 cm and HP 54 cm
    - 3 readmissions for pulmonary infection
    - Spastic diplegia with unstable walking, especially the right leg
    - Refractive problems (right eye), with discrete nystagmus
    - Never repeats any year at school, good student, practices football and has a low self esteem
    - IQ evaluation: Wisc4 68 (perc 39): Verbal comprehension 26 (perc 68), Perceptual reasoning 23 (perc 59), Working memory 14 (perc 31) and Processing Speed 5 (perc 6)
- Anatomic data at 15 years: Total volume of the corpus callosum and white matter using MRI and DTI

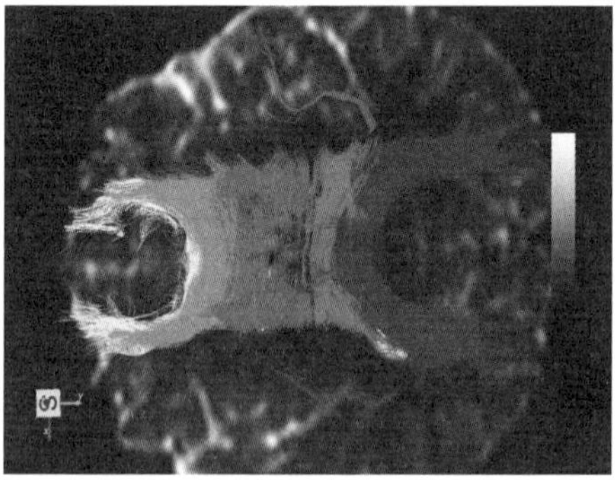

- TMS: Cerebral plasticity
  - Time of inhibition between hemispheres
  - Interhemispheric transfer of inhibition
  - Duration of inhibition between hemispheres
  - Ipsilateral responses (corticospinal) from the hemisphere stimulated
  - Basic excitability of each hemisphere (motor threshold)
  - Synchronization of brain cells (latency of motor responses)

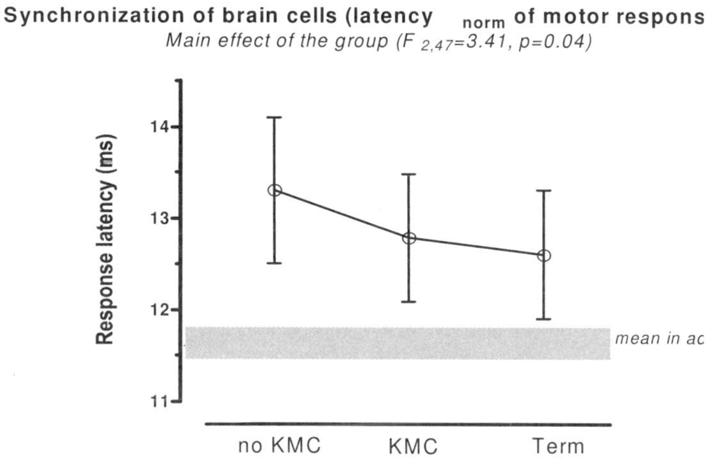

- fMRI: Cerebral activity

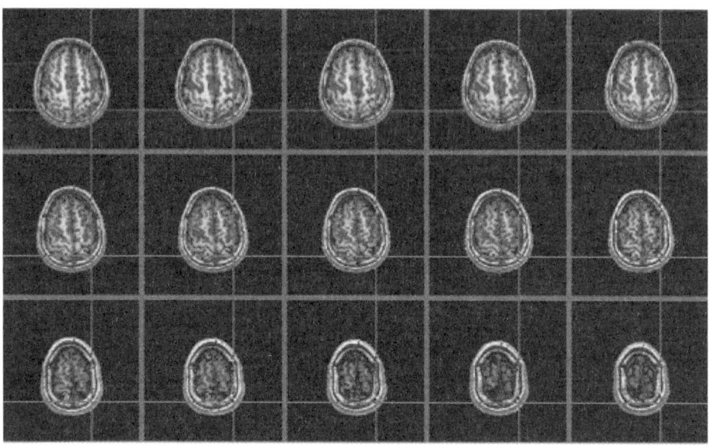

Conclusion:
Results from this pilot, hypotheses-generating exercise allow us to formulate the hypothesis that KMC intervention could be associated with better long term cognitive outcomes and might help to protect the immature brain and to promote better ex-utero maturation. Papers summarizing the results will be published before the end of the year.

These pilot activities should be extended to all subjects (500) who participated in the 1993-1997 KMC RCT. This Colombian cohort is unique, is still accessible and offers an exceptional opportunity to assess the long term impact of a complex, and promising intervention such as KMC.

SECOND PART:

Science of Medicine in the third Millennium: The evidence- based medicine approach

1. What is the reality?
   - Health care needs are unlimited
   - Meeting the needs is a challenge
     - Inequalities in access to health care
     - Limited resources for health care
   - By far, most of the health care expenditure is in the hands of physicians around the world.
   - Lack of efficiency should be regarded as a planetary emergency, jeopardizing the health and welfare of humanity as a whole.
   - Ineffective and potentially dangerous interventions are prevalent.
   - Choices need to be made in a scientific manner
   - False Causal Imputations (poor science):
     - Vaccines and Autism
     - High Power Lines and Child leukemia
     - Weak electromagnetic fields and immune suppression
     - Cellular Phones and CNS cancer

2. An example: the Retrolental Fibroplasia in the premature infant, a Case Study
   - 1940: The routine and unrestricted use of Oxygen in Preterm Infants are saving the infants

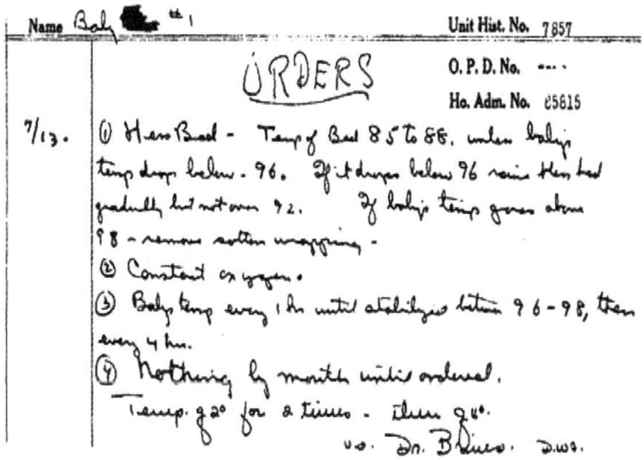
Name Baby [redacted] #1 | Unit Hist. No. 7857
ORDERS | O. P. D. No. ---
Ho. Adm. No. 25815

7/13. (1) Heat Bed - Temp of Bed 85 to 88. unless baby's temp drops below - 96. If it drops below 96 raise the bed gradually but not over 92. If baby's temp goes above 98 - remove cotton wrapping -
(2) Constant oxygen.
(3) Baby temp every 1 hr until stabilizes within 96-98, then every 4 hrs.
(4) Nothing by mouth until ordered.
Temp. of 92° for 2 times - then 94°.
v.o. Dr. Blues. D.W.S.

Facsimile of a medical prescription of continuous oxygen to a preterm twin Boston Lying-In Hospital July13, 1940

- A new disease is detected in 1941: A. Clifford (Boston) describes opacities in both eyes of preterm infants which corresponded to extensive vascular, hemorrhagic and fibrous tissue behind the lens and damaging the retina. It is the first report of Retrolental Fibroplasia (RLF) In 1945 Terry published the first series of 117 cases of RLF all affecting preterm infants
- The question was: there is a relation between Medicinal Oxygen and RLF?
  - Observational "Ecologic" studies highlighted that countries with unrestricted delivery of oxygen (USA and France) had higher rates of RLF than countries restricting the use of oxygen (UK and Australia)
  - This observation was contrary to physiological prevalent theories: neo-vascularization is a response to tissue hypoxia. O_2 administration ought to be protective.
  - 1950's: 2,500 blind preterm infants were produced per year
- Empirical testing the hypothesis of the "protective role" of O_2 in RLF
  - Baltimore 50's: Requested a US $4000 grant to the U.S. NIH to conduct a Randomized Controlled Clinical trial of high versus titrated O_2 in preterm infants
  - Initially rejected because it was an "unethical and scientifically unsound study"
  - 1952-1953: Was finally conducted with high resistance from hospital personnel
  - **Results on the incidence of RLF: High O_2 25%, Titrated (Low) O_2: 0%**
- Conclusion: It was enough to modify for ever the clinical practice? The answer is NO.

The RLF is currently in 2010 the main cause of blindness in Latin America in preterm infants!!!!!!

3. The Evidence-Based Medicine is urgently needed in contrast to the "Eminence" Based Medicine

- Actually there is a body of scientific knowledge (information, explanation & comprehension) produced through basic research (Theoretical or Experimental at Multi-level: Molecular, Cellular, Tissue, Organism; In-vitro/in-vivo with animal models) and Clinical/Epidemiological research (Problem-based; Realistic; Integrative; Interdisciplinary with humans)
- The Evidence-Based Medicine (EBM) is equivalent to "Science in Clinical practice" It Is a systematic approach to clinical problem-solving which integrates the best, current, valid and relevant evidences (produced by scientific research in human beings) in offering choices and recommendations that respects and involves the patients values and preferences
  - *(Sackett DL, Strauss SE, Richardson WS, et al. Evidence-based medicine: how to practice and teach EBM. London: Churchill-Livingstone, 2000)*

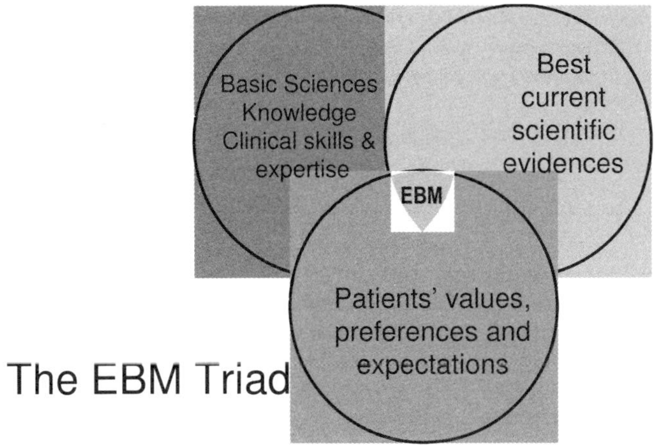

Armstrong, E.C. (2003) Harnessing new technologies while preserving basic values. Fam Sys & Health, (21)4, 351-355.

4. The KMC Intervention & Propagation is the perfect example of the evidence based approach
   - The Kangaroo Mother Care (KMC) method is a set of interventions for providing appropriate health care to preterm and/or LBW infants, based on the so-called kangaroo position (skin-to-skin contact). Since it was first described (Rey 1978) a considerable variability has developed about: a) definition of the target population and of the therapeutic goals, b) time for starting skin-to-skin contact, c) continuity and duration of the kangaroo position, d) feeding strategies, and e) discharge and follow up policies. There were an urgent need to standardize the intervention, based on scientific evidence that supports its benefits and limitations.
   - Research of the Kangaroo Foundation
     - Research on safety of KMC 1989-1992
       - Observational analytical study
     - Research on efficacy and safety of KMC 1993-1996
       - Randomized controlled trial
     - Enhancement of the intervention 1996-2010
       - Explicative research
       - Pragmatic research
     - Translation of research results into health interventions 1994-2010
       - National and international diffusion
       - Training in KMC Method
       - Resistances and quality improvement

- **Evidence based KMC guidelines with recommendations for KMC practice (2007-2009)**

We decided to develop a set of recommendations about the characterization and proper use of the different components of the KMC method, and to support each assertion with a systematic review of evidence.

A multidisciplinary group including authorities, users, parents of patients, and methodological and content experts worked between 2006 and 2007. After defining terms and characterizing components of the intervention, a systematic review of published literature (Medline, Lilacs, hand searching and review of previously compiled bibliographies) was conducted for identifying, appraising and summarizing the evidence regarding the effects, risks, expected benefits and limitations of each component. Evidence based assertions were widely discussed until consensus with each statement was achieved and were evaluated by external peers.

Standardized definitions of KMC and its component, variants, target population, indications and precautions were produced. An evidence-based foundation for each component was developed, identifying the strength of the evidence, the knowledge gaps, the areas of controversy and the needs for further research. Discussion of the basic aspects and the results of relevant research were added to each recommendation together with detailed descriptions of the appropriate implementation of KMC including practical tips. Extensive annexes include tables of evidence and additional information on related topics (mainly in nutrition and feeding strategies). The evidence based-KMC guidelines can be uploaded for free from the Kangaroo Foundation home page, in English and in Spanish. Although sound guidelines should be tailored to local needs and conditions, and this particular exercise is focused in the Colombian situation, it is reasonable to expect that many of the recommendations and certainly most of the evidence appraised will be a useful input for guidelines development elsewhere.

CONCLUSION

- Individual (clinical) and Collective (Public) Health care provision should be based on scientific relevant evidences.
- Not only basic, but clinical and epidemiological research produces evidences that should inform rational medical decision making.
- Physicians need to understand these principles of methodologically sound research to:
  - Be intelligent consumers of research results
  - Be able to collaborate in the production of sound and meaningful evidences
- There is evidence that EBM strategies (including use of Evidence-based Clinical Practice Guidelines) positively impacts health and rationalizes resource utilization, improving efficiency in health care delivery.

- EBM principles and practice should be widely disseminated and implemented as one effort to bring science into clinical practice, improve efficiency, equity and welfare.

Thank you for your time!

PERMANENT MONITORING PANEL REPORT ON LIMITS OF DEVELOPMENT

CHRISTOPHER D. ELLIS, CHAIR
School of Natural Resources & Environment, University of Michigan
Ann Arbor, Michigan, USA

The Permanent Monitoring Panel on the Limits of Development met from 9:30 to 16:40 on Thursday, August 19th. The scope of this year's meeting involved the idea of adaptive capacity from the point of view of urban settlements. The domains included expected forecasts of climate change, energy, water systems, land use change, infrastructure, buildings, agriculture, health of living organisms, and natural resource management. Papers ranged in depth from one domain alone to several domains. These included:

1. **Community Vulnerability, Adaptive Capacity and Resilience—Christopher D. Ellis**—This paper examined social vulnerability in the south eastern United States and its relationship to various natural hazards including those augmented by changes in climate. Ideas of resilience and adaptive capacity of communities were discussed in this context
2. **European Adaption and Mitigation Strategy on Climate Change—Hiltmar Schubert**—This paper discussed models of vulnerability and adaptation in European countries including issues of economic, environmental and social importance. It covered regions considered the most vulnerable, the sectors that are likely to be affected, and outlined four key pillars for addressing adaptation.
3. **The Connections between Environmental Vulnerability and Mobility in Africa—Mbareck Diop**—This paper summarized the vast body of knowledge being generated on climate change effects in Africa. It highlighted the relationship between climate change and mobility arguing the migration will be a necessary outcome of changes to climate.
4. **Climate Change Consequences in Developing Countries: The Mexican Case—González Pozo**—This paper covered the expected outcomes of climate change on Mexico. This included severe flooding inland due to increased storm frequency, intensity and duration, and along coastal areas due to sea-level rise. Other issues included hurricane damage and loss of biodiversity.
5. **Energy, Economy and Environment: The Situation in Argentina—Manolo Borthagaray**—This paper outlined an innovative approach to sustainable energy by combining wind energy, hydrogen production, and thermal power production. It suggested how social and economic benefits relate to the implementation of such a solution.
6. **Planning a Sustainable Development: The cases of ALAGO and AMEG municipal associations—Geraldo Serra—**This paper examined a model for engaging numerous communities over a large region in decision making processes aimed at long-term sustainable development. It adapted the UN

Human Development Index as a way to evaluate success of the plans. It concluded the communities are limited in their ability to expand sustainably unless they coordinate a long-term planning effort.

7. **Contribution of Satellite Earth Observations to Create a Tool for Long-Term Global Monitoring of the Global Vegetation Dynamics—Mikhail Antonovsky**—This paper outlined an approach to using time-series satellite imagery together with a normalized difference vegetation index to improve ecotype identification. This procedure holds significant promise as a way to monitor ecological responses to climate change.

The PMP members concluded that these topics were important and that monitoring of these topics should be continued. In addition several other topics were introduce for examination at next year's PMP meeting. These topics include:

1. Brownfield (contaminated site) Development (Pollution PMP)
2. Urban Sprawl and Greenfield Development (loss of resources)
3. Gated Communities and Class Separation (social implications)
4. Hazard Reduction and Recovery
5. Social Vulnerability

The PMP members are discussing these items over the coming year and will refine and shorten this list before a completed program is established. As a workout and to better prepare our discussion, members are encouraged to exchange views on the above subjects through e-mail. Members that intend to make a formal presentation of their viewpoints are required to inform the Chairman as soon as possible, in order to schedule it at the final program.

POLLUTION PERMANENT MONITORING PANEL—2010 ANNUAL REPORT

LORNE G. EVERETT
Chancellor Emeritus, Lakehead UniversityThunder Bay, Canada and L. Everett & Associates, LLC, Santa Barbara, USA

Pollution PMP—Workshops, Meetings, and Reports 1998–2010

1998	Workshop on Impacts of Pharmaceuticals and Disinfectant By-Products in Sewage Treatment Wastewater Used for Irrigation
1999	Seminar Session on Contamination of Groundwater by Hydrocarbons
1999	Workshop on Black Sea Pollution
2000	Seminar Session on Contamination of Groundwater by MTBE
2000	Workshop on Black Sea Pollution by Petroleum Hydrocarbons
2001	Workshop on Caspian Sea Pollution
2001	Seminar Session on Trans-Boundary Water Conflicts
2000	Workshop on Water and Air Impacts of Automotive Emissions in Mega-Cities
2003	Seminar Session on Water Management Issues in the Middle East
2003	Workshop on Monitoring and Stewardship of Legacy Nuclear and Hazardous Waste Sites
2005	Task Force Meeting on Groundwater Pollution Vulnerability Mapping of Sicily
2005	Workshop on Plastic Contaminants in Water
2005	Developed Groundwater Vulnerability Mapping Proposal of Sicily
2008	Resolving the Nuclear Waste Issue on the Road to Sustainability
2009	Green Chemistry Workshop
2010	Session on Energy & Pollution: Oil Spills and Deep Water Drilling
2010	Session on Pollution: Air, Water and Waste Scientific Criteria
2011	To Be Decided

http://www.federationofscientists.org/PMPanels/Pollution

POLLUTION PMP 2011 PROPOSALS

(1) MULTINATIONAL SUBSURFACE REPOSITORIES: RECENT DEVELOPMENTS AND PROPOSAL FOR 2011 SESSION

Charles McCombie, Arius Association, Switzerland

Energy PMP Potential Co-Sponsor

<u>New Global Interest in Nuclear Power:</u>

- 61 nations have requested support from IAEA to introduce the use of nuclear power, including:
  - Africa (20)
  - Latin America (12)

- Asia Pacific (20)
- Europe and FSU (9)

European Scale Example:

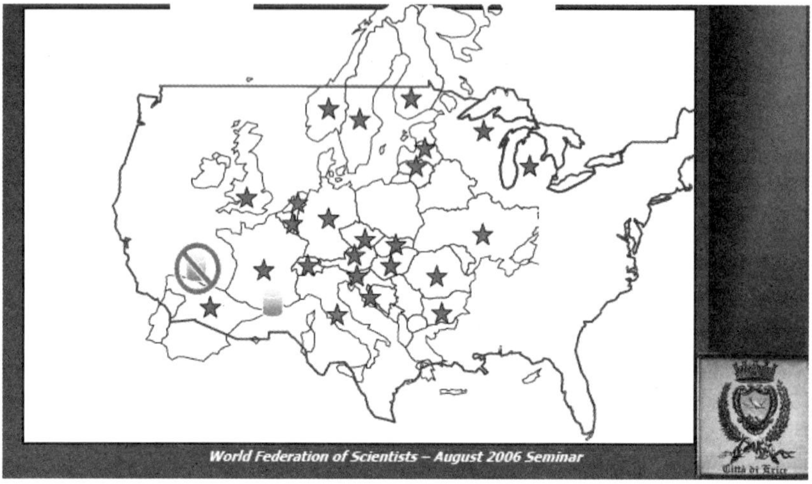

Proposed Participating Countries:

Europe:
Italy, Albania, Serbia, Portugal, Norway, Poland, Belarus, Estonia, Latvia, Ireland, Turkey
Middle East and North Africa:
Iran, Gulf states including UAE & Saudi Arabia, Yemen, Israel, Syria, Jordan, Egypt, Tunisia, Libya, Algeria, Morocco

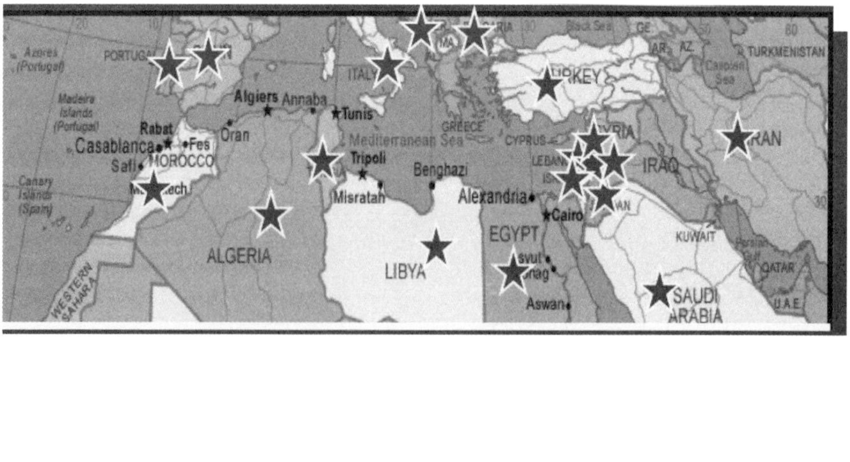

Multinational Solutions and Security

Limited numbers of facilities to be secured:

- Replace disconnected organisations
- Single facility easier to control and monitor

Earlier underground disposal for smaller nations:

- No realistic repository programmes at present

Enhanced security measures (engineered/institutional):

- Ensure highest standards of safety & security
- Encourage harmonisation of standards (e.g., EU: 15 NP States)

Enhanced safeguards oversight:

- Simpler surveillance

Improved financing arrangements:

- General economic advantages of sharing well-known
- Less chance of diversion of security funds

(2) CHEMICALS OF EMERGING CONCERN: PROPOSAL FOR 2011 SESSION

Frederick S. vom Saal, University of Missouri
Stefano Parmigiani, Universita di Parma

Chemicals of emerging concern (CECs): High volume, ubiquitous chemical contaminants in aquatic ecosystems with biological activity at very low concentrations (1 ng/L).

Sources of CECs: Pharmaceuticals, personal care products, cleaning agents, biofuels, food packaging and storage containers, building materials.

- CECs are not routinely monitored in water.
- CECs are not found in defined locations due to point-source emission

Many questions remain: Not all sources are known. Environmental fate is not known. Not all health effects are known.

Plenary Session and Workshop Proposal:

Exposure Pathways—Agricultural runoff, urban waste water, surface water, groundwater, and re-use water.

- What are the strategies of prevention and mitigation
- Impact on aquatic wildlife
- Impact on human health
- New methods of risk evaluation

(3) POLLUTION AND SCARCITY OF WATER: PROPOSAL FOR 2011 SESSION

Frank L. Parker, Distinguished Professor of Environmental and Water Resources Engineering, Vanderbilt University

Clean and Abundant water are Essential:

- Some of the poorest people in the world are also the unhealthiest.
- There are 1.1 billion people without access to improved water sources and 2.2 billion without basic sanitation.
- 1.4 million children die each year from preventable diarrheal diseases.
- Up to 50% of malnutrition is related to repeated and intestinal nematode infections due to unclean water.
- Every $1 invested in improved water supply and sanitation yields gains of $4 to $12 depending upon type of intervention.

What is Virtual Water?—The amount of water used to create a product.

Product	Gallons of Water Used to Produce one Pound of Item
Beef	1857
Sausage	1382
One Hamburger	634
Fresh Cheese	371
Cherries	185
Cup of Tea	9
Pair of Jeans	2900
Cotton T Shirt	766

Issues:

Terrorism and Water Infrastructure

What is at Risk? Water Resources may be potential targets for physical distribution, bio-terrorism and cyber attacks.

- 75,000 dams, thousand of miles of pipe, aqueducts, water distribution lines, and sewer lines; 160,000 public drinking water facilities and 16,000 publicly owned waste water treatment facilities.
- Other risk areas: Raw water systems, water intakes, water treatment, water distribution and interlocking dependencies.

How to Mitigate: Update and re-enforce infrastructures, improved monitoring, interconnecting water supply systems, implement protection measures against cyber and vandalism.

Potential Pollution and Scarcity of Water Topics for 2011 WFS Seminar, include:
- Overview and UN Global Water Partnership Program
- Water Resource Disputes
- Global Water Scarcity Issues/Sustainability
- Sanitation Problems
- Heath Issues and Infectious Disease
- Water Infrastructure and Terrorism

POLLUTION PMP MEMBERS

PMP Members/Associates:
- Dr. Gina Calderone—ECC, USA
- Dr. Lorne Everett—Haley & Aldrich, Inc.
- Distinguished Professor Frank Parker—Vanderbilt Univ. USA
- Dr. Richard Ragaini—LLNL, Retired, USA
- Dr. Charles McCombie—Arius Association, Switzerland
- Dr. Fred vam Saal—University of Missouri, USA
- Dr. Stephano Parmiagiano—University of Parma, Italy

Official Guests for August 2010 PMP Meeting:
- Dr. William Cosgrove—V.P. World Bank/World Water Council
- Dr. John Day—Professor Emeritus Louisiana State University
- Dr. LuAnn White—Tulane University Center for Applied Environmental Health
- Dr. Tom Isaacs—Stanford University, USA
- Dr. Thomas Grahame—U.S. DOE, Washington, DC, USA

Guests;
- Dr. Franco M. Buonaguro—Istituto Nazionale dei Tumori, Napoli
- Dr. Muheeb Awawdeh—Yarmouk University, Jordan
- Barbara Tannenbaum—Upton, New York

REPORT OF THE ENERGY PERMANENT MONITORING PANEL

WILLIAM FULKERSON, CHAIR
Institute for a Secure and Sustainable Environment, University of Tennessee
Knoxville, USA

JEF ONGENA
Plasmaphysics Laboratory, Ecole Royale Militaire
Brussels, Belgium

CARMEN DIFIGLIO
Office of Policy and International Affairs, U.S. Department of Energy
Washington, USA

The year's events caused significant changes in our planning. We had started off thinking about continuing our efforts begun in 2009 focusing on China and on sustainable well-being. This was to involve coordination with the Pollution, Water, Limits of Development, and Climate PMP's, and we thought to expand the activity this year to include India and maybe Vietnam. The China session in 2009 was successful, and we added Zhang Xiliang as a new member of the Energy PMP. However, the theme for the Seminars in 2010 was directed at the Role of Science for the 3rd Millennium, and our planning changed to accommodate it.

The Energy PMP organized three plenary sessions for 2010. These were:

- Passive Safety in Nuclear Power Plants, Session #2 requested by Professor Zichichi,
- Oil Spills and Deep Water Drilling, Session #3, organized by Carmen Difiglio with the Pollution PMP, and
- Science for Sustainable Energy Systems, Session #4.

In addition, Energy PMP Members helped attract several speakers to the Climate Sessions.

ENERGY PMP MEETING, AUGUST 19, 2010, DIRAC AUDITORIUM

Thirty-two people attended the meeting. The agenda is attached, as is a list of participants. Eight interesting talks were presented, and each deserves a line or two in this report, but the abstracts and/or visuals used in the presentations will be on the Energy PMP web site (http://energypmp.org/) maintained by Dick Wilson (wilson5@fas.harvard.edu) to whom the PMP is indebted.

Carmen Difiglio discussed the co-benefits of climate policies by reviewing a number of recent research studies. Difiglio described an International Energy Agency (IEA) study that analyzed how to achieve a 50% reduction in energy-sector CO_2 emissions by 2050. Policies would be needed to stimulate significantly increased energy efficiency and to decarbonize the power, industry and transportation sectors. Very significant additional investments would be required. However, based IEA's analysis, reduced fuel expenses would likely repay the required investment. Therefore, the co-benefits of climate policies would justify most of the policy actions without counting any benefits from reducing greenhouse gas concentrations. In addition to the financial payback, these measures also reduce sulfur, NO_x, PM, VOCs, mercury and other air pollutants resulting in improved health and other environmental benefits.

Akira Miyahara discussed basic science needs for advanced energy technologies being pursued in Japan. He introduced fusion science particularly the 14 MeV neutron sources being prepared in the framework of ITER-BA, but he emphasized more about needs of fusion materials science to meet the requirements of fusion commercial reactors. In fission Japan is still pursuing ion exchange enrichment that may have some advantages for proliferation reduction, U from seawater as described in 2009 by Tamada, and fast reactors with pyroprocessing of spent fuels. The focus is on proliferation resistance, safety and resilience (natural disasters). In addition, much attention is being paid to such diverse topics as forestry science and ultra high-field science using Exa Watt Laser.

Adnan Shihab-Eldin gave two papers. One covered clean energy options and strategies for oil producers. He gave a very comprehensive analysis of Gulf producing states that see efficiency, renewables, and nuclear as important particularly with rapidly rising electricity demand. Nuclear looks surprisingly good even compared to gas under low discount rates, but it has enormous baggage. Renewables, mostly solar, also look promising and targeted share of total electricity is in the range of 3 to 20% by 2020. A European super grid including the Gulf is an interesting prospect. Since most electricity will come from oil and gas CCS is being investigated particularly with respect to enhanced oil recovery.

Adnan's second paper was on world oil outlook from producer's perspective. Oil use declined by some 4.5 mbd since 2006, but is beginning to recover. Oil peaking analyses often ignore reserve growth. Oil is viewed as a black sheep because of climate change and security. No one knows future oil prices, but they may rise from about $70/b today to $120/b by 2030. Investments in advanced technology including renewables may bring oil prices below $60/b, however.

Dick Wilson gave a very interesting analysis of climate/energy legislation in the U.S. The Senate is not close to a bill, and it won't be pursued for the remainder of this year. That may mean not in the next two years since the Senate is likely to be much more Republican next term. Various proposed bills were discussed, but the Kerry-Lieberman proposal is a massive piece of legislation similar to the Waxman-Markey bill that passed the House more than a year ago. Much simpler legislation focusing on the front end of the fuel cycle never got anywhere.

Bill Fulkerson talked about the Asilomar, CA international conference on Climate Intervention Technologies (i.e., Geoengineering). Many notables including Nobel Laureate Paul Crutzen and the late Steve Schneider attended the conference. It was organized by Mike MacCracken. The Conference concluded with a set of rules for conducting climate intervention research. The Royal Society adopted a similar set.

Juan Manuel Borthagaray talked about transforming the wind swept deserts of Argentina into hydrogen generation sites using wind power. He speculated that it might be easier to transport hydrogen by truck than build power transmission lines. The hydrogen would supply electric generation in urban areas.

Jef Ongena and Akira Miyahara talked about the status of controlled fusion research including both magnetic and inertial. This followed the outline of their plenary session talk but with some more details. ITER is under construction at Cadarache, France. Laser inertial fusion experiments are underway at the National Ignition Facility at Livermore National Lab, and efforts to achieve ignition are expected by the end of the year or early in 2011.

DISCUSSION OF PLANS FOR 2011

Fulkerson led a discussion of possible topics and ideas for the Energy PMP for 2011. To the list of topics listed in the Agenda, the discussion identified more ideas including: a workshop on nuclear expansion around the world proposed by Bob Budnitz, Climate engineering e.g., extracting GHG's from the atmosphere, and a debate on climate sensitivity proposed by Chris Monckton.

The revised list was sent out via email to attendees who were asked to vote on their first, second and third choices. The list in the approximate order of voting is Table 1.

Table 1. Voting results for topics for 2011 (in approx. temporary order)

- Issues with Nuclear fission expansion
  - In depth discussion for a selected number of countries (~12)
- Pursuing silver bullets
  - The Innovation process
- Sustainable Transportation
  - Liquid fuels
  - Hydrogen
  - Energy storage
  - Electric vehicles
- Climate sensitivity debate—A debate between ICPP supporters and people with contrary views.
- Shale gas
- Protecting energy systems (i.e., hacking, incompetence, natural disasters, sabotage)
- Climate engineering
  - e.g., Extraction of GHG's from the air
- Continuous efficiency improvement
- New emerging issues, e.g., Gulf Oil spill
- Nuclear waste storage

- Nuclear Fusion
- Universal sustainable electrification

Dropped from the list before voting was the topic "China-India-Vietnam-etc. on sustainable energy and well-being."

The top two ideas on the list were almost identical in vote count, and these two were well above the next best on the list. Bob Budnitz proposed the first one on nuclear expansion and the second one on discovering silver bullets was suggested by Bruce Stram.

After the meeting Bob Budnitz provided the following description of his topic.

"The "emergency" is that unless nuclear power expands significantly not only in the developed world but also in the developing world, major energy problems will continue to occur worldwide from our increasing dependence on petroleum and natural gas. Today, about 25 countries in the developing world have announced their interest in installing nuclear plants for the first time—these are countries that never have had a nuclear power program, and in many cases do not today have the technical, infrastructure, or manpower capabilities to do so. The nuclear reactor/nuclear power community is both excited about the strong possibility of rapid expansion, and worried about whether many of these developing countries are capable of mounting a sensible program—a nuclear power program that can assure safety, security, and economic operation.

The purpose of organizing an extended workshop on this subject is to explore the many issues surrounding the possible expansion of nuclear power to developing countries. The principal questions to be addressed are:

- What are the conditions under which such expansion should be encouraged?
- Are there certain reactor and fuel-cycle technologies that are better suited to such countries, and other technologies that are not?
- Are there international institutional arrangements that need to be developed or enhanced to facilitate this expansion, but also to discourage it in countries that are not yet ready?

Invitees should include experts from the advanced countries that might supply the reactors, others from selected developing countries who have expressed interest, and still others knowledgeable about international arrangements (IAEA, OECD-NEA, World Nuclear Association, Nuclear Suppliers Group, etc.) who are in a position to advise on what arrangements are most likely to be workable. A workshop like this should have 20-25 attendees, and could not accomplish its objective in less than 3-4 full-uninterrupted days of discussion. A serious constraint to this idea is its expense. Cost sharing partners would need to be found."

Bruce Stram's idea of pursuing silver bullets that might substantially reduce the cost of low carbon energy services ranked almost as high in vote count as Budnitz's idea on nuclear expansion. The Stram proposal is attractive in the abstract because of the enormity of the potential gains in human welfare. But can the WFS play a role that would in fact supplement constructively the many energy R & D efforts, most governmentally driven, already underway or planned? The IEA estimates that these efforts are underfunded by a factor of 3 at least. Our first thought was to organize an extended workshop of representatives from nations with significant energy RD&D and focus on this gap issue. After discussion with Claude Manoli it was decided that such a workshop would be too grandiose and expensive for WFS to attempt. An alternative proposal is being conceived and discussed.

A debate on climate sensitivity suggested by Chris Monckton seemed to attract considerable interest throughout the Seminars, but it is not a direct responsibility for the Energy PMP to pursue. Also, Fulkerson pointed out that a debate can't happen if one side of the issue refuses to participate. It has been difficult to attract leaders of the climate modeling community to the Seminars. Subsequent to the Energy PMP Mtg. Fulkerson suggested by email that such a debate, if it were to be organized, should be focused on research ideas (i.e., measurements, analyses and "experiments") designed to increase understanding in the Galilean sense. The debate should be about the importance of these research ideas.

Also, subsequent to the Energy PMP meeting Carmen Difiglio suggested a joint Energy PMP/Pollution PMP plenary session on shale gas around the world. At the same time Lorne Everett suggested a joint effort on nuclear waste storage and sequestration. Both of these ideas will likely be pursued.

PMP REPORT FOR COSMIC OBJECTS

WALTER F. HUEBNER
Southwest Research Institute
San Antonio, USA

Progress in mitigation of potentially hazardous objects (PHOs: asteroids and comet nuclei that pass the Earth's orbit at a minimum orbit intersection distance (MOID) of 0.05 AU or less) was the starting point of our discussions. Several options for mitigations were discussed. Among them are countermeasures to deflect or destroy PHOs in space by means of (1) hypervelocity impactors, (2) conventional (chemical) explosives, and (3) nuclear explosives. The effectiveness of these impulsive measures were contrasted to slow, long-term deflection methods that require years or even decades of continuous action, such as gravity tractors, solar sails, mass drivers, the Yarkovsky effect, etc. Slow deflection methods also require matching the orbit of a rocket to that of the target PHO, similar to a rendezvous, to deliver and implement the mitigation procedure over a long period of time. Orbit matching will also require many years, if not decades, to accomplish. In addition to requiring identification of PHOs decades in advance of a potential collision with Earth, none of the slow deflection methods are sufficiently advanced in development for implementation. Our conclusion, in agreement with other studies[1,2], was that impulsive methods clearly were to be preferred. We also agreed that *mitigation*, which implies alleviation or lessening of effects, was not the best terminology to use. Instead we will refer to *countermeasures*. Clearly, we do not mean to *lessen* the effects of a collision of an asteroid or a comet nucleus with Earth, we want to *avoid* such a collision!

Evacuation of potentially affected areas has also been suggested. While evacuation should always be included in PHO mitigation issues, it does not prevent a collision and it is questionable how effective it can be. For example an impact in a sea or in an ocean near a coastline can produce a tsunami over very wide areas

Much progress has been made in detecting, tracking, physically characterizing, and cataloging near-Earth objects (NEOs: asteroids and comet nuclei that cross the Earth's orbit within 0.3 AU). PHOs are a subclass of these objects that pass the Earth within 0.05 AU.[3] Figure 1 shows the relative probability of a collision of a PHO as a function of its seize and the warning time. Warning time is the time from determining that an object may collide with Earth to the actual impact. Clearly, the small objects of Tunguska size are among the most common objects. The Tunguska object caused an explosion in the upper atmosphere over Tunguska (Siberia), on June 30, 1908, and destroyed an area of about 2,000 square kilometers. Note that the Tunguska object is now thought to have been only about 25 to 30 meters in effective diameter. There was no warning time.

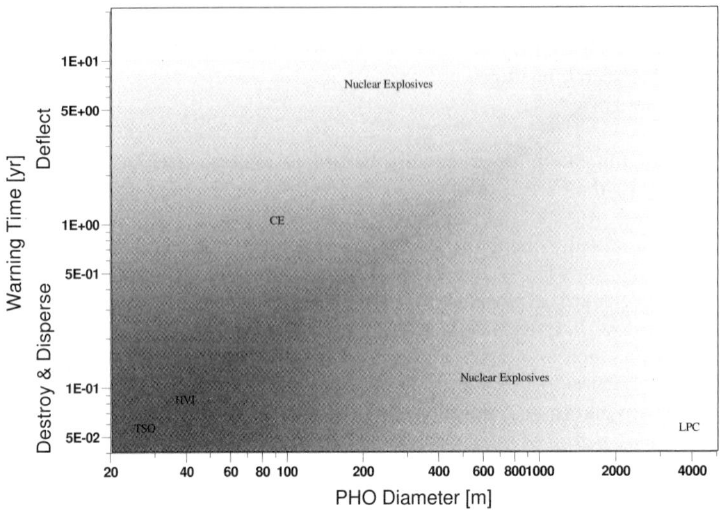

Fig. 1: Warning time versus seize of potentially hazardous objects. The shading indicates qualitatively the probability of an impact of an object on Earth: the darker the more likely a threat to Earth. TSO in the lower left corner indicates the Tunguska-seized objects. LPC in the lower right corner indicates long-period comet nuclei. The ordinate value of 5E-02 years is on the order of one week. Potential countermeasures are indicated as follows: HVI (hypervelocity impactors) can be used in a flyby mode for small objects (lower left corner), CE (chemical explosives may be used for somewhat larger objects if the warning time is on the order of a year or longer. Otherwise nuclear explosive have to be used, most likely in a standoff mode.

It is believed that about 90% of all objects larger than 1 km in effective diameter have been identified. New telescopes are coming on line that will have the power to find objects down to about 140 meters in effective diameter. It will take more than another decade to find 90% of objects larger than 140 m. Meanwhile, the smaller objects that will cause more regional or local devastation, rather than worldwide catastrophes, may not be found except very late on their collision course with Earth. If we want to defend against such objects we will have to be ready to launch countermeasures on very short notice, possible on a scale of a few weeks or even days. Impulsive countermeasures are the only ones that can be used in such a case.

Long-period comets are usually ignored in scenarios of defending Earth. They are usually large objects; tens of kilometers in seize, and come from the outer solar system, providing high velocities and therefore large kinetic energy. These comets are not confined to the ecliptic, they come from all directions and are therefore harder to find and are extremely rare. However, because of their high kinetic energy they are perhaps the most dangerous objects that could cause extinction of all human and mammalian life. We note here that several comets passed the Earth at close distances in the past several hundred years: Comet C/1983 H1 (IRAS-Araki-Alcock), which has an orbital period of 963.22 years, was discovered 27 April, 1983. It passed the Earth only two weeks later, on 11 May, 1983, at a distance of 0.0312 AU, thus, falling into the group of PHOs. In spite of its large seize, it was difficult to discover far in advance because it came at us from behind the Sun. Two other comets passed Earth at even closer distances: Comet D/1770 L1 Lexell passed Earth at a distance of 0.0151 AU and 55P/1366 U1 Tempel-Tuttle at a distance of 0.0229 AU. These data strengthen the reasons to develop countermeasures against such comets even when they appear with very little warning time.

We understand how to design and operate the three impulsive methods (hypervelocity impactors, conventional explosives, and nuclear explosives) mentioned above. What we do not know is their effectiveness on various classes of PHOs. The effects will be different on metallic (e.g., iron-nickel), stony, or carbonaceous asteroids and on comet nuclei. The relevant properties of these objects, e.g., their internal composition and structure are not well known. Some experiments with hypervelocity impactors (v > 10 km/s) and chemical explosives can be carried out in the laboratory with materials simulating asteroids and comet nuclei. One hypervelocity impact experiment has been carried out by NASA's Deep Impact mission, which released an impactor during a flyby of Comet 9P/Tempel 1 at a speed of 10.2 km/s relative to its nucleus to strike it on 4 July, 2005. Although this comet was not on a collision course with Earth, the mission might be considered one of the first hypervelocity impacts to test and develop such techniques. The effects of nuclear explosives can be modeled by sophisticated computer codes (see below).

Another topic of discussion, suggested by A. Zaitsev, concerns the political issues. Who will carry out the countermeasures? It was suggested that countermeasures be carried out under the auspices of the United Nations. Countermeasures using nuclear explosives can be developed and implemented, possibly jointly, by several countries in the eastern and western hemispheres. However, we will also need implementation in the southern hemisphere to be more efficient. An open letter addressed to the Secretary-General of UN Ban Ki-moon was presented to Prof. Zichichi for further consideration. In the letter we proposed a resolution, calling for the start of a practical realization of measures in defense of the Earth from asteroid-comet dangers.

Finally the discussions turned to the European efforts of the Don Quixote mission. A. Galvez pointed out that this concept is a test of a mitigation pre-cursor mission to further our understanding of asteroid impact hazards. It is a two-spacecraft mission consisting of Sancho, a science spacecraft, and Hidalgo, an impactor spacecraft. Sancho is intended to arrive at the target asteroid a few months ahead of Hidalgo to perform reconnaissance and measure size, shape, mass distribution, bulk density, large-scale mineralogical composition, the internal structure using seismology, the ratio of the moments of inertia, thickness of the regolith, and other regolith properties. Hidalgo will impact the asteroid. The impact will be observed by Sancho, which will remain with the asteroid for several months to provide information on the regolith properties, interior structure, and mechanical properties, in addition to making a direct measurement of the dynamical response (change in velocity, Δv, of the asteroid) to the kinetic impactor. The results will provide crucial information for all further development of countermeasure strategies, including numerical modeling. The mission has been put on hold.

ONGOING ACTIVITIES OF THE PMP

We continue to implement the database for geophysical and geological properties of near-Earth objects (NEOs). It can be accessed at http://neodata.space.swri.edu. It contains four major components:

1. Observational Data.
2. Material Properties Data.
3. Instruments and Mission Development.
4. Dissemination and Public Outreach.

We also participated in various international conferences at which papers about countermeasures against PHOs were presented and then published. In particular we mention the Tunguska conference held in Moscow[4] and the Asteroid Comet Hazard (ACH-2009) conference held in St. Petersburg.[5, 6, 7] The proceedings of this conference are in press and will be published by the Russian Academy of Sciences, St. Petersburg.

On our urging, several task groups have been established by the International Astronomical Union (IAU) to improve data from asteroids and comet nuclei. This effort is consistent with calling for international participation since the PHO threat is an international, i.e., worldwide problem.

A. Cellino pointed out the call for proposals by the European Union for efforts to participate in problems with countermeasures against PHOs. Proposals require institutions from three European Union member countries to submit proposals in which other countries should participate (see below).

Finally, discussions were carried out about improving risk analysis associated with impacts of asteroids and comet nuclei with Earth. It was emphasized that such analyses will have to be carried out carefully and that an analysis of sudden disastrous events such as the collision of an asteroid or comet nucleus with Earth and the associated enormity of destruction of infrastructure cannot be compared simply with risk analyses of other long-term world emergencies such as disease, poverty, lawlessness, and war. Comparisons with other sudden and unpredictable disasters such as earthquakes, volcanic activity, and floods are more appropriate. The major difference between the latter disasters and an impact by an asteroid or comet nucleus is that we have the technology to avoid impacts by PHOs.

NEW ACTIVITIES OF THE PMP

Before and after the seminar, we held a PMP meeting at Erice. Present were Alberto Cellino (Observatory Torino), Andres Galvez (Advanced Concepts and Studies Office, ESA), Catherine S. Plesko (Los Alamos National Laboratory), James D. Walker

(Southwest Research Institute, San Antonio), and Anatoly Zaitsev (Planetary Defense Center, Moscow). Discussions centered on how we should participate in the European Union call for proposals.

J. Walker pointed out the unique facilities at the Southwest Research Institute (SwRI) to accelerate masses of about 1 g to speeds exceeding 10 km/s. No other facility in the world can reach such velocities of 1 g masses. There are questions in the literature about the magnitude and importance of the momentum enhancement by the recoil of cratered material on the PHO during a hypervelocity impact. This facility could be used to clarify and quantify the enhanced momentum transfer during a hypervelocity impact on materials simulating asteroids and comet nuclei. At the same time impactor materials and shapes could be studied to maximize the effect.

As was pointed out above, the effects of nuclear explosives on an asteroid or comet nucleus has never been tested. However, as C. Plesko suggested, sophisticated radiation and neutron transport codes can be used to model such interactions. While radiation from a standoff nuclear explosion may be deposited in a PHO to a depth up to about 1 cm, neutrons can penetrate much deeper. Both, radiation and neutrons will heat the material and cause it to vaporize and impart momentum to the object during this ablation process. Since such computer simulation codes are classified as secret in all countries that have such capabilities, including Russia and the USA, it is not possible to describe any details. However, if two countries made such simulation calculations and then compared the results, e.g., the momentum transfer, confidence in the simulations would be gained, because the computer codes are totally independent of one another. We will search for European Union partners and suggest to them that our capabilities be included in their proposal. In the case of nuclear explosives this would mean contributions from both Russia and the USA.

REFERENCES

1. Near-Earth Object Survey and Deflection, Analysis of Alternatives (NASA White Paper to U.S. Congress, March, 2007.
2. Defending Planet Earth: Near-Earth Object Surveys and Hazard Mitigation Strategies: Final Report. Committee to Review Near-Earth Objects Surveys and Hazard Mitigation Strategies Space Studies Board; National Research Council of the National Academies. The National Academies Press, Washington, D.C. (2010).
3. 1 AU (astronomical unit) = 150,000,000 km. For comparison, the Earth – Moon distance is about 400,000 km.
4. Huebner, W.F., Johnson, L.N., Boice, D.C., Bradley, P., Chocron, S., Ghosh, A., Giguere. P.T., Goldstein, R., Guzik, J.A., Keady, J.J., Mukherjee, J., Patrick, W., Plesko, C., Walker, J.D., Wohletz, K. (2009) "A comprehensive program for countermeasures against potentially hazardous objects (PHOs)." *Solar Syst. Res.* **43**:348-356.
5. Zaitsev, A.V., Koroteev, A., Liaschuk, B., Popov, S. "The level of rapid response reaction of the planetary defense system." Proceedings of ACH-2009, A.M. Finkelstein, W.F. Huebner, V.A. Shor (Eds.), in press (2010).

6. Zaitsev, A.V., Klapovsky, A.A. "On the approach to form international-legal foundations to ensure planetary defense." Proceedings of ACH-2009, A.M. Finkelstein, W.F. Huebner, V.A. Shor (Eds.), in press (2010).
7. Huebner, W.F., Boice, D.C., Bradley, P., Chocron, S., Clement, R., Ghosh, A., Giguere, P.T., Goldstein, R., Guzik, J.A., Johnson, L.N., Keady, J.J., Mukherjee, J., Patrick, W., Plesko, C., Tapley, M., Walker, J.D., Weaver, R.P., Wohletz, K. "The engagement space for countermeasures against potentially hazardous objects (PHOs)." Proceedings of ACH-2009, A.M. Finkelstein, W.F. Huebner, V.A. Shor (Eds.), in press (2010).

ANNUAL REPORT PERMANENT MONITORING PANEL ON MITIGATION OF TERRORIST ACTS

ALAN LEIGH MOORE, JR.
The University of Alabama in Huntsville
Alabama, USA

1. Background
   a. The Permanent Monitoring Panel on Mitigation of Terrorist Acts (PMP-MTA) was established in approximately 2001 in parallel with the organization of the PMP on Motivation of Terrorist. Since that time the two PMP's have had a close working relationship, one focused understanding the issues associated with the creation of terrorists and their resulting activities, while the PMP-MTA is looking at the mitigation of the terrorist activities. This close working relationship will continue.
   b. Focus. The early primary focus of the PMP-MTA was understanding the evolving chemical, biological, radiological and nuclear (CBRN) posed by the terrorists, identifying and defining possible mitigation measures, and recommending actions that could be implemented by governments, private agencies and individuals to limit the effects of a terrorist event.
   c. Participation in the group has evolved as specially qualified experts have been invited to the meetings to address specific issues. A core set of individuals have sustained the effort and will reach out to experts with unique disciplinary expertise to support future activities.

2. Threat and Risks
   a. It is well understood that terrorists are watching the activities of targeted nations, listening and understanding their security measures, while developing "gaming changing" innovative approaches at security gaps that can be exploited with a high probability of success and with acceptable risk to the terrorists themselves.
   b. It is important to note the terrorist are capable of working within our countermeasures decision cycle, which is a significant concern. The terrorist organizations are generally small agile cells of individuals controlled by a limited and streamlined command and control structure, which is just the opposite of the bureaucratically structured organizations associated with government counter-terrorist organizations. The element of surprise and timing rests with the terrorist resulting in an increased risk for the target population.
   c. To counter the potential terrorist and terrorist acts, governments have spent significant funds to establish organization and implement public security measures. The costs of these activities have placed a heavy burden on a world economy still recovering from the financial meltdown of 2008. In Pakistan alone it is estimated they have lost approximately one-third of their Gross Domestic Product to countering terrorist activities.

3. Members of the PMP have indicated it is easier to obtain the materiel, construct, deploy and employ a WMD in lesser developed nations than in developed nations due to the developing nation's weaker security measures, weaker national command and control, greater threat of terrorist activity, relative political instability and reduced logistics (transportation of the device).

4. PMP-MTA Roadmap for the Future
   a. The terrorist threat is dynamic and evolving and will continue to do so for the foreseeable future. Today the fundamental science associated with CBRN threats and associated mitigation measures and understood, at least at the scientific level.
   b. Continuous activity will be required on the part of the PMP-MTA to detect, identify future terrorist asymmetrical threats such as unconventional agents (synthetics, genetically modified), deployed and employed in uniquely innovative applications with multiple non-linear potential outcomes. This will require, in addition to other steps, the revision and validation of models and predictability of measurable effects if we are to understand the potential impacts and mitigate terrorist acts.
   c. The assumption that terrorist are poor and ill-educated is not accurate. Some are well educated and employ tried and proven techniques in innovative ways to achieve their desired results as has been documented by the PMP on Motivation of Terrorists.

5. Mitigation of Gaps
   a. To mitigate the threat of terrorist acts will require an inter-disciplinary scientific approach. People working and thinking innovatively rather than bureaucratically. This will require the involvement of students, science fiction and thriller writers, and other creative thinkers, as well as the "Red Teaming" of results by inter-disciplinary group of subject matter experts.
   b. Consideration must be given to a mix of science and technology applications, and given appropriate funding, the development and use of 3-D computer modeling of scenarios/effects to predict the effectiveness of potential mitigation measures prior to a terrorist event.

6. Implementing Technology
   a. A concern has been the availability of supporting or implementing technology to affect the mitigation effort and to provide the command and control required in major population areas. Our latest survey has indicated advances have taken place in communications technology that can be implemented today by governments should they desire to do so.
   b. It should be noted that emergency response to major attacks has shown severe deficits in industrialized and developing countries alike. Some of the reasons for this deficiency are the lack of advanced planning, training of personnel and the implementations of appropriately integrated and interoperable communications means between national government authorities and first responders.

c. The PMP has noted the national and political will to implement solution sets are generally limited, take time and can be very expensive.
d. As noted above, no national or regional communications architecture exists today, much less inter-connectivity and interoperability to inform a worried public. Such a communications network is critical to retaining political stability while mitigating the terrorist act. However, with the establishment of a communications support system, special care in the design and architecture must be taken into incorporate cyber security measures. Failure to do so during the design phase could result in the system being taken off the air at a critical time, further enhancing the effects of the terrorist act and negatively impacting the mitigation measures.

7. Organizing for Success
a. To address the every evolving terrorist threat and innovative terrorist acts will require the development of a more comprehensive multi-disciplinary international team, as discussed below.
b. The innovative schedule of the terrorist is not constrained and continues at a significant rate. To keep up with the evolution of the threat the PMP-MTA is considering the conduct of multiple meetings a year. It will be necessary to conduct some of the meetings at politically neutral locations to allow access to wide range of scientific expertise.
c. The PMP-MTA is actively investigating an outreach effort to existing regional organizations whereby the PMP would act as scientific advisor or knowledge base for the organization. Such an activity, in addition to alternative funding, would have a number of advantages for all concerned as noted below.
d. "Science in Diplomacy": Science can provide advice to inform and support foreign policy objectives. "Diplomacy for Science": Diplomacy can facilitate international scientific cooperation. "Science for Diplomacy: Scientific cooperation can improve international relations."[1]

8. The Way Ahead
a. As noted earlier, it is critical the PMP-MTA build a stronger and broader multi-disciplinary (physicists, biologists, political scientists, psychologists, social scientist, physicians, etc.) scientific team focused on mitigating the primary consequences of terrorist acts. This may take place within the World Federation of Science in coordination with other PMP organizations or as an element of an out-reach program to existing regional organizations and agencies.
b. Communications
   – Architecture/cyber security development will be an important product that will evolve over the coming years. The architecture must be tailored to the governments and regions involved, but must be interoperable with commonly understood data sets, in a non-proprietary environment.

---

[1] The Royal Society and the American Association for the Advancement of Science, *New Frontiers in Science Diplomacy*, January 2010. Dr. Charles Ferguson presentation, Erice, Sicily, August 22, 2010.

- It will also be necessary to establish a PMP-MTA data repository. This may be at the World Federation of Scientists or as an adjunct to an existing internationally accessible data network.
- The expansion of the multi-disciplinary scientific team will result in an expanded self-organizing academic network which will further contribute to the effectiveness and utility of the PMP efforts.

c. National governments have spent extensive amounts of money to counter the terrorist threat and to mitigate the effects of an act if the counter-terrorist efforts were unsuccessful. Efforts will be taken to leverage existing national expenditures across CBRN mitigation efforts so as to enhance nation's return on investment.

d. The identification of mitigation opportunities and scientific approaches will be a limited value unless the involved government and first responders received validated technical training using approved scenarios. Such a training program and the periodic revisions thereto, must emphasize the scientific method.

9. Near Term Program Execution

a. The focus of the PMP-MTA for the next year will be on the development of meaningful measurable results in the identification of new and evolving threats, the defining of mitigation actions and opportunities, and the definition of mitigation communications and training needs.

b. Efforts will also be expended to identify and expand the scientific multi-disciplinary team/network and out-reach to existing regional organizations while establishing a web-based Center of Excellence.

c. To remain current and to track terrorist events in real time will require two or more face-to-face meetings per year with transnational scientific experts.

d. To fund the PMP-MTA scope of work will require the identification and development of supplemental funding sources. This will be a priority effort.

10. The PMP-MTA looks forward to a busy and meaningful year, while fulfilling and expanding its remit.

INFORMATION SECURITY PMP REPORT

HENNING WEGENER
Ambassador of Germany (ret.)
Madrid, Spain

JODY R. WESTBY, ESQ.
CEO, Global Cyber Risks LLC
Washington, USA

The work of the PMP during the 2009-2010 period was, in part, determined by the work taken in hand in the preceding period. The Group continued to expand its international networking and activities, mindful of the frontierless nature of its assignment. During the PMP meeting held on August 19th, the group received a report from Andrey Krutskikh on his work in the information security area with the United Nations First Committee.

The members of the PMP undertook to ensure the widest possible distribution of last year's Erice Declaration on Principles for Cyber Stability and Cyber Peace and carried forward its work on the Cyber Security Hard Problems List ("Hard Problems List"). The PMP reviewed some newly published documents on cyber security research areas and will update the Hard Problems List accordingly. A revised document will be circulated to PMP members and the document will be finalized by January 1, 2011. PMP members intend to make use of their international networks to bring the document to the attention of the international research community in order to purposefully advance research in areas where it is most necessary.

A major portion of the work of the Group was dedicated to drafting the various chapters of the book entitled *The Quest for Cyber Peace,* which is to be co-published with ITU Secretary General, Dr. Hamadoun Touré, and his organization. Select members of the PMP held a meeting in May in Geneva to advance work on the development of the book. During that same week, the PMP co-chairs represented WFS at the World Summit on the Information Society 2010. The PMP reviewed the initial draft of the book this week in Erice, with the active participation of the Secretary General. PMP members will revise their contributions and submit final text by October 31, 2010 and, following a final round of edits, will finalize the book by November 30, 2010. The book will be published in early 2011 in the six official languages of the UN.

Four members of the PMP were speakers at the EastWest Institute's (EWI) Worldwide Cyber Security Summit held in early May in Dallas, Texas. WFS entered into a Memorandum of Understanding with the EWI to collaborate on mutual cyber security initiatives and jointly published selected PMP documents, including the Erice Declaration of Principles for Cyber Security. The document, *Rights and Responsibilities in Cyberspace: Balancing the Need for Security and Liberty,* was released August 16, 2010.

SECTION 13

SEMINAR PARTICIPANTS

Professor Sergey V. Ablameyko	Belarusian State University Rector Minsk, Belarus
Lord John Alderdice	House of Lords London, UK
Professor Mikhail J. Antonovsky	Carbon Dioxide Division Institute of Global Climate and Ecology Moscow, Russia
Professor Richard Arnowitt	Department of Physics Texas A&M University College Station, USA
Dr. Muheeb Awawdeh	Department of Earth and Environmental Sciences Yarmouk University Irbid, Jordan
Professor William A. Barletta	U.S. Particle Accelerator School Department of Physics Massachusetts Institute of Technology Cambridge, USA
Prof. Juan Manuel Borthagaray	University of Buenos Aires Instituto Superior de Urbanismo Buenos Aires, Argentina
Dr. Vladimir B. Britkov	Institute for Systems Analysis Russian Academy of Sciences Moscow, Russia
Dr. Robert Budnitz	Lawrence Berkeley National Laboratory University of California Berkeley, USA
Dr. Franco Buonaguro	Istituto Nazionale dei Tumori "Fondazione G. Pascale" Napoli, Italy
Dr. Diego Buriot	(Former) Special Advisor Communicable Diseases UN World Health Organisation Lyon, France

Dr. Jacques Bus	DigiTrust.eu Brussels, Belgium
Dr. Gina M. Calderone	Hydrogeology Department ECC Marlborough, Massachusetts, USA
Professor Alberto Cellino	Pino Torinese Astronomical Observatory Pino Torinese, Italy
Dr. Nathalie Charpak	Kangaroo Foundation Bogotá, Colombia
Professor Luisa Cifarelli	Department of Physics University of Bologna Bologna, Italy
Professor Andrea Contin	Department of Physics University of Bologna Bologna, Italy
Dr. William J. Cosgrove	Ecoconsult Inc. Québec, Canada
Professor Pierre Darriulat	Vietnam Auger Training Laboratory Institute for Nuclear Science & Technology Hanoi, Vietnam
Dr. John W. Day	Dept. of Oceanography and Coastal Sciences Louisiana State University Baton Rouge, USA
Dr. Socorro de Leon-Mendoza	Jose Fabella Memorial Hospital Neonatology Unit Manila, Philippines
Dr. Carmen Difiglio	Office of Policy and International Affairs U.S. Department of Energy Washington, USA
Dr. Mbareck Diop	Science & Technology Advisor to the President of Senegal Dakar, Senegal

Dr. Barbara Dirac-Svejstrup	Cancer Research UK London Research Institute Hertfordshire, UK
Dr. William L. Duax	Hauptman-Woodward Medical Research Institute Buffalo, USA
Professor Robert Duncan	University of Missouri Columbia, USA
Professor Merab Eliashvili	Theoretical Physics Department A. Razmadze Mathematical Institute Tbilisi, Georgia
Professor Christopher D. Ellis	School of Natural Resources & Environment University of Michigan Ann Arbor, Michigan, USA
Professor Christopher Essex	Department of Applied Mathematics University of Western Ontario London, Canada
Dr. Lorne Everett	Chancellor Emeritus, Lakehead University Thunder Bay, Canada and L. Everett & Associates, LLC Santa Barbara, USA
Dr. Charles D. Ferguson II	Federation of American Scientists Washington DC, USA
Professor William Fulkerson	Institute for a Secure and Sustainable Environment University of Tennessee Knoxville, USA
Dr. Bertil Galland	Writer and Historian Buxy, France
Dr. Andres Galvez	Future and Strategic Studies Office (DG-PF) European Space Agency Paris, France
Professor Richard L. Garwin	Thomas J. Watson Research Center IBM Research Division Yorktown Heights, USA

Prof. Alberto González-Pozo	Theory and Analysis Department Universidad Autónoma Metropolitana Mexico D.F., Mexico
Mr. Thomas Grahame	Office of Fossil Energy U.S. Department of Energy Washington DC, USA
Dr. Balamurugan Gurusamy	ERE Consulting Group Sdn Bhd Subang Jaya, Malaysia
Professor William Happer	Cyrus Fogg Brackett, Physics Department, Princeton University Princeton, USA
Dr. Adalat Hasanov	Azerbaijan National Academy of Sciences Baku, Azerbaijan
Dr. Udo Helmbrecht	European Network & Information Security Agency Heraklion, Greece
Professor Pervez Hoodbhoy	Physics Department Quaid-i-Azam University Islamabad, Pakistan
Professor Walter F. Huebner	Southwest Research Institute San Antonio, USA
Dr. Daniel Ingersoll	Oak Ridge National Laboratory Oak Ridge, USA
Professor Tom Isaacs	CISAC Stanford University Stanford, USA
Dr. Hisham Khatib	World Energy Council Amman, Jordan
Dr. William Kininmonth	Australasian Climate Research Kew Victoria, Australia
Professor Toichiro Kinoshita	Cornell University Ithaca, USA

Dr. Edward Kobal	Slovenian Scientific Foundation Ljubljana, Slovenia
Dr. Vasiliy Krivokhizha	International Department Federal Assembly of the Russian Federation Moscow, Russia
Dr. Andrey V. Krutskikh	Chief of Dept. for Security & Disarmament Affairs Ministry of Foreign Affairs Moscow, Russia
Dr. Valery P. Kukhar	Ukrainian Academy of Sciences Laboratory of Fine Organic Synthesis Kiev, Ukraine
Professor Askar Kutanov	National Academy of Sciences National representative to the NATO SPS Programme Bishkek, Kyrgyzstan
Professor Tsung-Dao Lee	Department of Physics Columbia University New York, USA
Professor Axel Lehmann	Institute for Technical Computer Sciences German Armed Forces University München Neubiberg, Germany
Dr. Sally Leivesley	Newrisk Limited London Managing Director London, UK
Dr. Giovanni Levi	Centre National de Recherches Scientifiques Evolution des Régulations Endocriniennes Paris, France
Professor Richard S. Lindzen	Earth, Atmospheric & Planetary Sciences Massachusetts Institute of Technology Cambridge, USA
Professor Bruno Maraviglia	Specialisation in Sanitary Physics School Università degli Studi di Roma "La Sapienza" Rome, Italy

Professor Sergio Martellucci	Faculty of Engineering Università degli Studi di Roma "Tor Vergata" Rome, Italy
Dr. Daniel Masiga	International Centre of Insect Physiology and Ecology (ICIPE) Nairobi, Kenya
Dr. Charles McCombie	McCombie Consulting Gipf-Oberfrick, Switzerland
Professor Stephen McIntyre	University of Guelph Toronto, Ontario, Canada
Professor Akira Miyahara	National Institute for Fusion Science Tokyo, Japan
The Viscount Monckton of Brenchley	Carie, Rannoch, Scotland
Dr. Alan Leigh Moore Jr.	The University of Alabama in Huntsville Alabama, USA
Dr. André Naudi	Director of Finance and Human Resources (Former) CERN, Geneva, Switzerland
Professor SunKun Oh	Department of Physics Konkuk University Seoul, Korea
Dr. Jef Ongena	Plasmaphysics Laboratory Ecole Royale Militaire Brussels, Belgium
Professor Frank Leon Parker	Environmental Engineering Vanderbilt University Nashville, USA
Professor Stefano Parmigiani	Evolutive and Functional Biology Università di Parma Parma, Italy

Dr. Cathrine S. Plesko	Applied Physics Division Los Alamos National Laboratory Los Alamos, USA
Dr. Tom Quirk	Institute of Public Affairs Melbourne Victoria, Australia
Professor Aleksander K. Rebane	Physics Department Montana State University Bozeman MT, USA
Dr. Anita Ronne	Institute of Legal Science, Section B Copenhagen, Denmark
Professor Zenonas Rudzikas	Lithuanian National Academy of Sciences Vilnius, Lithuania
Dr. Juan Ruiz	San Ignacio Hospital Santafé de Bogotá, Colombia
Mr. Ivan Sandrea	Vice President of Strategy for International Exploration & Production Statoil London, UK
Professor Hiltmar Schubert	Fraunhofer-Institut für Chemische Technologie, ICT Pfinztal, Germany
Professor Geraldo Gomes Serra	NUTAU University of São Paolo São Paulo, Brazil
Dr. Nir J. Shaviv	Hebrew University of Jerusalem Mount Scopus Campus Jerusalem, Israel
Dr. Adnan Shihab-Eldin	Kuwait Foundation for the Advancement of Sciences Safat, Kuwait
Professor S. Fred Singer	Science & Environmental Policy Project Arlington, USA

Dr. Annette Sobel	University of Missouri Vice President Columbia, USA
Dr. Tanju Sofu	Engineering Analysis Department Nuclear Engineering Division Argonne National Laboratory Illinois, USA
Professor Friedrich Steinhäusler	Physics and Biophysics Division University of Salzburg Salzburg, Austria
Dr. Bruce Stram	Element Markets Houston, USA
Professor Gheorghe Stratan	National Institute of Physics & Nuclear Engineering Bucharest-Magurele, Romania
Professor Kyle Swanson	Department of Mathematical Sciences University of Wisconsin-Milwaukee Milwaukee WI, USA
Professor Michael J. Tannenbaum	Brookhaven National Laboratory Physics Department Upton, New York, USA
Dr. Hamadoun I. Touré	Secretary General International Telecommunications Union Geneva, Switzerland
Professor Anastasios A. Tsonis	University of Wisconsin-Milwaukee Department of Mathematical Sciences Milwaukee, USA
Professor Vitali Tsygichko	Institute for System Studies Russian Academy of Sciences Moscow, Russia
Professor Jan Veizer	University of Ottawa Department of Earth Sciences Ottawa, Ontario, Canada

Dr. Valery Vengrinovich	Institute of Applied Physics of Belarus National Academy of Science Minsk, Belarus
Dr. James D. Walker	Southwest Research Institute Engineering Dynamics Department San Antonio, USA
Dr. Henning Wegener	Ambassador of Germany (ret.) Information Security Permanent Monitoring Panel, World Federation of Scientists Madrid, Spain
Dr. Jody Westby	Global Cyber Risk LLC Information Security Permanent Monitoring Panel, World Federation of Scientists Washington, USA
Dr. LuAnn White	Tulane Center for Applied Environmental Health Tulane University New Orleans, USA
Dr. Crispin Williams	CERN Geneva, Switzerland
Professor Richard Wilson	Department of Physics Harvard University Cambridgc, USA
Professor Ada Yonath	The Kimmelman Centre for Biomolecular Structure and Assembly Weizmann Institute Rehovot, Israel
Professor Yuan Daoxian	UNESCO International Research Center on Karst Guangxi, P.R. China
Dr. Anatoly Zaitsev	NPP “Planetary Defense Center Moscow, Russia
Dr. Lyudmila Zaitseva	Division of Physics and Biophysics University of Salzburg Salzburg, Austria

Professor Xiliang Zhang	Institute of Energy, Environment & Economy Tsinghua University Beijing, P.R. China
Professor Antonino Zichichi	CERN, Geneva, Switzerland and University of Bologna, Italy and Centro Enrico Fermi Erice, Italy

SECTION 14

ETTORE MAJORANA ERICE SCIENCE FOR PEACE PRIZE

Pontifical Academy of Sciences

The Vatican, 25 November 2009

Why Science is Needed for the Culture of the Third Millennium

PREMIO ETTORE MAJORANA—ERICE—SCIENZA PER LA PACE' 2008

PROGRAMME

Pontifical Academy of Sciences
The Vatican, 25 November 2009

MORNING SESSION—*AWARD CEREMONY* 10.30-12.30

H.E. Mons. Marcelo Sánchez Sorondo	*Greetings and Messages*
Professor Antonino Zichichi	*The Erice Science for Peace Prize Awards and Motivations*
Professor Herbert Aaron Hauptman Represented by Professor William Duax	Hauptmann Woodward Medical Research Institute, Inc.
Professor David Hunter Hubel	Harvard Medical School – Neurobiology
Professor Robert Huber	Max Planck Institute for Biochemistry
Professor Bengt Samuelsson	Karolinska Institute
Professor Honglie Sun	Chinese Academy of Sciences
Professor Ada Yonath	Kimmelman Centre—Weizmann Institute

AFTERNOON SESSION—***SCIENTIFIC SESSION ON THE WFS PLANETARY EMERGENCIES***

The World Federation of Scientists Interdisciplinary Group 16.00-19.00

Professor Antonino Zichichi	*Why Science is Needed for the Culture of the Third Millennium*

Various Interventions
General Discussion and
Closing Remarks

PARTICIPANTS

Professor Tleuken Akhmetov	Ministry of Energy, Industry and Trade Adviser of the Minister Astana City, Kazakhstan
Lord John Alderdice	House of Lords London, UK
Professor Mikhail J. Antonovsky	Institute of the Global Climate and Ecology Carbon Dioxide Division Moscow, Russia
Professor Albert Arking	Department of Earth and Planetary Sciences John Hopkins University Baltimore, USA
Professor Richard Arnowitt	Department of Physics Texas A&M University College Station, USA
Professor Rinaldo Baldini	Centro Enrico Fermi, Director Rome, Italy
Dr. Edoardo Babusci	Ministero della Sanità Rome, Italy
Professor Antonio M. Battro	Battro & Denham Buenos Aires, Argentina
Avv. Alessandro Carnevali	Consiglio dell'Ordine degli Avvocati di Roma Rome, Italy
Dr. Nathalie Charpak	Kangaroo Foundation Bogotá, Colombia
Avv. Valerio Cianciulli	Consiglio dell'Ordine degli Avvocati di Roma Rome, Italy
Professor Luisa Cifarelli	Department of Physics University of Bologna Bologna, Italy
Dr. Socorro de Leon-Mendoza	Jose Fabella Memorial Hospital Neonatology Unit Manila, Philippines

Dr. Carmen Difiglio	Office of Policy and International Affairs U.S. Department of Energy Washington, USA
Dr. Mbareck Diop	(Former) Science & Technology Advisor to the President of Senegal Dakar, Senegal
Dr. Barbara Dirac-Svejstrup	Cancer Research UK London Research Institute Clare Hall Laboratories Hertfordshire, UK
Professor William L. Duax	Hauptman-Woodward Institute State University of New York Buffalo, USA
Professor Merab Eliashvili	Theoretical Physics Department A. Razmadze Mathematical Institute Tbilisi, Georgia
Professor Christopher Essex	Department of Applied Mathematics University of Western Ontario London, Ontario, Canada
Dr. Bertil Galland	Writer and Historian Buxy, France
Professor Stefan Hadjitodorov	Bulgarian Academy of Sciences Scientific Secretary Sofia, Bulgaria
Dr. Adalat Hasanov	Azerbaijan National Academy of Sciences Baku, Azerbaijan
Professor David Hunter Hubel	Harvard Medical School Neurobiology Boston, USA
Professor Robert Huber	Max Planck Institute for Biochemistry Martinsried, Germany

Dr. Hisham Khatib	World Energy Council Amman, Jordan
Dr. Edward Kobal	The Slovenian Scientific Foundation Ljublyana, Slovenia
Professor Valery Kukhar	Laboratory of Fine Organic Synthesis Bioorganic Chemistry and Petrochemistry Ukrainian Academy of Sciences Kiev, Ukraine
Dr. Askar Kutanov	President-rector of Academy of Management under the President of the Kyrgyz Republic Bishkek, Kyrgyzstan
Mr. Claude G. Manoli	World Federation of Scientists Secretary General Geneva, Switzerland
Professor Bruno Maraviglia	Scuola di Specializzazione in Fisica Sanitaria Università degli Studi di Roma "La Sapienza" Rome, Italy
Professor Larry McLerran	Brookhaven National Laboratory Upton, USA
Dr. Akira Miyahara	National Institute for Fusion Science Tokyo, Japan
Dr. André Naudi	(Former) Director of Finance and Human Resources, CERN Geneva, Switzerland
Dr. Jef Ongena	Plasmaphysics Laboratory Ecole Royale Militaire Brussels, Belgium
Professor Vytautas Ostasevicius	Chairman of the Division of Technical Sciences The Lithuanian Academy of Sciences Vilnius, Lithuania
Professor Gennady Palshin	ICSC-World Laboratory Ukraine Branch Kiev, Ukraine

Professor Frank Leon Parker	Environmental Engineering Vanderbilt University Nashville, USA
Professor Aleksander K. Rebane	Physics Department Montana State University Bozeman MT, USA
Dr. Anita Ronne	Associate Professor in Energy Law Institute of Legal Science, Section B Copenhagen, Denmark
Professor Zenonas Rudzikas	Lithuanian Academy of Sciences Vilnius, Lithuania
Dr. Fiorella Ruggiu	Ettore Majorana Foundation and Centre for Scientific Culture Erice, Italy
Professor Bengt Samuelsson	Karolinska Institute Stockholm, Sweden
Mgr. Marcelo Sánchez Sorondo	Pontifical Academy of Sciences, Chancellor Vatican City State (Holy See)
Professor Mikhail Shaposhnikov	Ecole Polytechnique Fédérale de Lausanne Lausanne, Switzerland
Dr. Adnan Shihab-Eldin	Kuwait Foundation for the Advancement of Sciences Safat, Kuwait
Professor Honglie Sun	Vice President Chinese Academy of Sciences Beijing, P. R. China
Professor Kyle Swanson	Department of Mathematical Sciences University of Wisconsin-Milwaukee Milwaukee WI, USA
Professor Jan Szyszko	University of Warsaw Warsaw, Poland

Professor Michael J. Tannenbaum	Brookhaven National Laboratory Physics Department Upton, USA
Professor Anastasios Tsonis	University of Wisconsin-Milwaukee Department of Mathematical Sciences Milwaukee, USA
Avv. Emanuele Verghini	Consiglio dell'Ordine degli Avvocati di Roma Rome, Italy
Dr. Henning Wegener	(Former) Ambassador of Germany WFS Information and Security PMP Chairman Madrid, Spain
Professor Richard Wilson	Harvard University Department of Physics Cambridge, USA
Professor Maw-Kuen Wu	Institute of Physics Director Academia Sinica Taipei, Taiwan
Professor Ada Yonath	The Kimmelman Centre for Biomolecular Structure and Assembly Weizmann Institute Rehovot, Israel
Professor Antonino Zichichi	CERN, Geneva, Switzerland and University of Bologna, Italy and Centro Enrico Fermi, Italy